LEAN CONSTRUCTION

Lean Construction

Edited by
LUIS ALARCÓN
School of Engineering, Catholic University of Chile, Santiago, Chile

A.A.BALKEMA/ROTTERDAM/BROOKFIELD/1997

Published by
A.A. Balkema, P.O. Box 1675, 3000 BR Rotterdam, Netherlands (Fax: +31.10.4135947)
A.A. Balkema Publishers, Old Post Road, Brookfield, VT 05036-9704, USA (Fax: 802.276.3837)

ISBN 90 5410 648 4

Contents

Applications

Tools

Quality management

Preface

In the last two decades, great improvements in performance have been observed in manufacturing. In particular, lean automobile industry is now using less of everything: half the manufacturing space, half the human effort in factory, half the product development time, half the investments in tools. In general, significant improvements in all performance indicators have been observed simultaneously, challenging classic paradigms. All these improvements have not been the product of a radical or sharp change of technology but the result of the application of a new production philosophy which leads to "Lean Production". The new production philosophy is a generalisation of such partial approaches as JIT, TQM, time-based competition, and concurrent engineering. Its adoption is expected to change almost every industry bringing revolutionary changes to the way we work. So far, in construction, lean production is little known but several companies have started to explore applications of the concepts of lean production to construction. Even if only a small fraction of the gains observed in manufacturing were realised in construction, the incentive to apply these concepts would be tremendous.

The new production philosophy recognises two types of activities in a production system: conversions activities which add value to the material or piece of information being transformed into a product and flows (inspection, waiting, moving), through which the conversion activities are bound together but which do not add value. The improvement of non value adding flow activities should primarily be focused on improving reliability if not reducing or eliminating them, whereas conversion activities should be made more efficient. In construction, management attention has been focused on conversion processes and flow activities have not been controlled or improved, leading to uncertain flow processes, expansion of non value-adding activities, and reduction of output value. The opportunities of improvement are enormous. During the last four years an increasing number of researchers have joined efforts to investigate the implications of lean production to construction. They have shared their views and experiences with people from the industry, suggested new approaches to lean construction and worked to advance a new theory of production in construction.

This book summarises the new and evolving conceptualization of lean construction by collecting the work developed by members of The International Group on Lean Construction (IGLC) during the last three years. The authors, who are from different backgrounds and include people from the industry and the academia, have covered theoretical aspects as well as relevant areas for lean

IX

construction, such as performance measurement, improvement tools, implementation issues, and case studies. The result is a challenging exchange of ideas and experiences which include stories of success and also some of failure.

Luis F. Alarcón
Pontificia Universidad Católica de Chile
Santiago, Chile
November, 1996

Acknowledgements

There is some key people who must be acknowledge for their contribution to the development of research on Lean Construction. The pioneer work of Lauri Koskela, from VTT, Finland, was an important milestone in developing a stream of research on Lean Production applied to Construction. In 1992, Lauri wrote an inspiring report on Lean Construction during his visit to Stanford University. Then, upon his return to Finland, he organised the First Conference on Lean Construction which was held in Espoo, Finland, in 1993. The Second Conference on Lean Construction was hosted by the Universidad Católica de Chile, in Santiago, Chile in 1994. Glenn Ballard, from the University of California, Berkeley, and Gregory Howell from the University of New Mexico, Albuquerque, have been also key contributors and enthusiastic promoters of the research on Lean Construction. Greg and Glenn hosted the Third Conference on Lean Construction which was held in Albuquerque, NM, USA, in 1995.

The Corporación de Investigación de la Construcción of the Chilean Chamber of Construction is acknowledge for its support to the organization of the Conference on Lean Construction in Chile and for supporting the research in this area, in the Construction Engineering and Management Program at the Universidad Católica de Chile.

Lean production in construction*

LAURI KOSKELA
VTT Building Technology, Espoo, Finland

ABSTRACT: In manufacturing, great gains in performance have been realized by a new production philosophy, which leads to 'lean production'. This new production philosophy is a generalization of such partial approaches as JIT, TQM, time-based competition, and concurrent engineering. In construction, lean production is little known. The concepts, principles and methods of lean production are reviewed, and their applicability in construction is analyzed. The implications of lean production to construction practice and research are considered.

1 INTRODUCTION

The chronic problems of construction are well-known: low productivity, poor safety, inferior working conditions, and insufficient quality. A number of solutions or visions have been offered to relieve these problems in construction. Industrialization (i.e. prefabrication and modularization) has for a long time been viewed as one direction of progress. Currently, computer integrated construction is seen as an important way to reduce fragmentation in construction, which is considered to be a major cause of existing problems. The vision of robotized and automated construction, closely associated with computer integrated construction, is another solution promoted by researchers.

Manufacturing has been a reference point and a source of innovations in construction for many decades. For example, the idea of industrialization comes directly from manufacturing. Computer integration and automation also have their origin in manufacturing, where their implementation is well ahead compared to construction.

Currently, there is another development trend in manufacturing, the impact of which appears to be much greater than that of information and automation technology. This trend, which is based on a new production philosophy, rather than on new technology, stresses the importance of basic theories and principles related to production processes (Shingo 1988; Schonberger 1990; Plossl 1991). However, because it has been developed by practitioners in a process of trial and error, the nature of this approach as a philosophy escaped the attention of both professional and academic circles until the end of 1980's.

In construction, there has been rather little interest in this new production philoso-

*Presented on the 1st workshop on lean construction, Espoo, 1993

1

phy. The goal of this paper is to assess whether or not the new production philosophy has implications for construction. The paper is based on a more detailed study (Koskela 1992a).

2 LEAN PRODUCTION

2.1 *Origins of lean production and the new production philosophy*

Since the end of 1970's, a confusingly long array of new approaches to production management has emerged: JIT, TQM, time based competition, value based management, process redesign, lean production, world class manufacturing, concurrent engineering.

After closer analysis, it transpires that the above mentioned management approaches have a common core, but view this from more or less different angles. This common core is made up by a conceptualization of production or operations in general; the angle is determined by the design and control principles emphasized by any particular approach. For instance JIT stresses the elimination of wait times whereas TQM aims at the elimination of errors and related rework but both apply this angle to a flow of work, material or information.

Thus, a new production philosophy is emerging through generalization of these partial approaches, as has been suggested recently by various authors (Schonberger 1990; Plossl 1991). The new production philosophy, regardless of what term is used to name it (world class manufacturing, lean production), is the emerging mainstream approach practised, at least partially, by major manufacturing companies in America, Europe and Japan. The new philosophy has already had a profound impact in such industries as car manufacturing and electronics. The application of the approach has also diffused to fields like customized production, services, administration and product development.

The conception of the new production philosophy evolved through three stages: It was viewed as a tool (like kanban and quality circles), as a manufacturing method (like JIT) and as a general management philosophy (referred to, for example, as world class manufacturing or lean production). The conceptual and theoretical aspects of the new production philosophy are least understood. However, without conceptual and theoretical understanding the application of methods is bound to remain inefficient and haphazard.

In Figure 1, an attempt for a consolidation of the new production philosophy is presented. The various levels are analyzed in the following.

2.2 *Conceptual framework*

The core of the new production philosophy is in the observation that there are two aspects in all production systems: Conversions and flows. While all activities expend cost and consume time, only conversion activities add value to the material or piece of information being transformed into a product. Thus, the improvement of non value adding flow activities (inspection, waiting, moving), through which the conversion activities are bound together, should primarily be focused on reducing or eliminating

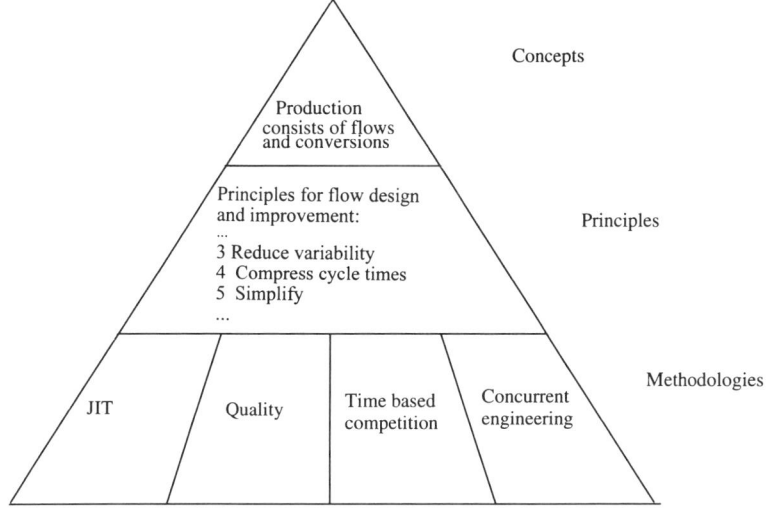

Figure 1. Different levels of the new production philosophy.

them, whereas conversion activities should be made more efficient. In design, control and improvement of production systems, both aspects have to be considered. Traditional managerial principles have considered only conversions, or all activities have been treated as though they were value-adding conversions.

Due to these traditional managerial principles, flow processes have not been controlled or improved in an orderly fashion. We have been preoccupied with conversion activities. This has led to complex, uncertain and confused flow processes, expansion of non value-adding activities, and reduction of output value.

Material and information flows are thus the basic unit of analysis in the new production philosophy. Flows are characterized by time, cost and value.

2.3 *Principles*

In various subfields of the new production philosophy, a number of heuristic principles for flow process design, control and improvement have evolved. There is ample evidence that through these principles, the efficiency of flow processes in production activities can be considerably and rapidly improved. The principles may be summarised as follows (Koskela 1992a):

1. Reduce the share of non value-adding activities (also called *waste*);
2. Increase output value through systematic consideration of customer requirements;
3. Reduce variability;
4. Reduce cycle times;
5. Simplify by minimizing the number of steps, parts and linkages;
6. Increase output flexibility;
7. Increase process transparency;
8. Focus control on the complete process;

9. Build continuous improvement into the process;
10. Balance flow improvement with conversion improvement;
11. Benchmark.

In general, the principles apply both to the total flow process and to its subprocesses. In addition, the principles implicitly define flow process problems, such as complexity, intransparency or segmented control.

Experience shows that these principles are universal: They apply both to purely physical production and to informational production, like design. Also, they seem to apply both to mass production and one-of-a-kind production.

2.4 *Methodologies and tools*

Among the methodologies for attaining lean production are the following most important:
 – Just in time (JIT);
 – Total quality management (TQM);
 – Time based competition;
 – Concurrent engineering;
 – Process redesign (or reengineering);
 – Value based management;
 – Visual management;
 – Total productive maintenance (TPM);
 – Employee involvement.

Most of these methodologies have originated around one central principle. Even if they usually acknowledge other principles, *their approach is inherently partial.* Thus, for example, the quality approach has variability reduction as its core principle. Time based management endeavors to reduce cycle times. Value based management aims at increasing output value.

In the framework of all these methodologies, useful techniques, tools and procedures have been developed. For example, such techniques as quality circles and the 7 quality tools (fishbone diagram, Pareto-diagram etc.) are used in TQM.

2.5 *Comparison between conventional production and lean production*

What is the conventional production philosophy being now replaced by the new philosophy? It is the paradigm of industrial mass production, which evolved in the beginning of this century. The most important differences between the conventional and the new philosophy are summarized in Table 1.

The results of the implementation of the conventional and the new production philosophy are schematically illustrated in Figure 2.

Conventional production is improved by implementing new technology, primarily in value adding activities, to some extent also in non-value adding activities (like automated storages, transfer lines and computerized control systems). However, with time, the cost share of non-value adding activities, which are not explicitly controlled, tends to grow: production becomes more complex and prone to disturbances.

In lean production, non value-adding activities are explicitly attended. Through measurements and the application of the principles for flow control and improve-

Table 1. The conventional and the new production philosophy.

	Conventional production philosophy	New production philosophy
Conceptualization of production	Production consists of conversions (activities); all activities are value-adding	Production consists of conversions and flows; there are value-adding and non-value-adding activities
Focus of control	Cost of activities	Cost, time and value of flows
Focus of improvement	Increase of efficiency by implementing new technology	Elimination or suppression of non-value adding activities, increase of efficiency of value adding activities through continuous improvement and new technology

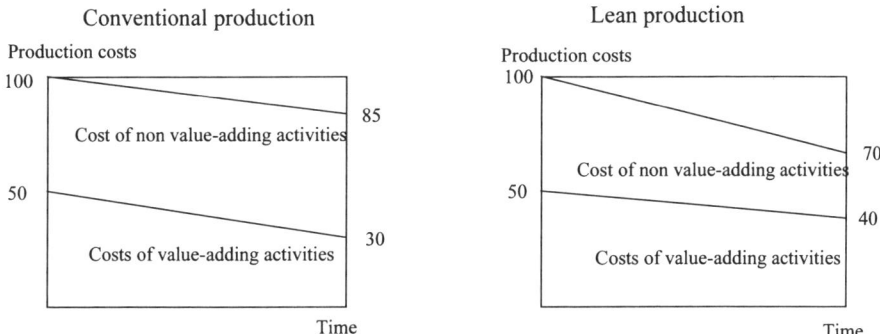

Figure 2. Conventional and lean production: Focus of development efforts.

ment, it is possible to initially reduce the costs of non value-adding activities considerably. Value adding activities are first improved through internal continuous improvement and finetuning of existing machinery. Only after these improvement potentials are realized, major investments in new technology are considered. The implementation of new technology is easier in lean production, because less investments are needed and the production is better controlled. Thus, after the initial phase, increase of efficiency of value adding activities should also be more rapid in lean production than in conventional production.

3 LEAN PRODUCTION IN CONSTRUCTION

3.1 *Preliminary implementation*

In the construction industry, the overall diffusion of the new philosophy seems to be rather limited and its applications incomplete. Quality assurance and TQC have been adopted by a growing number of organizations in construction, first in construction material and component manufacturing, and later in design and construction. The new approach, in its JIT-oriented form, has been used by component manufacturers, for example in window fabrication and prefabricated housing.

Why has the diffusion of the new production philosophy been so slow in construction? The most important barriers to the implementation of these ideas in construction seem to be the following:

– Cases and concepts commonly presented to teach about and diffuse the new approach have often been specific to certain types of manufacturing, and thus not easy to internalize and generalize from the point of view of construction;

– Relative lack of international competition in construction;

– Lagging response by academic institutions.

However, it seems that these barriers are of a temporary nature. On the other hand, the slow diffusion is not explained by an inadequacy of the new philosophy with respect to construction. This is justified by following analyses of waste and peculiarities in construction.

3.2   *Waste in construction*

To what degree do the problems associated with the conventional production view, as observed in manufacturing, also exist in construction? If the flow aspects in construction have been historically neglected, it logically follows that current construction would demonstrate a significant amount of waste (non value-adding activities). Thus, it is appropriate to check whether the existing information supports this hypothesis.

There has never been any systematic attempt to observe all wastes in a construction process. However, partial studies from various countries can be used to indicate the order of magnitude of non value-adding activities in construction. The compilation presented in Table 2 indicates that a considerable amount of waste exists in construction. However, because conventional measures do not address it, this waste is invisible in total terms, and is considered to be unactionable.

3.3   *Problems of construction are caused by neglect of flows*

Analysis (Koskela 1992a) shows that, as in manufacturing, the conceptual basis of construction engineering and management is conversion or activity oriented. The construction process is seen as a set of activities, each of which is controlled and im-

Table 2. Waste in construction: Compilation of existing data (Koskela 1992a).

Waste	Cost	Country
Quality costs (non-conformance)	12% of total project costs	USA
External quality cost (during facility use)	4% of total project costs	Sweden
Lack of constructability	6-10% of total project cost	USA
Poor materials management	10-12% of labor costs	USA
Excess consumption of materials on site	10% on average	Sweden
Working time used for non-value adding activities on site	Appr. 2/3 of total time	USA
Lack of safety	6% of total project costs	USA

proved as such. Conventional managerial methods, like the sequential method of project realization or the CPM network method, deteriorate flows by violating the principles of flow process design and improvement. They concentrate on conversion activities. The resultant problems in construction tend to compound and self-perpetuate. In project control, fire-fighting current or looming crises consumes management resources and attention so totally that there is little room for planning, let alone improvement activities. As a consequence, there is considerable waste in construction.

3.4 *New conceptualization of construction*

Following the lead of manufacturing, the next task is to reconceptualize construction as flows. The starting point for improving construction is to change the way of thinking, rather than seeking separate solutions to the various problems at hand.

Thus, it is suggested that the information and material flows as well as work flows of design and construction be identified and measured, first in terms of their internal waste (non value-adding activities), duration and output value. For improving these flows, it is a prerequisite that new managerial methods, conducive to flow improvement, are developed and introduced. Such methods have already been developed to varying degrees. Not unexpectedly, they try to implement those flow design and improvement principles which are violated by the respective conventional method. However, lacking a sound theory, these efforts have remained insufficient.

Generally, taking flows as the unit of analysis in construction leads to profound changes of concepts and emphasis.

3.5 *Construction peculiarities*

Construction peculiarities refer especially to following features: one-of-a-kind nature of projects, site production, and temporary multiorganization. Because of its peculiarities, the construction industry is often seen in a class of its own, different from manufacturing. These peculiarities are often presented as reasons when well-established and useful procedures from manufacturing are not implemented in construction.

Indeed, these peculiarities may prevent the attainment of flows as efficient as those in stationary manufacturing. However, the general principles for flow design and improvement apply for construction flows in spite of these peculiarities: *construction flows can be improved*. Consider, for example, the one-of-a-kind nature of construction projects. The same peculiarity is shared by many – if not most – product development projects in manufacturing. However, it has been possible to shorten the development time and to improve the output quality in such projects by implementing principles of the philosophy.

On the other hand, these construction peculiarities can be overcome. Initiatives in several countries, like 'sequential procedure' in France, 'open building system' in the Netherlands and 'new construction mode' in Finland try to avoid or alleviate related problems:

– One-of-a-kind features are reduced through standardization, modular coordination and widened role of contractors and suppliers;

– Difficulties of site production are alleviated through increased prefabrication, temporal decoupling and through specialized or multi-functional teams;

– The number of temporary linkages between organizations is reduced through encouragement of longer term strategic alliances.

However, the *elimination of construction peculiarities is not any solution itself*: It just brings construction to the same level as manufacturing. Unfortunately, a large amount of waste also exists in manufacturing before process improvement efforts begin. Thus, only a starting point for effective process improvement is provided.

Thus it is concluded that the construction peculiarities do not diminish the significance of the new philosophy for this industry.

4 IMPLICATIONS

The implications of the new production philosophy for construction will be far-reaching and broad, as they are in manufacturing. The renewal of manufacturing has been realized in a feverish burst of conceptual and practical development. This might also happen in construction.

4.1 *Implications for academic research*

Current academic research and teaching in construction engineering and management is founded on an obsolete conceptual and intellectual basis. It is urgent that academic research and education address the challenges posed by the new philosophy. The first task is to explain the new philosophy in the context of construction. Formalization of the scientific foundations of construction management and engineering should be a long term goal for research.

4.2 *Implications for major development efforts in construction*

Current development efforts like industrialized construction, computer integrated construction and construction automation have focused primarily on the efficiency of value-adding and to some extent also non value-adding activities. They have to be redefined in order to acknowledge the needs for flow improvement. For example, the following guidelines for construction automation can be derived from the principles presented above (Koskela 1992b):

– Automation should be primarily focused on value-adding activities;

– Construction process improvement should precede automation;

– Continuous improvement should be present in all stages of development and implementation of automation.

4.3 *Implications for the industry*

Every organization in construction already can initially apply the generic principles, techniques and tools of the new production philosophy: defect rates can be reduced, cycle times compressed, and accident rates decreased. Examples of pioneering companies show that substantial, sometimes dramatic improvements are realizable in a few years after the shift to the new philosophy. Given the presently low degree of penetration, there are ample opportunities for early adopters to gain competitive benefits.

However, for continued progress, new construction specific managerial methods and techniques are needed; presumably they will emerge from practical work, as occurred in manufacturing.

ACKNOWLEDGEMENTS

Reprinted from *Automation and Robotics in Construction X*, 1993, pp 47-54, with kind permission from Elsevier Science, Amsterdam, Netherlands.

REFERENCES

Koskela, L. 1992a. *Application of the New Production Philosophy to Construction.* Technical Report No. 72. Center for Integrated Facility Engineering. Department of Civil Engineering. Stanford University. 75 p.

Koskela, L. 1992b. Process Improvement and Automation in Construction: Opposing or Complementing Approaches? *The 9th International Symposium on Automation and Robotics in Construction, 3 -5 June 1992, Tokyo. Proceedings.* pp. 105-112.

Plossl, G.W. 1991. *Managing in the New World of Manufacturing.* Prentice-Hall, Englewood Cliffs. 189 p.

Schonberger, R.J. 1990. *Building a chain of customers.* The Free Press, New York. 349 p.

Shingo, S. 1988. *Non-stock production.* Productivity Press, Cambridge, Ma. 454 p.

What do we mean by lean production in construction?*

BERT MELLES
Delft University of Technology, Delft, Netherlands

1 GENERAL INTRODUCTION

This chapter will discuss lean production in construction. The primary goal of lean production is to avoid waste of time, money, equipment, etc. (Japanese: Muda) (Shingo 1992). Everything is focused on productivity improvement and cost reduction by stimulating all employees.

Koskela (1993) gave an overview of waste in construction. He found results of 6 to 10% of the total project costs in Sweden and the USA.

Investigations in construction companies in the Netherlands (source: INFOCUS Management Consultants) did give the same results. Quick scans gave a result of failure costs (costs to restore failures) of at least 6% of the project costs!

Lean production is a philosophy to decline the waste in production companies. Some elements of this philosophy are used already in construction. We discuss the principles and experiments.

2 BASIC PRINCIPLES OF LEAN PRODUCTION

In fact lean production does not include really new principles of management techniques. It only combines existing principles in a new day. The primary goal of lean production is to avoid waste of time, money, equipment, etc. (Japanese: Muda) (Shingo 1992).

Everything is focused on productivity improvement and cost reduction by simulating all employees. This implicates that everything is done to bring the pain to them who created it. In fact a lot of scheduling and decision problems can be avoided by creating lateral relations between task groups, without managing them in a hierarchical way. Let everybody manage his own problems and don't create new problems by managing the problems of somebody else! Lean production is invented in Japan. Especially the implementation of Toyota is very famous. A wide spectrum of techniques ca be discussed as part of lean production. The most important instruments are:

1. Multifunctional task groups;
2. Simultaneous engineering;
3. Kaizen;

*Presented on the 2nd workshop on lean construction, Santiago, 1994

4. Just-in-time-deliveries;
5. Co-makership;
6. Customer orientation;
7. Information, communication and process structure.

All these aspects can be discussed separately but in fact they have a lot in common. The use of all these instruments together are the basis of lean production concepts. Authors from different fields of management science support the importancy of different techniques. In fact there is no pure lean production concept.

Let's discuss the most important instruments of lean production.

Multifunctional task groups
Many authors agreed that the instrument of multifunctional task-groups is one of the most important instruments of lean production. Instead of homogeneous task groups a multifunctional task groups produce a number of different products. This makes it possible to produce a more complex or more completed product with one production unit. It transfers the maximum number of tasks and responsibilities to those workers actually adding value. In the mean time an accurate response to market developments ca be achieved by flexible deployment of personnel (Womack et al. 1990). In multifunctional task groups workers do not have to wait to each other. It also does not give stocks. To achieve the principle of multifunctional task groups personnel has to be trained intensively in recombining thinking and doing (Kenward 1992).

Simultaneous engineering
Today technology changes rapidly. This reduces the lifecycle of products. For this reason a reduction of product development time is essential. Simultaneous engineering can achieve this. By using simultaneous engineering the design and manufacture of the product are no longer separated, physically and time-wise, but integrated and synchronized, through face to face co-operation between designers and producers in a product development team. Direct communication and co-operation can reduce the development period of products significantly (factor 2 to 3). Simultaneous engineering reduces muda by avoiding miscommunication between engineering and production. Within simultaneous engineering also market research is incorporated. This reduces the development of products which are not liked by the clients.

Kaizen
Kaizen is Japanese for permanent and stepwise quality improvement. Kaizen stimulates personnel at all levels in a company to use their brains to reduce costs. In fact Kaizen requires permanent new ideas for cost reduction. In some cases this implicates a strict demand from the management to all production units to create a new idea each week.

A good implementation of Kaizen implicates cost reduction and zero defects in final products. It is obvious that Kaizen reduces muda (Imai 1993). Kaizen demands employee involvement.

Just-in-time deliveries
Just-in-time is a concept for good-flow control. It stimulates reduction of stocks of material by providing goods when and in the amounts needed (Ohno & Mito 1988).

Traditional good-flow oriented control concepts are managing the stock. Instead, primarily short-term decisions are made based upon the current demand for products. New subassemblies are made only immediately before they are actually needed. The ultimate result is that only extremely small subassembly inventories are needed. Traditional inventory control is based upon detailed scheduling techniques (demand for parts is 'pushed'). With JIT, the actual production of new subassemblies is initiated based upon the demand for products which are really need (the 'pull' approach). Transparent production control (visual management) is important. Stock of materials is seen as muda.

The implementation of JIT needs reliable production (zero defects) and good (and steady) relations with suppliers.

Long term relationships with suppliers (comakership)
The basic idea of comakership is to create co-operation with your suppliers (Womack et al. 1990). This means e.g.:
– Mutual technology transfer;
– Mutual openness;
– Mutual management support;
– Mutual declining of stock;
– Mutual sharing of profits.

Long term relationship with suppliers stimulates a relation which is founded on co-operation instead of conflicts. Disturbances in relations causes muda.

Customer orientation
The entire company must be focused on the client (Womack et al. 1990). Client-supplier-relations are very important internal as well as external. Good communication with your client declines problems. As a result this declines muda.

Information, communication and process structure
Lean production demands a transparent organization (Koeleman 1991). A transparent and flat organization implicates better information and communication, internal as well as external. A simple organization structure makes it easier to communicate. A transparent organization makes is easier to have an overview of consequences of control actions. It is obvious that bad communication declines muda.

3 RESULTS OF EXPERIMENTS WITH INSTRUMENTS OF LEAN PRODUCTION IN CONSTRUCTION IN THE NETHERLANDS

In this section we give an overview of some experiments with instruments of lean production in The Netherlands (Botermans 1994).

Multifunctional task groups
Five years ago some experiments were carried out to use multiskilled gang of workmen in housing. Since the sixties all gang were specialized. A large amount of work was subcontracted. The amount of failures was tremendous, because of the fact that every gang only looked after his own production. The trends of using multi-skilled

gang of workmen were good, but full implementation caused radical changes in organization. One of the consequences was that homes should not finished via a construction project bases at a construction site and using specialized work crews (such as kitchen installers, electricians, plumbers) but rather from a central yard using all-round work crews (Melles & Wamelink 1993). In fact the construction companies did not have the attitude to realize this innovative ideas although the results of the experiments were profitable.

Simultaneous engineering and customer orientation
In fact a lot of engineering activities in construction do have some aspects of simultaneous engineering. During the last years in several design and construct projects simultaneous engineering is used. The results were not always profitable. De Ridder (1994) designed new organization structures for this kind of projects. To implement these ideas a company needs a good system of management procedures. These systems are missing very often. The relation between construction company and client is poor managed (on both sites!).

Kaizen
The first approach to implement Kaizen in construction companies was in the early eighties. The so-called MANS-philosophy (MANS = new style management) created a temporary innovative action. The problem was how to communicate the new ideas in the organization. Another problem was created by the fact that it was difficult to stimulate the employees to improve permanently. MANS died in silence (Melles & Wamelink 1993).

The second attempt was implemented in total quality management. This attempt is still going on but seems to have more success. We will discuss this later.

Just-in-time deliveries and long-term relations with suppliers
Also these experiments (especially in building) were carried out for one project only. A real long-term relationship with suppliers did not exist. All experiments were more or less focused on using detailed delivery schedules. In fact this is in contradiction with the pull-orientation of the JIT-concept.

Information, communication and process structure
During the last years the importance of good and clear information flows is understood. Some research projects were carried out on this subject (Melles & Wamelink 1993). Good communication protocols are impossible if an organization does not have a system of management procedures.

4   APPLICABILITY OF LEAN PRODUCTION IN CONSTRUCTION

The basic idea of lean production is very simple. Keep your production system and production organization simple and avoid waste. Stimulate your employees to improve their own production process. If you want to avoid complex information systems the best way to create good communication within a complex organization is to create bilateral relations between different task groups (e.g. engineering and production units) and to give task groups responsibilities (Galbraith 1973). Employees in such an organization have to change their attitude. The management has to create the management frame (what production units do we have, what products do we make, etc.). After that the production units will manage themselves.

In fact the most important goal of lean production is to change the attitude of all employees of a company. In our view Kaizen is the most important instrument of lean production. All other instruments are logical implications of the change in attitude. For example, simultaneous engineering is a logical conclusion of the change in attitude. If we like to make the total production process transparent, if we like to simplify the communication structure, if we like to avoid stock of subassemblies we have to think about the production during engineering activities and reverse about engineering during production.

Up until now most experiments with lean production in construction were focused on implementing one instrument. Most instruments do have overlap with each other instruments, but in fact they can be implemented as stand-alone instruments. For example, it is possible to implement just-in-time deliveries (in a primitive way) on a construction site, without using simultaneous engineering or multifunctional task groups.

Only Kaizen really stimulates all other instruments (including instruments like bench marking which are not mentioned above).

The problem with the Japanese version of Kaizen is that it is developed for the Japanese culture. The mentality of a country and its people is founded in historical events. Neither the Japanese society nor the economic structure of Japan is the same as in Western countries. In Japan the company demands ideas for improvement. Everybody is proud of and loyal to his company and his part of the production process. In fact there is a very emotional relation to the company. In Western countries the attitude to the own company is less emotional.

Kaizen is invented for Japanese companies. This is why Western version of Kaizen has been developed. In fact it is part of total quality management, based on certification according the ISO-9000. Total quality management includes quality assurance and quality improvement. Within the quality assurance manual the system of thinking and acting with quality improvement (Kaizen) has been described. The quality system as well as the implementation can be checked based on the ISO-9000. Such a total quality management philosophy can be externally checked and certified. In the Dutch building industry this seems to give good results. Ten years ago another temptation to implement the basic ideas of Kaizen (MANS-experiment) failed because there was no external check (Melles & Wamelink 1993).

Total quality management is not concerned with only one aspect of the company. It is an integral concept for all units. This makes TQM the integrator of instruments in lean construction. Implementation of TQM based on ISO-9000 is possible in construction companies in Western countries turns out to create a real change in attitude of employees. The real change in attitude can be discovered in general one and a half year after certification (this is 3 to 4 years after the start of the total quality management program!) (source: INFOCUS Management Consultants).

5 CONCLUSIONS AND LESSONS FOR IMPLEMENTATION OF LEAN PRODUCTION IN CONSTRUCTION

In fact the instruments of lean production are not new. The attitude to make it possible to use all these instruments together seems to enforce a real change. If we do not change the attitude of all employees of a construction company we can forget real implementation of lean production. Beside that we need a system of good management procedures to assure good implementation of new ideas. This is the reason why we think that total quality management, based on ISO-9000 is essential to create an environment in which other instruments of lean production can be worked out. If we start with the other instruments they all have a very temporary character.

REFERENCES

Botermans, D.J.M. 1994. *Ook de bouwvakker aan de lijn?* (in Dutch). Catholic University Brabant.
Galbraith, J.R. 1973. *Designing Complex Organizations.* Addison-Wesley Publishing Company, Reading.
Imai, M. 1993. Kaizen, Kaizen.
Kenward, M. 1992. The Fine Art of Mass Production. *New Scientist* 18 July, New York 1992.
Koeleman, H. 1991. *Interne communicatie als managementinstrument. Strategie, middelen, achtergronden* (in Dutch). Bohn Stafleu Van Loghum, Houten/Zaventem.
Koskela, L. 1993. Lean Production in Construction. *Proceedings International Symposium on Automation an Robotics in Construction, Houston* 1993.
Melles, B. & J.W.F. Wamelink 1993. *Production Control in Construction, Different Approaches to Control, Use of Information and Automated Data Processing.* Delft University Press.
Ohno, T. & S. Mito 1988. *Just-in-time for today and tomorrow.* Productivity Press, Cambridge.
Ridder, H.A.J. de 1994. *Design and Construct of Complex Civil Engineering Systems, A new Approach to Organization and Control.* Delft University Press.
Shingo, S. 1988 or 1992. *Non Stock Production.* Productivity Press, Cambridge.
Womack, J.P., D.T. Jones & D. Roos 1990. *The Machine That Changed The World.* Harper Perenial, New York.

Lean production theory: Moving beyond 'can do'*

GREG HOWELL
Department of Civil Engineering, University of New Mexico, Albuquerque, USA

GLENN BALLARD
Department of Civil Engineering, University of California, Berkeley, USA

1 INTRODUCTION

Lauri Koskela (1992) identified the first task for academics 'is to explain the new philosophy in the context of construction' and this is first objective here. The second is to provide a foundation to understand the contributions of Glenn Ballard which follow. The chapter first discusses changes in the construction industry to suggest why a new (or for that matter any) production theory is required. The extent of the uncertainty experienced on projects leads to yet another comparison between manufacturing and construction. A new understanding of the construction process is offered. Next the concepts of flows and the role of 'lean production theory' (LPT) is examined. The chapter closes with a reflection on the mental models which support current thinking.

One caution, our perspective is drawn from experience in petrochemical and process piping projects. While there appear to be many parallels with experience in other project types, the specific thinking and applications occurred primarily in this industry segment.

2 UNDERSTANDING THE CONSTRUCTION CONTEXT FOR LPT

Significant gains in manufacturing are reported from implementation of LPT in industry. Koskela (1992) identifies the overwhelming dominance of conversion thinking in construction and argues for replacing the conversion model with a flow/conversion model in order to reduce waste. Unfortunately, the foundations of conversion thinking are not clearly explained so its nature must be derived from the tools, techniques, contracts and organizational forms in use. Perhaps the heavy emphasis on the 'critical path method' (CPM) as beginning and ending of planning best exemplifies the conversion theory in practice. Designed for relatively slow, simple and certain projects, these tools, techniques, contracts and organizational forms are inadequate to manage let alone improve practice on quick, complex, and uncertain projects (Laufer et al. unpubl. paper). A trend toward complex, uncertain and quick projects is obvious in the petrochemical business and apparent in other segments.

Competition is becoming intense as constructors try to find new ways to reduce

---

*Presented on the 2nd workshop on lean construction, Santiago, 1994

costs even as projects become more difficult. Experiments with various forms of TQM, partnering, constructability provide some improvements but no consistent pattern or theory has yet emerged. The development of LPT in manufacturing appears related to changes in the competitive environment which are similar to those being experienced in construction. In construction, as in manufacturing, the changed project environment is the driving force behind the need for new understanding. We should not forget that the impact of LPT in manufacturing extended well beyond the shop floor.

There must be many partial explanations for the persistence of conversion thinking if it is as inadequate as we suggest. Let us offer a few. It is relatively easy to contract for the purchase of a thing and relatively difficult to contract for behaviour (MacNeil 1974). Commercial contract law for the purchase of goods tends to govern the rules applied to construction. Hence we have a continuing focus on contract while projects fail because of lack of teamwork – a behaviour issue.

A second reason may be the apparent efficiency of using a single set of tools for a number of functions. Wouldn't it be wonderful if work could be completely coordinated by a schedule which also provided updated forecasts for senior management, limited claims, and could be broken into smaller plans to direct specific activities? It would be wonderful but no such tool exists – despite the claims of CPM software salesmen. In our experience, it is impossible to show all of the logic constraints with CPM. Further CPM is inadequate in the face of complex resource constraints as Prof. Fondahl himself noted in his early work.

A third set of reasons is suggested by what happens when conversion thinking represented by CPM doesn't work. The typical response in the face of inadequate performance has been to blame the problem on unmotivated or untrained users. To even suggest inadequacy is to provoke strong emotional reactions. After they subside, the problem remains that conversion thinking is inadequate in the face of quick uncertain and complex projects.

Finally, perhaps conversion thinking persists because no adequate alternative has been proposed and the environment of construction projects really has changed, that is, the pressure for completion on uncertain projects has increased dramatically in the last few years.

Current thinking, resting on the needs of a different era, is both unable to deliver significant breakthroughs, and is itself far more damaging than previously understood.

3 THE SITUATION AND ITS IMPLICATIONS

A review of data on the state of uncertainty at the beginning of the construction phase is instructive. Data in Figure 1 shows the state of uncertainty at the beginning of typical construction projects as reported by about 175 project managers representing a broad spectrum of project sizes and types. The data confirms that significant uncertainty is to be expected even as late as the start of construction (Howell & Laufer 1993).

Data in Figure 2 was collected from managers of similar projects. Here the managers reported on their most recent projects as opposed to their 'typical' projects as

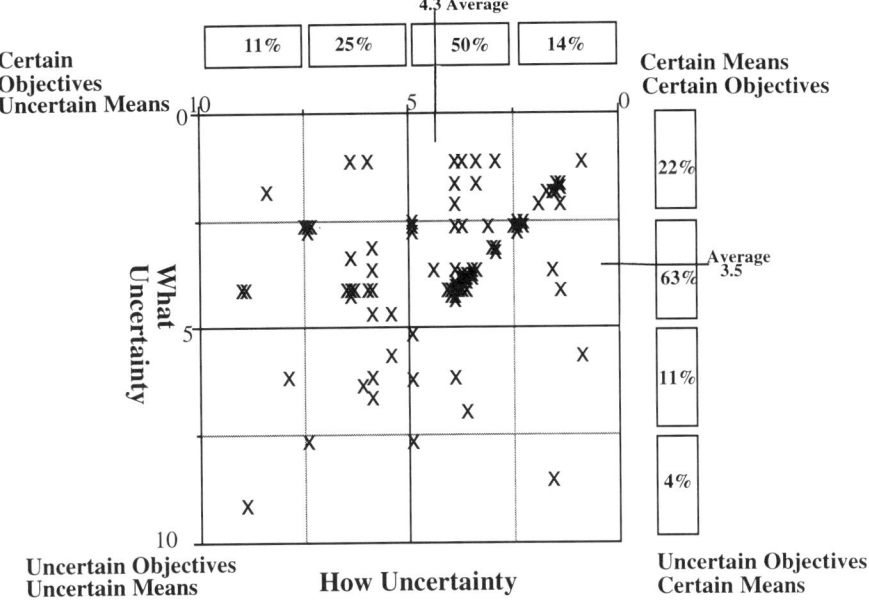

Figure 1. Assessment of uncertainty at the start of construction: Typical projects.

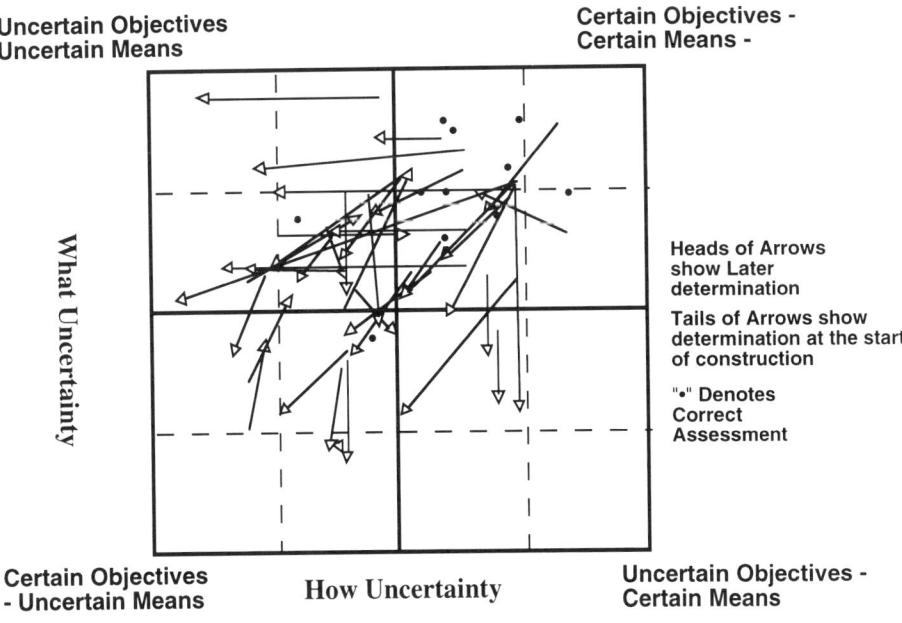

Figure 2. Assessment of uncertainty at the start of construction: Most recent project.

in Figure 1. The managers were asked to use a 'T' to locate where they thought the project was when construction began and an 'R' where it really was once they understood the situation.

This pattern is even more disturbing and compelling. In 85% of the projects, the manager underestimated the extent of uncertainty. The problems they didn't know about were bigger than the problems they knew about.

Consider the waste of proceeding with detailed planning and mobilization on such an unstable basis. If project management accepts real responsibility for project success, the misperception must be rooted in the way planning is conducted. The degree of instability suggests an overwhelming tendency to optimistic evaluations of project circumstance. These evaluations drive managers to plan in greater detail than supported by their information. This persistent optimism suggests either a genetic predisposition on the part of construction planners or a defect in current design of planning systems. Whatever the cause, managers are tending to focus on planning to a fine level of detail far too soon. Focus on technique without an accurate diagnosis of the situation doesn't make much sense. Information must be collected and verified.

In addition to the data on uncertainty, the pressure to reduce project durations is clear. The CII is conducting research in the area and the need for faster completion is widely reported. Recent interviews with superintendents leave little doubt about the increased urgency and complexity of projects in their charge.

LPT in construction must come to grips with the entire design and construction process because increasingly complex projects are being urgently pressed forward under greater uncertainty. Field operations can be improved using LPT principles but even they occur in a different context from manufacturing production. A comparison with manufacturing shows the key feature which distinguishes construction from manufacturing is the extent of uncertainty evident throughout the phase (Table 1).

Table 1. Context of manufacturing and construction production.

	Start of manufacturing production	Start of construction in the field
What	Highly defined	Evolving as means refines ends
How	Highly defined. Operations plan is in great detail based many trials. Primary sequence of major tasks is inflexible, interdependencies are documented and analyzed. Positions in process determine required skills	Partly defined but details un-examined. Extensive planning remains as situation evolves. Primary sequence only partly determined by hard logic but may change. Interdependencies due to conflicting measurements, shared resources, and intermediate products only partly understood. General craft skills to be applied in a variety of positions
Assembly objectives	Produce one of a finite set of objects where the details of what and how are known at the beginning of assembly	Make the only one. The details of what and how are not completely known at the beginning of assembly
Improvement strategy	Rapid learning during the first units preparing for production runs	Rapid learning during both planning and early sub-assembly cycles

In important ways, the life of a construction project is similar to the product development stage in manufacturing. Because a construction project is analogous to the preparation of a prototype, completing the construction phase is better understood as one of the preliminary steps leading to the 'production' which occurs once the facility is completed rather than as manufacturing production exposed to the elements. Reducing uncertainty related to 'what and how' defines the process of 'building a prototype in place'. The challenge for LPT is to reduce waste through bringing stability to the planning process as 'what and how' are refined.

Stability is a key aspect of LPT in manufacturing. There the idea is to minimize input variations so non-value adding steps or flow related activities could be eliminated from the process. Managing flows in construction is more difficult than in the production phase of manufacturing because there is uncertainty both in what is to be accomplished and in the provision of requirements for assembly.

Current construction thinking tends to deny the existence of uncertainty or to suggest it is some sort of moral failure.

– 'If the owner would make up their mind – once and for all, we could do our job.'; or

– 'If the process design engineers would...' and on down the chain.

Once the reality of uncertainty is accepted, a construction project becomes less the transmission of unambiguous orders from the owner to the worker and more a series of negotiations. The object of these negotiations is the rapid reduction of uncertainty. Anything which inhibits these negotiations adds waste. This is true whether the negotiation is between project objectives and means during project definition and design or the more constrained negotiation between shoulds and cans which occurs as foremen prepare weekly work plans.

It is time to examine the concept of flows in relation to the reduction of uncertainty.

4 FLOWS RECONSIDERED

The idea of flows of materials and information from one conversion process to another is quick to grasp. Work in a factory or on a site can be thought of in terms of the movement of materials and information (stuff) through 'input – process – output/input – process – output' chains. Stabilizing work in these chains reduces waste. It requires managing the timing and sequence of the flow of stuff, and assuring it meets downstream requirements.

This simple I/O model is adequate for field assembly operations but is not sufficient for understanding the flows involved in the planning process. Minimizing uncertainty in the flow of decisions and information required in planning is as important as minimizing uncertainty in the flow of stuff. To visualize the flows associated with planning, we propose to expand the horizontal I/O model to include the concepts of directives, i.e. the vertical flow of instructions or standards, the plans for the process at hand. Similar ideas have been expressed by $IDEF_0$ and SanVido but have been thoroughly developed by Talley & Ballard (1990).

In a sense, plans are directives produced by a planning process. They tell the next level what 'should' take place. Inputs such as materials to the work processes determine what 'can' be done. Thus there are two different kinds of flows – one of the plans which become more narrow as the assembly process nears, and stuff which is used in the assembly process. Uncertainty may be transmitted to the work site through either flow. Stabilizing work flow, the subject of Chapter 11 (pp. 101-110), proposes a technique for shielding the workers from uncertainty in both plans and stuff. This is the first step in waste reduction and it provides a basis for further improvements.

Reducing the waste occasioned by the flow of stuff is closely tied to the development of plans. Stable plans both rest on firm upstream assumptions (or premises) and have been tested against the availability of resources. Reducing the variation in the flow of both plans and stuff is the topic of Chapter 10 (pp. 93-100).

LPT, as we understand it, reduces waste by rapidly reducing uncertainty. The implementation strategy is to stabilize work flow by shielding, reduce in-flow variation, then better match labor to available work, and finally improve downstream performance. This strategy both solves problems on projects and clarifies our understanding of LPT. Once this approach is adopted it becomes clear that current management techniques inject uncertainty into the project. Examples will be offered as time permits.

The immediate goal of LPT should be to bring stability to the process by more efficient 'negotiations' between ends and means at every level. Activities such as partnering and constructability which are considered partial implementations of LPT exemplify the negotiation aspect of construction. Important work remains in learning to package and planning to the right level of detail so plans remain in force and stable despite environmental changes. Conversion thinking offers little advice on how to package work so that activities may proceed independently.

5   MOVING BEYOND 'CAN DO'

'Can do' the slogan of the SeaBees of the US Navy summarizes the underlying mental model of most constructors. Ambiguous as it may be, 'can do' is an answer to an assignment. It means, 'no matter what the problem or situation, you can count on me to get the job done.' (no wonder they chose 'can do'.)

A new answer, 'won't do' is possible under LPT because it makes explicit the criteria for decision making. As we develop our understanding of LPT in construction we will confront the underlying thinking of an industry built on 'can do'. Real information on the performance of planning and resource systems can only be available when those charged with planning and doing the work can say 'won't do.' Having the right to say 'no' makes real commitment possible. I am not saying people are allowed to say no on a whim, rather that they are *required* to say no when asked to act beyond the limits of established criteria. This sounds a lot like Ohno's radical decision to allow workers to stop the production lines.

Current management planning and controls systems rely on two unspoken assumptions:

1. The last planner (who you will meet shortly) will always select work in the 'correct' order to achieve project objectives; and

2. Last planners lack the intelligence to manipulate the cost/schedule system for their own short term ends.

In effect we believe they do not know how to protect themselves by selecting the easy work when pressed to increase productivity or production or loose their job.

In short, current management approaches are built on and entice dishonesty. We cannot improve performance unless new thinking exposes the contradictions and weaknesses in our underlying mental models and injects certainty and honesty into the management of projects. It is simple in concept and not hard in execution once we take the challenge of no longer accepting 'can do' when 'won't do' is appropriate. Only then will we have the consistent feedback needed for rapid learning (Senge 1994).

6 WHERE DO WE GO FROM HERE?

Simple, certain and slow jobs hardly pose a challenge. Those best able to manage complex uncertain and quick projects will claim the future. Tools developed to enhance performance on these projects will prove useful on all projects but the aim of LPT should be to help those managing in turbulent situations. Since construction projects are really vast product development processes seldom repeated by the same group of people, and there are so many of these projects, LPT drawn from construction should prove useful across the spectrum of product development efforts.

The idea that LPT ideas drawn from construction could be valuable in other arenas is at first surprising – we tend to think of ourselves as primitive compared to manufacturing. Perhaps our field operations are primitive in comparison with the auto factory of today. This chapter has argued that this is an incorrect comparison. The better comparison is with the product development phase. Here we may not look so bad. It is worth noting that Gilbreth and Ohno, two seminal thinkers in industrial or manufacturing engineering began their careers working in construction.

One caution, we must avoid the tendency (particularly among academics) to deny the nasty uncertainty of the real world. We must avoid the temptation to becoming manufacturing engineers who attempt to change circumstance to fit a theory which is useful in a more stable arena. Rather we must develop our own unique approach to managing all of the flows occasioned by the complex negotiation between ends and means. Once we bring stability to the work environment through better planning, we can turn to the details of methods analysis and there utilize similar principles to those applied at the project level.

REFERENCES

Koskela, L. 1992. *Application of new production theory in construction.* Technical Report No. 72
     CIFE, Stanford University.
Howell, G.A. & A.L. Laufer 1993. *Uncertainty and project objectives.* Project Appraisal.
Laufer, A.L., Denker, G. & V. Shenhar (unpubl. paper). *Simultaneous management.*
MacNeil, I.R. 1974. *The many futures of contract.* University of Southern California.
San Vido, V. Various papers in ASCE journals.
Senge, P. 1994. The Fifth Discipline Fieldbook.
Talley, J. & H.G. Ballard 1990. Work Mapping Package.

Pattern transfer: Process influences on Swedish construction from the automobile industry*

J. BRÖCHNER
Royal Institute of Technology, Stockholm, Sweden

ABSTRACT: Sweden differs from other Nordic countries in having a large share of manufacturing accounted for by automobile production. In this paper, it is shown that since the 1910s, the process of manufacturing and in particular manufacturing cars has served as a paradigm for process change in Swedish residential and commercial construction. Government policies on construction and joint action by the construction industry have been influenced explicitly by features of the car design, manufacturing and marketing processes. Direct transfer and influence has occurred in contractual relations between owners in the automobile industry and contractors. Over the years, supplier relations or customer relations in the industry have formed patterns for changes in the construction process: Standardization of components for mass production, functional design logic, limited customization, mass marketing and recently EDI links to suppliers.

1 INTRODUCTION

For many years, car makers have been invoked as models for the reengineering of a conservative building industry, perceived to remain stubbornly at a pre-industrial stage and bogged down by obsolete practices inherited from medieval craftsmen. Cars and residences have in common they are about the most costly and durable goods that an average household may consider buying. Both have to fit the human scale, and thus neither can follow very far the path of microminiaturization that has led to such startling increases in productivity and decreases in prices associated with electronics.

But do practices diffuse from car manufacturers to construction contractors? What features migrate, if any, and which are the mechanisms in these cases? The issues to be dealt with here in the Swedish context are mainly two:

1. How is the car industry influence transmitted to the construction industry?
2. Which are the lessons perceived to be learnt from the car industry?

Answers to both these questions have changed over the years.

*Presented on the 3rd workshop on lean construction, Albuquerque, 1995

2 WAYS OF TRANSMISSION

Fundamentally, a number of mechanisms for technology diffusion from car manufacturing to residential and commercial construction can be envisaged:

– A car producer as owner commissions buildings to be produced with construction process changes inspired by its primary production process for vehicles;

– Construction companies search actively for automotive paradigms;

– Government research funds are channelled into industrialization of construction, using car production as a pattern;

– Construction companies recruit white- and blue-collar employees, intentionally or unintentionally, with production experience from the car industry;

– Joint ownership of car related business with construction related business;

– Owners outside the industry point to the automotive paradigm and exert pressure on construction companies to apply practices according to the paradigm.

Sweden differs from other Nordic countries in having car producers within its borders. This raises expectations of direct influences on the Swedish construction sector. Nevertheless, it should be said that although much smaller in terms of population than most other car producing countries, the level of per capita car production is comparable to that of much larger producing countries. Table 1 shows how per capita car production varies over a number of countries. Swedish and US figures for 1990 almost coincide.

3 PARADIGMATIC ASPECTS OF CAR PRODUCTION

3.1 *The Ford model T Era: Standardized mass production*

When Gunnar Asplund and his co-authors, being all the major proponents of architectural modernism in Sweden, published their highly influential 'acceptera' (accept!) in 1931, they pitted the teachings of the (internationally not very widely known) Swedish sociologist Gustaf Steffen against Henry Ford, with Steffen in the role of the observer and with Ford seen as the doer (Asplund et al. 1931).

However, Steffen must be credited with pioneer status, given his analysis more than a decade earlier of how the production methods of industrialism had failed to transform European residential construction, and in particular so in Sweden (Steffen 1918). In his analysis, which he wrote as a member and initiator of the first Swedish Royal Commission on Housing, Steffen noted that only in America would you find factory mass production of components to be assembled – or replaced – for home

Table 1. Car production (per capita) in 1990.

Country	Cars produced per capita
Sweden	0.025
Japan	0.080
USA	0.024
France	0.058

Source: UN Statistics.

construction. With hidden reminiscences of his earlier interest in Ruskin's teachings, he did voice some doubts as to the aesthetic consequences; nevertheless, his plea for prefabrication is clear. His paradigms are not presented in detail and tend to belong to 19th century industrialism: Railroads and steamboats, although there is a sweeping gesture towards 'the whole range of later technical improvements'. Steffen was born in 1864, the year after Henry Ford, and had published his first analysis of industrialism and the Worker Question already in the 1880s.

The 'acceptera' chapter on Industrial Housing Production – Standardization begins with a reflection on the fact that any worker can afford a bicycle. This the authors ascribe to machine work and industrial organization, and they continue to stress that this is the case for most of our commodities, from pins to cars. They refer to the principle of reproduction, implying mass or series production, and find that this necessitates a reduction to a limited number of fixed types, or in other words standardization. Photographs serve to illustrate their concept of types: A Gothic church, a bookshelf, a Doric temple and a 1930s car are reproduced in 'acceptera'.

The impact of standardization on construction activities is expressed by Asplund et al. (1931) as implying a minimization of site work through prefabrication. Variety in construction will be based on a limited number of basic components or types in the market. Photographs from Frankfurt construction are used to illustrate this. When discussing large-scale production, the authors indicate the possibility of using more efficient site equipment for assembly and transports. Not all features are inspired by car manufacturers, it must be allowed: Building components can be manufactured on site using available local materials, they suggest. Teams with interior specialization could be moved from house to house as in the Frankfurt experiment. Identical workers or teams repeat the same work process for each house, which is supposed to lead to an increased work rate and increased precision in the work performed. As to personalization, 'acceptera' takes a dark view of car vendors: 'Or who believes, although the opposite is claimed by smart advertisements, that he for instance can buy a personalized car or a personalized textile for a suit... ?'

At that time, when referring to car production, the immediate Swedish paradigms were probably far less important to the authors than the images transmitted by means of leading foreign modernists, in particular ideologues such as Walter Gropius and le Corbusier. These emphasize a complex of production efficiency and product quality. First, they asserted that they had solved the problem of how to construct residential buildings using a dry method ('la maison *à sec*', le Corbusier 1930) and also that they had acted as the manufacturers of automobiles and railroad rolling stock. Secondly, in 1928, Gropius found a graphical presentation showing price increases in North America since 1913, namely 50% for Ford, 78% for the car industry and 200% for houses, where as the general cost of living had risen to 150% (le Corbusier 1930). This divergent pattern was explained by the perfection of methods of mass production among car manufacturers.

It is not only influences transmitted by thinkers on the European continent which can be traced in 'acceptera', but also models from US sociologists. Since the authors refer to the original Middletown study elsewhere in their book, it is probable that they had been influenced by the heavy criticism levied against the local construction industry in Middletown/Muncie: 'Standardized large-scale production, the new habit in industry that makes Middletown's large automobile parts shops possible, is coming very slowly in the complex of tool-using activities concerned with making house; the building of homes is still largely in the single-unit handicraft stage'. (Lynd & Lynd 1929). Something similar applied to Swedish residential construction at that time.

One of the more influential architects behind 'acceptera', Uno Åhrén, had direct experience from working with the automobile industry just before the book went to press. His responsibility had been the Ford Motor Company assembly and repair factory in the Free Port of Stockholm, built in 1930/1931 and considered a break-through project in Sweden. Brunnberg (1990) in her study of the industry architecture of modernism in Sweden points out the influence of Ford's chief architect through many years, Albert Kahn, and his floor layouts and innovative use of new materials in the service of car production.

3.2   *The BMC 1000 Era: Product development*

The principles of mass production as advocated in the 1930s would not be applied extensively to residential production in Sweden until the 1960s, when the annual volume went up rapidly. Although high-rise apartment buildings construction with large prefabricates stood for a smaller proportion after the early 1960s, in 1965, the decennial million programme was declared by the government, based on support for mass production. The objective of one million new homes was met, but criticism of stereotyped environments produced grew more vocal. Could monotony be broken, and would the car manufacturers provide another paradigm?

The Swedish Industries' Building Study Group (1969) used an innovations study published by the Royal Swedish Academy of Engineering Sciences two years earlier where 'the British car BMC 1000 is used to exemplify active, product-yielding research and development work'. The example shows how systematic development work based on well-known techniques results in better products being offered to the consumers. Also underlined was how R&D leads to results when transformed through product development into successful products, and 'the decisive part played by companies as regards development and progress'. One consequence was emphasis on functional design logic and pressure on the formulation of building codes so that performance requirements would be used more frequently, leaving room for product innovation in the construction industry.

3.3   *Quality assurance: Volvo as customer*

The type of industrial thinking represented by the million programme was seldom focused on details of product quality or on responsiveness to owner and user needs. Rising maintenance costs and owner dissatisfaction were often discussed but with no obvious solution. The Volvo Group Headquarters outside Gothenburg (Petersson 1984) was however a fresh initiative. Here, the largest car manufacturer in Sweden required that the contractor should present a system for quality assurance. Although there were precedents in the construction of nuclear energy plants during the 1960s, no contractor had any other experience of quality assurance systems. Interesting is that the owner (Volvo) was unwilling to impose on contractors its then current quality systems for suppliers of car parts. Instead, when pressed by contractors, Volvo refrained, and as a consequence, the successful contractor, F.O. Peterson, had recourse to the at that time unique Canadian standard for quality assurance and had to adapt this standard independently to construction purposes (Augustsson 1995).

3.4 *Electronic data interchange: SAAB and suppliers*

Interest in and the introduction of Electronic Data Interchange (EDI) in the Swedish construction sector have clearly been influenced by the relations between suppliers and producers in the car industry. The fact that SAAB requires suppliers to communicate through the medium of ODETTE has been important.

4 FROM MASS TO LEAN PRODUCTION: CURRENT VIEWS

4.1 *Remnants of the mass production paradigm*

The traditional interest in the mass production paradigm is still alive. Johansson & Snickars (1992) mention development phases and periods of change in the automobile industry, starting with the analysis performed by Altshuler and others published in the MIT 1984 report on the Future of the Automobile. Johansson & Snickars (1992) claim that the Swedish construction industry had developed up to 1950 to the point where automobile production found itself before the Ford model T Era, i.e. before 1910-1920. It is especially economies of scale and the advent of mass production in housing during the 1960s that Johansson & Snickars (1992) discuss.

Supply chain coordination is an issue which includes both the old concern with prefabrication and other aspects. In the 1994 PEAB Annual Report, the CEO of one of the largest contractors in Sweden says 'Volvo or SAAB would hardly buy e.g. pipes in ten metre lengths to be cut and bent so that they fit the car. But that is the way we work on a construction site'. Such observations remain commonplace and are basically along the lines of the reasoning first formulated earlier in the century.

4.2 *Focus on time*

It is only with the 1990s that the focus on time as a metric for production arrives in the construction sector, first as part of the Skanska 3T program (Ekstedt & Wirdenius 1994; Bröchner 1994). Here, the lag between acceptance of time focus in the manufacturing industry and in construction is minimal. The theme is often repeated during the first half of the 1990s: Redtzer (1994) refers to the automotive industry and its manufacturing of components and systems to cut construction times by half, which would give at least 20% lower construction costs.

4.3 *Personalized products*

More surprising is the claim that the present car industry provides greater scope for personalization of products than does construction. A recent trade magazine article (Redlund 1995) on future residential construction starts by noting that the car industry delivers its products with specialized details and equipment, according to the wishes of the individual customer, and then ventures to ask whether future homes can be preordered in a similar way and adapted to the needs of the resident. This is very far from 'acceptera' in 1931, when car producers were seen as paradigmatically uninterested in individual whims.

4.4 *Learning from lean production: Arcona AB*

The construction management activities of Arcona AB are of special interest because they were until recently carried on in parallel with another company in the group being agent for German and Japanese cars in Sweden. However, with the possible exception of the Nissan transplant factory in Sunderland, UK, top Arcona managers are reluctant to identify direct influences from the car industry on their methods for construction management (Birke & Jonsson 1995). As external sources of ideas in addition to the car industry, Arcona does acknowledge the ABB Group and its T50 program for reducing cycle times by half. Effects of the T50 program are spread by ABB subsidiaries in their role as suppliers to firms such as Arcona. T50 is important for its customer orientation and its hunt for time. Another source of inspiration is producers of luxury yachts, where just-in-time procedures are based on traditional craftsmanship, well organized traditional craft work being appreciated as a form of lean production. However, the main source of inspiration is claimed to be the day-to-day activities and experiences in construction projects.

For Arcona, focus on time has been the main principle, discovered as the key to rational production. Process orientation has been seen as essential, implying that non-value-adding activities should be squeezed out. There should be raised precision in all that is done, so as to minimize surprises. This means using the best competence available and dependable partners as strategic suppliers, who should participate from the outset. In this manner, the basis for continuous change is laid, including technology change. Everything should be subordinated to relations so as to escape from the traditional division into opposite parties being kept apart by strict boundaries. In short, all firms involved should profit from being less costly in the process. Ideally, there should be incentive agreements tied to all interfirm relations. According to the 'less of everything' principle, small organizations in a small concentrated core group which (*pace* IT) should sit together in the literal physical sense is desirable; also that this organization is retained from project to project, again with as few people as possible being involved. 'If people always do right, fewer people will be needed'.

5 CONCLUSIONS

In spite of the frequency with which car manufacturing has been invoked as a paradigm for construction, there have only been a few clear-cut cases of direct transfer of ideas and work practices from the car side, and then only when a car manufacturer such as Ford or Volvo has acted as customer to the construction industry. Perhaps the more efficient ways to influence construction has been through the medium of gen eral interest in foreign cultures of manufacturing, historically appearing on the scene in the order of US and Japanese influences on the Swedish industrial imagination. To take only the two most prominent, focus on time and on stronger relations with key suppliers are transforming all branches of industry and consequently also how large contractors choose to operate. The time lag between the rest of industry and construction is not long nowadays; the slowness to conform has probably to be explained by peculiarities in the IT support for most types of construction activities.

REFERENCES

Asplund, G., Gahn, W., Markelius, S., Paulsson, G., Sundahl, E. & Åhrén, U. 1931. *Acceptera* (Accept, in Swedish). Stockholm: Tiden.

Augustsson, R. 1995. Personal communication.

Birke, H. & Jonsson, J.E. 1995. Personal communication.

Brunnström, L. 1990. *Den rationella fabriken* (The rational factory, in Swedish). Umeå: Dokuma.

Bröchner, J. 1994. Organizational adaptations to strategic developments in construction companies. *Proceedings of the A.J. Etkin International Seminar on Strategic Planning in Construction Companies, Haifa, 8-9 juni 1994*, pp. 35-49. Haifa: National Building Institute, Technion.

Corbusier le, E.J. 1930. Analyse des éléments fondamentaux du problème de la 'maison minimum' (Analysis of fundamentals of the Minimal Home Problem, in French). In: *Die Wohnung für das Existenzminimum*, pp. 20-29. Frankfurt: Englert & Schlosser.

Ekstedt, E. & Wirdenius, H. 1994. Enterprise renewal efforts and receiver competence: The ABB T50 and the Skanska 3T cases compared. Paper presented at the IRNOP Conference, Lycksele, Sweden, March 22-25.

Gropius, W. 1924. Wohnhaus-Industrie: Ein Versuchshaus des Bauhauses (Residential industry: An experimental Bauhaus house, in German). *Bauhausbücher*, Vol. 3. Munich: Albert Langen. (Tr. in Scope of Total Architecture. New York: Harper & Brothers 1955).

Johansson, B. & Snickars, F. 1992. *Infrastruktur: byggsektorn i kunskapssamhället* (Infrastructure: the construction sector in the Knowledge Society, in Swedish). The Swedish Council for Building Research, T33:1992. Stockholm.

Lynd, R.S. & Lynd, H.M. 1929. *Middletown: A Study in Contemporary American Culture*. New York: Harcourt, Brace & Co.

PEAB Årsredovisning 1994. (Annual Report) (p. 6).

Petersson, C.-G. 1984. Volvo HK ett pilotprojekt för kvalitetssäkring (Volvo HQ, a pilot project for quality assurance, in Swedish). *Byggmästaren*, No. 5, pp. 37-38.

Redlund, M. 1995. Framtidens bostäder kräver trendnissar (Future homes need trendies, in Swedish). *Byggindustrin*, No. 19, pp. 26-28.

Redtzer, U. 1994. Vi bygger dyrt med förlegade metoder (We build expensively with obsolete methods, in Swedish). *Dagens industri*, February 1.

Steffen, G. 1918. *Bostadsfrågan i Sverige* (The Housing Question in Sweden, in Swedish). Bostadskommissionens utredningar, Vol. IX. Stockholm.

The Swedish Industries' Building Study Group 1969. *The New Building Market: Product Responsibility, Competition, Continuity*. Stockholm: Byggförlaget.

The knowledge process*

DEBORAH J. FISHER
University of Mexico, Albuquerque, USA

ABSTRACT: This chapter describes the knowledge process and how it is a part of lean construction principles. The knowledge process is then further viewed within the context of two current research projects being conducted by the department of civil engineering at the University of New Mexico. One research project, funded by the National Cooperative Highway Research Program (NCHRP), is entitled 'Constructibility review process for transportation facilities,' whose purpose is to develop a formalized constructibility review process for state highway agencies. The other research program, sponsored by the Construction Industry Institute (CII), is entitled 'Modeling the lessons learned process,' and has as its purpose to develop a formalized lessons learned process for CII member companies to follow, in order to implement knowledge. Both of these research projects are viewed in the broader context of knowledge management and current organizational learning theory as it applies to lean construction.

1 INTRODUCTION

Before I begin this chapter, I must first defend my basic premise that the knowledge process is part of lean construction. Knowing that the definition of lean construction is still being argued, this becomes no less a challenge. If you use Womack's definition of *lean production* as a starting point for the definition of *lean construction*, then the principles would be (Womack 1990):
  – Teamwork;
  – Communication;
  – Efficient use of resources;
  – Elimination of waste;
  – Continuous improvement.
  Certainly the learning organization demonstrates these abilities, regardless of whether you are in a production organization (i.e. manufacturing) or a project organization (i.e. construction). Senge refers further to learning organizations and the 'disciplines of theory, methods, and tools representing bodies of actionable knowledge (Senge 1994)'. Both of these authors substantiate the importance of knowledge and learning as an important aspect of lean production and therefore of lean construction.

*Presented on the 3rd workshop on lean construction, Albuquerque, 1995

It is a commonly known fact that we have entered the age of information. In fact, one could say, to quote vice president Al Gore, that we are in the age of 'exformation' because of the deluge of knowledge that is available to all on internal, corporate computer networks, as well as externally on the world wide web. Peter Drucker (1994), writing in The Atlantic Monthly, describes a social transformation that is dramatically changing the socioeconomic makeup of the United States as the 'rise of the knowledge worker'. This worker will need both formal education and a habit of continuous learning (Drucker 1994). Corporations are finding that they must incorporate both continuous improvement and organizational learning in order to improve business results and compete in a global economy (Gupta & Fisher 1994).

2 CONSTRUCTIBILITY REVIEW PROCESS

Constructibility is defined by CII as 'the optimum use of construction knowledge and experience in planning, design, procurement, and field operations to achieve overall project objectives (CII 1986)'. The use of the word 'knowledge' implies that it can be thought of as is a subset of a body of knowledge acquired during the construction phase of a project. This could include both knowledge that has been collected historically with experience, or current new knowledge being acquired and demonstrated in the form of best practices.

The NCHRP has funded a research project with the purpose of developing a formalized constructibility review process (CRP) for the state highway agencies to utilize, in order to implement constructibility into their project development processes (PDP's). In this research project, a tool similar to business process reengineering (BPR) was utilised to model the CRP. The specific process modeler that was used was IDEF0, a modeling technique that was developed originally by the US Air Force. This technique formalizes a process by identifying the primary functions of the process and including, inputs, outputs, constraints and resources associated with each function (see Fig. 1).

As with any process modeler, the intent is to model the process as it now occurs (known as the 'as is' process) and then to reengineer that process in the form of a 'would be' model, attempting to improve the process by eliminating non-value added functions. This is a basic premise of the new production philosophy (Koskela 1992). The only problem with applying this technique to this project was that an 'as is' CRP was non-existent for transportation agencies. In a survey of all 50 state transportation agencies, it was discovered that only 23% of the agencies had a formalized CRP process, and upon further investigation of these 'formal' processes, the definition of 'formal' was found to be quite subjective. Most of these 'formal' processes lacked distinct functions or steps that lead the user through the implementation process (Anderson & Fisher 1994). This is not surprising, since CII stated that the number one barrier to constructibility was the perception that companies were already doing that (CII 1993). Therefore, the researchers superimposed the CRP over the standard transportation agency project development process (PDP) as the solution to process modeling (see Fig. 2).

As a result of modeling, a useful, formalized CRP process resulted that is cur-

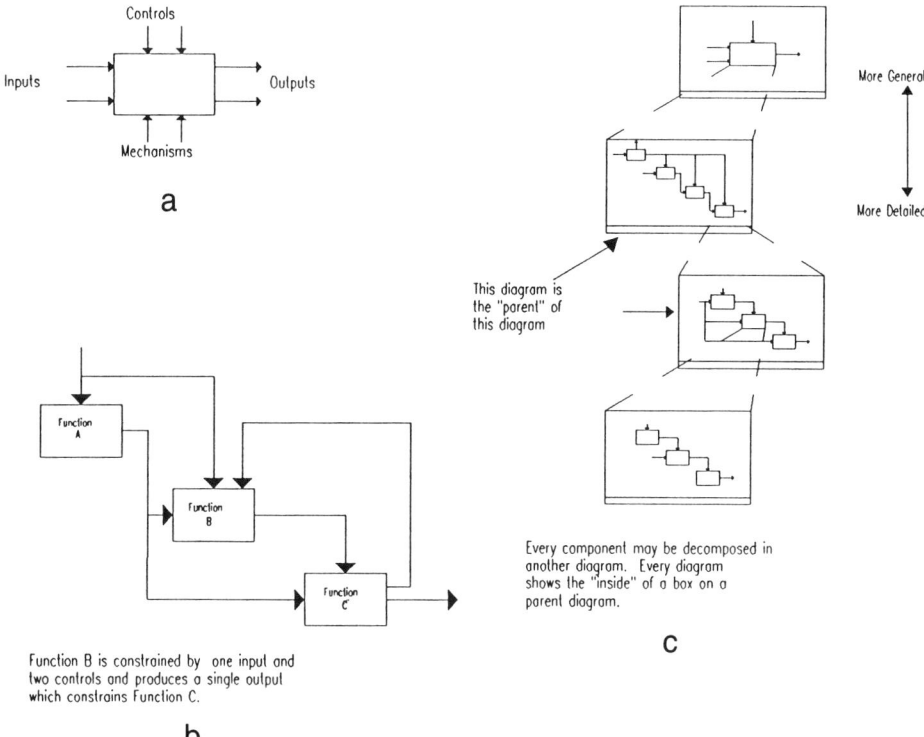

Figure 1. IDEFO function modelling: a) Function box and interface arrows, b) IDEFO model structure, c) Constraint diagram.

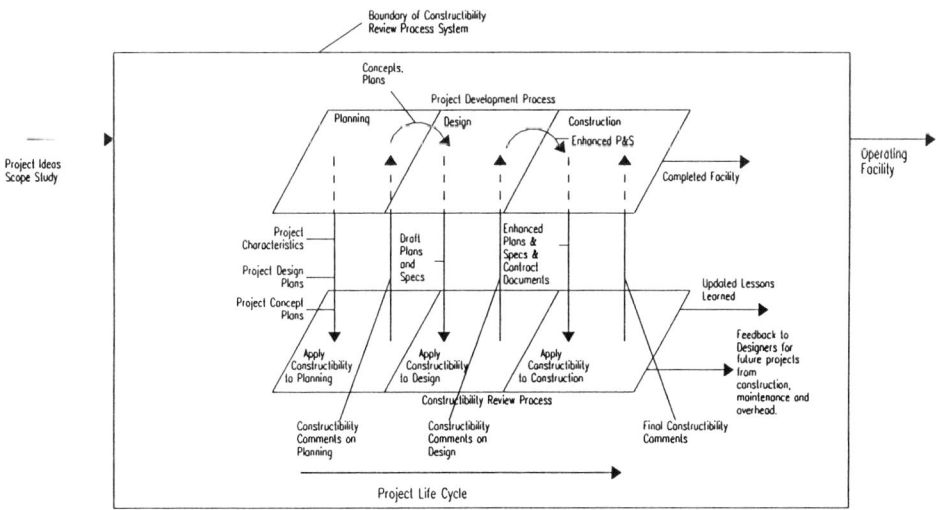

Figure 2. Linking constructibility functions to project development process.

rently being structured into the form of a workbook to train state transportation agencies on how to implement constructibility into their standard PDP's. However, perhaps more insight was gained by the researchers in identifying certain paradigm shifts encompass lean construction principles. Paradigm shifts resulting from this research were (Anderson & Fisher 1994):

- Existence of agency policy for constructibility;
- Use of project constructibility processes for design and construction;
- Contract strategy;
- Use of a constructibility consultant/engineer;
- Use of lessons learned;
- Use of constructibility implementation tools;
- Use of constructibility team;
- Enhancement of plans and specifications and contract documents;
- Feedback from maintenance and operations.

Even though it was beyond the scope of this project, these paradigm shifts indicated that we began to look outside of the project boundaries and see the importance of organizational culture and what changes are necessary for learning to take place.

3 MODELING THE LESSONS LEARNED PROCESS

The construction industry struggles with its ability to capture the 'lessons learned' from its projects and activities for the benefit of future, similar work. Very often, the knowledge gained on a particular project is lost with the changing or leaving of the people who worked on the project. This problem occurs throughout project execution, but often is most evident during the later phases (i.e. construction and operations) when design is well complete. Owners and contractors must depend on job end reports and/or rapid communication to transfer lessons learned from project to project. In today's fasttrack project environment, this is virtually impossible without a formal, systematic process that is to some extent automated. For these reasons, CII has just begun a research effort this year to develop a lessons learned model for use by its member companies, in order to implement lessons learned into the earlier phases of a project so that mistakes are not repeated from project to project. This model should increase the widespread use of lessons learned from previous projects as a tool for continuous improvement.

In an initial survey, 45% of CII member companies state that they have a formalized lessons learned process. As with the NCHRP project, we are in the process of investigating these 'formal' existing processes, as well as identifying other best practices, such as the Martin Marietta model illustrated in Figure 3a and 3b (Sidell 1993). We have learned, so far, that most companies are using some form of knowledge sharing, such as Lotus Notes or the World Wide Web. We anticipate that some hybrid formal process will be developed as a result of this research project for dessemination to CII member companies.

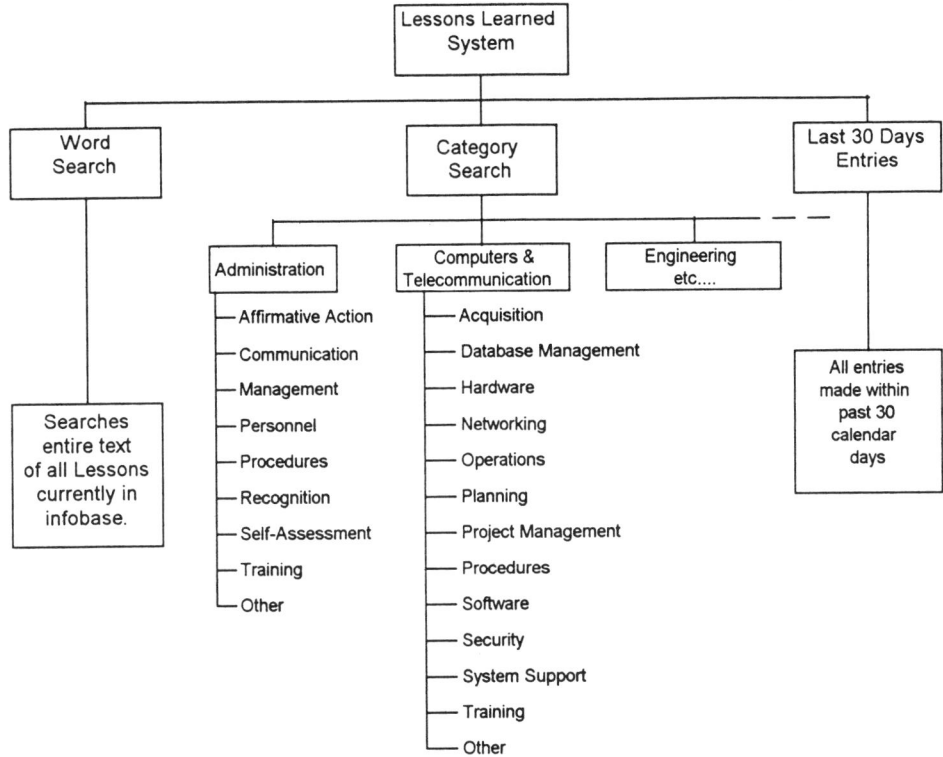

Figure 3a. Lessons learned organization.

4 ISSUES IN KNOWLEDGE MANAGEMENT

If you view the knowledge process as containing the three areas illustrated in Figure 4, you will see that at the centre of the process is the knowledge itself, that is contained within the knowledge management process of collecting, analyzing, and implementing this knowledge into some sort of form or process that the organization can use. At the recent Knowledge Imperative Symposium in Houston, TX, sponsored by Arthur Anderson and the American Productivity and Quality Centre, we learned that these two inner circles only represent about 10 to 20% of the knowledge process. By far, the more difficult issue to address, and the one with far greater potential, is the outer circle of this figure, that is to say the organizational learning culture. Perhaps this is because of what Senge says 'the organization continually becomes more aware of its underlying knowledge base – particularly the store of tacit, unarticulated knowledge in the hearts and minds of employees (Senge 1994)'.

So if at the heart of lean construction is the management of the knowledge process, what solutions can be applied to research projects at UNM, in order to make the types of improvements in construction that Womack (1990) espouses in lean manufacturing? Solutions are found in the following issues, summarized from the recent knowledge symposium that address how to implement the knowledge process into

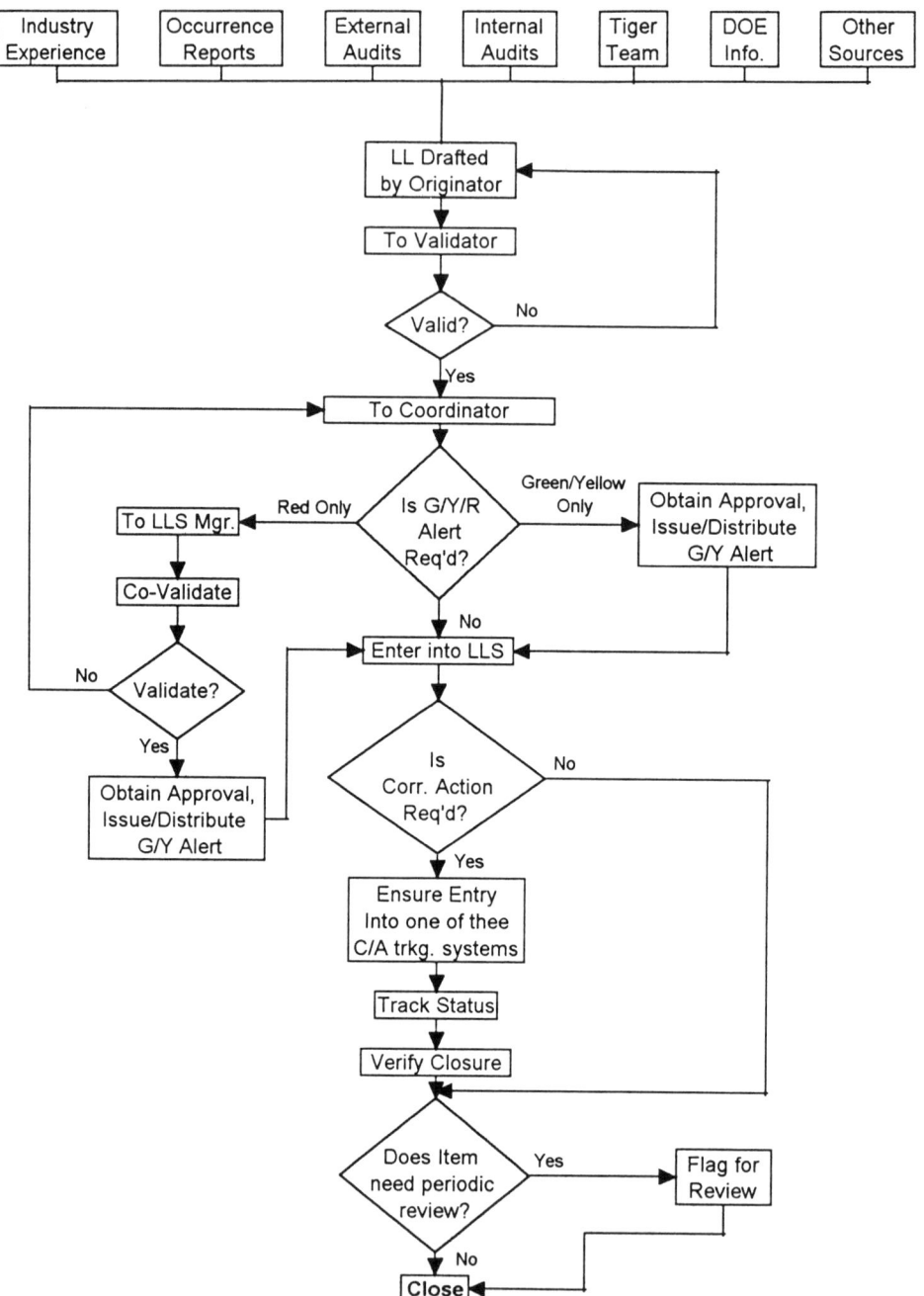

Figure 3b. Lessons learned system flow.

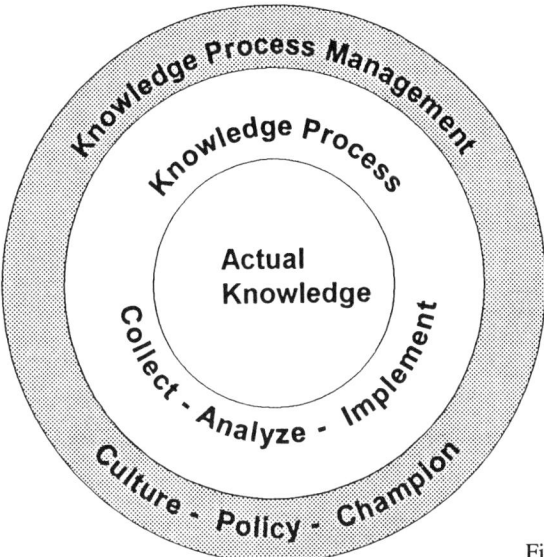

Figure 4. Knowledge process.

the organizational culture. Issues are divided into the three categories of process, cost/benefit, and environment (Knowledge Imperative Symposium 1995).

Process:
 – A commonly understood framework for capturing knowledge must be established;
 – Everyone should be a knowledge leader;
 – Most valuable application is networked word processing (Lotus Notes, World Wide Web);
 – People need training in how to add value to information to make knowledge;
 – There is some confusion as to where knowledge management should reside (R&D, human resources, legal dept., information systems, etc.);
 – OK = $(P + I)^s$ where OK = organization knowledge; P = people; I = information; + = technology; s = sharing;
 – Databases are not magic solutions;
 – Technology only enhances the knowledge process;
 – Build an organizational culture that fosters the knowledge process and then change the metrics to correspond to this culture.

Cost/benefit:
 – Knowledge workers need time and incentives to share personal knowledge;
 – Resources must be allocated to support an integrated knowledge sharing environment or it will fail;
 – Benefits of knowledge management needs to be quantified to management (ROK);
 – Need short-term results using a long-term strategy;

– Knowledge management is expensive, but so is ignorance (3.5-10% of revenues);

Organization environment:
 – An organization must make a cultural change to support knowledge sharing;
 – The process needs a visible, approachable champion;
 – Behaviours must be changed before technology will work;
 – Replace the title of 'chief knowledge officer' to 'facilitator' or 'enabler' to make them more approachable;
 – Knowledge does not equal power or job security;
 – Knowledge tends to be shared more in a decentralized company;
 – Many barriers exist to sharing knowledge (i.e. systematic process, competition, lack of leadership, lack of recognition, resistance to change, etc.).

5 CONCLUSIONS

Organizational learning and knowledge sharing are part of lean construction. This paper emphasizes the application of the lean construction concept of the knowledge process to two UNM research projects to enhance their effectiveness. By far, the greatest payoffs in improving the efficiency of the construction industry and having a competitive organization is with the knowledge process at the organization level. Detailed processes and technology will occur at the project level more successfully once an organizational climate has been established.

Researchers on the NCHRP project were beginning to recognized these facts when we discovered the paradigm shifts that came out of process modeling. These will be highlighted in the executive summary portion of our workbook that will be completed next spring and presented at our constructibility workshop next fall here in Albuquerque. It is hoped that we will follow suit on the CII research project and identify not just detailed processes of collection, analysis, and implementation, but look further to the organizational climate that will foster the success of the process.

REFERENCES

Anderson, S. & D.J. Fisher 1994. Constructibility Review Process for Transportation Facilities, Interim Report, NCHRP No. 10-42, Texas A&M Research Foundation, College Station, Texas.
Construction Industry Institute 1986. *Constructability: A Primer.* Publication 3-1, University of Texas at Austin.
Construction Industry Institute 1993. *Constructability Implementation Guide.* Publication 34-1.
Drucker, P. 1994. The Age of Social Transformation. *Atlantic Monthly.* USA.
Gupta, V.K. & D.J. Fisher 1994. Achieving World-Class Status Through Organizational Learning. *Project Management Journal* 25(3): 16-23. Project Management Institute, USA.
Knowledge Imperative Symposium 1995. Sponsored by Arthur Anderson and the American Productivity and Quality Center, Houston, TX.
Koskela, L. 1992. Application of the New Production Philosophy to Construction. Technical Report No. 72, CIFE, Stanford University.

Mayer, R.J. (ed.) 1992. IDEF0 Function Modeling; A Reconstruction of the Original Air Force Wright Aeronautical Laboratory Technical Report AFWAZ-TR-81-4023. Knowledge Based Systems, Inc., College Station, Texas.

Senge, P.M. 1994. The Fifth Discipline Fieldbook. Doubleday, Inc., New York, NY.

Sidell, S.A. 1993. Lessons Learned: It's the Right Thing to Do. *ASQC Quality Congress Transactions, Boston*, p. 167-173.

Womack, J.P. 1990. *The Machine That Changed The World*. Harper Perennial, Inc., New York, NY.

Identifying and monitoring key indicators of project success*

RALPH D. ELLIS Jr
Department of Civil Engineering, University of Florida, Gainesville, USA

1 INTRODUCTION

Recently we have seen a growing interest in new innovations in construction management. This search for improvement is driven by a long list of industry wide problems. Major failures in project performance include cost overruns, delays in planned schedules, quality problems, and an increase in the number of disputes and litigation. As a result, construction professionals are looking for new (or at least different) ways of managing the construction process.

This search for new paradigms encompasses all aspects of the construction process. Alternatives to the traditional 'low bid' procurement system are being tried. Quality management is almost becoming a religion. Project management computer software is continually being revised. 'Partnering' is rapidly on its way to becoming an industry standard in the US.

The author is convinced that the path to success lies in improving our ability to quantify and measure project performance. The identification, quantification, and measurement of key indicators of project performance is a major challenge facing the construction industry. This chapter will present a few examples from industry of recent attempts to expand our performance measurement systems.

Although it is true that today we do have a global construction market, many local differences in procedures and techniques must be recognized. This chapter deals specifically with the US construction market. However, it seems likely that the concepts presented may be applicable to the construction management process in general.

2 EXPANDING CONTRACTOR SELECTION SYSTEMS

2.1 *The traditional low bid system*

Competitive bidding is deeply rooted in the American tradition. For example laws requiring competitive bidding have been in existence in New York state since 1847 (Harp 1988). The basic concept is that awarding to the low bidder provided public protection from corruption and provided the project at the lowest cost. Today the low

*Presented on the 1st workshop on lean construction, Espoo, 1993

43

bid method of competitively selecting contractors remains the predominant system used in the United States. Alternatives such as negotiated contracts sometimes occur in the private sector but public procurement is almost exclusively based on the low bid method.

Over the years some small modifications to the original low bid method have occurred. Pre-qualification of bidders is now common. The term 'responsible bidder' has been added to the statutes governing public construction awards. However, the basic 'low bid' concept remains intact today.

The low bid system offers one single advantage, objectivity. Since price is the only selection criteria, all subjective evaluation is eliminated. For many participants this is reason enough for continuing the low bid price only system. However, the low bid system has many disadvantages. Clearly, bid price is not the only critical measure of project success. Other factors such as project time, quality, and safety are often equally important. In fact, low bid price may be negatively correlated with these additional performance measures. The low bidder may not be the best overall performer and may not have accurately estimated the project cost.

Competition and objectivity are both positive aspects of the low bid system. What is needed is a selection methodology that includes other factors in addition to price, and still retains the objectivity and competitive aspects of the low bid system. This requires identification and quantification of the additional parameters.

2.2 *Cost-time or A + B bidding system*

Several state highway agencies in the US have begun to experiment with a new method of bidding utilizing both price and time as a basis of award. In this system, the bidder submits both a price bid and a time bid. The time bid represents the total project time proposed by the contractor. Time is quantified by the owner in terms of cost or value per day. For a highway project, this typically involves calculating the highway road user cost including fuel and other operational cost. In general, the time value represents the owner's calculation of the cost of each additional day of project duration. Bids are compared by adding the bid price to the total time value.

For example, a project has been assigned a US$5,000 per day time value. A contractor proposes a bid price of US$4,500,000 and project time of 200 days. The total bid cost for award comparisons would be calculated as follows:

Bid price	US$4,500,000
Time cost 210 days × US$5,000 =	US$1,050,000
Total bid cost	US$5,550,000

Award is made to the bidder with the lowest combined bid price and time cost. The contract price is set at the contractor's bid price and the specified project time is the time proposed by the contractor.

Table 1 presents the results of a time-cost bid for a highway project in Mississippi. The time value for this project was established by the owner as US$7,000 per day. Award was made to bidder A who had the lowest overall cost but not the lowest bid price.

Initial results of the time-cost bidding system are very encouraging. An evaluation

Table 1. Results of bid tabulation using the cost/time concept (source: Herbsman & Ellis (1992)).

Bidder (1)	Bid cost base (2)	Days bid (3)	Time value (4)	Total amount (5)
A	US$15,636,180.56	450	US$2,250,000.00	US$17,886,180.56*
B	US$16,070,558.46	426	US$2,130,000.00	US$18,200,558.46
C	US$15,628,815.06	523	US$2,615,000.00	US$18,243,815.06
D	US$16,231,527.80	646	US$3,230,000.00	US$19,461,527.80
E	US$15,835,768.22	780	US$3,900,000.00	US$19,735,768.22

*The lowest combined bidder.

Table 2. Summary of case study results (source: Herbsman & Ellis (1992)).

Case study number (1)	State (2)	Successful bid price (US$) (3)	Savings to owner (US$) (4)	Times savings to owner (days) (5)
1	Delaware	3,034,765	250,000	50
2	Kentucky	17,886,181	1,387,635	219
3	Mississippi	4,721,599	166,331	49
4	Kentucky	16,329,262	2,885,000	577
5	Kentucky	12,583,349	315,000	63
6	Kentucky	9,186,877	1,620,000	324
7	Kentucky	18,554,123	715,000	143
8	Delaware	2,306,380	175,000	35
9	Maryland	35,087,606	0	0
10	Missouri	1,637,015	(460,000)	(23)
11	Georgia	1,361,009	(147,000)	(21)
12	Texas	39,833,648	150,000	30
13	Texas	39,781,121	300,000	60
14	Texas	15,867,833	55,000	11

of 14 projects indicated significant performance improvement (Ellis & Herbsman 1990). Table 2 presents a summary of the 14 projects studied. Using the cost-time bidding resulted in significant cost and time savings in 11 of the 14 projects. Savings to owner were calculated taking into account the contractor's price and proposed time as compared to owner's estimate of cost and normal time.

Cost-time bidding is an example of how we can expand the traditional basis for contractor selection. By broadening the selection criteria, we have improved project performance. Other factors such as quality might also be included. All that is required is that they be objectively quantified and correlated to a common cost base (Herbsman & Ellis 1992).

3 EXPANDING MEASURES OF PROJECT PERFORMANCE AND COMPENSATION

3.1 *Traditional approach*

Historically, the construction delivery format has been rather simple. Typically, the contractor produces the product which is then inspected. Payment for the product follows acceptance. Non-conformance results in a demand for rework and non-payment or occasionally a negotiated reduction in payment.

The product or project is incrementally accepted or rejected based upon conformance to technical specification. Other measures of performance which relate to the process itself are neglected or at least assigned to a distant secondary position of importance. Why must technical compliance be the only measure of performance? It is true that compliance with final completion times is enforced by the assessment of liquidated damages. However, early completions are rarely rewarded. Furthermore, during the performance interval from start to specified completion date the contractor is largely free to perform at will. Judgement with regard to performance time is suspended until the specified completion date.

It seems that there should be other performance indicators which are important to project success. If so, these additional measures of performance should be identified, quantified and measured as a part of the construction management process.

3.2 *Quantifying and measuring quality*

The quantification of quality is one of the largest challenges facing the construction industry. However, significant advances have been made with the adoption of statistical acceptance procedures. Statistical based acceptance specifications provide an objective format for measuring product quality. Equally important, they also can provide a procedure for adjusting payment with regard to measured quality.

We are beginning to see quality components such as concrete strengths, pavement thickness, and base densities tied to a quality index pay factors. In certain cases, the contractor has been allowed to bid on the quality performance. For example, one of the important measures of highway quality is conformance to specified profile measured in inch/mile. Some state highway agencies have allowed for a contractor bid on pavement profile. A bid of 8 inch/mile would be better than a bid of 13 inch/mile. The owner must establish the cost value of profile conformance in terms of cost for inch/mile.

3.3 *Adding additional performance measures: A case study*

A particularly innovative example of expanding project performance measures can be seen in a recent power generation project in Florida. The project consists of the construction of a new 40-megawatt cogeneration plant. The owner, a local electric utility, placed great importance on a number of performance measures in addition to cost. Selection of the construction contractor included a careful prequalification process, scope review, and competitive price proposals.

What makes this project particularly unique and interesting is that a number of performance measures were worked into the bid proposal and are used to measure performance during the project.

Table 3 provides a listing of the agreed upon performance measures and the project goals assigned to each. On this project the contractor's fee or profit is contingent upon meeting these performance measures which include safety, quality, schedule, craft control, and environmental awareness.

Table 3. Cogeneration project: Performance measure schedule.

Performance measure	Value (US$)
Safety (40%)	
– Incident rate (max 6.9%) 40%	76,560
– Frequency rate (max 3.25%) 30%	57,420
– Severity rate (max 73.55%(30%	57,420
Quality (20%)	
– Weld rejection (max 5%)20%	19,140
– NCR's (max 2/mo) 40%	38,280
– Failure of regulatory inspection (max 0) 20%	19,140
– Ultimate submittal (max 3/mo) 20%	19,140
Schedule (20%)	
– Attain milestones	95,700
Craft control (10%)	
– Verifiable complaints (max 0)	47,870
Environmental awareness (10%)	
– Citations/reportable incidents (max 0)	47,850
Total fee	478,500

Table 4. Cogeneration project: Incentive fee report card.

Incentive fee report card	Results this period	Possible value (US$)	Valued earned this month (US$)
Period from 31 May 1993			
Period to 30 June 1993			
Safety (40%)			
– Incident rate (max 6.9%) 40%	0%	8,000	8,000
– Frequency rate (max 3.25%) 30%	0%	6,000	6,000
– Severity rate (max 73.55%) 30%	0%	6,000	6,000
Quality (20%)			
– Weld rejection (max 5%) 20%	1.8%	2,000	2,000
– NCR's (max 2/mo.) 40%	0	4,000	4,000
– Failure of regulatory inspection (max 0)	0	2,000	2,000
– Ultimately submittal (max 3/mo.) 20%	0	2,000	2,000
Schedule (20%)			
– Attain milestones	OK	10,000	10,000
Craft control (10%)			
– Verifiable complaints (max 0)	0	5,000	5,000
Environmental awareness (10%)			
– Citations/reportable incidents (max 0)	0	5,000	5,000
Total		50,000	50,000

Each month the contractor and the owner jointly evaluate the contractor's project performance. If the contractor falls below the performance standard, a reduction in profit for that period occurs. Table 4 is a copy of the Incentive Fee Report card which is used to determine the fee payment earned.

Table 5. Owner performance measures report.

Item	Current concern		
	Yes	No	N/A
Permits			
Redesign			
Timely submittal review			
Material readiness			
Late design release			
Late changes			
Poor teamwork			
Lack of communications			

So far, this system appears to have worked well. The project is in the eight month of a 12-month performance period. Only one minor shortfall with regard to safety has occurred so far.

It is important to note that all of the performance measures are expressed in quantifiable terms. The measures used are entirely objective. It is the opinion of the project team that subjective measures would be unworkable.

Also of interest is the fact that the owner's performance is also measured. However, the owner performance indicators do not relate to the contractor's payment and are much less objective. Table 5 provides a listing of owner related measures. Performance problems in any of these owner related areas are noted each month. Even though no adjustment in payment occurs, at least the owner's performance is noted and recorded each month.

Overall, the project appears to be successful from both the owner and contractor point of view. On schedule completion is anticipated and both parties are satisfied with the cost performance.

4 SUMMARY AND CONCLUSIONS

Traditional simplistic measures of project performance such as the 'low bid price only' bidding system are in many ways obstacles to improving construction productivity. What we do not measure and quantify, we do not do well. Measurement and accountability are prerequisites of productivity improvement.

Today, the construction industry is looking for innovative ways to improve project performance and productivity. A few examples of innovative methods for expanding project performance measures can be found.

However, much work needs to be done in this area. If performance measures are to be usable, they must be quantifiable and objective. More research needs to be done on quantification methods and on correlating performance indicator values with project success.

REFERENCES

Afferton, K., Freidenrich, J. & Weed, D. 1992. Managing Quality: Time for a National Policy. *Transportation Research Record 1340*, Transportation Research Board, Washington, D.C.

Byrd, L.G. 1989. Partnerships for Innovation: Private Sector Contributions to Innovation in the Highway Industry. *National Cooperative Highway Research Program Synthesis of Highway Practice 149*, Transportation Research Board, Washington, D.C.

Ellis, R.D. & Herbsman, Z. 1991. Cost Time Bidding Concept: An Innovative Approach. *Transportation Research Record 1282,* Transportation Research Board, Washington, D.C.

Harp, W.D. 1988. Historical Background of the Low Bid Concept. *Task Force on Innovative Contracting Practices,* Transportation Research Board, Washington, D.C.

Herbsman, Z. & Ellis, R.D. 1992. A Multi-Parameter Bidding System An Innovation in Contract Administration. *Journal of Construction Engineering and Management, ASCE*, Vol. 118, No. 1, New York.

Yarbrough, R.L. 1990. *In Search of Performance Excellence: Moving Away from Method Specifications*. Focus Strategic Highway Research Program, Washington, D.C.

Modeling waste and performance in construction*

LUIS F. ALARCÓN
Department of Construction Engineering and Management, Catholic University of Chile, Santiago, Chile

ABSTRACT: The introduction of new production philosophies in construction requires new measures of performance. Traditional models offer only a limited set of measures. However, it is possible to use old concepts in implementing new approaches to construction performance improvement. In this chapter, several models are briefly reviewed and examples are given on how a traditional model and a new model can be used to predict and measure performance at the site and project level. At the site level, an example is shown on how work sampling techniques are applied to measure different waste categories in construction. At the project level, a model to evaluate the effect of management strategies on a flexible set of performance elements is introduced.

1 INTRODUCTION

Modeling and evaluation of performance in construction projects has been a challenge for the construction industry for decades. Several models and procedures have been proposed for the evaluation of project performance at the site and project level. Some of these models focus on prediction of project performance while others focus on measuring, however, most of them limit their analysis to a number of measures such us cost, schedule, or productivity (usually labor productivity). The application of new production philosophies in construction requires the evaluation of new measurements (Koskela 1992), such as waste, value, cycle time or variability. However, traditional control systems, and models are not appropriate to measure such performance elements. Nevertheless, some of the concepts developed in previous research can be applied in modeling new performance elements for construction.

In this chapter, traditional performance elements are discussed together with new elements required for continuous improvement. Several models are briefly reviewed and examples are given on how some traditional and new models can be used to predict and measure performance at the site and project level. At the site level, an example is shown on how work sampling techniques are applied to measure different waste categories in construction. At the project level, a model to evaluate the effect of management strategies on a flexible set of performance elements is introduced.

*Presented on the 1st workshop on lean construction, Espoo, 1993

2 PERFORMANCE ELEMENTS

The word 'performance' involves all aspects of the construction process. Performance as applied to on-site activities or associated activities is a broad, inclusive term, encompassing four main elements, namely, productivity, safety, timeliness, and quality (Oglesby et al. 1989). When applied in its more general definition to on-site and off-site activities it involves additional aspects. One author (Sink 1985) has characterized performance, in a broad definition, as seven criteria or elements on which management should focus its effort: (1) effectiveness; (2) efficiency; (3) quality; (4) productivity; (5) quality of work life; (6) profitability; and (7) innovation. More specifically, these can be described as:

1. *Effectiveness.* A measure of accomplishment of the 'right' things: a) On time (timeliness), b) Right (quality), c) All the 'right' things (quantity). Where 'things' are goals, objectives, activities and so forth;

2. *Efficiency.* A measure of utilization of resources. It can be represented as the ratio of resources expected to be consumed divided by the resources actually consumed;

3. *Quality.* A measure of conformance to specifications. In construction projects, quality has two dimensions: The first and overall one is that of the completed project functioning as the owner intended; the second concerns the many details involved in producing this result;

4. *Productivity.* Theoretically this is defined as a ratio between output and input. According to the Bureau of Labor Statistics, a measure of productivity is more specifically an expression of the physical or real volume of goods and services (output) related to the physical or real quantities of input (e.g. labor, capital and energy). In the context of the construction industry, the output is the structure or facility that is built or some component of it. The major input into the construction process includes work force, materials, equipment, management, energy and capital. Labor productivity is also a measure of efficiency but, because of the labor intensive nature of construction, it is treated as a separate dimension. Productivity is primarily measured in terms of cost;

5. *Quality of work life.* A measure of employees' affective response to working and living in organizational systems. Often, the management focus is on insuring that employees are 'satisfied', safe, secure and so forth;

6. *Innovation.* This is the creative process of adaptation of product, service, process or structure in response to internal as well as external pressures, demands changes, needs and so forth;

7. *Profitability.* This is a measure or a set of measures of the relationships between financial resources and uses for those financial resources. For example, revenues/costs, return on assets and return on investments.

Koskela (1992) has recently pointed out the need for new measures for construction, to stimulate continuous improvement. Some proposed measures are the following:

– Waste: Number of defects, rework, number of design errors and omissions, number of change orders, safety costs, excess consumption of materials, etc.;

– Value: Value of the output to the internal customer;

– Cycle time: Cycle time of main processes and sub processes;

– Variability: Deviations from the target, such as schedule performance.

The definitions above are examples of the type of elements which should be used to describe performance. However, the evaluation of performance in construction generally concentrates on only some aspects of performance by using measures that reflect a partial picture, usually profitability and productivity. The need for including each of the aspects discussed above in the evaluation of construction performance has been discussed by Maloney (1990). Profitability and productivity are necessary, but not sufficient, conditions for survival in the construction industry. There are other important measures for getting improvement in construction.

The problem of performance evaluation is a multiattribute or multicriteria one. No two organizations or managers will equally weight individual measures. Also, different managers probably will use different performance elements. Therefore a model for evaluation or prediction must have the flexibility to include the individual organizational objectives in the evaluation process. It also must have the ability to examine the effect of changes in those objectives in the evaluation process.

3 SOME CURRENT PERFORMANCE MODELS

Several authors have tried to build conceptual models of the construction process by explaining different aspects of project performance. Some of these models and methods and their potential uses in research are briefly discussed in the following sections.

3.1 *Measurement models*

A Business Roundtable Report (BRA 1982) published in 1982 showed that no satisfactory measures of aggregate construction productivity were available and that new indexes and data collection procedures should be developed. Kellogg et al. (1981) proposed a holistic model, called the hierarchy model of construction productivity, that could be the basis of a cognitive plan to solve the problems of pulling together all the diverse elements of the construction industry and permit the 'total study' of total factor productivity of the industry. This model defines and measures the factors and elements that influence construction productivity at each level of the construction process; but the broad scope and simplistic form limit its application as a site model for construction productivity. On the other hand, there is a need for greater use of site productivity measurement systems that may allow owners and contractors to monitor and improve productivity. Thomas & Kramer (1988) have studied several procedures to measure productivity on site and have developed recommendations to use effectively these procedures to monitor and improve productivity. Tucker et al. (1986) have implemented a Petrochemical Model Plant database to be used as a baseline measurement of productivity for the industry. In the future, periodic updates of the database would help assess the impacts of different actions on productivity.

Work study-based models have been extensively used to indirectly measure labor productivity. Delay models use stopwatch techniques to record productive time and delays occurring during the day. Activity models use work sampling techniques to categorize activities observed into productive, supportive or idle times. The validity of some of these models to measure construction productivity has been recently questioned (Thomas et al. 1990) because of the difficulties found in supporting the assumptions that link their result to contruction productivity. This aspect is discussed further in Section 4.

3.2 *Productivity prediction models*

There is no standard method for predicting productivity of construction work. Many estimating manuals provide guidance to account for different conditions. Some estimators simply take data from similar work in one area and assume it is directly applicable to a project in a new area. Others apply a 'gut feel' factor based on expert's judgment. Generally, the main variables considered to determine productivity are: the number of direct minutes available per hour, the characteristics of worker population and the rate of work during direct work times. Some authors have provided quantitative information on the effects of different factors on labor productivity (Dallavia 1952; Edmonson 1974; Neil 1982; Neil & Knack 1984). They have proposed methods to predict productivity including the effects of factors like weather, location, labor availability and design. In general, using tables, graphs, and judgment in some cases, corrections are made to an initial productivity value to get the predicted value. These techniques are based on a mathematical analysis that attempts to provide a more formal treatment to this problem. Several other authors have developed adjustment factors based on similar models (Brauer 1984; Lorenzoni 1978; Riordan 1986; CORPS 1979). Borcherding & Alarcón (1991) reviewed near fifty publications that provide quantitative relationships on construction productivity. These continuous efforts to develop quantitative methods during the last four decades show the importance and the difficulties of this task.

3.3 *Productivity theory factor model*

Prediction models such as Dalavia's try to predict average productivity throughout the activity from a contractor's perspective. The factor model, from a research perspective, tries to predict average productivity during much shorter periods when a particular set of conditions exists. The theory says that the work of a crew is affected by many factors that may lead to random and systematic disturbances to performance (Thomas & Yiakounis 1987). The cumulative effect of these disturbances is an actual productivity curve that may be irregularly shaped and difficult to interpret. However, if these disturbances can be mathematically discounted from the actual productivity curve, one is left with an ideal productivity curve (Fig. 1). This curve is a smooth one consisting of a basic performance allowance, plus a component resulting from improvements in repetitive operations. The shape and magnitude of the ideal productivity curve is a function of a number of factors that reflect the site environment, construction methods and constructability aspects. Based upon design requirements and construction practices, it is theorized that this curve can be established prior to commencing the work. If cause-effect relationships are known, then actual productivity can be predicted as a function of the number of units produced or as a function of time. The model contains systematic, random and time-dependent variables.

This model can be used as a framework for quantifying the effects of various fac-

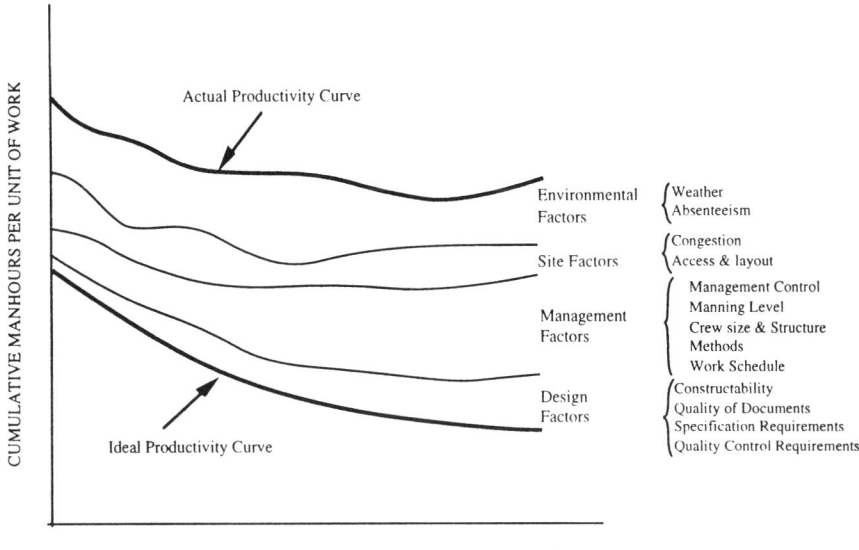

Figure 1. Factor model of construction productivity (Thomas & Yiakounis 1987).

tors on construction productivity. It provides procedures for collecting data on a daily basis and methodologies for combining data from different activities to account for the learning-curve effect. These procedures may allow the successive collection of new information for future research and use the information already obtained to develop more reliable mathematical models using different approaches.

3.4 *Conceptual construction process model*

Sanvido (1988) developed a conceptual model of the management functions that are required to improve the productivity and performance of site construction operations (Fig. 2).

The basic features of the model are: (1) definition of the basic tasks of the craftsworkers and the input resources required; (2) identification of interrelationships between different functions involved in supporting the field construction process and specification of rules to govern their performance; (3) definition of the scope and boundaries of the on-site construction process; and (4) categorization of external influences on the construction process that are beyond the control of the site personnel. Figure 2 illustrates some of these concepts.

The model represents a structure of functions to be performed on a project. The author claims that projects that function closer to the ideal case specified by the model perform better in terms of schedule, cost and quality than those which are further from the ideal situation. The author presents a methodology to use this model to improve the productivity of construction projects. Sanvido's approach can serve both planning and monitoring roles by permitting comparisons between the ways responsibilities should be and actually have been assigned on a project.

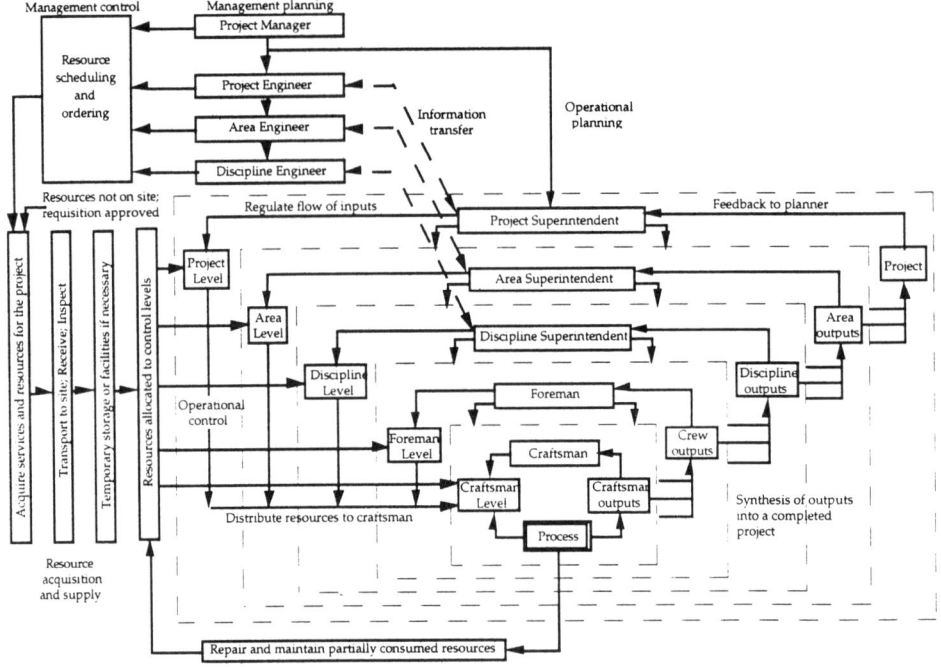

Figure 2. Interplay among authority, responsibility, and communications on every work-face task on a large construction project (Sanvido 1988).

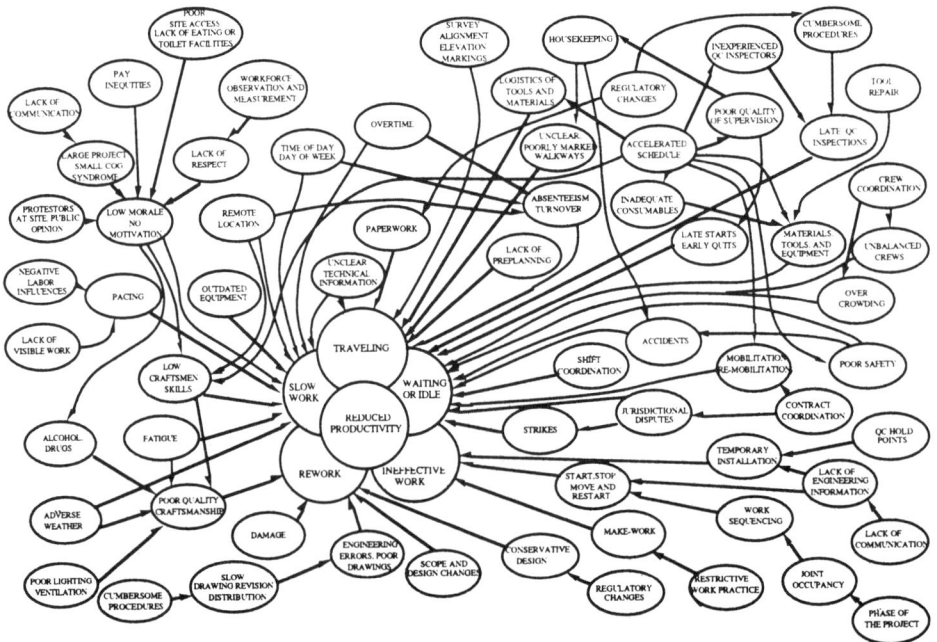

Figure 3. Sources of reduced productivity (Borcherding et al. 1986).

3.5 *Causal models*

Borcherding et al. (1986) provide an interesting qualitative model to identify causes of reduced productivity in construction work as shown in Figure 3. Productivity loss on large complex construction projects is explained using five major categories of unproductive time: (1) waiting or idle, (2) traveling, (3) working slowly, (4) doing ineffective work, and (5) doing rework. The reason why the crafts workers produce less output per unit of time is relegated to one of these basic nonproductive activities. The activities are affected directly or indirectly by several other factors. The diagram illustrates the complexity of the interactions of the numerous factors that affect productivity.

Some authors have attempted to develop causal models of construction productivity to help researchers understand the factors and interactions that affect performance. Maloney and Fillen have developed a model, based on the expectancy theory of human motivation, that shows the relationship between incentives, efforts, barriers and crew performance (Maloney & Fillen 1985). Shaddad proposed a causal model to conceptualize the influence of different management subsystems on construction productivity (Shaddad & Pilcher 1984). The model includes a detailed classification of factors affecting productivity and intermediate relationships.

4 TRADITIONAL AND NEW TOOLS TO EVALUATE PERFORMANCE

This section uses two examples to illustrate the application of a traditional and a new model to evaluate performance in construction. The first example shows how the measurement technique of work sampling is used to measure waste and detect opportunities for improvement at the construction site. The second example introduces a new performance modeling technique, at the project level, to evaluate the impact of management actions on project performance.

4.1 *Measuring waste at the construction site*

This example is based on the work sampling technique, that measures time engaged in various activities (Thomas 1981). Work sampling has been used as an indirect measure of productivity but is considered, at best, a surrogate measure of productivity since there is no measure of output. Two assumptions are made to link work sampling results to labor productivity:

1. It is assumed that reducing delays and waiting time will increase productivity;
2. It is assumed that productive time is related to output and productivity.

These assumptions have been shown insupportable for most construction operations (Thomas et al. 1990) and work sampling has been almost dismissed as a model to measure construction productivity. Nevertheless, Maloney has proposed the use of work sampling as a tool to determine the presence of 'organizationally imposed constraints', within a framework for analysis of performance (Maloney 1990).

The experience of several years carried out by a professional team from the Catholic University of Chile (CUCH), with more of 10,000,000 sq. ft. of building construction, has shown that work sampling is an effective tool to promote improvement in construction. The fact that there is no direct correlation between work sampling results and construction productivity does not reduce its potential as a diagnosis and measuring tool for performance improvement. This technique can be used to measure directly different waste categories that are necessary for continuous improvement.

The work sampling model used by the CUCH team focus on specific categories of contributory and noncontributory work as an diagnosis tool to detect sources of reduced performance and to measure improvement. Table 1 shows an example work sampling report used for this purpose.

The way this information is structured allows detection of otherwise difficult to detect waste sources. For instance, Table 1 shows that carpenters spend 24% of their time doing contributory work such as transportation of materials; this is an important waste category that can be defined as follows: 'unskilled work performed by skilled labor'.

The report structure distinguish among specialty crews activities and locations, to help management to compare performance between crews and work areas, facilitating identification of sources of reduced performance. Table 2 shows the aggregate result for activities.

Table 3 shows an analysis of contributory work by specialty, the number over the diagonal is the current week average, the number below is the previous week result. This type of information allows management to take actions and monitor results for evaluation.

Some example actions taken based on this information are directed to remove 'organizational constraints' or 'minimize waste', for example: reduce traveling time by providing portable showers or bathrooms, provide materials or tool storage next to the construction site; modify site layout, provide unskilled workers to support skilled workers; provide transportation crews to eliminate traveling and transportation time, and many other practical actions to reduce waste.

Observation of the evolution of the different performance elements over a period of time allows measurement of 'variability', which has been suggested as a necessary measure for improvement in construction (Koskela 1992). Figure 4 shows the graphical evolution of the classical work categories for work sampling: productive time, contributory work and idle time. In fact, the experience of the CUCH team suggest that a correlation may exist between 'variability' in productive work and construction productivity. It has been observed that when variability is reduced in

Table 1. Work sampling results by specialty.

	Carpentry			Helpers			Steel work			Concrete work		
	PW	CW	Idle	PW	CW	Idle	PW	CW	Idle	PW	CW	Idle
26/07/93 (M-M)	67	19	15	13	56	31	74	21	5	62	14	24
28/07/93 (W-A)	54	25	21	39	33	28	58	23	19	69	19	13
29/07/93 (T-M)	62	22	16	42	39	19	70	20	10	62	14	24
02/08/93 (M-M)	53	32	15	14	61	24	65	28	8	63	26	11
Average	59	24	17	27	47	26	67	23	10	64	18	18
Old average	52	25	23	24	49	28	68	16	16	54	9	37

PW = productive work; CW = contributory work.

work sampling measures, better productivity is reached at the construction site. This hypothesis is currently under examination.

Work sampling used as described plays the role of a sensitive monitor of project performance. Even though it may not reflect productivity levels, it is very effective in detecting changes in work conditions that are usually associated with inefficiencies and waste sources. In this way it help management to immediately focus attention on problem areas and provide information for decision making.

Foreman Delay Surveys (Tucker 1982) is another traditional tool that has been successfully applied to measure waste in construction. They can complement work sampling results to supply a more precise diagnosis of the causes of delays and job interruptions and provide valuable information to keep track of the improvement effort.

Table 2. Work sampling results by activity.

	Superstructure			Finishing			Installations		
	PW	CW	Idle	PW	CW	Idle	PW	CW	Idle
26/07/93 (M-M)	51	27	21	67	19	14	–	–	–
28/07/93 (W-A)	54	24	21	68	20	13	66	14	20
29/07/93 (T-M)	59	22	19	74	12	14	74	15	12
02/08/93 (M-M)	48	32	20	65	23	12	75	6	19
Average	53	26	20	68	18	13	71	12	17
Old Average	49	25	26	54	20	16	56	18	25

PW = productive work; CW = contributory work.

Table 3. Contributory work distribution.

Speciality	Contributory work categories												Average
	Tran. −5 m		Tran. +5 m		Cleaning		Instruction		Measure-ment		Others		speciality
Carpenter	4	4	2	3	0	0	5	6	10	5	3	6	24% 25%
Helpers	20	20	12	21	14	5	1	3	0	0	0	0	47% 49%
Steel work	14	8	2	3	0	0	7	2	0	2	0	1	23% 16%
Concrete	7	7	3	1	0	0	3	1	3	0	3	0	18% 9%
Earthworks	5	5	4	4	0	0	3	3	1	1	0	0	13% 13%
Plaster	5	9	3	0	0	0	2	1	3	4	1	4	15% 18%
Stucco	5	9	3	0	0	0	2	1	3	4	1	4	15% 18%
Roofing	–	–	–	–	–	–	–	–	–	–	–	–	– –
Electrical	18	7	0	20	0	0	0	0	0	0	0	0	18% 27%
Sanitary	0	0	15	25	0	0	0	0	8	0	4	0	27% 25%
	New/Old		New/Old		New/Old		New/Old		New/Old		New/Old		New/Old
Total	8	7	7	10	3	2	3	3	3	3	1	3	25% 27%

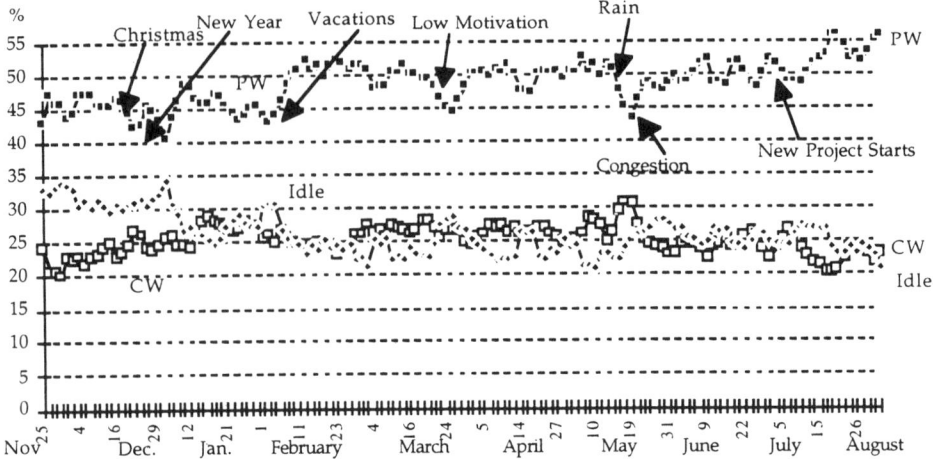

Figure 4. Evolution of work sampling results.

4.2 *A general performance model*

This is a methodology for evaluating performance at a project level. It is a new tool, but it is partially based on concepts proposed in some of the performance models reviewed in this chapter: causal structures, simplified models, qualitative and quantitative relationships for prediction of performance.

The GPM was developed by Alarcón & Ashley (1992), working with the Construction Industry Institute's (CII) Project Team Risk/Reward (PTRR) Task Force. It is a performance modeling methodology for application to individual projects which combines experience captured from experts with assessments from the project team. The methodology consists of a conceptual qualitative model structure and a mathematical model structure. The conceptual model structure is a simplified model of the variables and interactions that influence project performance. The mathematical model uses concepts of cross-impact analysis (Honton et al. 1985) and probabilistic inference to capture the uncertainties and interactions among project variables.

Project options such as organizational design, incentive plans, and team building alternatives are incorporated into the model knowledge base. The GPM allows management to test different combinations of project execution options and predict expected cost, schedule, and other performance impacts. Project performance can be modelled using multiple performance elements defined by the user and a flexible weighting procedure for evaluation.

4.2.1 *Model structure*

The simple model structure, shown in Figure 5, is used to capture, store and link the special expertise of many different parties in the construction industry. The modularization of the knowledge allows for independent elicitation of knowledge from the most qualified experts in specific areas. The model combines the client's preferences, or weights, toward outcomes such as schedule or budget with the special insight of the project team charged with the design, procurement and construction of the facility. Important project management expertise is drawn from CII experts who have

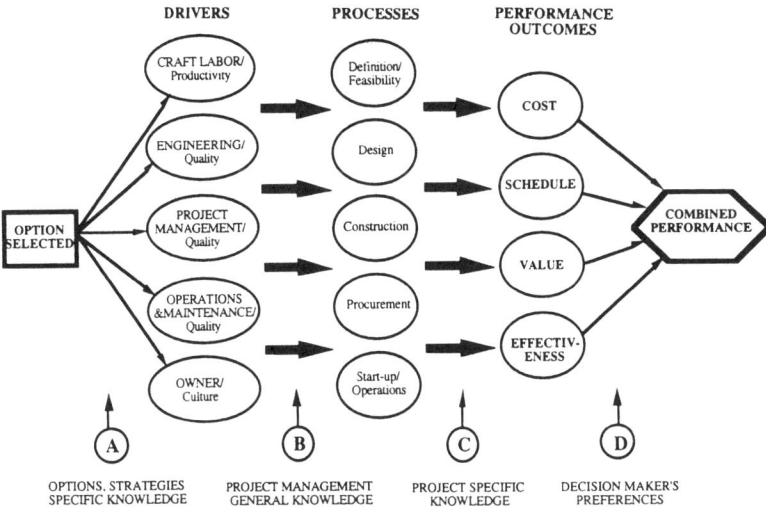

Figure 5. GPM structure and knowledge inputs.

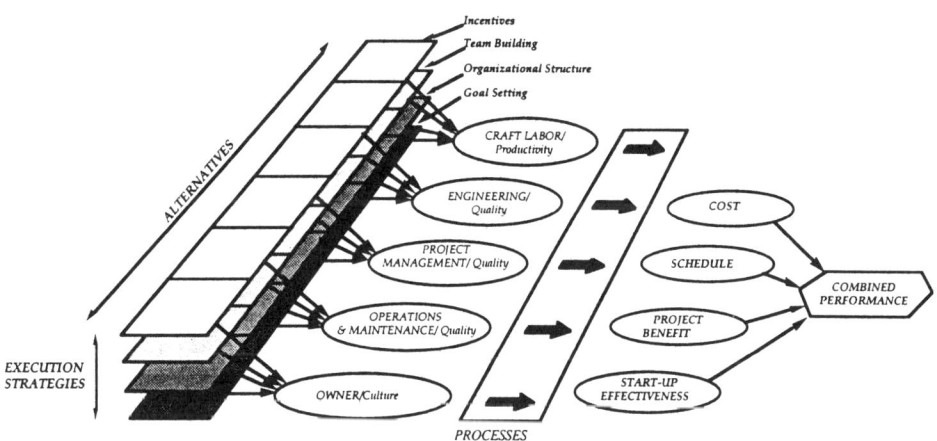

Figure 6. General performance model.

judged how the people can drive the processes toward improved performance. Finally, the expertise of the specialists in incentives, or team building, or perhaps partnering, is used to determine how such management actions motivate people.

The model can evaluate execution strategies, individually and combined as shown in Figure 6. Starting with the left-hand side of the model, each layer represents a single option set such as incentives, team building or project organizational structure. There are many alternatives and combinations of alternatives for each option set. A specific incentive plan can be combined with a specific team building program and choices among the other option areas to form an execution strategy.

Following the options there is a set of variables that is directly affected by project options; these variables are called drivers. Each combination of alternatives within an option set is assessed as to its probable impact on drivers. Drivers, in turn, propagate these effects through interactions among themselves and with processes. Drivers include the field labor constructing the project, engineering personnel involved in all design and specification activities, the project management team, key operating and maintenance individuals, and, of course, the client. One way to visualize these people drivers is on the basis of to what degree they achieve their full potential and have the maximum impact on quality or productivity.

The project processes included in the model mirror the typical time phases of a project: (1) definition/feasibility, (2) design, (3) procurement, (4) construction, and (5) start-up/operations. Assessments of how each process will likely impact each outcome are made by the project team on a project specific basis. Direction and magnitude of this effect are both assessed.

The right-hand side of the GPM shows a 'combined performance' box. This measure is obtained by combining several performance outcome measures using weights elicited from top management. The performance outcomes are the true results of the model. The analysis approach developed allows for the comparison of multiple objectives and the client organization's top management has the flexibility of determining and modifying these objectives by selecting new performance measures. Example performance measures used in this study are project cost, project schedule, the on-going value or contribution of the facility to the firm and the effectiveness with which the facility is placed into operation. The last two performance measures are examples of new measures necessaries for the evaluation in the new production philosophies.

4.2.2 *Analysis capabilities*

The computational scheme utilized within the model allows all possible execution strategies to be compared on a relative basis. Preferred strategies are ranked either on the basis of combined performance or on any single chosen criterion. Figure 7 shows a comparison of the benefits of different organizational structures for an hypothetical project.

Analyses can be extended to other options or to a combination of options for the project. Figure 8 shows a comparison of strategies which combine three options: organizational structure, team building and incentive plans. In addition, sensitivity analysis can be performed for selected alternatives to determine the robustness of any highly ranked strategy, and the drivers or processes with greater impact. The causal structure can be reviewed in detail to gain a better understanding of the performance impact mechanisms.

The outputs of this model are predictive, quantified comparisons of project execution strategies in terms of the outcome measures and detailed qualitative and quantitative explanations of the causal interactions. The idea behind this model seems appropriate to evaluate project performance from a general system analysis perspective, to get a better understanding of the global effect of management actions.

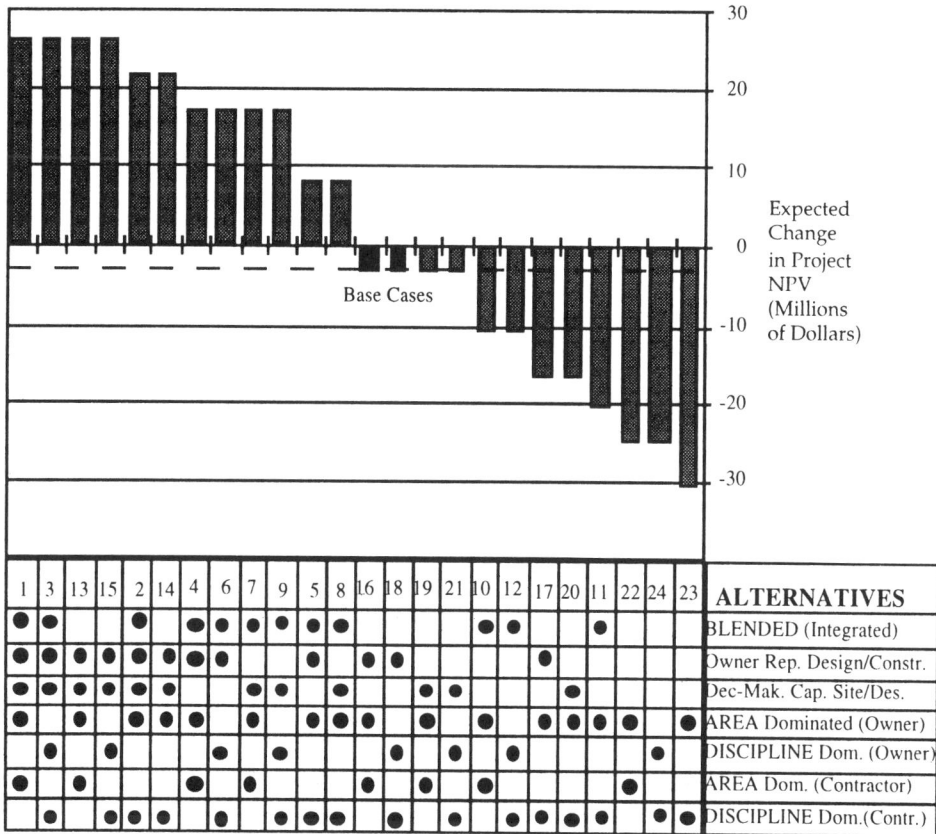

	1	3	13	15	2	14	4	6	7	9	5	8	16	18	19	21	10	12	17	20	11	22	24	23	ALTERNATIVES
	●	●		●			●	●	●	●	●	●					●	●		●					BLENDED (Integrated)
	●	●	●	●	●	●	●	●				●		●	●				●						Owner Rep. Design/Constr.
	●	●	●	●	●	●			●	●		●				●	●			●					Dec-Mak. Cap. Site/Des.
	●		●		●	●	●		●		●	●	●		●		●		●	●	●	●		●	AREA Dominated (Owner)
		●		●				●		●				●	●		●					●			DISCIPLINE Dom. (Owner)
	●		●				●		●				●		●		●				●				AREA Dom. (Contractor)
		●		●	●	●		●			●	●	●		●		●	●	●	●	●		●	●	DISCIPLINE Dom.(Contr.)

Figure 7. Benefits of organizations.

This modeling methodology can be easily adapted to model project performance for continuos improvement. In fact, Ashley & Teicholz (1993) have recently propose to use this methodology as a basis to predict the impact of project management actions on industrial facility quality. The effect of project options such as project organization, contractual conditions or data integration on a set of performance outcomes, which capture a comprehensive evaluation of quality/performance, could be evaluated using this approach.

5 SUMMARY AND CONCLUSIONS

The models discussed in this chapter provide concepts that could be applied when introducing new production philosophies in construction. It is important to build a bridge between traditional and new developments in construction performance improvement. The examples shown in this chapter illustrate how traditional concepts and tools can be valuable aids in developing continuous improvement efforts.

Before attempting improvement it is necessary to establish a framework for meas-

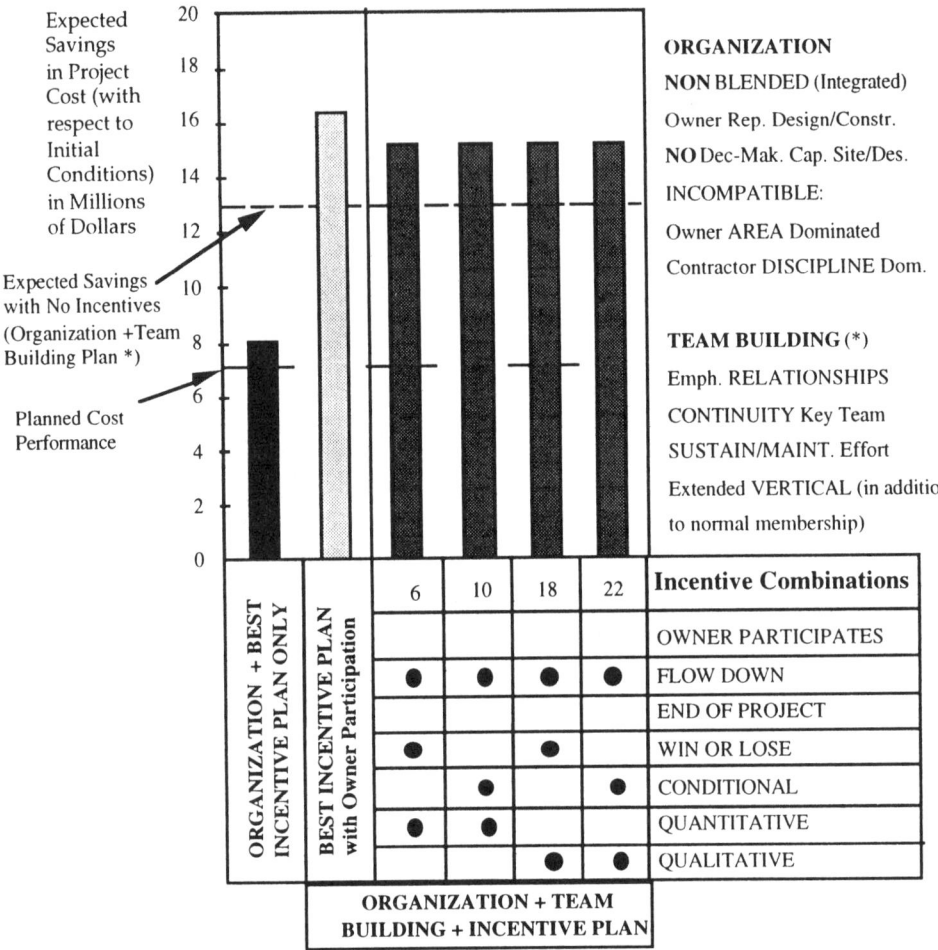

Figure 8. Combined effects: Incentive plan, organization, and team building.

urement and evaluation. Measurement and performance evaluation models have developed standards that, if adopted, may allow improved quality of the information available for decision making and research.

Prediction models provide systematic procedures to account for the different factors that affect productivity. The Productivity Theory Factor Model (Thomas & Yiakounis 1987) provides both standard procedures for collecting information and a more rigorous way of accounting for productivity factors. The Conceptual Construction Process Model (Sanvido 1988) shows a different perspective on the problem that explains more thoroughly the functional relationships and influences that affect productivity.

Causal models provide a qualitative model structure to explore actions that can affect productivity and to understand the mechanisms which produce the results. Borcherding's model (Borcherding et al. 1986), in particular, provides an interesting classification of sources of reduced productivity which closely resemble waste categories.

Work sampling models can be a valuable tool to measure waste, variability and other performance elements on the construction site. Its application by personnel adequately trained, to promote continuous improvement, can provide additional value to the application of this simple tool. The benefits of direct observation, as a result of this approach, can help to detect additional sources of reduced performance.

In general, most of the models discussed in this chapter focus on a limited number of performance elements, usually one or two, and they are restricted at some particular project phase or level. The General Performance Model provides a more global perspective including multiple performance elements and integrate project phases and levels. It integrates the complexity of the causal structure within the procedures used for prediction of performance to help explain the factors and interactions that affect performance in a meaningful way.

REFERENCES

Alarcón-Cárdenas, L.F. & Ashley, D.B. 1992. Project Performance Modeling: A Methodology for Evaluating Project Execution Strategies. A Report to the Construction Industry Institute, The University of Texas at Austin, Source Document 80.

Ashley, D.B. & Teicholz, P. 1993. Prediction of Integration Impacts on Engineering-Procurement-Construction (EPC) Processes and Industrial Facility Quality – a Proposal to the National Science Foundation.

Borcherding, J.D. & Alarcón, L.F. 1991. Quantitative Effects on Construction Productivity. *The Construction Lawyer*, Vol. 11, No. 1. American Bar Association, USA.

Borcherding, J.D., Palmeter, S.B., & Jansma, G.L. 1986. Work Force Management Programs for Increased Productivity and Quality Work. *EEI Construction Committee Spring Meetings*.

BRA 1982. Construction Productivity Measurement. Report No. A-1. Business Roundtable, New York, N.Y.

Brauer, R.L. 1984. *AFCS Climatic Zone Labor Adjustment Factors*. US Army Corps of Engineers, Construction Engineering Laboratory, Champaign, Illinois.

BRSR 1983. *More Construction for the Money: Summary Report of the Construction Industry Cost Effectiveness Project*. Business Roundtable. New York, N. Y.

Dallavia, L. 1952. *Estimating Production and Construction Costs*. Dallavia CO, Houston.

Edmondson, C.H. 1974. You Can Predict Construction Labor Productivity. *Hydrocarbon Processing*, pp. 167-180.

Gregerman, I.B. 1981. Worker Productivity-Characteristics and Measurement. *AACE trans.* I.2.1-I.2.5. American Association of Cost Engineers, USA.

Honton, E.J., Stacey, G.S. & Millett, S.M. 1985. *Future Scenarios: The BASICS Computational Method*. Battelle, Columbus Division, Ohio.

Horner, R., Talhouni, B. & Whitehead, R. 1987. Measurement of Factors Affecting Labour Productivity on Construction Sites. *CIB W-65 The Organization and Management of Construction 5th International Symposium*, London.

Howell, G.A. & Casten, M. 1983 (in-house publication). *Making Jobs Run Better*. Howell Assoc.
Judson, A.S. 1982. The Awkward Truth About Productivity. *Harvard Business Rev.* 82(5): 93-97.
Kellogg, J.C., Howell, C.G. & Taylor, D.C. 1981. Hierarchy Model of Construction Productivity. *Journal of the Construction Division, ASCE*, vol CO1, Proc. Paper 16138, pp. 137-152.
Kern, D.R. 1982. Engineering and Construction Projects, The Emerging Management Roles. *Proceedings of the Specialty Conference, ASCE,* New Orleans, Louisiana.
Koehn, E. & Brown, G. 1985. Climatic Effect on Construction. *Journal of Construction Engineering and Management, ASCE*, Vol. 111, No. 2.
Koskela, L. 1992. Application of the New Production Philosophy to Construction. Technological Report No. 72, CIFE, Stanford University.
Kuipers, E.J. 1977. Field Confirmation of Environmental Factors for Construction Tasks. Technical Report P-78, US Army Construction Engineering Research Laboratory, Champaign, Illinois.
Leonard, C.A. 1987. The effect of Change Orders on Productivity. *The Revay Report*, Revay and Associates Limited, Vol. 6, No. 2.
Lorenzoni, A.B. 1978. Productivity... Everybody's Business and It Can be Controlled!'. *Transactions 5th International Cost Engineering Congress, Utrecht*, The Netherlands.
Maloney, W.F. 1990. Framework for Analysis of Performance. *Journal of Construction Engineering and Management*, Vol. 116, No. 3. ASCE, USA.
Maloney, W.F. & Fillen, J.M. 1985. Valance of and Satisfaction with Job Outcomes. *Journal of Construction Engineering and Management*, Vol. 111, No. 1. ASCE, USA.
Neil, J.M. 1982. Labor Productivity, Chapter 8. *Construction Cost Estimating for Project Control.* Prentice-Hall, Inc., Englewood Cliffs, N. J.
Neil, J.M. & Knack, L.E. 1984. Predicting Productivity. *American Association of Cost Engineers (AACE), Transactions* H-3. USA.
Oglesby, C.H., Parker, H.W. & Howell, C.A. 1989. *Productivity Improvement in Construction.* Mac Graw Hill Book Company.
Riordan, B. 1986. Labor Productivity Adjustment Factors: A Method for Estimating Labor Construction Costs Associated with Physical Modifications to Nuclear Power Plants. Cost Analysis Group, Office of Resource Management, US Nuclear Regulatory Commission, Washington D.C.
Sanvido, V.E. 1988. Conceptual Construction Process Model. *Journal of Construction Engineering and Management, ASCE,* 114(2): 294-310.
Shaddad, M.Y.I. & Pilcher, R. 1984. The Influence of Management on Construction System Productivity – Towards a Conceptual System Causal Research Model. *CIB W-65 The Organization of Management of Construction 4th international Symposium*, Waterloo, Ontario, Canada.
Sink, D.S. 1985. *Productivity Management: Planning, Measurement and Evaluation, Control and Improvement.* John Wiley & Sons.
Thomas, H.R. Jr. 1981. Construction Work Sampling, Department of Civil Engineering, The Pennsylvania State University, University Park, Pa.
Thomas, H.R. Jr. & Kramer, D.F. 1988. The Manual of Construction Productivity Measurement and Performance Evaluation, A Report to the Construction Industry Institute *(CII)*, from Pennsylvania State University.
Thomas, H.R. Jr. & Yiakounis, I. 1987. Factor Model of Construction Productivity. *Journal of Construction Engineering and Management, ASCE,* 113(4): 623-639.
Thomas, R.H., Maloney, W.F., Horner, R.M. W., Smith, G.R., Handa, V.K. & Sanders, S.R. 1990. Modeling Construction Labor Productivity. *Journal of Construction Engineering and Management, ASCE*, vol. 116, No 4.
Tucker, R.L. & Scherer 1986. The CII Model Plant. A Report to The Construction Industry Institute (CII), The University of Texas at Austin.
Tucker, R.L. 1982. Implementation of Foreman Delay Surveys. *Journal of Construction Engineering and Management,* ASCE, 108(4): 577-591.

Characterization of waste in building construction projects*

ALFREDO SERPELL
Department of Construction Engineering and Management, Catholic University of Chile, Santiago, Chile

ADRIANO VENTURI & JEANETTE CONTRERAS
School of Civil Construction, Catholic University of Chile, Santiago, Chile

ABSTRACT: Information obtained from productivity consulting services provided to a substantial number of building construction projects from 1990 to 1994, has been processed and organized with the purpose of studying the types of construction waste in this kind of construction projects. From the results obtained in this study, this chapter presents:
1. A classification of construction waste and their main causes, and
2. Statistics of the frequency and relative importance of these types of waste.

1 INTRODUCTION

Since the beginning of a construction job, project/site management has to deal with many factors – most of them caused by their own actions or inaction – that negatively affect the construction process, producing different types of waste that can convert a good 'to be' project into a bad 'it was' project. In most cases, construction managers do not know or recognize the factors that produce waste nor have they measurements of their importance. It can be said that most of the factors are not easily visible. Thus, the identification of these factors and their causes, and the measurement of their importance is a useful information that would allow management to act in advance to reduce their negative effects.

During the last 5 years, consulting services have been provided by a consulting group at the Department of Construction Engineering and Management of the Catholic University of Chile, to more than 40 construction sites that have started programs to improve their site performance, most of them successfully. Seventeen of these construction sites were high rise building construction projects and were selected for studying purposes.

This consulting experience has generated a great amount of information on wasted resources and time in building construction projects. The information obtained has been processed to study the types of waste that have occurred during construction work, and their relative importance.

*Presented on the 3rd workshop on lean construction, Albuquerque, 1995

2 CHARACTERISTICS OF THE SAMPLE AND INFORMATION

The sample implicated 17 buildings constructed by seven construction companies in the Metropolitan Region of Chile, around the capital city of Santiago, from 1990 to 1994. Table 1 shows a summary of the building's sample.

During the consulting/observation time, information was generated mainly in the form of reports that were delivered to site managers once a week. The structure of these reports consisted of two sections:

1. Quantitative section, based on work sampling results.

2. Qualitative section, where an analysis of the quantitative data and general observations regarding work conditions were presented, and corrective actions proposed.

The quantitative section is the most important part of the report because it works as detection tool of waste sources. Work sampling results in this section have been divided as follows:

– General activity levels, using the categories of productive work, contributory work and non-contributory work for the complete job site;

   – Activity levels of company personnel and subcontractors' personnel separately;

   – Activity levels of each of the most critical trades.

The categories of contributory work and non-contributory work have been further subdivided into the following subcategories included in the reports:

   – Contributory work: Transporting, cleaning, receiving instructions, measuring, and other specific activities;

   – Non-contributory work: Waiting, idle time, travelling, resting, and reworking.

This relative detailed information facilitates the identification and attention to the most significant waste sources, and becomes the basis for analysis of improvement

Table 1. Characteristics of the sample.

Number	Type of building	Number of stories	Total gross area (m^2)	Time of observation (months)
1	Office	33	90,000	19
2	Office	25	60,000	6
3	Office	23	30,000	5
4	Housing	20	30,000 (app.)	7
5	Housing	19	7,000	14
6	Hotel	18	15,000	13
7	Housing	17	14,000	10
8	Housing	17	13,000	13
9	Office	16	33,000	7
10	Hotel	16	10,000	13
11	Housing	16	10,000	9
12	Housing	15+15	16,800	5
13	Housing	15	9,000	8
14	Housing	14	10,000	11
15	Housing	14	8,000	13
16	Housing	14	7,000	8
17	Housing	12	6,000	8

actions to reduce any identified waste. Supporting information from visual observations and, sometimes, from other parallel studies, is also incorporated in the qualitative analysis.

The information provided by work sampling presents some limitations that should be considered. The most important are as follows:

– It does not show clearly and precisely the origin of waste;

– It only measures work time utilization, but it doesn't directly provide production rates, which should be measured in other ways;

– It doesn't provide measurements of waste of materials or equipment.

Despite these limitations, the information obtained using the described approach has been extremely useful to achieve the stated objectives. The application of judgement and experience allows consultants to identify and detect situations that are producing waste and to point out the main causes that are producing them.

3 BACKGROUND ON THE CONSTRUCTION PRODUCTION PROCESS

Figure 1 presents the construction production process on which this work has been based. The proposed model is an open and dynamic system inside an environment that conditions its status and behaviour. Part of this environment is controllable by the system but other factors are outside of its control.

The main and most critical components of the construction process as portrayed in the figure, are:

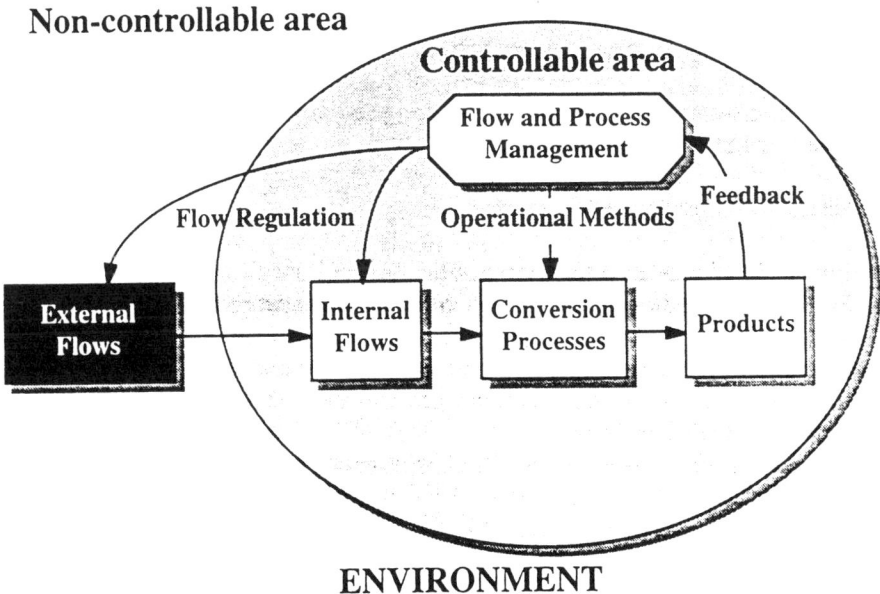

Figure 1. Construction production process model.

a) Flows and conversion management: Responsible of making the decisions that define the performance of the system;

b) Flows: Are the inputs to the system and can be separated in two types, resources (labor, materials and construction equipment), and information. These flows contemplate all activities since the need is defined and the elements arrive to the conversion location. There are controllable and uncontrollable flows. Examples of the first type are the flows of materials or instructions from a warehouse or management respectively, to the workplace. Uncontrollable flows are: Suppliers' provision of resources and design information;

c) Conversion activities: The processes that transform the flows into finished and semi-finished products. The methods used in this activities are decided by the flows and conversion management;

d) Products: The results of conversion activities.

The flow and process management, is the function that puts the system into action through three major actions:

1. Resources and information flow regulation, including: Allocation of resources, defining quantity and specifications; planning and coordination; distribution.

2. Design of work methods, looking for optimization of activities' execution.

3. Monitoring and controlling of system activities.

Construction waste is then produced during the construction process due to several causes as shown in the following section.

4 CLASSIFICATION OF CONSTRUCTION WASTE

This study deals with the identification of the most relevant factors that produce waste of productive time in building construction works. This waste comes from flow activities, conversion activities and management activities. Their occurrence is generally manifested by two common construction situations: work inactivity and ineffective work. Figure 2 shows the classification adopted in this study after reviewing the information collected from the building construction projects.

The classification shown in the figure presents some limitations that should be considered:

– Slow work: This waste of time is related to the efficiency of processes, construction equipment, and personnel. Then it is difficult to measure it because it is first necessary to know the optimal efficiency that could be reached, which is not always possible;

– Rework: Not always is the result of labor ineffectiveness; uncontrolled problems like weather conditions, earthquakes, etc. also result in rework, although they can be prevented.

5 CLASSIFICATION OF WASTE CAUSES

The most important causes of wasted time identified by the study were classified as shown in Figure 3.

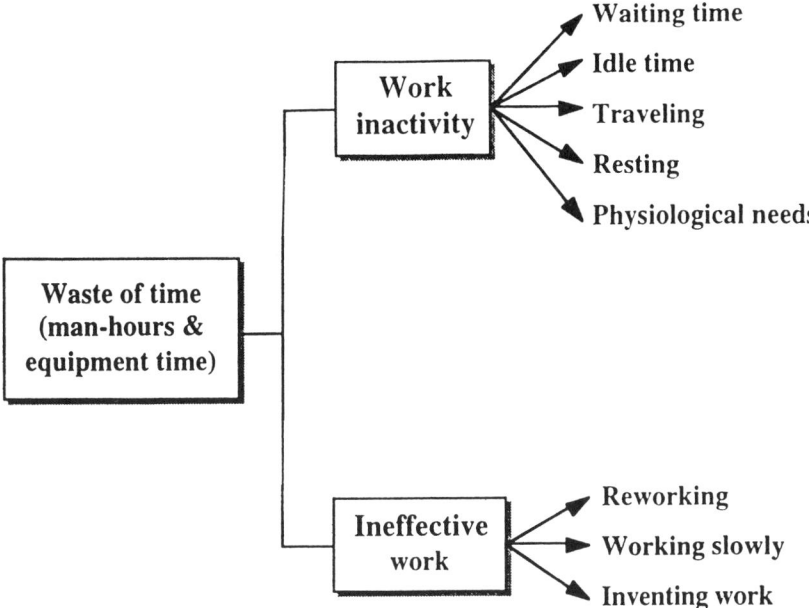

Figure 2. Categories of wastes of productive time.

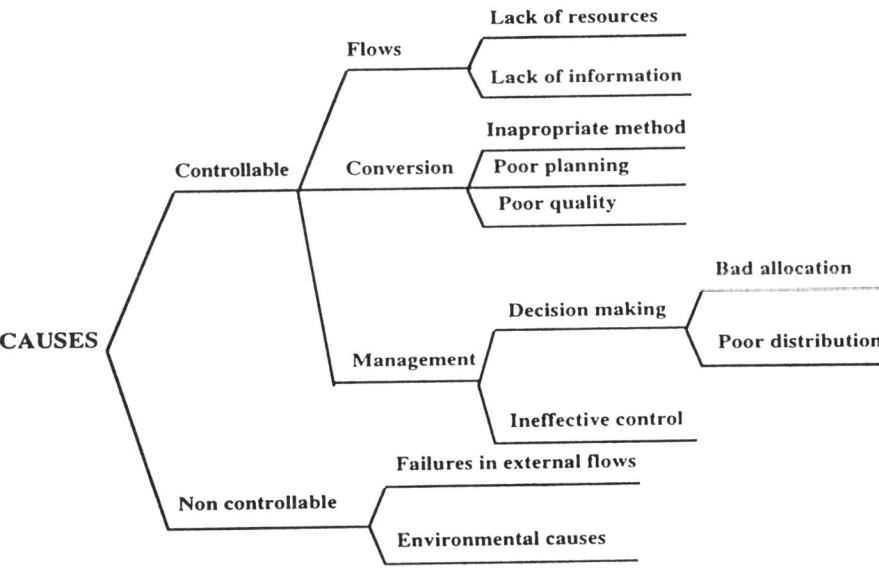

Figure 3. Classification structure of causes of wasted time.

5.1 *Controllable causes associated to flows*

The principle flow causes were as follows:

a) *Resources*

– Materials: Lack of materials at the work place; materials are not well distributed; inadequate transportation means;

– Equipment: Non availability; inefficient utilization; inadequate equipment for work needs;

– Labor: Personal attitudes of workers; stoppage of work.

b) *Information*

– Lack of information;

– Poor information quality;

– Timing of delivery is inadequate.

5.2 *Controllable causes associated to conversions*

The following causes were identified:

a) *Method*

– Deficient design of work crews;

– Inadequate procedures;

– Inadequate support to work activities.

b) *Planning*

– Lack of work space;

– Too much people working in reduced space;

– Poor work conditions.

c) *Quality*

– Poor execution of work;

– Damages to work already finished.

5.3 *Controllable management related causes*

The following causes were identified:

a) *Decision making*

– Poor allocation of work to labor;

– Poor distribution of personnel.

b) *Supervision*

– Poor or lack of supervision.

Regarding the non-controllable causes, the majority of them were associated to suppliers' and designers' performance. Also, there were some causes related to the environment, like weather conditions and festivities.

6 ANALYSIS AND RESULTS OF THE STUDY

This section presents a summary of the most relevant results obtained from the analysis of information.

6.1 *General distribution of work sampling*

The average distribution of working time of the 17 observed building is shown in Figure 4. The minimum value of productive work was 35% and the maximum was 55%. Regarding the non-contributory work category, the minimum value was 18% for the same building that has the top productive value, and the maximum was 31%, corresponding to the building with the lowest productive value. Finally, the extremes for the contributory work were a minimum of 24% and a maximum of 34%.

The differences between the best and the worst building are very important. Analysis are being done to find out what were the factors that might explain these differences.

6.2 *Types of wasted time measured by non-contributory work*

As shown in Figure 5, the main subcategories of non-contributory time which explain 87% of the total value of this category are: waiting time, idle time and travelling, in that order. A second order subcategory is the time spent in resting. An unex-

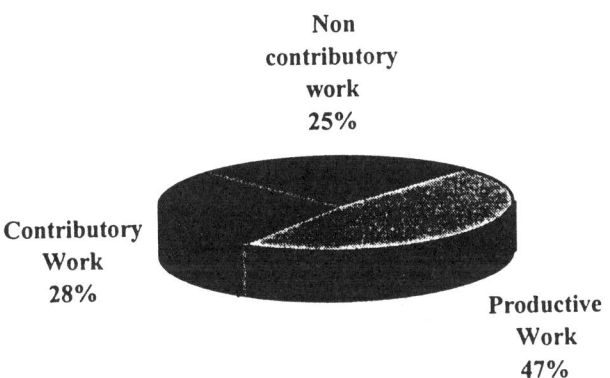

Figure 4. General distribution of work categories.

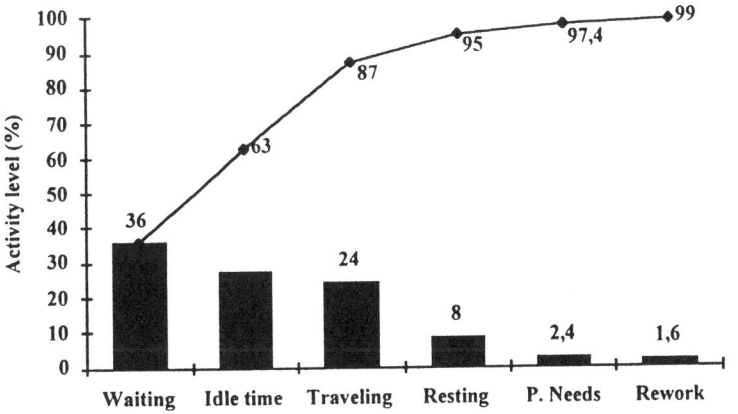

Figure 5. Distribution of non-contributory time.

pected result is the low value of rework, a factor that has been observed more critical that is shown here. One possible explanation of this outcome is the fact that work sampling is not the most appropriated tool to detect and measure this type of waste.

The causes that were found accountable of the waiting time, are presented in Figure 6. The graphic displays the total frequency of occasions where each factor was identified as responsible of waiting conditions. Overmanning clearly was the most critical factor, which is consistent with other observations. There is a general agreement that Chilean building construction projects normally have more people than needed, especially unqualified people. Foremen like to have a surplus of people to face potential risks in their work.

Figure 7 shows the factors that are responsible for idle time. The most important factors by far are: lack of supervision and overmanning. The lack of supervision has two components:
– Workers stop when they are not controlled; a cultural and educational related problem;
– Workers cannot work because they are waiting for instructions that supervisors should give them.

Overmanning also produces idle time when there are more workers than work to do, a common problem as mentioned before.

Finally, the main causes of travelling time are presented in Figure 8. Again, overmanning appeared as the most frequent factor, followed by lack of supervision, workers' attitude, materials supply and site working conditions.

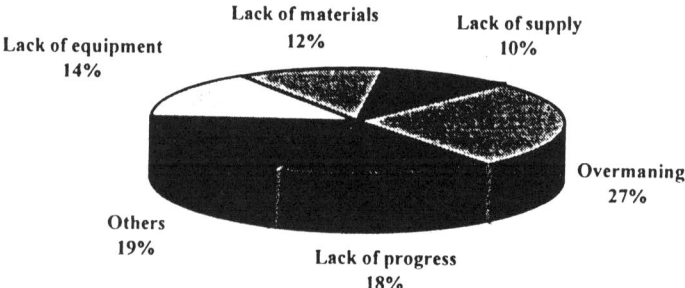

Figure 6. Causes of waiting time.

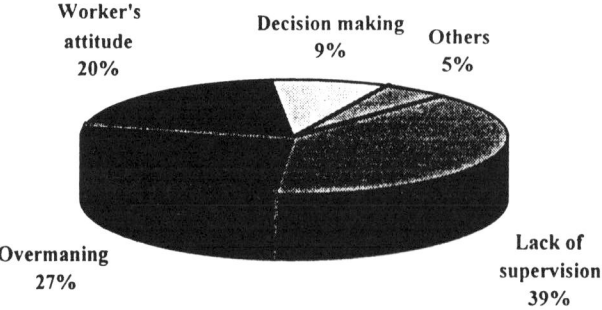

Figure 7. Causes of idle time.

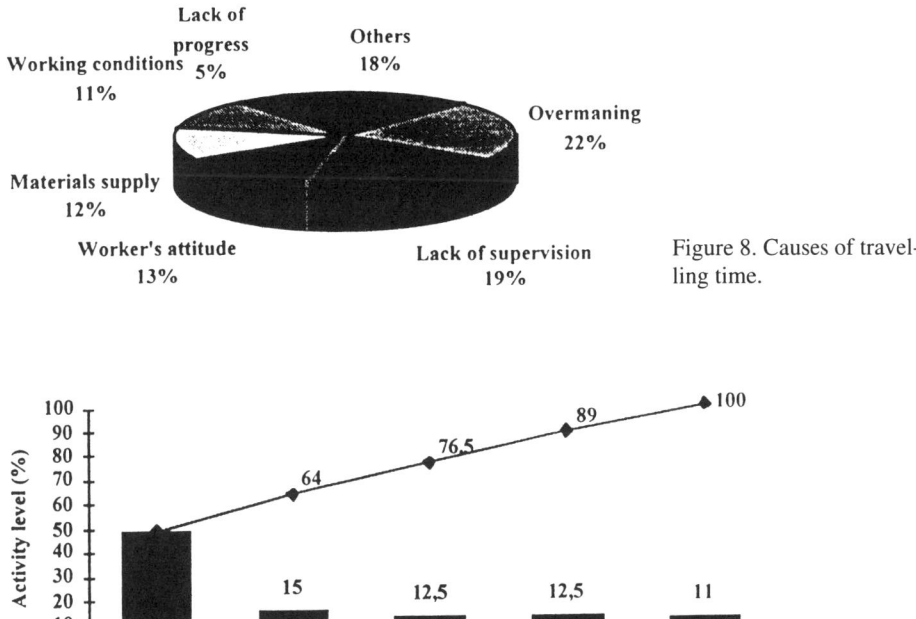

Figure 8. Causes of travelling time.

Figure 9. Distribution of contributory work time.

6.3 *Wasted time related to contributory work*

Figure 9 describes the main categories of contributory work found in the study. Although it is not possible to completely eliminate contributory work time, its reduction allows an increasing of productive time and thus it is highly convenient. Transporting is responsible for almost 50% of the total category, a very interesting outcome, that confirms previous observations performed at many building construction sites.

The factors that account for most of the transportation time are presented in Figure 10. Inadequate transportation methods is the most significant factor, being responsible of almost 40% of the transportation cases. In many cases, qualified personnel (carpenters, plumbers, electricians) dedicate a considerable amount of time transporting their materials and tools instead of having laborers or less qualified personnel doing this job. Poor distribution and the lack of transportation equipment also are important sources of transportation time.

Inadequate distribution of materials is a situation that has been observed in almost every construction project. Long transportation distances due to deficient layout of temporary facilities, extra movements of materials and equipment because of lack of planning of their initial unloading positions are just two of many other similar cases of transportation waste.

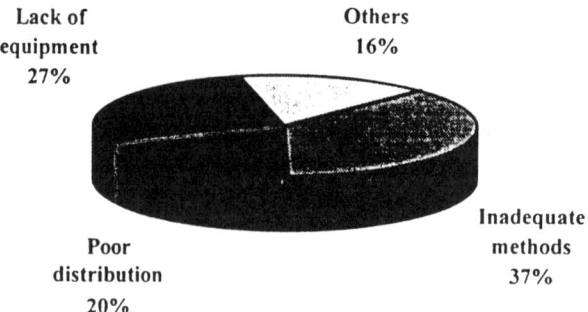

Figure 10. Causes of transportation time.

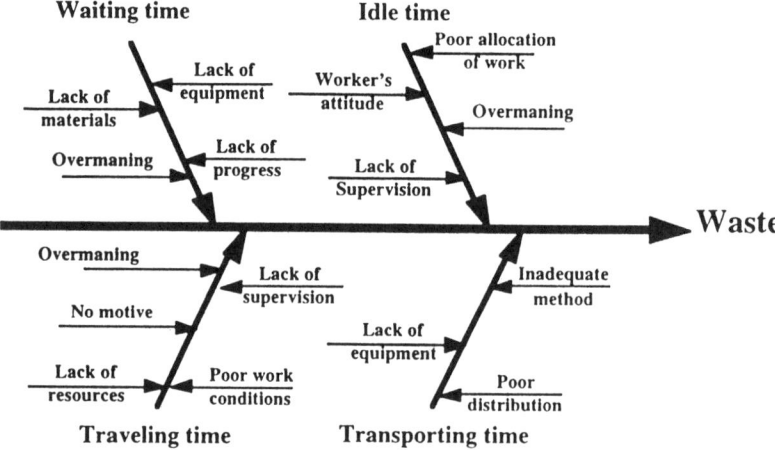

Figure 11. Cause-effect diagram for the main causes of construction waste.

6.4 *Summary of findings*

Work time distribution in the building construction projects considered in this study, demonstrates that around 53% of the total working time is dedicated to non productive activities. Many different factors are the sources of these activities. A summary of the most relevant ones identified in this study is shown in Figure 11. Many others were found to be significant but less important and have not been reported here.

7 CONCLUSIONS

The information provided in this chapter can be very valuable for site managers in many ways. First, they will be able to inform themselves about the main waste factors in construction work. Thus they may become better informed managers, who know what are the problems they face, a requisite to solve any problem. Second, they can then focus their attention on these potential risks of unproductive time and act effectively to reduce both the risks and their impacts.

Planning is the key managerial function that should be used to be effective in reducing or eliminating these waste factors. Most of the factors shown in Figure 11 are clear demonstrations of a lack of adequate planning. Resources not available, supervision not available or inadequate, poor layout distribution, overmanning, lack of progress, poor allocation of work, are all planning deficiencies.

As reported by Serpell et al. (1995) project planning faces several problems in construction. The most frequently are related to a poor definition of job objectives, insufficient use of computer planning systems and the reduced availability of planning and control data. What might be the main problem is the lack of time that site managers have to plan. Generally, they are assigned to the project team a short time before the project starts. Later, after the project is under execution, traditional priorities and habits restrict their available time for planning. These problems has also been addressed by Howell & Ballard (1994).

Then, to be effective in addressing these problems, planning should be an activity of every site manager, from project managers to foremen. Although professionals normally have received planning instruction, general foremen and foremen have not. Then we need to educate and train our first line managers, to make them able to carry out effective planning.

The major planning focus at this level should be on short-term planning. Most of the wastes that take place at construction sites on day to day operations are the result of lack of effective short-term planning. A simple, but effective tool to carry out short-term planning has been used successfully in construction works as shown by Serpell (1993).

Finally, this work, still underway, contributes in that it is a systematic attempt to observe and measure wastes in construction processes. Outcomes of the study will help in the understanding of the nature of the construction production process. This understanding is necessary to create a much needed theoretical framework of construction.

ACKNOWLEDGEMENTS

The authors gratefully acknowledge the Corporación de Investigación de la Construcción (Construction Research Corporation), for providing funding to support this research effort.

REFERENCES

Howell, G. & Ballard, G. 1994. Implementing Lean Construction: Stabilizing Work Flow. *2nd. Annual Conference of the International Group of Lean Construction*, Santiago, Chile.

Serpell, A., Crovetto, J. & Seymour, D. 1995. A Current Vision of Construction Management Practices in Chile. Unpublished paper submitted to the CIB Working Commission 65 Organization & Management of Construction Symposium to be held in Glasgow, Scotland, 1996.

Serpell, A. 1993. *Construction Operations Management* (in Spanish), Ediciones Universidad Católica de Chile, Santiago, Chile.

Lean construction and EPC performance improvement*

GLENN BALLARD
Department of Civil Engineering, University of California, Berkeley, USA

The conversion process model encourages suboptimization. One notorious example results from an owner insisting on keeping design costs below a certain percentage of total installed cost. A second example is the harmful practice of buying on price tag rather than cost to use, a practice much derided by Deming (1986). A third example is the practice of controlling manpower quantities as a sole or primary means for achieving schedule.

The mismanagement encouraged by the conversion model clearly can infect any or all of the phases and functions involved in a construction project; the design phase, the procurement phase, and the construction phase proper. This paper explores the consequences for performance improvement strategies of displacing the conversion model with lean construction concepts and principles.

The inadequacy of the conversion process model is especially apparent in regard to projects in which engineering, procurement, and construction overlap in time, i.e. 'fast track' projects, the norm in industrial plant construction. How is it reasonable to restrict design cost to a maximum percentage without taking into account the downstream impacts of design quality on plant performance and cost to construct?

Strange as it may seem, such suboptimization is the rule rather than the exception. Our ways of thinking about and managing EPC projects appear to have been formed in the 'good old days' when each phase was performed sequentially. The designer produces the design, including equipment and material specifications. Equipment and material are purchased and delivered to the plant site. A construction contractor is selected to assemble the equipment and material into the desired facility in accordance with the drawings and specifications from the designer. This non-overlapping sequence encouraged the misconception that each phase could be considered separately, without regard to interdependencies and trade-offs.

The shift to concurrent design, procurement and construction strains the assumption of independence, especially because of the obvious need to integrate the work activities of each phase within a single unified schedule. The examples from the automobile industry of lean design, lean supply, and lean manufacturing are now joining up with the already strained assumption of independence. Old concepts and new fight for supremacy as we move from the old paradigm to the new.

Fast track EPC projects provide an experimental laboratory for redesigning how we manage work and people. The results can then be adapted for projects in which

*Presented on the 1st workshop on lean construction, Espoo, 1993

the phases occur sequentially. In fact, several US industrial contractors have launched efforts to substantially reduce time and cost of EPC projects, recognizing the need for breakthrough both in performance and in management thinking.

These initiatives pose interesting questions for the development of 'lean construction':
 – What is the optimum investment program?
 – How to take into account the interdependency of functions?
 – Is it smart to invest in the front end and realize the gains on the back end? If so, how?
 – How to bring the entire process under control: The prerequisite for breakthrough to new performance standards?

Goals
EPC projects: 30% reduction in duration; 30% reduction in total installed cost.
Stand alone projects: Apply redesigned processes and modify as needed.

Goals will be achieved within 3 years; goal achievement will be demonstrated on projects; and each participating business unit will designate one or more pilot projects for experimentation.

Performance improvement strategy
In one case, the strategy is to pursue changes in three steps, beginning with controlling each function, then investing in design quality, supplier delivery, and construction cycle time reduction, in that order (Table 1).

Engineering is a supplier to construction; supplying drawings and specifications. Engineering supplies procurement with requisitions detailing what equipment, fabricated items, and bulk materials need to be purchased and what specialized services need to be contracted. External suppliers and service providers are suppliers to construction. Construction is obviously a supplier to the owner of the facility. The model for understanding the interdependency of these functions is the simple supplier/customer model.

In accordance with that model, it is apparent that the quality of design (the quality dimension important to construction is design constructability, document clarity and consistency, dimensional accuracy, etc.; not the performance capability, operability, or maintainability of the plant) and the reliable delivery of resources to the construction site are critical inputs to the construction work process.

Table 1. Performance improvement strategy.

Engineering	Procurement	Construction
1. Reduce delivery variation	1. Develop needed tools	1. Minimize the impact of erratic resource delivery and poor quality.
2. Improve the quality of engineering deliverables	2. Reduce delivery variation and the cost of equipment, materials and services	2. Reduce construction costs
3. Reduce delivery cycle time	3. Reduce delivery cycle time	3. Reduce construction duration

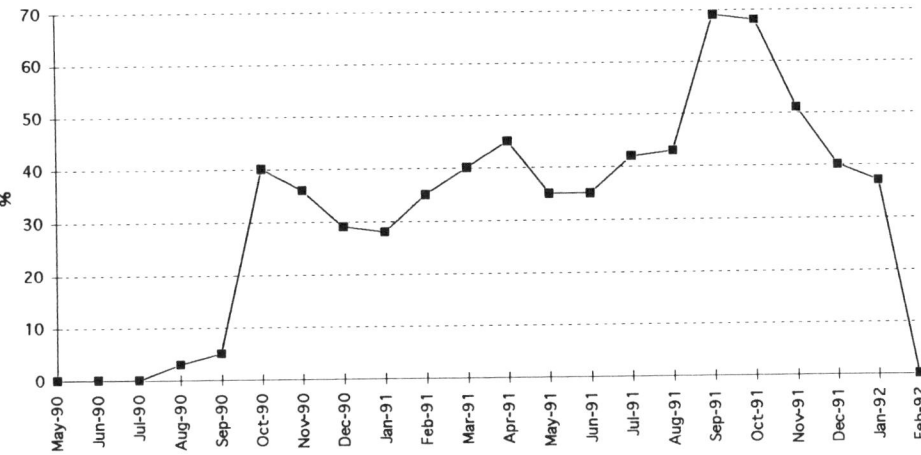

Figure 1. Project A % milestones missed.

Figure 2. Planning system elements.

What happens when the delivery to construction of an external resource is erratic? Figure 1 displays the percentage of drawings and specifications that were issued past scheduled milestone dates. On average, more than 30% of engineering deliverables were behind schedule. What's more, the average days late was 56.

Interestingly, the project was completed on time, on budget, and to the entire satisfaction of the owner. Does that mean the late delivery of engineering drawings and specifications had no impact on the project? I suggest that current schedules and budgets assume that such poor performance will occur. In short, current standards of performance include tremendous amounts of waste.

In order to explain the importance of reducing variation in resource delivery, I must introduce a few concepts. First, the essential elements of a planning system are those that determine what *should* be done, what *can* be done, and what *will* be done. In my experience in the construction industry, we usually do a good job with 'should', a mediocre job with 'can', and a very poor job with 'will'.

The front line supervisor, whether foreman, squad boss or purchasing supervisor, does his planning based on information he receives regarding 'should' and 'can'. His job is to approximate 'should' within the limits of 'can', thus producing assignments that are practical, and providing reliable input for the planning of interdependent work processes (Fig. 2). Where the commitment to what will be done is made within the organization, and how much in advance of the plan period, varies from project to project. When directives are contradictory or resource information is faulty, supervisors give up on crew level planning, coordination disappears, and performance dete-

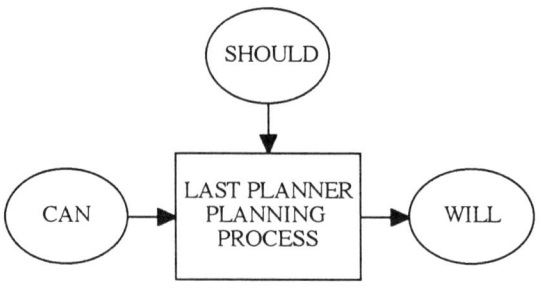

Figure 3. Crew level planning.

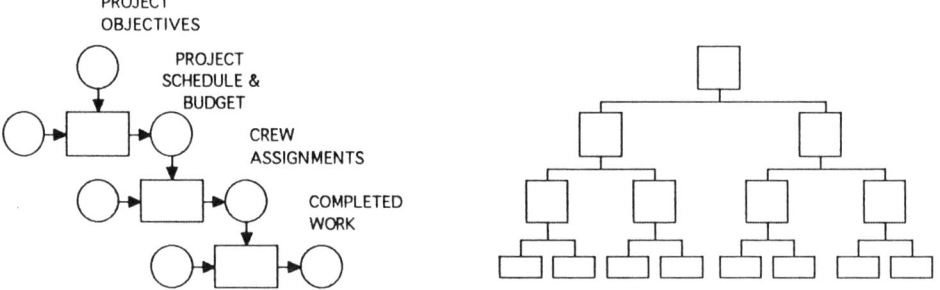

Figure 4. Last planner (the concept).

riorates. Assignments become mere expressions of the original schedule, untempered by resource availability. Too frequently, craftsmen spend most of their time trying to determine whether or not the assignment can be carried out.

It is obviously preferable to produce good assignments, and that means enabling the Last Planners in each constituent organization.

The Last Planner (Fig. 3) is last in a chain of planners, each providing directives ('shoulds') to the next. Construction is complex. Planning is not done by one person or group at one time. It is distributed throughout the organization and over the life of a project. The 'last planner' is the one who produces directives that drive direct work processes, not other planning processes; i.e. assignments. If the planning system fails to produce good assignments, it does not matter how good the upstream planning was. Those plans never get realized.

I am now attempting with several clients to shift the control focus closer to root causes by monitoring the 'percent plan complete' at the Last Planner level. Initial measurements show percentages from 35% to 50% (Fig. 4).

As we analyze and act on the causes for incompletion, those percentage rise, and productivity and production rise along with them.

Eliminating root causes can require the redesign of planning and control systems, coaching and training of planners, redistribution of information, as well as improvements in the physical distribution of resources such as materials, tool and equipment.

In the case illustrated here, insufficient and inadequate materials account for 40% of plan failures. Consequently, we chose to work first on the materials system (Fig. 5).

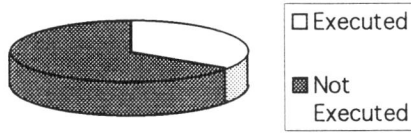

Figure 5. Percentage of daily plans executed.

Table 2. Performance improvement strategy.

Engineering
1. Reduce delivery variation: – Release time for investment in performance improvement by reducing supervisory firefighting time, maintain workable backlog, assign only from backlog, provide accurate and timely information on status of prerequisites – Understand the actual logic of past work processes and develop model logics, identifying appropriate and inappropriate workarounds – Chart work flow and eliminate queues between operations – Streamline interface with suppliers – Reduce inspection time by eliminating repetitive errors – Get the facts about defects, rework, delays, schedule conformance, constructability, design quality

Returning to the strategy for EPC performance improvement, it is thought necessary to first bring all work processes under control, in order to provide the conditions in which substantial improvements can be made. In this instance, 'in control' means able to keep commitments and standardization of processes; especially as regards delivery of inputs to downstream customers. The strategy for achieving control is to identify last planners, clarify their role and expectations, and enable them to be successful.

Step 2 for engineering is to improve quality. Step 3 is to reduce the duration of the design phase. Quality comes first because of its impact on the time and cost of procurement and construction.

Let's look more closely at engineering's Step 1 (Table 2). 'Release time for investment...' is a way of shielding the Last Planner from an erratic flow of resources, and may appear to be in contradiction with the principles of lean construction. Shingo (1981) declares the goal in his term 'non-stock production', while I am advocating the deliberate creation of inventory surge piles. The goal is the same, but to get there requires freeing management time and energy for improvement. The strategy is perhaps counter-intuitive, but, inverting Lenin, one step backward does lead to two steps forward. Reducing uncertainty at the Last Planner level reduces the amount of time supervisors spend hustling resources and fighting fires, and results in more in-process inspection of work, more coaching and innovation. We cannot reasonably expect those on the firing line to work on making improvements as long as they must struggle to simply get the work done at all.

When I first began research with this contractor, they were consistently unable to meet design schedules. Investigation revealed poor goal setting as one cause. Senior managers made whatever commitments were needed to win work; admittedly with some reliance on historical project durations. Often there would be meetings early in the project in which engineering management (project, process, mechanical, civil, structural, piping, electrical, and instrumentation) would develop a design schedule. The first attempt would be much too long, so each manager would cut some constraints. 'You don't need certified vendor data to estimate horsepower requirements! That's what you get paid for!' A rerun of the logic network would reveal a shorter duration, but still too long, so more constraints would be cut. Eventually, the managers went beyond their ability to understand and manage the risks resulting from their decisions. The consequence was an erratic flow of work, a deterioration of detailed planning, and gross schedule overruns. Construction was usually able to absorb the late and out-of-sequence delivery of drawings and materials to the job site, but at a tremendous cost.

To make better decisions about process logic, the contractor has begun an analysis of past projects grouped by type of facility being designed. While producing a road map to guide decision-making, they are also collecting work around strategies, i.e. strategies for managing the risks attendant upon cutting constraints (Fig. 6). One example is redundancy; e.g. size all starters at the top end to accommodate the greatest possible horsepower range.

A related issue here is the harmful substitution of progress planning and control for schedule planning and control. In industrial facility design, the control focus is not on producing the right output at the right time, but on producing the right amount of output, usually measured in terms of earned man-hours. Perhaps the erratic flow of work caused by pressure for speedier production has led people to give up. Unfortunately, controlling progress rather than schedule makes it even more difficult to achieve schedules.

Turning now to Procurement, the first step is to develop the tools needed for performance improvement (Table 3). The second step is to reduce delivery variation and cost. The third step is to reduce the time required to produce and provide goods and services.

Again, duration is the last target, and again for the same reason. The greater impact on total installed cost and total project duration is from erratic, unreliable delivery. In addition, the reduction of uncertainty is the precondition for improvement. What's more, the reduction of procurement durations requires prior investment in technology and in relationships.

This contractor has invested in an integrated materials management system. That means they can generate purchase orders directly from requisitions, the purchase orders show up on the Expediting and Supplier Quality modules, and the same data supports job site receiving and issue. The intent is to avoid hand-offs across functional boundaries as a product progresses through the project cycle, all the way to installation and use (or disposition of surplus).

□ Insufficient Material
▨ Inadequate Materials
■ Insufficient Manpower
▢ Insufficient Time
▨ Poor Communication

Figure 6. Causes for non-compliance with plans.

Table 3. Performance improvement strategy.

Procurement

1. Develop needed tools:
– Integrated project materials system: An integrated database supporting all materials management functions from quantity takeoff to disposition of surplus
– Supplier quality process: Evaluation and ranking of suppliers; reduce the number of suppliers; exercise a bias for proximity in supplier selection; blend with clients' preferred supplier lists, especially for alliances or partnering arrangements
– Develop supplier alliances: Exchange equity/take other steps to unify interests
– Get the facts about: The quality and timeliness of directives provided to suppliers; schedule performance of suppliers; quality performance of suppliers (defects in delivered products or services); lead times required by suppliers

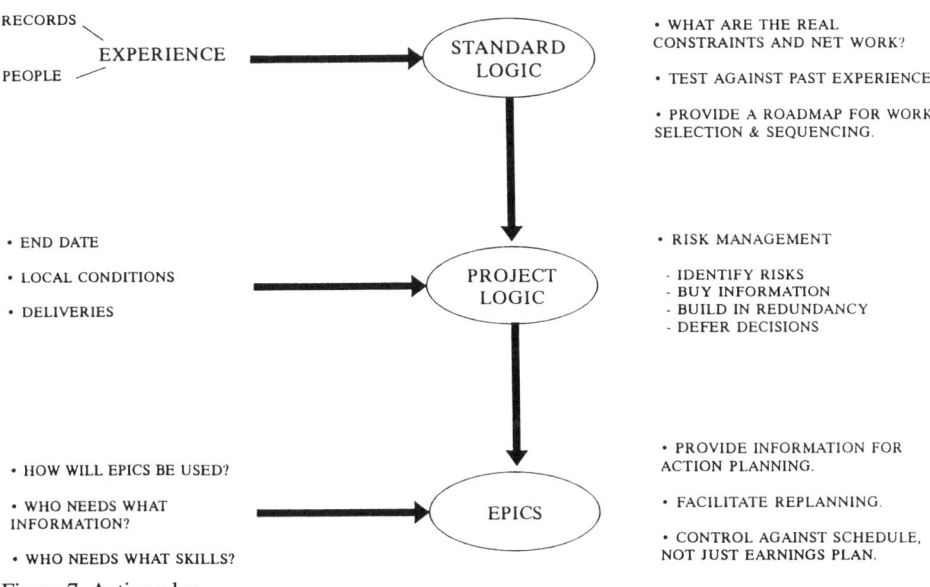

Figure 7. Action plan.

They are also implementing a supplier quality process and developing supplier alliances. Last but not least, they are beginning to get the facts regarding procurement processes and supplier performance.

One example is this control chart showing the delivery of structural steel over a nine month period. The range is from 13 weeks early to 11 weeks late. This reduced craft productivity, not only for structural ironworkers, but also for the electricians, who were forced to pull wire and cable within a shorter duration because prerequisite work was not completed on time (Fig. 7).

Like engineering and procurement, construction's first task is to minimize the impact of poor quality and timeliness of inputs from others (Table 4). Again, the last target is reduction of durations, which requires for substantive gains prior improvement in the flow of design documents and materials. While waiting for such im-

Table 4. Performance improvement strategy.

Construction
1. Minimize the impact of erratic resource delivery and quality: – Stabilize the work process by maintaining workable backlogs and assigning only from backlogs – Eliminate delays for materials, tool, equipment, and crew interference – Make the foreman a manager: require detailed planning, provide information regarding what ought to and can be done, require analysis of deviations from performance standards, select and train for success – Get the facts about delays and rework

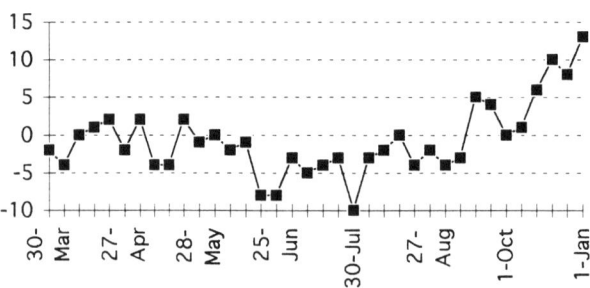

Figure 8. Steel delivery performance-scheduled delivery versus actual.

provements, construction can capitalize on a more stable environment and tackle cost reduction in Step 2.

The key here is to make the foreman a manager rather than a gang pusher. As discussed previously, that involves enabling the foreman to act as a Last Planner.

Actually, Step 1 also includes some cost reduction, i.e. reduction of delays, which reduces non-productive time and improves productivity. When the flow of drawings and materials is off schedule, productivity will deteriorate. However, the damage can be minimized by insuring that assignments are material-sound and can be done.

The magnitude of delays on large industrial projects is indicated in Figure 8, with approximately 25% of direct labor time lost, and that was on projects at the good end of the range (this data is from craftsman questionnaire surveys).

I use a systems graphic language I call 'workmapping' to display the interdependence of processes. The focus of this one is field planning, shown producing assignment that drive direct work, which in turn consumes resources including labor, materials, tools and information. Also shown are the control/breakthrough diamonds representing processes that answer the question, 'Do we need to change?' Change can be a return to an existing standard of performance (control) or the introduction of a superior performance standard (breakthrough) (Fig. 9).

Moving to Step 2, this lists some key actions Engineering needs to take (Table 5). Obviously, the supervision time released in Step 1 needs to be invested. The choice made here is to invest in error reduction and in determining customer wants.

Customer wants are not restricted to the proximate customer, usually someone from the client's home engineering office. Also included are the construction contractor, and client marketing, financial, operations and maintenance personnel.

Value engineering can play a helpful role if modified to address not only the cost

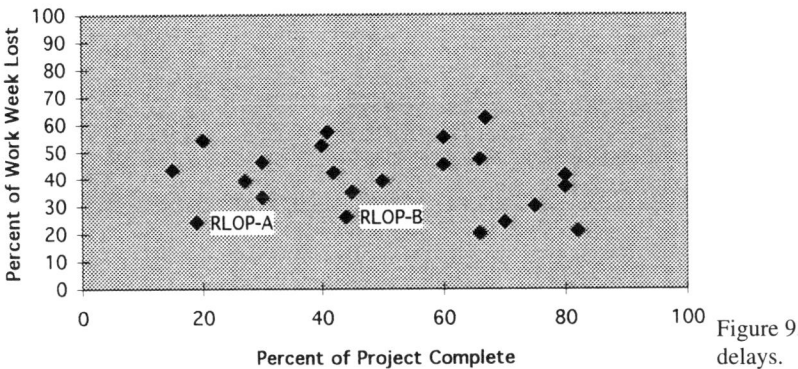

Figure 9. Summary of delays.

Table 5. Performance improvement strategy.

Engineering
2. Improve the quality of engineering deliverables: – Invest released supervision time in coaching to reduce errors resulting from ignorance, lack of skill or inattention – Analyze repetitive errors back to root causes and eliminate – Constantly monitor customer needs and wants – Apply value engineering techniques to evaluate and improve design components as regards performance (functionality), constructability, operability, maintainability, life cycle costs and time to market – Bring supplier design forces into the project design team – Eliminate unnecessary specifications, improve the quality of directives to suppliers – Train design engineers in constructability criteria, rotational assignments, field trips and video reports

to accomplish necessary plant functions, but the other phases and dimensions of the entire plant life cycle as well.

One more aside: There is a great temptation to seize on the one change that will yield a competitive advantage. Unfortunately, there is no one thing. There are many changes needed and they are themselves interdependent, so cannot be successfully implemented out of sequence. This is very frustrating to most American managers. The solution is to actually understand how work processes fit together within a supplier/customer framework.

Procurement has two tasks in Step 2: To reduce delivery variation and to reduce cost (Table 6). Both are pursued by implementation of the tools developed in Step 1, i.e. the integrated database and the supplier quality process.

This is an area where the construction industry can most completely (although till not entirely) imitate manufacturing. According to the book, *The Machine That Changed the World*, even US manufacturing needs to redesign contractual relationships to align supplier and user interests to continuous improvement. That is more difficult in construction, but still can be done. For example, fabricators are usually inadequately compensated for production and delivery that interrupts cost-saving long production runs. One part of the solution is to change contracts so they are

Table 6. Performance improvement strategy.

Procurement
2. Reduce delivery variation and the cost of equipment, materials and services: – Improve baseline agreements and contracts to reflect process capabilities and organizational interests, incorporate incentives for continuous performance improvement – Integrate supplier and user work processes, e.g. the vendor print cycle – Create intermediate mechanisms for detailed integration of work processes beyond baseline agreements and master schedules – Help suppliers reduce their production costs – Minimize receiving inspection at job sites – Reduce purchase price by volume purchases, selecting suppliers based on past performance and prior relationships

Table 7. Performance improvement strategy.

Construction
2. Reduce construction costs: – Reap the benefits of better constructability – Shift work to shop conditions – Structure onsite fabrication as closely as possible to shop conditions – Maintain tools and equipment for zero downtime – Develop craftsmen and crews with multi-craft capabilities – Experimentally improve crew mix on repetitive operations – Reduce rework within field control by completing installations the first time, identify and learning from installation errors as quickly as possible, tracking repetitive errors back to causes and eliminating those causes – Reduce the number of organizational layers – Increase average crew sizes to adjust for reduced supervisory time spent firefighting – Reduce double handling of materials by placing prefabricated process equipment directly onto its foundation, by placing fabricated items such as piping and structural steel within crane reach of final positions – Avoid assembly operations in dangerous or hard-to-reach locations, e.g. preassemble pipe rack bents on the ground and lift into position – Improve installation methods by training supervisors and engineers in methods design, by providing staff support and equipment such as video cameras, by collecting suggestions, and by running experiments to breakthrough to superior methods and performance standards – Reduce redundant inspections, break the chain of repeated rework/inspection by assembling all involved to witness repairs and approve on the spot

compensated. Another part is to change fabrication shop processes to reduce the cost of producing shorter runs. For the most part, US suppliers are still in the mass production mode. Large owners and contractors can facilitate their transition to lean production.

Step 2 in construction is focused on cost reduction, and assumes the stabilized environment created in Step 1 (Table 7).

I have been working closely with several contractors during the construction phase of EPC industrial projects. My first observation has already been stated several times regarding the disruption and waste caused by poor quality and delivery of off-site resources, specifically design documents and permanent plant materials and equipment. A second observation is that the construction industry is in some ways not completely shifted from craft to mass production – much less to lean production. The absence of industrial engineers from project sites and the lack of standard work methods is one sign of the dominance of the craft production model – no little assisted by the contention that each facility is unique. On the other hand, the industry has followed the mass production model in its extensive division of labor; a phenomenon even more characteristic of merit shop projects than of union projects. The worker who can move from carpentry to pipefitting is extremely rare, as is the worker who can build formwork and do finish carpentry. The trend has been to assign specialists to each type of work, and attempt to coordinate the work activities of these specialists by increasing the number and layers of supervision. Consolidating specialty functions into single points of responsibility and eliminating management layers is much needed.

Another area of opportunity is inspection and rework. No one, as far as I know, really understands what percentage of craft man-hours is devoted to rework of various kinds. My guesstimate is 25%. Much of that will be eliminated by releasing supervisors to supervise. In addition, there must be aggressive steps taken to accelerate enable these changes, it will be necessary to convert quality inspectors from policemen into teachers, and to convince those making errors that it is safe to report them; i.e. that the root cause of errors will be eliminated, and not the messenger.

At this point in time, the strategy for Step 3 is least developed. These are some of the components (Table 8).

Concurrent engineering is the primary model and engine for reducing project duration. In addition to the aspects developed in manufacturing, the construction industry needs to understand how to execute the work of interdependent engineering disciplines simultaneously, as well as simultaneously addressing all life cycle design criteria. One idea is to apply both technological and organizational tools developed

Table 8. Performance improvement strategy.

Engineering	Procurement	Construction
3. Reduce delivery cycle time: – Concurrent engineering: Development of principles for the design of industrial facilities analogous to DFA and DFM, further development of cross-functional teams, change to strong form of project management where not yet implemented – Systematic development and sharing of workarounds across projects – Establish partnering relationships with clients	3. Reduce delivery cycle time: – Joint streamlining of interdependent processes – Implementation of construction triggers (releases) for resource delivery by suppliers just when needed – Reduce transport distances by selection of local suppliers – Optimum use of blanket purchase orders to minimize acquisition cycle time	3. Reduce construction duration: – Advance mobilization within constraint of available backlog – Reduce backlog quantities to reflect more reliable deliveries – Divide construction sites into subprojects and facilitate cross functional team approach to craft/craft and direct hire/sub-contractor integration

in manufacturing, i.e. electronic data interchange and cross-functional teams. Even though engineers are assigned to large projects under the control of strong project managers, in a task force mode, there remain tremendous problems coordinating across disciplines. The contractor implementing the strategy I have presented may experiment with mixed teams, with joint responsibility for a set of interdependent deliverables, and considerable autonomy at managing the internal interfaces.

Procurement can reduce the time required for acquisition of resources by eliminating wasted time in information flows, reducing transport distances by selection of local suppliers (or the more expensive use of local staging areas), and by the use of blanket purchase orders that get some steps in the cycle done ahead of time.

In addition, procurement must work with construction on timing of deliveries. The goal is for construction to release resources for delivery just when needed. This reduces on-hand inventory, space requirements, and double handling when equipment and materials can be placed directly into final position off delivery vehicles.

As delivery variation declines, construction can reduce the size of backlogs required to initiate work, thus advancing construction mobilization. In addition, there will remain opportunities for squeezing time from construction processes, primarily

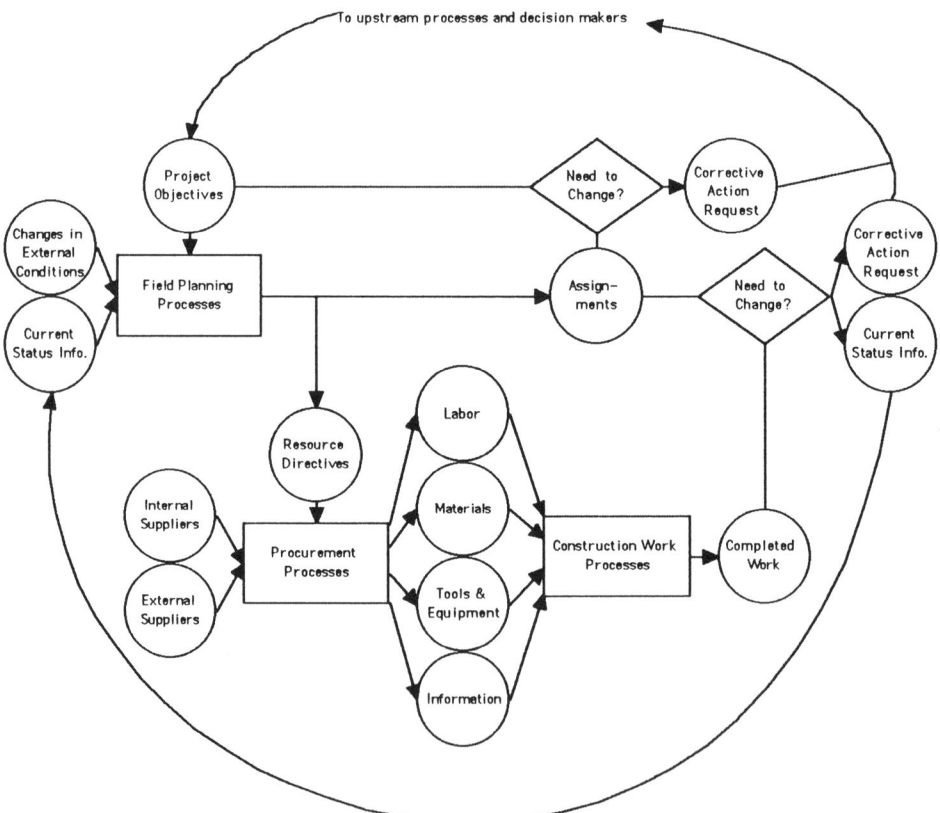

Figure 10. Field planning system-generic workmap.

from better coordination of interdependent crafts. Better coordination can be achieved by the same strategy used in engineering, i.e. subprojects with cross-functional teams. In cases where a substantial part of the work is subcontracted, the contractual relationship will require redesign to facilitate teamwork.

To conclude: the conversion process model would have us attempt to achieve performance improvement on EPC projects by separately reducing the cost and time of engineering, procurement and construction, without regard to their interdependencies. The lean construction model facilitates EPC performance improvement by revealing those interdependencies.

The strategy pursued by one US industrial contractor begins by implementing the Last Planner concept as a means of shielding the direct work force from variation in resource flow. This releases energy and time for reducing quality and delivery variation, i.e. investment is made in design quality, supplier delivery and construction cost reduction. This provides the basis for substantial reductions in project duration, largely from accelerating site mobilization as deliveries become more reliable and inventory requirements fall.

REFERENCES

Deming, W.E. 1986. *Out of Crisis*. MIT Centre for Advanced Engineering Study: Cambridge, Mass.

Shingo, S. 1981. *Study of Toyota Production System*. Japan Management Association.

Implementing lean construction: Reducing inflow variation*

GREG HOWELL
Department of Civil Engineering, University of New Mexico, Albuquerque, USA

GLENN BALLARD
Department of Civil Engineering, University of California, Berkeley, USA

In many circumstances variety is the spice of life. But it is a bitter herb when you are trying to complete a complex and uncertain fast track project. Significant variations occur at every stage in the construction process. Plans change and materials are late. In compressed circumstances, variation becomes more apparent and critical as it exposes the interdependence between activities. Once the work environment is stabilized through modifying the planning system, it becomes possible both to reduce variation in flows and to work behind the shield to improve downstream operations. Suggestions for research and improved practice are offered specifically regarding the management and reduction of flow variation.

1 UNDERSTANDING AND RESPONDING TO VARIATION

Variation in delivery is a fact. Responding to variation is a major aspect of Lean Production Theory (LPT). Buffers between operations are an important tool because they allow two activities to proceed independently. Variations in output from upstream operations does not limit the performance of the downstream operation. Buffers can serve at least three functions in relation to shielding work by providing a workable backlog:

1. To compensate for differing average rates of supply and use between the two activities;

2. To compensate for uncertainty in the actual rates of supply and use;

3. To allow differing work sequences by supplier and using activity.

As valuable as buffers are, they are expensive, hard to size, and hardly an optimal solution. The costs associated with buffers include storage space, double handling, inventory management, loss prevention, buffer fill time, and idle inventory. Buffers are hard to size because the actual supply and use rates are unknown and they vary. Ohno (1987) recognized that buffers are hardly an optimal solution and admonished management to cease reliance on them. As he said, 'you must drain the water from the river to see the rocks'.

Oddly, the flow management problem in construction is more difficult because we seldom are allowed to build buffers. Great pressure for immediate production is

*Presented on the 2nd workshop on lean construction, Santiago, 1994

93

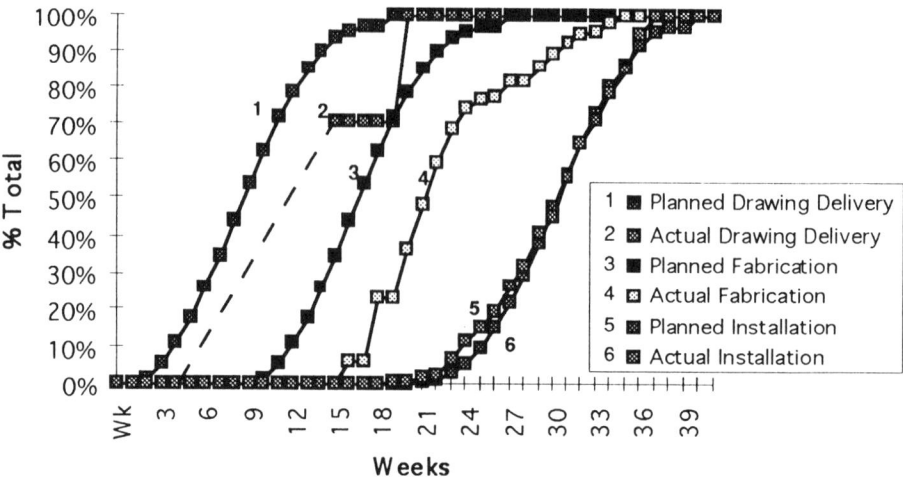

Figure 1. Planned and actual performance of engineering, fabrication and installation.

transmitted to users by the cost/schedule system, the requirement for cash flow, and the early-start mentality of project managers. As a result, planning is inhibited because the actual rates of consumption are unknown because 'normal rates' include delays due to waiting for resources. As a consequence, our advice to establish buffers at first appears contradictory. But in order to find the uninhibited use rates we must make it possible to work without interruptions. An example is in order.

Figure 1 traces the planned and actual delivery of isometric drawings from the engineer to the fabrication shop, the planned and actual fabrication and delivery of pipe to the site, and the planned and actual installation rates. Each stage except the last shows a high degree of variation from the planned rates of provision.

This project was built under fierce time constraints. Even so, the installation contractor held to their policy of *not* starting installation until 85% of the pipe, structural steel and equipment was on site. They believe, and their balance sheet supports, that they can work extremely efficiently and quickly by waiting for the backlog to develop. Oddly, this company only accepts lump-sum contracts so they will not be forced by the owner into inefficient practices. Their policy is to avoid growth so they only bid on enough work to keep their backlog nearly filled with 'high' quality projects. While the strategy works for this contractor, others often proceed with work before a workable backlog exists.

Figure 2 shows the relationship on 4 projects between buffers of pipe on hand at the beginning of installation and project performance. Project 2 shows similar performance to Project 1 (described in Fig. 1). Interestingly these projects are almost complete opposites in every other aspect including size, location, union/non union labor, contract form, constructability reviews etc. Despite their differences, management in both cases made explicit decisions to establish buffers and neither gave in to tremendous pressure for partial releases of work or starting before enough work was available so the downstream operation could be expected to run steadily to completion. The value of buffers of resources is understood throughout the construction

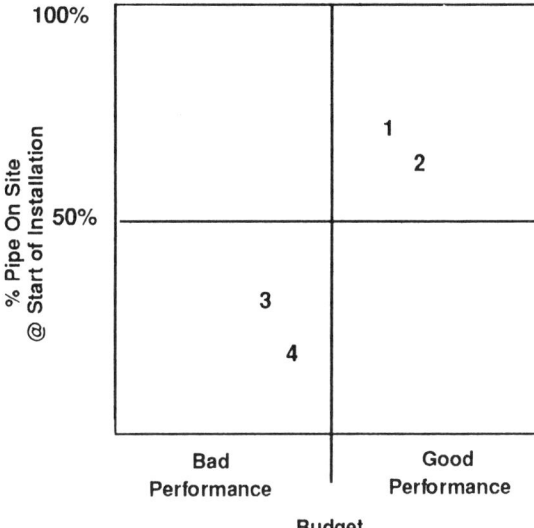

Figure 2. Early data relating performance and piping buffers.

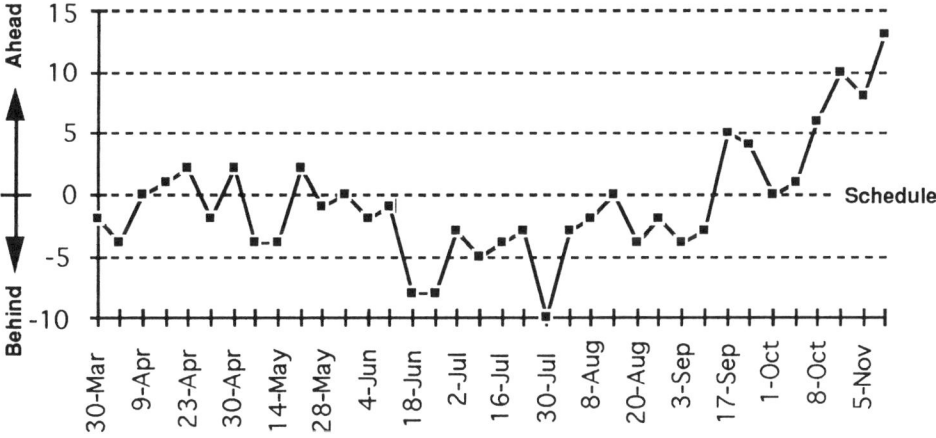

Figure 3. Actual delivery of steel compared to schedule.

organization. They occur at all levels in response to variations. We often find middle level supervisors hoarding materials, tools and equipment in the face of strong pressure to release them to others for use. As a result they have learned clever techniques for hiding resources and will sometimes lie to maintain concealment.

Figure 3 provides an example of an attempt to provide a buffer which failed because variation had been underestimated. Structural steel was delivered to a jobsite over an eight month period, from April to January. The steel had been divided into delivery sequences to match the contractor's installation schedule. The data points on the chart show in weeks when sequences were delivered relative to their scheduled delivery dates. Points above the line represent early deliveries and points below the line represent late deliveries. The level of variation shown on the chart subverted the contractor's plan to work from a three week backlog of steel-in-hand. The contractor had underestimated the degree of variation and uncertainty in the flow of that resource.

Decisions were taken to reduce labor when the amount of steel on site was evaluated and the degree of unpredictable delivery documented. The owner and contractor then took the logical step and dispatched an engineer to expedite fabrication. They found the hold up was in paint completed steel.

2 VARIATION IN PLANS

Plans change. The extent of uncertainty was documented earlier on (see first chapter Lean production theory: Moving beyond 'can do', Figs 1 and 2). The current practice is to update the plans to reflect the current status and to reissue the schedule with strong letters urging improved performance. The discussion here is illustrated in Figure 4 and follows the strategy of improving the stability in planning from the bottom up. The first phase is to shield work execution and then to move up to the adjusting mechanism and finally to rethinking initial planning.

3 DEVELOPING WORKABLE BACKLOG FOR THE LAST PLANNER

Gathering and assuring a backlog of stuff is a complex inventory management prob-

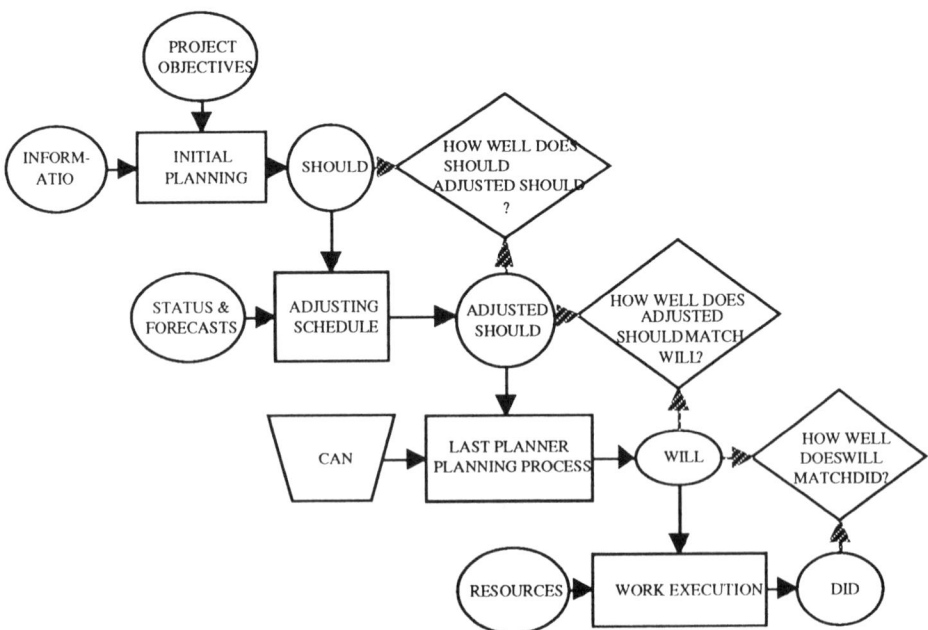

Figure 4. Expanded planning system.

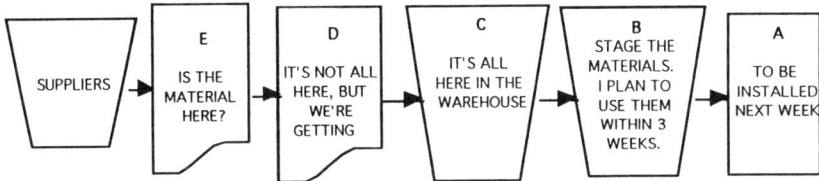

Figure 5. Managing workable backlog.

lem. The difficulty comes from the unpredictable rates of use and supply of specific resources. In our experience, the resource supply or materials group views their task as one of supplying resources to the crew. As important as this is for doing work, it overlooks the task of providing information to planners on what can be done. It often appears that the material systems in these companies were designed on the premise that everything would work as it should. In fact, even the last planner faces the same task of matching should with can, breaking the operation into pieces, and deciding the appropriate level of detail and update frequency.

An example of managing workable backlog illustrates the idea and shows that a contractor need not be totally at the mercy of the third party engineer to deliver drawings and materials in accordance with the project schedule (Fig. 5).

A strategy for using resource buffers appropriate to construction-only (but still fast track) projects is illustrated above. The sequence of steps on the flowchart show how the piping craft controlled the flow of materials onto and within the jobsite. Piping supervision had access to computer terminal listings of piping drawings, with the typical progressing milestones displayed: fabrication, erection, connection, trim, punch. From schedules, models and marked up drawings, the supervisors decided what and how much work needed to be done 4 weeks in the future and marked the selected milestones with an 'E'. The same database was accessible by materials management, who took 'E's' as a directive to get the associated materials into the warehouse within two weeks. Materials managers first allocated existing inventory to the priorities, then purchased/expedited missing materials. Once an item was materially sound, the appropriate milestone was marked by materials management with a 'D', indicating to piping supervisors that the material required to do that work was in the warehouse, reserved for that use, and ready for issue to the field. The week before they needed to work that material, piping supervisors placed a 'C' in the appropriate milestone nodes, thus telling the warehouse to send it to field storage, at which time it was marked a 'B'. Piping foremen selected work to be done next week from 'B's', marking those selected as 'A'. When a milestone was marked with a letter, the man-hours budgeted for that work were automatically summed for each drawing and for the total items selected, so the planner could know if the right amount of work had been selected.

The consequence of this approach to planning is that people only commit to doing work that can be done. Work is selected for material-soundness, available materials are allocated to plan priorities, and expediting is directed to support plan priorities that are very specific. Secondary benefits include elimination of separate progress

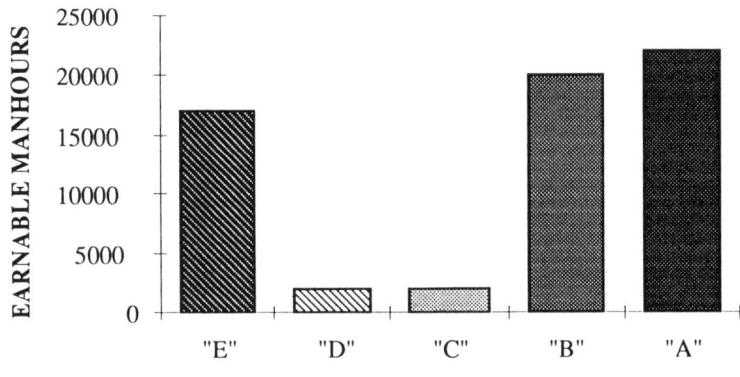

Figure 6. Quantifying workable backlog.

reports and material requisitions. Perhaps equally important is the backlog system gives management a significant degree of forward control.

Figure 6 shows the actual level of backlogs at each station early in the implementation of the planning phase. With this data, it is no longer necessary to wait for cost reports to trickle in after being massaged by middle management to see the problem and begin to take action.

4   THE ADJUSTMENT PROCESS

The adjusting process is the next level in the planning system. Current adjustment approaches fail to provide useful plans in part because the primary task of the planning and controls group is to prepare forecasts for senior management. Unfortunately, since corporate rules force the planners to put the best face on the situation, these forecasts are seldom prepared using detailed updates on resource availability. Many practitioners will take strong exception to this claim. However most plans rest on a policy of including all future activities as if the resources for all were 100% certain of arriving to support the plan. A review of the current approach to the 90 day look ahead schedule used by a major contractor makes the point.

The 90 day look ahead schedule is a drop-out from the project schedule, sometimes at a higher level of detail. A scheduled activity may not be deleted from the 90 day even when the management knows it cannot be done because that appears to reduce the pressure both upstream and downstream to meet contractual commitments and milestones. When they know something cannot be done, the status is communicated informally to craft planners, so they know not to expect on-time delivery.

We suggest the opposite: allow scheduled activities onto the 90 day *only* if managers are morally certain that it will enter into workable backlog as scheduled. Each week, slide the 90 day window forward one week and review the status of all scheduled activities that fall within the window. If an activity is questionable, it falls out of the 90 day, gets flagged for expediting attention, and reappears on the 90 day when its on-time delivery is assured or the schedule has been adjusted to accommodate a later delivery.

Picture a line running from 0% to 100% probability of on-time delivery. We now adjust the schedule only when probability nears 0%. Instead we should begin adjusting the schedule when probability departs from 100%. Defining risk and agreeing on trigger levels of probability may seem difficult, but in fact must occur if we are to find a way to provide both the earliest warning of impending variation and the information on performance needed to improve the planning system.

5 INITIAL PLANNING

Let us return to Figure 4 to better understand the stages of planning. At the top, consider the obstacles to improving the stability of initial plans. The directives to the initial planning typically include explicit project objectives with an implication that the objectives should remain stable. At this stage, (as in all following stages) control is achieved and variation reduced by careful attention to the stability of inputs. Initial plans are based on the answers to questions like; 'How is the market trending? What is happening to the process technology? What is our available cash flow?' Since none of these questions can be answered with absolute certainty and changes in the answers can have dramatic effect on project success, the best advice falls in two categories:

1. Monitoring the basis for the plan: This means carefully identifying key assumptions or premises and assigning the responsibility for monitoring and early detection of changes to specific individuals. This applies the principle of reducing inflow variation at the outset of the project. Significant research is required to better understand how to diagnose the situation in terms of uncertainty, how to break projects into pieces, how to set the level of planning detail, and how to determine the appropriate update frequency. Similar research would be useful at each of the following planning stages. Practitioners are advised to explicitly consider these questions when planning the planning process (Laufer & Howell 1994);

2. Carefully testing the objectives against means for achievement. Objectives are only fixed when the means have been carefully examined. This applies the principle of matching 'should' with 'can' at the outset.

6 SUMMARY

Flow variation can be reduced by stabilizing all functions through which work flows from concept to completion. Better understanding uncertainty, suppliers and customers can eliminate the causes and so reduce variation in shared processes. In addition planning systems must be redesigned to include a level for adjusting 'should', so operations can better match 'should' with 'will'. The next step in implementation of lean construction is to work behind the shield on improving performance, taking full advantage of the reduction in variation and uncertainty thus far achieved.

REFERENCES

Ohno, T. 1987. *Toyota Production System.* Productivity Press.
Laufer, A. & G. Howell 1994. Construction planning: Towards a new paradigm. *PMI Journal.*

Implementing lean construction: Stabilizing work flow*

GLENN BALLARD
Department of Civil Engineering, University of California, Berkeley, USA

GREG HOWELL
Department of Civil Engineering, University of New Mexico, Albuquerque, USA

1 IMPLEMENTING LEAN CONSTRUCTION ON FAST TRACK, COMPLEX PROJECTS

1.1 *Introduction*

Lean construction has at least two foci that distinguish it from traditional construction management. One focus is on waste and the reduction of waste. Breaking from the conversion process model, and reconceiving production processes in terms of Koskela's flow process model (Koskela 1992) reveals the time and money wasted when materials and information are defective or idle. Instead of simply improving the efficiency of conversion processes, the task is extended to the management of flows between conversions. Consequently, in addition to its focus on waste, lean construction also focuses on managing flows, and to do so, puts management systems and processes into the spotlight along with production processes.

Flow management is a much more difficult task on complex, fast track projects such as refineries, chemical plants, food processing plants, paper mills, etc. (Ballard 1993). These projects have long, complicated supply chains, many players, typically are under pressure to hit market windows for product, and are subject to multiple, extensive process design changes motivated by the opportunity to make much more money than is lost through disruption of construction. In this environment, traditional approaches to construction management fail miserably. The conversion process model conceals everything that needs to be revealed; particularly the design of systems and processes to manage work and work flow.

1.2 *Implementation strategy*

This chapter proposes a way of implementing lean construction on complex, fast track projects. The first step is to stabilize the work environment by shielding direct production of each component function from upstream variation and uncertainty management has not been able to prevent. Once that shield is installed, it becomes possible to move upstream in front of the shield to reduce inflow variation, and to move downstream behind the shield to improve performance (Fig. 1). In the follow-

---

*Presented on the 2nd workshop on lean construction, Santiago, 1994

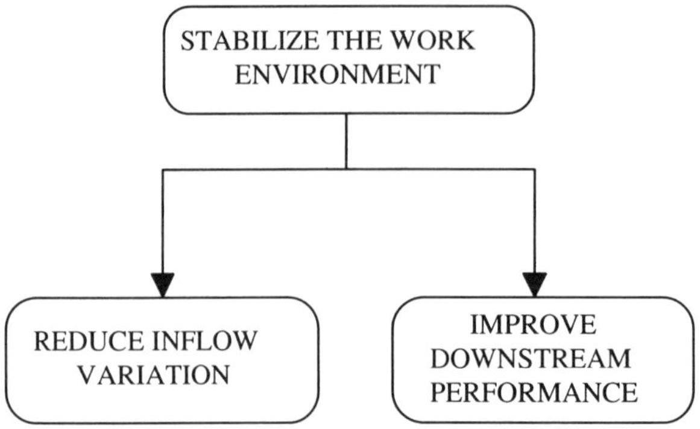

Figure 1. Stabilize the work environment.

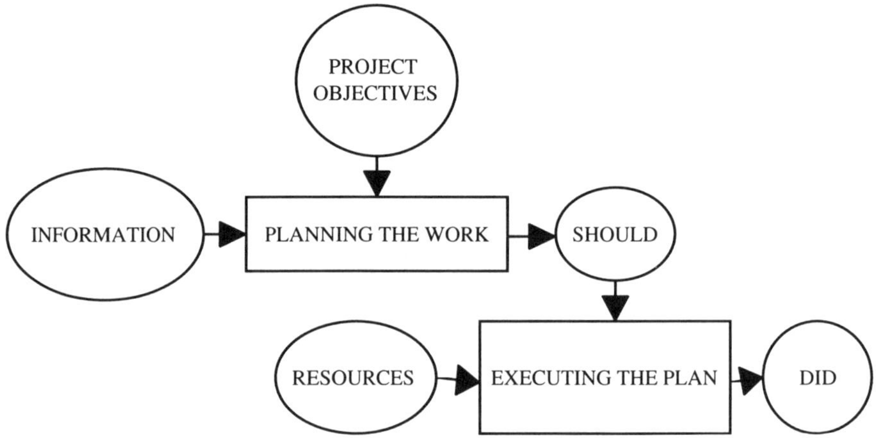

Figure 2. Traditional management practices.

ing pages and chapters, each of these steps will be explained, with primary emphasis in this presentation on installing the shield.

2 TRADITIONAL MANAGEMENT PRACTICES

2.1 *Traditional planning*

The construction industry devotes tremendous energy and resources to planning projects and developing the schedules, budgets and other requirements that collectively tell project personnel what they *should* do. Project management thereafter monitors and enforces conformance of *did* to *should*. Planning at the beginning of the project is replaced by control during project execution (Fig. 2).

Everything works fine until someone drops the ball, then a chain reaction takes place. A vendor fails to return certified data on time, causing design to slip, leading to a delay in fabrication, and late delivery to the jobsite. Or, a new process technology emerges late in the game, but offers an opportunity to shave 10% off the cost per unit of product. Or, the market window is advanced to meet an unexpected competitive challenge.

As slack disappears from the schedule, more and more pressure is put on everyone in the chain to produce more, faster. This usually makes things worse rather than better. Working hand to mouth and betting on the come results in ever more nonproductive time, demoralizes supervision, and directs energy and attention toward getting stuff to work with rather than learning how to do work better and faster.

If this traditional approach to planning worked perfectly, *did* would always match *should*. Indeed, a major E/C contractor's project management policy includes the following statement:

'The project management team is responsible for finding methods of meeting the control budgets and schedule rather than justifications for not meeting them.'

This tells project management that there are no legitimate reasons for failing to meet control budgets and schedule. The result is failure to identify and act on reasons why planned work does not get done, and failure to learn and improve. It is assumed that all work and resources can be coordinated by schedules, and that inability to perform to schedule are rare or evidence of lack of commitment.

2.2 *Measuring match of 'did' with 'should'*

Actual measurement reveals that what actually gets done differs from what is supposed to be done roughly 1/3 of the time. The data shown in Figure 3 are from one of many studies. On 5 construction projects, scheduled activities amounted to 625 during the study period, of which 227 activities were not completed as scheduled. The percentage of planned activities completed was 64%; i.e. the percentage of planned activities not completed was 36%. This kind of data suggests that the lack of fit be tween what we *should* do and what we *can* do is substantial and systemic, and that we must learn how to manage in such conditions (Fig. 4).

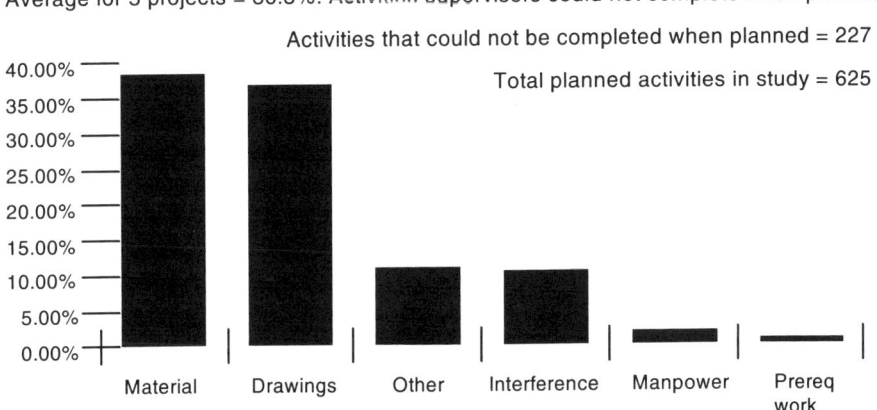

Figure 3. Construction last planner: Comparing 'should' with 'did'.

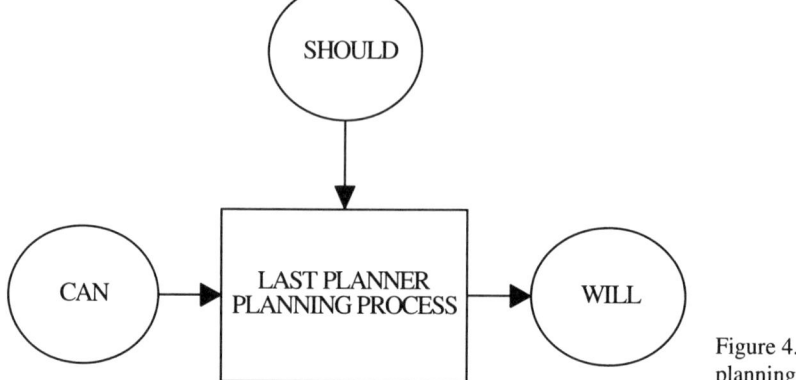

Figure 4. Last planner planning process.

Returning to the chart, note that almost 80% of the misses were due to lack of materials and drawings; i.e. resources provided to the construction phase by design or suppliers. Relying on schedules to coordinate work flow through the process does not have a good track record.

3 THE LAST PLANNER

3.1 *Should-can-will*

Decisions regarding what work to do in what sequence over what durations using what resources and methods are made at every level of the organization, and occur throughout the life of the project. Ultimately, some planner produces assignments that direct physical production. This 'last planner' (Ballard 1994) is last in the chain because the output of his/her planning process is not a directive for a lower level planning process, but results in production (Fig. 4).

Stabilizing the work environment begins by learning to make and keep commitments. Last planners can be expected to make commitments (*will*) to doing what *should* be done, only to the extent that it *can* be done. Expressing this as a rule, we might say: Select assignments from workable backlog; i.e. from activities you know can be done.

When this rule is not observed, direct workers inherit the uncertainty and variation of work flow we have not prevented. The result is a high percentage of non-productive time and a demotivated work force less and less willing to fight through these obstacles.

3.2 *Quality characteristics of weekly work plans*

Quality characteristics of weekly work plans:
- Work is selected in the right sequence;
- The right amount of work is selected;
- The selected work can be done.

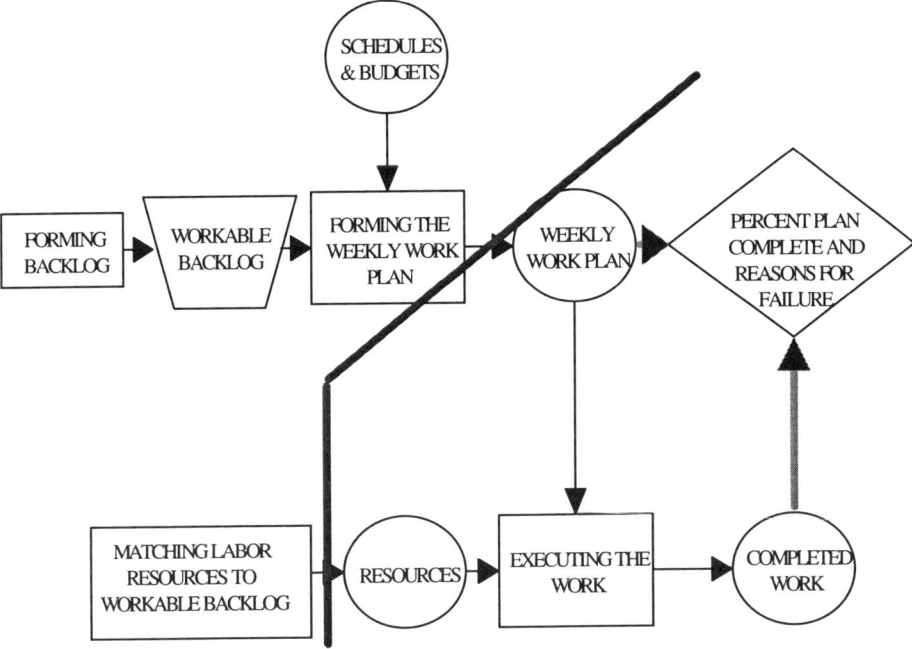

Figure 5. Developing a weekly work plan.

In addition to selecting practical assignments, there are two other primary quality characteristics of the commitment level of planning, which I will express here as 'weekly work plans' (Fig. 5). 1) The right sequence of work is to be selected; i.e. work in the sequence that best moves the project towards its objectives. Sequencing directives are provided by schedules developed to coordinate work flow and production activities. Sequencing decisions can also be made by last planners based on their intimate knowledge of working conditions and constructability issues. As an example, the project schedule may sequence installation of piping by reference to specific areas into which the total project has been divided. Even assuming that fabrication and delivery of piping materials occurs as scheduled, it is usually advantageous to allow the foreman or superintendent some latitude in sequencing the work within an area. When deliveries do not occur on schedule, you may choose to have a piping engineer do the detailed sequencing rather than the foreman, but someone has to make those decisions locally, with intimate knowledge of the work to be done and the conditions in which it is to be done.

The second quality characteristic is 2) The 'right amount' of work is selected; i.e. that amount of work that uses your labor and equipment capacity as directed by the schedule. An interesting issue here is measuring the productivity and progress that will be achieved if a plan is fully executed; i.e. if all planned activities are completed. By reviewing and signing off on plan quality beforehand, management validates plan quality and can then focus on controlling execution of the plan. Please note that this is how we think now of the entire project, but our attempts to control against *should* disregard the mismatch between *should* and *can*, which vitiates management control.

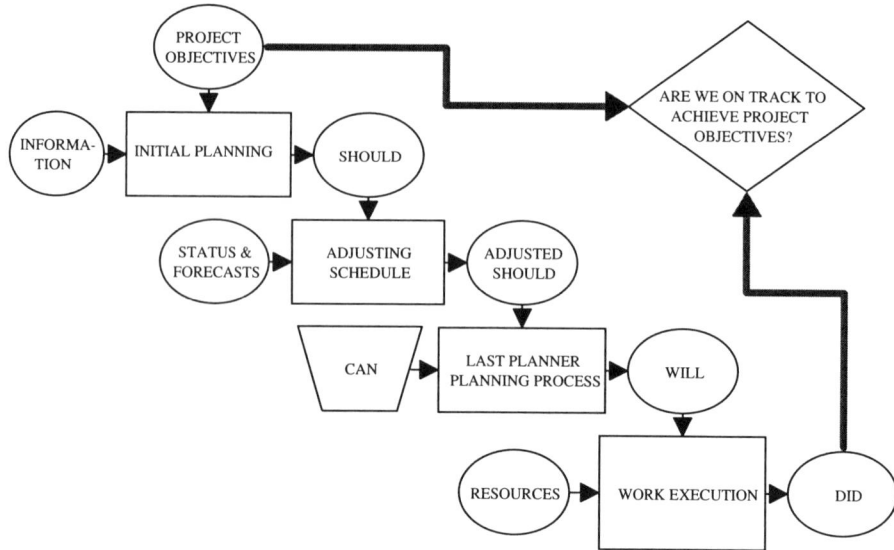

Figure 6. Results-oriented control.

3.3 *Shielding production*

The planning system needs additional levels in order to better manage uncertainty and complexity. The first level to be added is the level of last planner commitment, the implementation of which shields the direct work force from upstream variation and uncertainty (Fig. 6).

Shielding occurs in part simply from selecting only assignments that can be successfully completed, assignments for which all materials are on hand and all prerequisite work is complete. But in order to be able to select assignments from workable backlog, we need a process for forming backlog. And in order to avoid a mismatch between labor force and work flow, we need some way of matching labor and labor-related resources (tools, construction equipment, temporary facilities, etc.) to the work flow into backlog. Lastly, in order to perfect the shield, we must measure the degree of fit between *did* (completed work) and *will* (weekly work plan), identify the root causes of failures to complete planned work, and eliminate those causes to prevent repetitions.

4   MEASURING AND IMPROVING PERFORMANCE

4.1 *Measuring project performance*

The second level to be added to the planning system is devoted to adjusting *should* to better match *can* and *will*. That is discussed in the chapter on reducing in-flow variation. Attend now to the issue of measuring and improving performance. As explained earlier, we focus measurement on results and compare actual to desired results to see if we are on track to achieve project objectives. This results-oriented control is intended to reveal problems so they can be solved, thus keeping the project on track. In fact, this approach to control is too abstract to identify what needs to be changed in order to improve (Fig. 6).

4.2 *Measuring system performance*

To determine where to intervene, it is necessary to focus measurement and control on system components or levels. In the case of planning systems, that means measuring the match between output and directives at each level and understanding the root causes of mismatches. The match of *will* and *did* is measured by percent plan complete (Ballard et al. 1994). The match of *adjusted should* and *will* can also be measured and improved, as can the match of *should* and *adjusted should* (Fig. 7).

4.3 *Improving PPC*

The starting point for improvement in planning is measuring the percentage of planned activities completed (PPC), identifying reasons for non-completion, and tracing reasons back to root causes that can be eliminated to prevent repetitions (Fig. 8).

Measuring PPC allows us to distinguish between failures rooted in plan quality and failures to execute plans. Currently that distinction cannot be made because the quality characteristics of plans are not made explicit, and it is assumed that all failures are execution failures. Our findings suggest that the vast majority of failures to complete planned work are rooted in the quality of plans. Consequently, planning

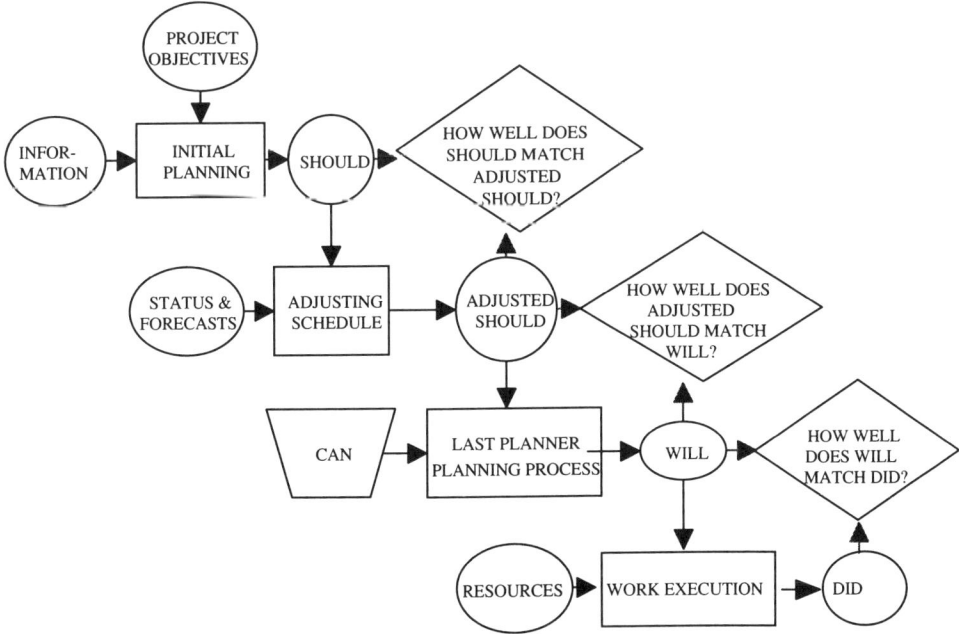

Figure 7. Focusing control on system components.

PERCENT PLAN COMPLETE

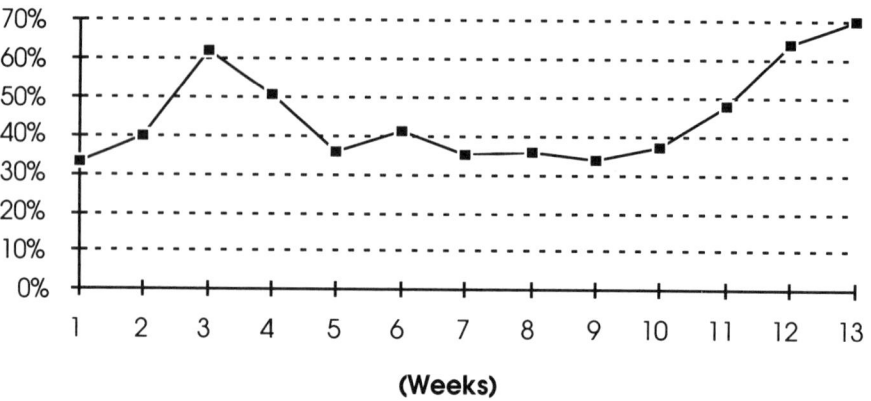

Figure 8. Measuring percent plan complete.

system performance at the commitment (*will*) level can be improved by such actions as educating planners, improving the supply or quality of planning information, clarifying or modifying directives, etc.

5 BENEFITS OF STABILIZING THE WORK ENVIRONMENT

5.1 *Benefits of shielding*

Shielding direct workers from upstream variation and uncertainty has tremendous benefits, not least of which is injecting certainty and honesty into the work environment, as opposed to unreliability and dishonesty. We do what we say we are going to do, at least week by week. Our suppliers do the same. Others can count on us. We can count on others. There is no blind pressure for production, and there is a commitment to learning and improving, so there is no reason to conceal the facts and every reason to reveal them.

Shielding promotes accountability because expectations can be met, and failures to meet expectations can be understood and acted upon.

Control improves because we have the facts about causes and capabilities, and can easily see the quality of planning and execution at the foreman and crew level.

The reduction in system noise facilitates performance improvement. Confusion and ambiguity are minimized. We know what numbers mean.

Non-productive time falls. Less paid labor time is spent waiting on or hunting for something to work with, or moving to alternative work.

Especially front line supervisors feel as though a burden was removed from their shoulders. Not having to explain away performance results over which they have no control, not having to spend time trying to keep their people busy-this releases time and energy for improving performance.

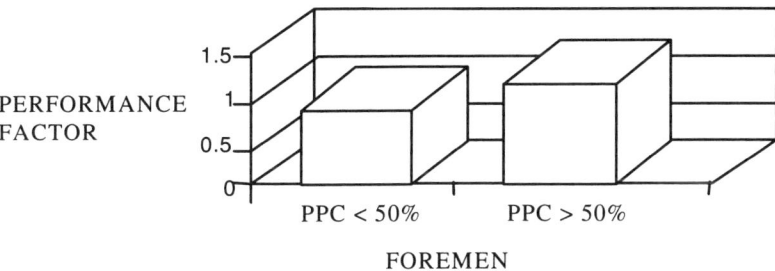

Figure 9. PPC and productivity.

5.2 *PPC and productivity*

An indication of the benefits to be achieved by stabilizing work flow is shown on Figure 9. On one major industrial project, piping foremen were divided into two groups:

1. Those whose PPC was below 50%; and
2. Those whose PPC was above 50%.

The productivity of the second group was 30 points higher than the first group. In other words, if the first group was performing at 15% over budget, the second group was performing at 15% under budget.

6 SUMMARY

As currently designed, planning systems do not shield direct production from upstream variation and uncertainty. The result is longer durations and higher costs. To reduce project duration and costs, direct production must first be shielded by the introduction of a near-term, commitment level of planning, with explicit plan quality characteristics.

In addition, processes must be installed to identify workable backlog, to match labor to work flow into backlog, and to measure and improve the match of *did* and *will*.

Stabilizing the work environment through the implementation of these planning processes results in substantial performance improvement (e.g. 30% improvement in productivity) and creates the conditions for even more substantial improvement.

The next steps are to move upstream to reduce in-flow variation, and to move downstream to improve performance behind the shield.

REFERENCES

Ballard, G. 1993. *Lean Construction and EPC Performance Improvement.* Lean Construction Workshop, Espoo, Finland.

Ballard, G. 1994. *The Last Planner.* Northern California Construction Institute Monterey, California.

Ballard, G., Howell, G. & Kartam, S. 1994. Redesigning Job Site Planning Systems. In: *Proceedings of the American Society of Civil Engineers Conference on Computing in Construction*, Washington, D.C. June, 1994.

Koskela, L. 1992. Application of the New Production Theory to Construction. *Technical Report* No. 72, Centre for Integrated Facilities Engineering, Stanford University.

Implementing lean construction: Improving downstream performance*

GLENN BALLARD
Department of Civil Engineering, University of California, Berkeley, USA

GREG HOWELL
Department of Civil Engineering, University of New Mexico, Albuquerque, USA

1 INTRODUCTION

1.1 *Overview*

The conversion process model would have us attempt to achieve performance improvement on complex, fast track projects by separately reducing the cost and time of engineering, procurement and construction, without regard to their interdependencies. The lean construction model facilitates performance improvement by revealing those interdependencies. This not only avoids suboptimization. The logic of lean construction implementation requires a certain sequence of initiatives, which progressively reveal additional opportunities for improvement.

Ohno & Shingo (1987) two of the principal architects of the Toyota Production System, argue persuasively that manufacturing be conceived in two complimentary but different ways (Fig. 1).

1. As a process, i.e. the course of events through which material is changed into a product; and

2. As an operation, i.e. the course of events through which man and machine work on the product.

They also argue that process must be balanced and managed prior to addressing operations. We are following Ohno & Shingo when we advocate implementation of lean construction in three phases, beginning with stabilization and reducing in-flow variation (process), and finally turning to operations.

1.2 *Realizing potential gains*

It is often possible to reduce the cost or time of operations taken singly on the order of 25-50%. However, it is not so easy to realize those gains in actual cost or time savings. Excess manning of operations may signal failure to balance flows. Lack of discipline in planning and execution will vitiate attempts to implement improvements. Stabilization is a prerequisite for operations improvement (Fig. 2).

Once the work environment is stabilized, substantial improvement in operations becomes possible, and the potential gains can be realized in cost or time savings.

Reducing in-flow variation brings additional benefits and opportunities for im-

*Presented on the 2nd workshop on lean construction, Santiago, 1994

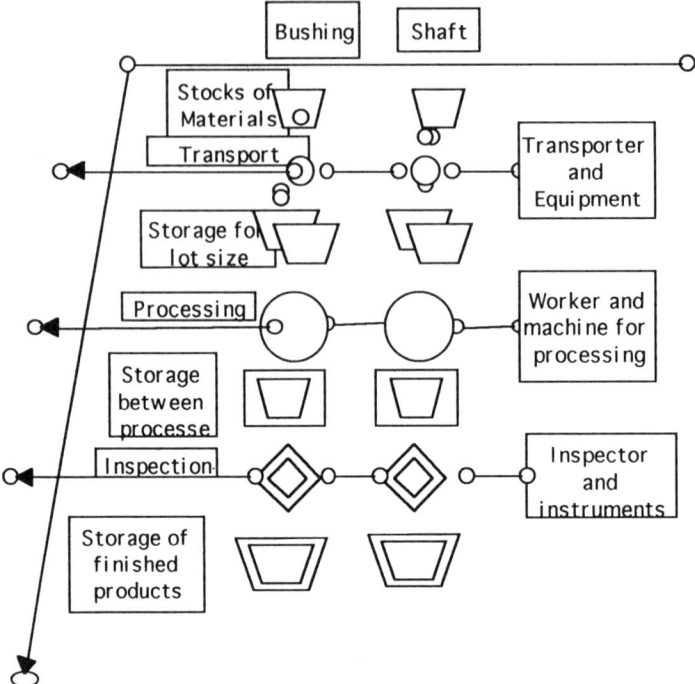

Figure 1. Adapted form Shingo.

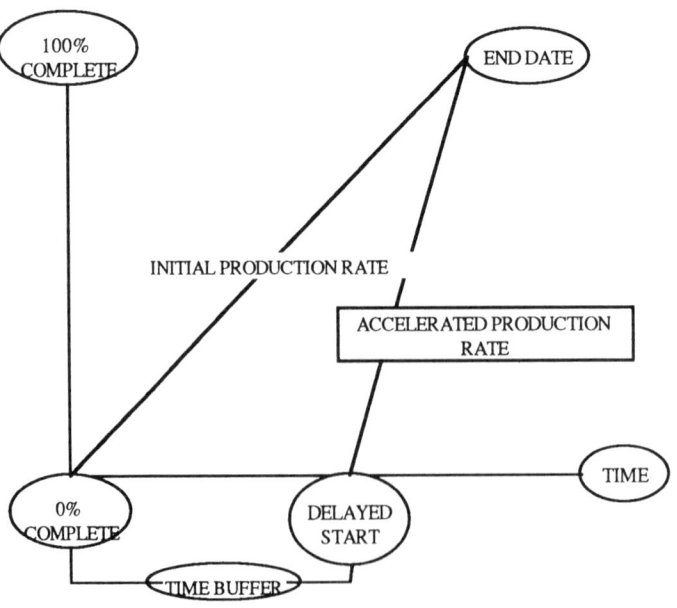

Figure 2. Wait to start, then go faster.

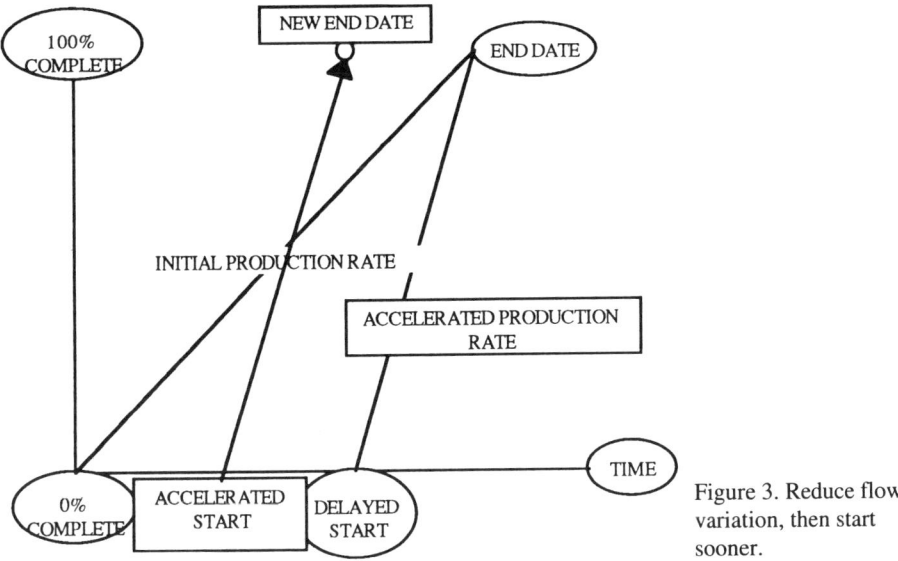

Figure 3. Reduce flow variation, then start sooner.

proving operations. An immediate benefit is illustrated in Figure 3. As delivery variation declines, so does the size of backlogs required to initiate work without risk of interruption, thus advancing phase initiation, e.g. construction mobilization. With a more certain in-flow of work, more optimum sequences can be selected and better matching of labor resources can be done. Further, there will remain opportunities for squeezing time from processes occurring within engineering, construction, etc. – primarily from reducing operation cycle times and from better coordination of interdependent disciplines or crafts.

1.3 *Are foremen the 'last' planners?*

It now becomes possible to improve performance by changing the way work is done, as opposed to managing the conditions in which it is done. However, that does not mean turning immediately away from planning to execution. Flow management extends into the different project functions, as well as between them. Consequently, planning also must be extended downwards. In addition, planning addresses both process flow and operations design.

Even when PPC (Percent Planned Activities Completed) analysis reveals no plan quality failures from upstream planning processes, there may remain plan quality failures. The weekly work plan is not the last plan, so strictly speaking, the producer of the weekly work plan is not the last planner. Planning goes on at every level of the organization until work has been executed. The distinction between plan quality failures and execution failures is relative to the division of planning responsibilities made at each level of the planning system (Fig. 4).

In construction, the effective point of intervention has proven to be the weekly work plan, because that is where work is selected and commitments are made, and the key to stabilization/reduction of uncertainty is improving the ability to keep

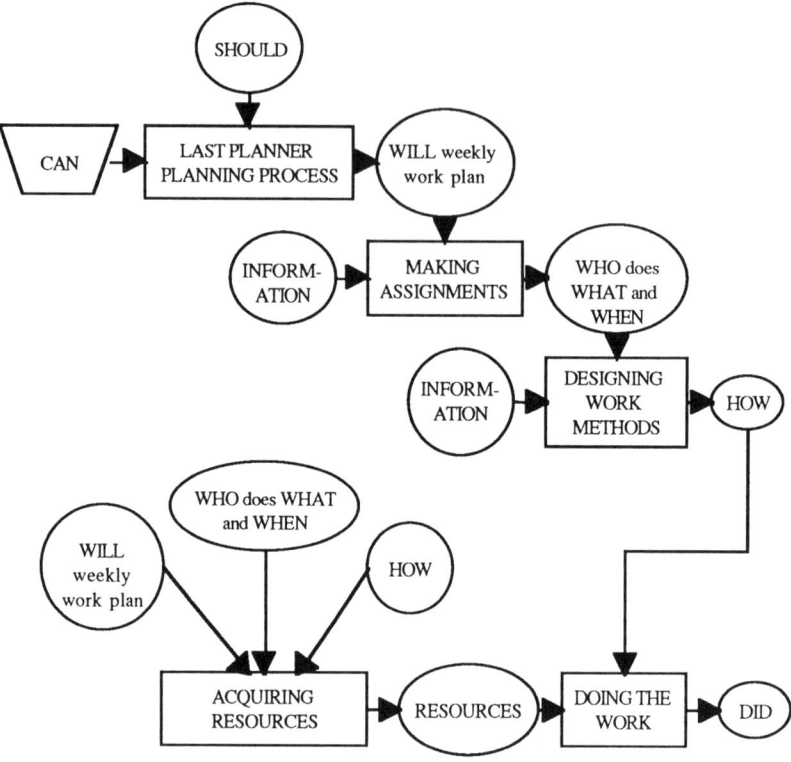

Figure 4. Are foremen the last planners?

commitments through better selection of work to be done. Nonetheless, planning goes on beyond the selection of work to be done next week, often by foremen, sub-crew and individual craftsmen as they produce assignments, daily plans and work methods (Howell et al. 1993). Consequently the first task is to follow the planning system through its final levels, assessing and improving plan quality.

2   WORKING THE PLAN

2.1 *Performance standards*

First there is the issue of working the plan; i.e. making the plan the standard of performance for work execution. This is an issue because many instances have been observed where weekly work plans are not used to develop assignments, or the assignments do not include the goals necessary to be achieved in order to complete planned work. This is obviously in part a function of the degree of definition in the weekly work plan itself. When a bar chart is produced showing the durations and sequencing of assignments by sub-crew for each day of the week, simply providing the bar chart to the sub-crews may be an effective method of driving work execution with the

Table 1. How foremen set standards.

– Keep busy
– Do as I would
– Make the craft standard
– Make the budget: average for aggregate work
– Make the target: budget adjusted for work mix and conditions

weekly work plan. However, having such a bar chart and using it may be two very different things.

In some sectors of the construction industry, it has been customary for upper levels of supervision and management to monitor productivity and progress, while simply urging lower levels and direct workers to push and work harder. Indeed, many 'good' foremen only control against inactivity, and have no explicit goals at all. In some cases, they are so absorbed in removing obstacles that little time or energy remains. In other cases, they assume that eliminating constraints that cause inactivity produces optimum performance, so setting goals would be irrelevant.

Another group of foremen control against standards formed from their personal experience: 'How long would it take me to do that work?' With them, we move from activity levels to assignment durations as the locale in which standards are established – definitely a step forward. The next group establishes expected assignment durations from budget unit rates or craft rules of thumb that serve the same purpose. As an example, I know whole tribes of pipefitters in the south-eastern United States who figure everything will be all right if they can average 10 linear feet of pipe per crew member per day – and they are usually right!

The best way to set standards is to draw them from the specific work to be done. That is the intent of the rule to 'select the right amount of work to be done next week'. Craft rules of thumb or budgeted unit rates are not themselves the standards directly governing assigned work. They inform the planner regarding the aggregate quantities and costs, which the planner adjusts in application to a specific mix of pipe material types and sizes, in different locations, some behind obstacles, etc.

The planner may also adjust the standards to reflect broader performance goals such as completing the piping work 15% under the direct labor budget.

2.2 *Goal setting theory and worker motivation*

Research on the motivational impact of goals on performance contributes these two important findings (Locke & Latham 1990).

1. Given capability, difficult, specific goals lead to better performance than 'do your best' goals;

2. As tasks become more complex, performance becomes less dependent on mere motivation and more dependent on the quality of task strategies, i.e. work methods.

The most popular way of setting goals in assigning work assumes that 'do your best' is the best that can be done. It is also widely held that motivation is more important than the quality of planning and know-how. It is also important to note that there is no support in research for the widely-held belief that participation in goal setting increases motivation to achieve goals. We suspect that understanding goals is more important for performance than participating in goal setting. That does not mean workers have no role in planning. However, it is not increasing motivation, but improving plan quality that is the reason for involving direct workers in planning; especially in planning how to do the work.

3 IMPLICATIONS FOR PROJECT CONTROL

3.1 *Identifying variances*

We believe that the current approach to cost and schedule control in the construction industry is fundamentally flawed in some respects. It is assumed that aggregate averages (budget unit rates) are applicable to each component, and variances are measured from those aggregate averages (Fig. 5). Superintendents and foremen have been well indoctrinated in the importance of productivity, but the attempt to assess performance against false standards inevitably has bad consequences.

1. Near impossibility of identifying a real variance when examining short-term results, and the resultant focus on blaming rather than analyzing; and

2. Failure to direct actual production towards project objectives through the provision of reasonable goals; and

3. Waste of craft supervision's energies and time spent selecting work and shaping reports so they appear to be working as closely as possible to aggregate averages, and thus avoid 'drawing fire'.

The change needed is to decide before work begins if the right work has been selected in the right amounts, given the work mix and conditions. Plan execution is the right standard of performance for a week. Its measures are Percent Planned Activities

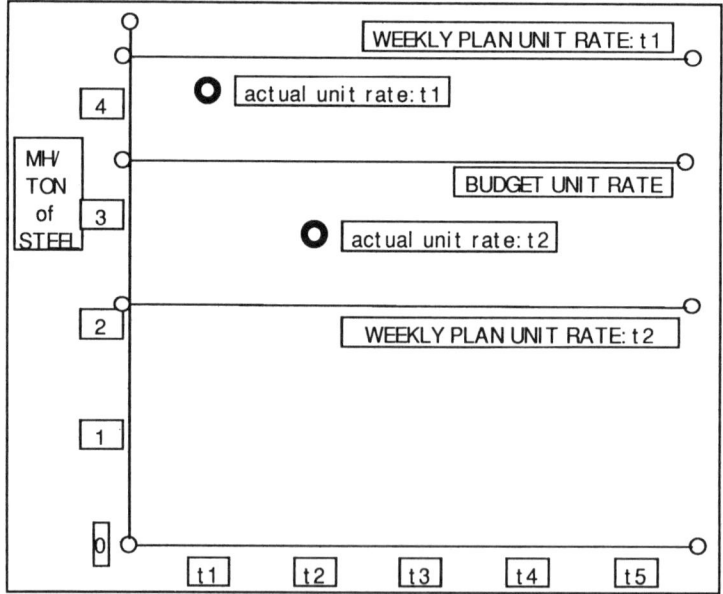

Figure 5. What variances are significant?

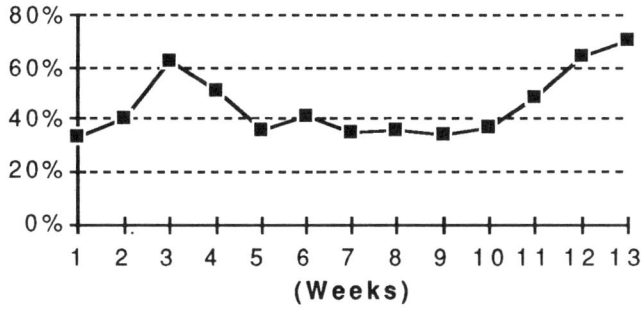

Figure 6. Percent plan complete.

Completed and Planned Productivity. Budget unit rates are the right standard of labor consumption for a project, and serve as means of calculating and assessing PPAC and planned productivity. When you shift your control focus to adjusted standards, variances can be identified and can be analyzed using reasons why planned work was not done, and you have a chance of improving performance and avoiding repetitive errors. With the prevailing approach to controls, its use of false standards and inability to identify variances, performance improvement is accidental.

3.2 *Proactive control of plan quality*

When the field planning system begins to work properly, you will know before-the-fact if labor is matched with work to be done, if planned productivity is reasonable, if the right work is being done in the right amount, etc. After-the-fact analysis can then focus on percent planned activities completed, and can identify when a performance variance (positive or negative) is due to plan quality or to execution. Variances will then be assessed relative to planned performance, not against aggregated standards, thus providing a more efficient variance analysis than one which begins by discounting most variances as apparent rather than real, and providing a more effective analysis because the standard has already been adjusted to reflect work mix and difficulty. Incidentally, this way of operating removes incentives for crafts to move man-hour charges around to reduce apparent variance, and so improves accuracy (use of statistical analysis to interpret variation is also sadly missing, but unfortunately cannot be addressed in this chapter from lack of time).

Controls should be focused on improving PPAC (Percent Planned Activities Completed), improving on-time resource deliveries, and matching labor to resource deliveries (Fig. 6). Now control consists of determining if the total project and its component parts (by discipline and area) are on schedule and budget, but generates little if any information useful in determining causes of variance or acting on them.

4 REMOVING OBSTACLES

4.1 *Reasons why planned work does not get done*

When attempts are first made to shield direct production from inflow variation, the primary reason for failing to complete planned work is usually lack of materials.

Once it has become second nature to select assignments from workable backlog, missing materials begins to disappear as a reason, and is replaced by items more within local control, such as coordinating interdependent work activities and shifting priorities (Fig. 7).

Identifying and eliminating root causes is analogous to stopping the manufacturing production line, something we rarely if ever do in construction.

4.2 *Resource utilization studies*

Acquiring and coordinating the use of shared resources is one of the major planning activities that must occur between committing to a weekly work plan and executing that plan. Some shared resources are coordinated by means of scheduling; e.g. tower cranes. When it is not feasible to schedule the use of shared resources, other means are needed to match supply with demand.

Utilization rates have proven useful as partial measures of that match. In the case of light, mobile cranes, some slack is needed for travelling and to accommodate uneven demand. (Although, Ohno & Shingo would likely chastise our faintheartedness in pursuing waste reduction). Sampling-based measurement of utilization shows the extent to which current capacity is being absorbed, and provides data to consider alongside complaints and requests for more.

It is obvious that low utilization can co-exist with high delays. This occurs when the allocation process is ineffective, or when user planning is poor. Low lead time and lumped requisitions signal poor planning. Tracking and eliminating such practices can allow reducing the number of items of equipment, maintaining a high utilization rate, and *reducing* craft delays. On a project from which data was taken, equipment leasing and labor costs were 15% under budget.

4.3 *Performance-based budgeting of support crafts*

Another promising approach to managing shared resources is to determine resource sizing by reference to maintaining a service level. An example is scaffolding during

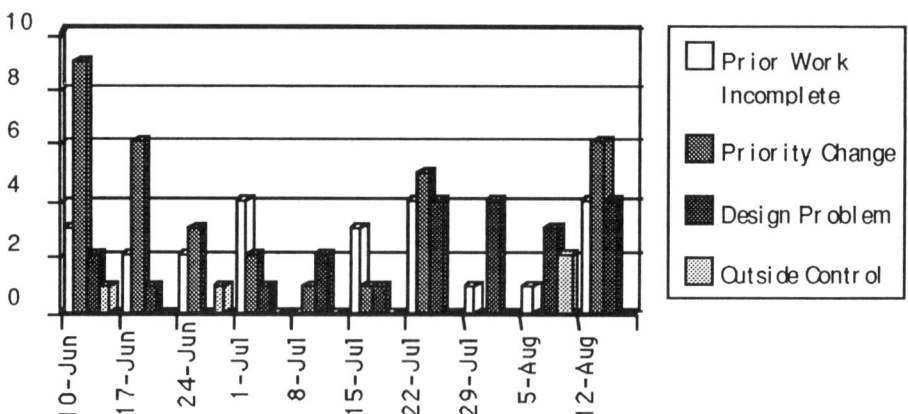

Figure 7. Reasons why planned work was not done.

intensive piping work on industrial projects. Scaffolding becomes a support craft and has often been sized as a certain percentage of total primary craft labor hours. On one recent project, the scaffolding 'budget' was 15% of piping labor hours, but we concluded the project at under 12%, and saved a pile of money.

What we did was make a deal between scaffolding and piping. If piping foremen would provide at least 48 hours lead time in their requests to have scaffolding erected, scaffolding labor force and materials would be adjusted to maintain a 48 hour response time. Lead time and response time were monitored and made available to both scaffolding and piping. This information helped management make the right decisions about the amount of investment needed in scaffolding resources, as well as the corrective action needed in pipe foreman planning.

5 CHANGING HOW WE DO THE WORK

5.1 *First run studies*

The way work is done is called a work method. All kinds of work can be represented in methods charts and can benefit from redesign and streamlining, especially with the involvement of a team consisting of representatives of the different functions involved. Administrative and resource management processes should be first on the list because they support all different types of production processes. Planning systems and processes have already been singled out for special attention. Other administrative processes that are likely candidates include RFIs and change orders. Resource management processes that should be examined include small tool supply and distribution, construction equipment supply and distribution, purchasing, requisitioning, delivery/material handling, hiring and in-processing, etc.

As soon as craft work is ready to begin, the first run of each type of craft operation should be examined in detail, ideas and suggestions solicited from all parties, and experiments performed to explore alternative ways of doing the work. The result will be a performance standard, for both results and the methods for achieving them. This standard is best used not as a rigid procedure, but as a challenge to meet or beat the best done thus far.

As was mentioned at the beginning of this presentation, it is often possible to reduce the time or cost of operations, or to improve their safety or quality on the order of 25-50%. Simple but powerful tools can easily be taught to craftsmen and craft supervisors, including process charts for representing work methods and systematically seeking ways of eliminating, reducing or making concurrent the steps represented on those process charts. It is certainly true that such training occurs too infrequently, but that signals the fact that the problem is not technical but philosophical.

Theory comes before policy and policy comes before training. Hopefully the development of lean construction theory will soon move us into the policy phase. As regards policy, we recommend adaptation and use of the traditional PDCA cycle (Fig. 8) developed in the 1930's by Walter Shewhart, perhaps better known as W. Edwards Deming's mentor.

Plan:
1. Select work processes to study;

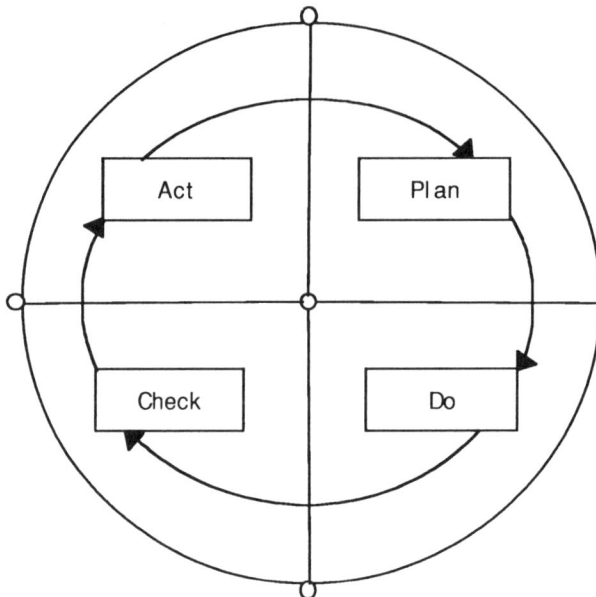

Figure 8. PDCA cycle.

2. Before the first run of each process, assemble people with input or impact;

3. Bar chart the work process steps;

4. Brainstorm how to eliminate, reduce or overlap process steps;

5. Check process designs for safety; anticipate hazards and specify preventions;

6. From past experience, list probable errors and specify preventions;

7. Assign optimum labor, tool and equipment resources.

Do:

8. Try out your ideas on the first run.

Check:

9. Describe and measure what actually happens: process steps, sequences and durations; errors, omissions and rework; accidents, near misses and hazards; resources used (labor, tools, equipment, support crafts, etc.); outputs.

Act:

10. Reconvene the team, including those who actually did the work, review data and share ideas. Continue until opportunity for improvement is exhausted;

11. Communicate the improved method and performance as the standard to meet or beat.

The intent is to thoroughly plan and study first runs of major operations, using past studies as guidelines, and producing standard work methods designs for use on the project. This experiment-based approach produces a tested method that can be taught to all crews, thus reducing cost, errors and accidents. Use the standard PDCA cycle for improvement, involve everyone who could help or hinder, establish standard methods to be met or improved upon on this project and on future projects. First run studies can be a substantial part of a contractor's investment in innovation, and an arena within which direct workers can make a significant contribution. When these studies are a routine part of your business, it will be easy to experiment with new work process designs, new technologies and tools, new crew mixes, etc. Once workers see that you are interested in finding better ways of doing work, they will develop and share their ideas.

An important feature of first run studies is to integrate all performance dimensions into work process design, with safety first, then quality, time and cost. You can apply the same control and improvement strategy to this level of planning as we have recommended for the selecting of work for weekly work plans. Plan as well as you can, then analyze actual performance to distinguish between plan quality and plan execution failures, so each can be addressed and improved.

5.2 Inspection and testing

5.2.1 What is rework?
Rework has traditionally been defined in terms of errors requiring work to be redone. A typical list might include design errors or changes, vendor errors, and installation errors. We suggest adding two more items to the list: incompletions and rehandling (see Table 2).

Incompletions occur when a task that should have been completed was not. This occurs quite commonly, sometimes from the habit or tradition of dividing work into ever smaller parts, and sometimes from the lack of materials or completed prerequisites. In either case, there is a cost of incomplete work; e.g. the costs associated with returning and gaining access, which is often more difficult after other items have been installed. There is also the cost of replanning and supporting additional assignments. And finally, there is the risk of misinformation between those who did the original work and those who are completing it.

I was perversely delighted to see that Toyota had to fight through the perception that early deliveries were on-time (Shingo 1981) We struggle so hard in construction to get what we need to do work that it may seem foolish to complain about something coming before scheduled arrival. Nonetheless, there is a substantial cost involved in early deliveries. Attention is usually focused on the cost of capital tied up in unneeded inventory, but the greater cost is rehandling. Think what we could save if we were able to place all equipment, steel, and piping directly into final position off the delivery vehicle, thus avoiding off-load, reload, haul to work area.

5.2.2 Linking the planning and inspection systems
The construction industry has practiced the 'inspect quality into the product' approach rather than designing quality into the work process. One of the primary focus

Table 2. Components of rework.

Traditional components	Proposed additions
Design errors	Incomplete installations
Design changes	Rehandling materials
Vendor errors	
Installation errors	
Damage by other crafts	

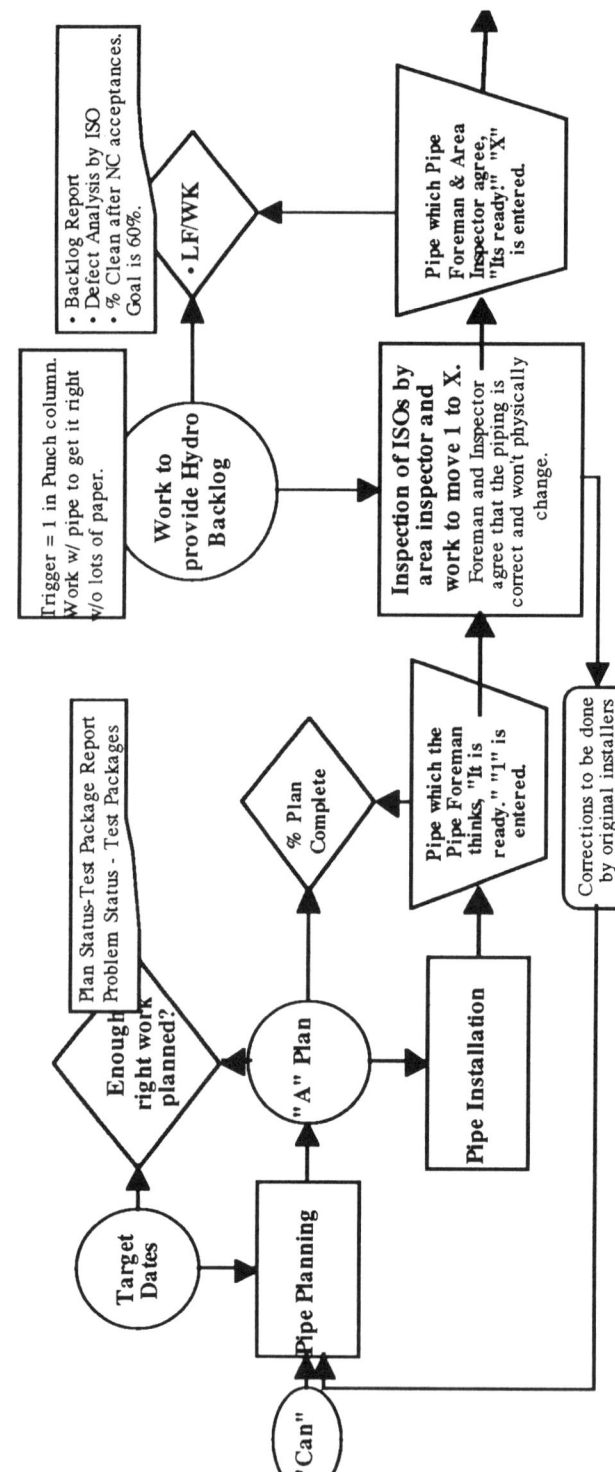

Figure 9. Inspection and turnover.

of process improvement activities should be the inspection processes associated with each craft and its work. The objective is to reduce the time and cost of inspection by increasing the frequency of getting it right the first time. The amount of time spent doing work over has never been accurately measured, but can safely be said to be substantial, absorbing perhaps as much as 25% of paid labor time. Much of this time is expended after work is reported to be complete, i.e. during the punch-out and correction phase.

We have had some success in piping using the following approach. First of all, we stopped putting pressure on hydrotest to test x number of feet per week. Instead, the decision was made to try and get it right up front; both in terms of production quantities and quality of work. A vital part of that effort was assigning inspectors to work informally with piping as work was being installed, so those doing the work would get feedback as quickly as possible, and would be able to make corrections. This effectively substituted in-process inspection for the traditional end-of-line inspection. Figure 9 is a small part of a larger flow chart describing the piping inspection and hydrotesting process.

Not only did production quantities increase by 50% (linear footage of test package piping turned over for hydrotest each week), but over the life of the last unit, repetitive errors dropped from one every 21 feet in an earlier unit to 1 error every 42 feet, a 50% reduction in error rate.

By the end of the project, 80% of test packages were passing final inspection 'clean'. If that level can be achieved upstream, all later inspections can be eliminated. Owner companies with which we have worked now intend to eliminate one of the two end-of-the-line inspections, and ultimately involve their people doing the in-process inspection.

We have an opportunity to develop more beneficial ways of involving foreman and crew members in reducing rework. Among the many possibilities to explore: provide craftsmen laminated lists of things to check before walking away from an installation, find a way to 'stamp welds' applicable to other types of work, provide better and more timely feedback regarding repetitive errors, and improve the in-process inspection process.

6 CONCLUSION

6.1 *Engineering, procurement and construction*

This chapter has focused on improving performance on the construction site. Obviously, other locations will also enjoy opportunities as a consequence of stabilization and reduction of in-flow variation. In Engineering, concurrent engineering is the primary model and engine for reducing project duration. In addition to the aspects developed in manufacturing, the construction industry needs to understand how to execute the work of interdependent engineering disciplines simultaneously, as well as simultaneously addressing all life cycle design criteria. One idea is to apply both technological and organizational tools developed in manufacturing, i.e. electronic data interchange and cross-functional teams. Even though engineers are assigned to large projects under the control of strong project managers, in a task force mode, there remain tremendous problems coordinating across disciplines. Contractors implementing the strategy we have presented will experiment with mixed teams, with joint responsibility for a set of interdependent deliverables, and considerable autonomy at managing the internal interfaces.

Procurement can reduce the time required for acquisition of resources by eliminating wasted time in information flows, reducing transport distances by selection of local suppliers (or the more expensive use of local staging areas), and by the use of blanket purchase orders that get some steps in the cycle done ahead of time. Vendors and fabricators will benefit from the increased certainty in work flow to them, and will be able to more reliably serve the needs of their downstream customers.

In addition, procurement must work with construction on timing of deliveries. The goal is for construction to release resources for delivery just when needed. This reduces on-hand inventory, space requirements, and multiple handling when equipment and materials can be placed directly into final position off delivery vehicles.

6.2 *Review*

The fast track, complex projects of the manufacturing and process plants sector offer some of the greatest challenges to the construction industry. It is vital that we learn how to manage in conditions of rapid change and uncertainty because those conditions are becoming the norm for all types of construction. Lean production concepts and techniques offer help in meeting those challenges.

The first step in applying lean production to construction is to shield direct production from variation and uncertainty in the flows of directives and resources (Fig. 10). The second step is to reduce flow variation. The third step is to improve performance behind the shield; i.e. to improve operations within a context of managed flows.

Many opportunities exist for improving operations, but the logical starting point in each case is extending the planning system down to execution, first addressing the making of assignments (goal setting, division of labor), then acquisition and management of shared resources, and finally the design of work methods.

We advocate involving direct workers in the planning and execution of experiments in work methods design, a practice we have titled First Run Studies. We also advocate extending the concept of rework to include incompletions and rehandling, in order to draw attention to the benefits of task completion and the avoidable costs of rehandling materials.

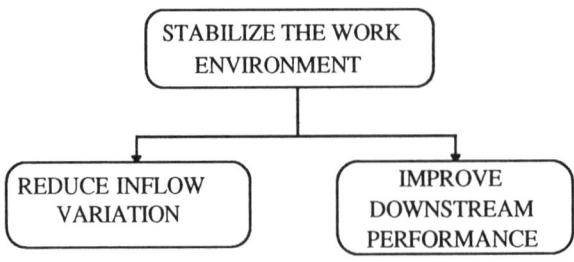

Figure 10. Improvement strategie.

Implementation of lean production concepts and techniques in the construction industry is the way to the future, but following that path requires letting go of traditional thinking. We hope to have shown how to make the first steps on the way.

REFERENCES

Howell, G., Laufer A. & Ballard G. 1993. Interaction Between Subcycles: One Key to Improved Methods, *ASCE Journal of Construction Engineering and Management*, Vol. 119, No. 4.
Locke, E. & Latham, G. 1990. *Goal Setting Theory*. Prentice Hall.
Ohno, T. 1987. *Toyota Production System*. Productivity Press.
Shingo, S. 1981. *Study of Toyota Production System*. Japan Management Association.

Identification of critical factors in the owner-contractor relation in construction projects*

LUIS FERNANDO ALARCÓN, PATRICIO VENEGAS & MARIO CAMPERO
Department of Construction Engineering and Management, Catholic University of Chile, Santiago, Chile

ABSTRACT: This article presents a modelling exercise in which the effects of certain characteristics of risk in owner-contractor relations are evaluated in relation to the results of construction projects. It uses a conceptual model based on contractors' experience in public works and a mathematical model which integrates the assessments of the modelling participants.

Characteristics of the relation such as good faith, the quality of design received by the contractor, the inspection regime stipulated by the owner, and the quality of payments, among others, were analyzed to predict the effects that different scenarios would have on cost, schedule and overall quality of projects.

The authors' analysis of the results, indicates that contractors' efforts to understand and solve problems should focus on those aspects which are of major importance to the desired results, namely:

1. Contract flexibility should reflect the quality of design, i.e. flawed design requires greater flexibility; and

2. Increased confidence in owner-contractor relations forms a work team with better inspection values.

1 INTRODUCTION

The current dynamism of the construction industry obliges contractors and owners to work closely together to develop and complete their projects. But contractors work with changing conditions and different project owners. It is difficult to forecast the results of specific interactions on the finished project. An adequate understanding of these possibilities is key to facing new challenges in construction.

Working with the Committee of Public Works of the Chilean Chamber of Construction, the authors used an analysis methodology, developed in previous research, to forecast and evaluate the effects that certain characteristics would have on projects developed by a public works department (Alarcón & Ashley 1992).

This methodology uses two basic structures. One is a conceptual model which identifies important variables and interactions in the owner-contractor relation during the construction process and estimates their influence on the success of a finished project.

*Presented on the 2nd workshop on lean construction, Santiago, 1994

127

The second structure is a mathematical model for quantitative analysis. This structure uses a stochastic model to process interactions and uncertainties among the variables of the conceptual model.

Applied, the model produces a comparative analysis which reflects the characteristics of an owner and the implicit risk to the contractor for the results of a project.

2 MODEL STRUCTURE

In carrying out a project, contractors normally experience an intense interaction with the project owners. It is well-known that this interaction creates the flow of information needed to develop a project which satisfies the pre-established objectives of both parties. However, contractors lack detailed information about how the characteristics of their relations with owners affect their internal operations. These characteristics are called 'risk characteristics' in this paper. An accurate evaluation illustrates the importance of the relation in project development and suggest guidelines to assure a loss-free project.

The following model is simplified. It integrates experience and opinion from actors whose knowledge permits the quantification of risk in owner-contractor relations. The model assumes that the objectives of contractors in any project are tied to schedule, cost and quality. Finally, the experience of construction company owners is used to determine the effects that the specific characteristics of a project owner have on the development of a project.

The structure of the model is shown in Figure 1. This structure shows four levels (from left to right).

2.1 *Risk characteristics of the owner-contractor relation*

These characteristics were divided in two groups seen as distinct: 'initial conditions' and 'conditions of interrelation'. Each group is described below. To simplify the analysis, to each characteristic we assigned only two possible conditions, one of which is operative in the model.

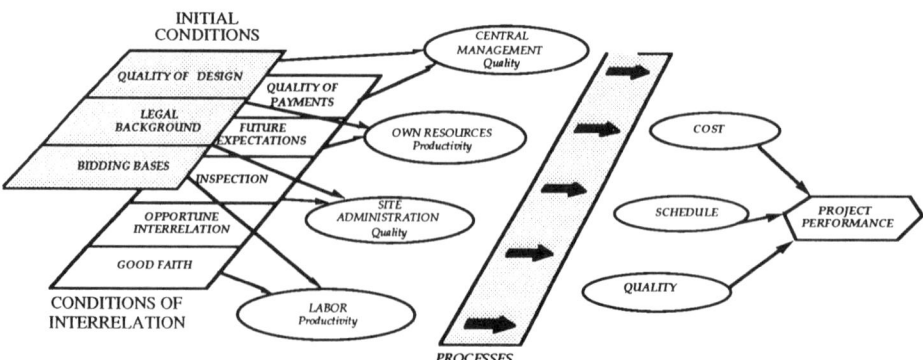

Figure 1. Simplified conceptual model.

Initial conditions

These are characteristics 'married' to the project. They are the formal framework of the relation and are very difficult to alter once the project is under way. In this group of characteristics the following variables were considered:

Quality of design. This characteristic refers to the contractor's appreciation of the design presented by the project's owner, represented in drawings and specifications. Quality of design considers aspects such as: constructibility, expectations for change, finishing level, clearness and completion of design. The quality of design can be 'good' or 'bad'.

Legal background. This characteristic refers to the legal environment in which the project develops. It considers juridical processes, arbitration systems, contract systems, and other legalities that could stimulate or inhibit the contractor's ability to complete the project. The legal background may be 'stimulating' or 'inhibitory'.

Bases of bidding. This characteristic contemplates the bidding guidelines that the owner sets for the project. For the contractor, these bases may be good or bad to the extent that they represent more or less risk. The bases of bidding can be 'good' or 'bad'.

Conditions of interrelation

These conditions are less formal and more subjective than the preceding ones and reflect personal assessments of relations between the owner and the contractor during the realization of the project. In this group of characteristics we considered the following variables.

Good faith. This characteristic refers to the confidence the parties to the project have in each other. For the contractor, good faith may be present or not, according to his level of confidence in the owner. Good faith, in the modelling, 'exists' or 'does not exist'.

Opportune interrelation. This characteristic indicates the efficiency, measured as speed of response, in communications between the parties when they need to exchange information about the project. Opportune interrelation 'exists' or 'does not exist' in the modelling.

Inspection. This characteristic refers to the diligence and regularity of the inspection of the project as determined by the owner. Inspection may be 'stimulating' or 'inhibitory'.

Future expectations. An owner may raise the contractor's expectations of future projects. For the contractor, these signals may be negative or positive. Future expectations can be 'good' or 'bad'.

Quality of payments. The contractor may be more or less comfortable with the owner's payment system depending on factors such as procedures for payments, timing and reliability in committed payments, etc. The quality of payments can be 'good' or 'bad'.

Each combination of states approximates the character of the contractor's relation with a specific owner. This relation is evaluated from the probable effects that a particular scenario will have over the directly affected variables.

2.2 *Directly affected variables*

These variables are directly affected by the characteristics of the owner-contractor relation, and spread their effects inside the contractor's organization, thus affecting his management. Among these, we consider:

Central management/quality. This variable refers to the administration in the main office of the construction company. It considers aspects such as administrative efficiency and the ability to respond positively and as a team to the conditions imposed by the project.

Craft labor/productivity. This variable refers to the labor employed in the construction of the project. This variable takes account of work force size and levels of skill.

Own resources/productivity. This variable pools the contractor's material resources involved in the project. It considers existing equipment, tools and financial resources, their efficiency in use and their ability to respond to unforeseen circumstances.

Site administration/quality. This variable represents the on-site administration team in charge of scheduling and coordinating the physical materialization of the project.

2.3 *Processes*

For the model, the processes are variables conditioned by the directly affected variables, that directly affect the contractor's results. The Committee selected planning, supply, construction, and control as process variables.

Planning and scheduling. This process refers to the scheduling and coordination of the different activities and resources used in the materialization of a project, in light of the goals pre-established for it.

Procurement. This process includes all work needed to acquire and supply material, human and other resources required by the project.

Construction. This process refers to the physical materialization of a construction project and includes functions such as operations administration, contract administration and operations execution.

Control. This process verifies the routine and timely completion of goals preestablished for the project. It includes budget control and quality control.

2.4 *Results*

Results are measures of performance that quantify the effects of owner-contractor relation on the project. The measures used are cost, schedule, and quality.

Cost. The contractor's total cost of a project, including direct and indirect costs of the project. Cost was defined as the value of the offer minus the expected returns from the materialization of the project. It is measured in pesos ($).

Schedule. The time elapsed from when the contractor is awarded the work until it is completed and accepted by the owner. Schedule is measured in months.

Quality. The contractor's level of satisfaction when the work is finished. Quality considers the adherence to project specifications and the presence or absence of deviations from the contractor's initial construction program.

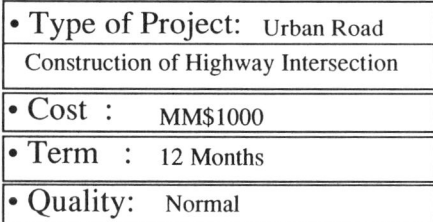

Figure 2. Definition of a base project.

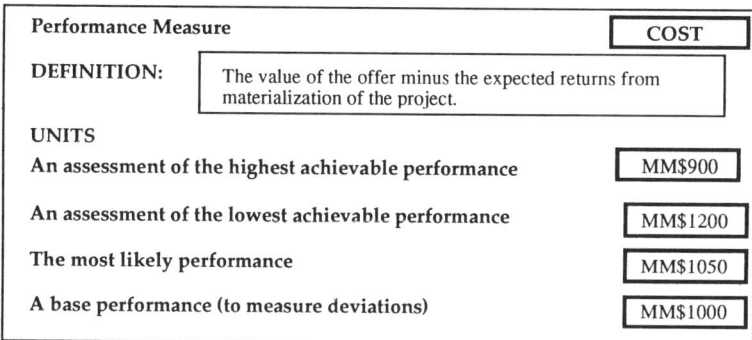

Figure 3. Input required for individual performance measure.

3 DEFINITION OF A BASE PROJECT

To exercise the conceptual model described above, we defined a base project to measure the effects that specific characteristics of risk to the contractor would have on results. For this purpose, we used an urban road system project: The construction of a vial node with a cost of MM$1000. Figure 2 gives the basic information for the selected project.

To complement this information, variables such as performance measures and the ranges of these variables were elaborated by estimating pessimistic, optimistic and more probable values in a scheme similar to that used by the PERT system. This information incorporates the uncertainty of these variables in the mathematical model by adjusting a probable distribution to the data. Figure 3 shows the information for the cost of the project which is one of the three measures of performance defined in the conceptual model. Similar information was prepared for the other two measures of performance.

4 INTERACTION AMONG THE VARIABLES

The knowledge about interactions among the different variables of the model is consolidated in a 'cross-impact matrix'. For the example under discussion, Figure 4 shows a part of this matrix that corresponds to the effects of the directly affected variables on the processes of the project. The elements of this matrix respond to the

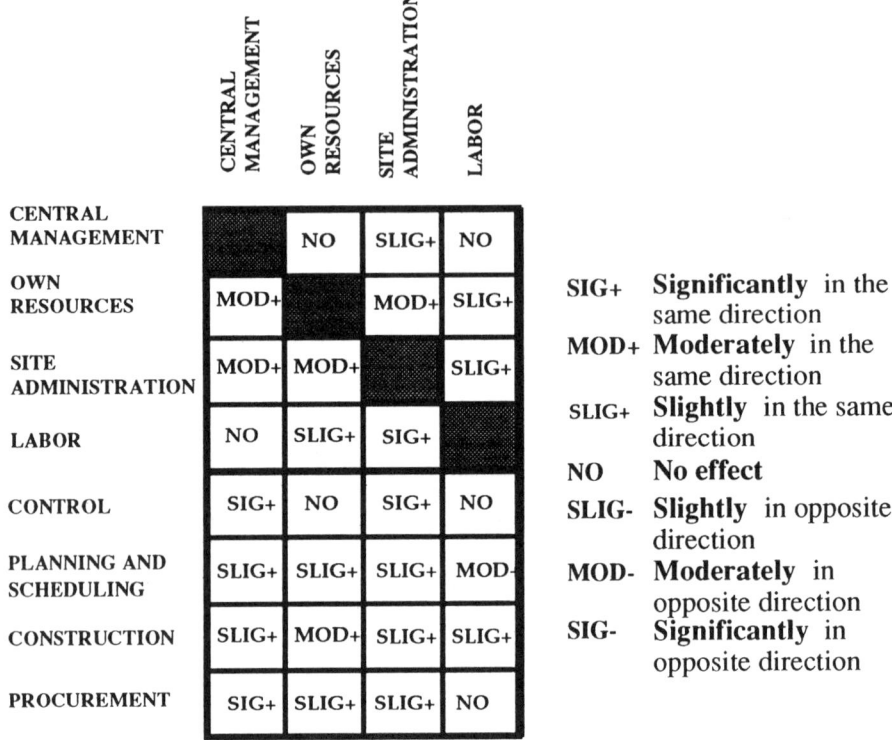

Figure 4. Impact of directly affected variables.

question, 'If changes in the column states occur, how are the row states affected?'. The answer indicates the intensity and direction of the effects according to the scale shown. For example, Figure 4 indicates that an improvement in site administration significantly affects (by intensity) labor productivity resulting in an improvement (in the same direction), an impact with the assigned value SIG+. This way of identifying impacts among the variables is a simplification of the procedure used in 'cross-impact analysis' (Gordon et al. 1986), the technique employed by the mathematical model to facilitate the process of user's modelling. This information is later converted to the numerical format required by the formalism of cross impact. Only direct impacts are set out in this matrix; indirect impacts are captured through the interaction among the variables.

In the same way, we built a matrix that, according to the personal perceptions of the members of the committee, represents the effects how processes of a project have on its results and how the processes affect each other.

5 ANALYSIS OF RESULTS

Using the data we collected, we proceeded with simulations using a mathematical model to process the interactions and uncertainties among the variables of the conceptual model and to give quantitative results as detailed below.

5.1 *Analysis of individual options*

The conceptual model evaluated the effects which specific risk characteristics in the owner-contractor relation would have on the performance of a project. The results point to the outstanding variables in the relation with the owner of a project.

5.1.1 *Conditions of interrelation*
Figure 5 shows the effects on the schedule of a project that some characteristics of the relation with an owner would have for the condition of bad quality of payments. This graph shows that, in the more positive scenarios 1-4 of these combinations, (with the common characteristics bad quality of payments, good faith and stimulating

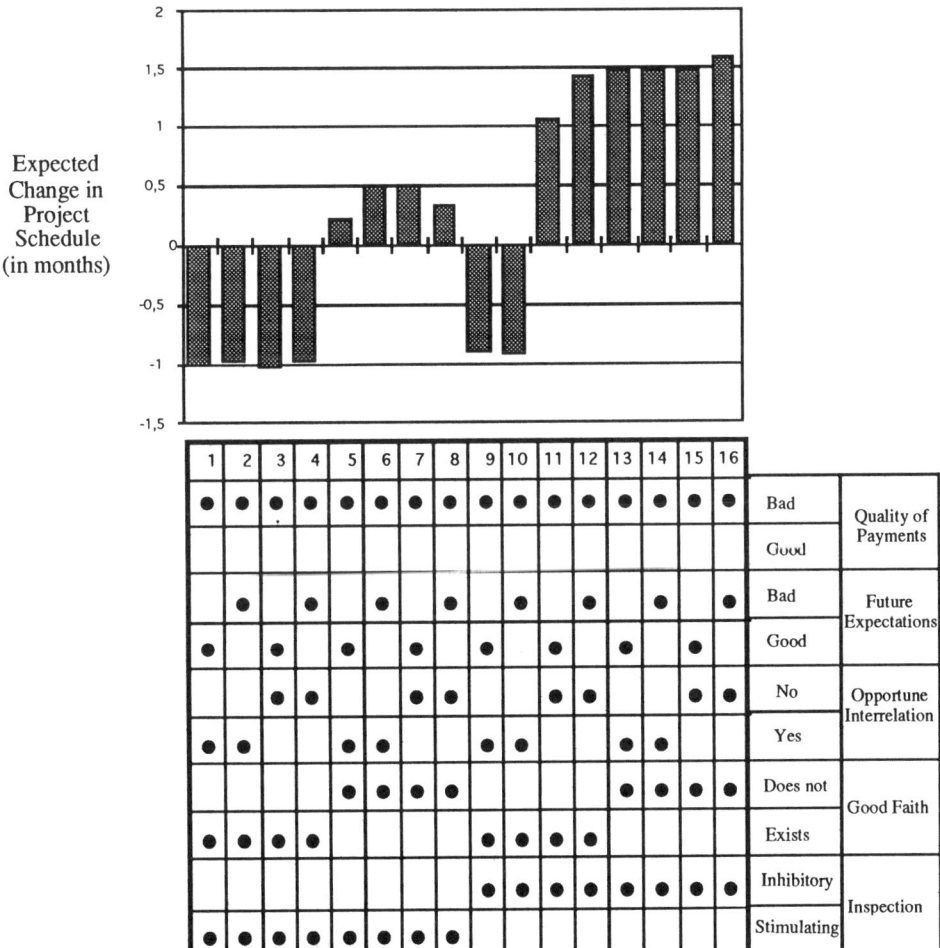

Figure 5. Effects of the conditions of the interrelation on the project schedule.

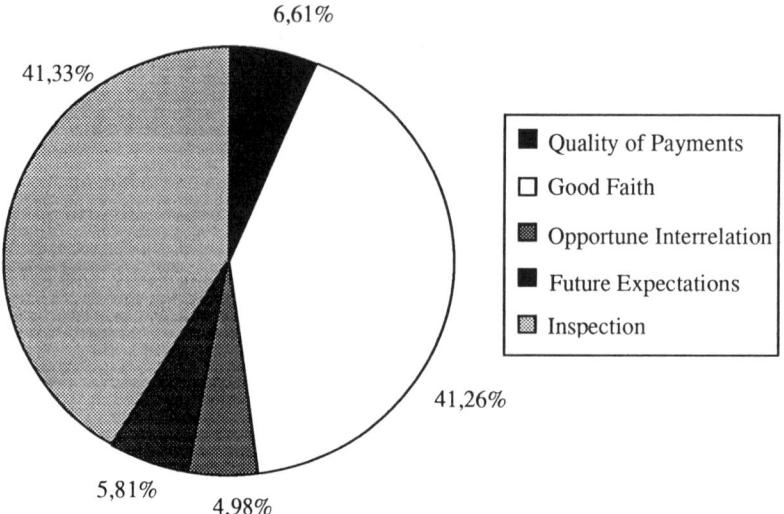

Figure 6. Relative importance of interrelation characteristics on the project schedule.

inspection), it is possible to shorten the schedule by almost one month from what was planned. In the most negative cases (11-16), (with the common characteristics bad quality of payments, absence of good faith and inhibitory inspection), the effect on the schedule of the project could be up to 1.5 months more than what was planned.

We underline the results of scenarios 9 and 10. The conditions of interrelation in those scenarios seem to be very positive for the schedule of the project. This happens when good faith and opportune interrelation are present, and independent of future expectations, stimulating inspection and good quality of payments. However, an exhaustive analysis of other intermediate scenarios permits the observation that, in general, stimulating inspection and good faith in the relation seem to be the most important variables affecting the schedule of the project.

From this analysis we may build a graph like Figure 6 to show the average relative importance of conditions of interrelation on the schedule of the project. Obviously, good faith and inspection are the most important characteristics. Quality of payments, future expectations and opportune interrelation are of much less importance.

5.1.2 *Analysis of conditions of specific interrelation*

Figure 7 shows the sensitivity of costs to negative conditions of relation (scenario 16 in Fig. 5) with the following characteristics: bad quality of payments, bad future expectations, inhibitory inspection, no opportune interrelation and non-existent good faith. We see that by moving from an inhibitory to a stimulating inspection, the contractor may save approximately 4%; the saving is 2% if good faith exists, and if the quality of payments improves, it may produce a cost reduction of 1%.

We should highlight the dynamics of good faith, where timing is critical. Good faith assumed at the bidding or planning stage (savings of 2%) must be operative at the time of construction and requires the efforts of both owners and contractors to create and maintain confidence.

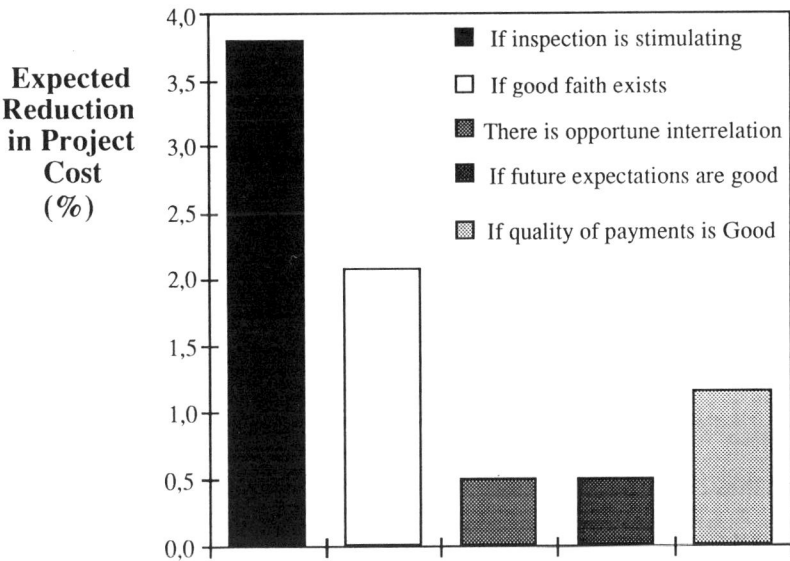

Figure 7. Cost sensitivity to conditions of interrelation in the worst-case scenario.

5.2 *Analysis of initial conditions*

Figure 8 shows the effects on project costs of the initial conditions of the project. This figure shows that in the more positive scenarios (1, 2 and 3) of these combinations including good quality of design, cost savings may be close to 2%. In the most negative cases (6, 7 and 8) of these combinations, with bad quality of design, the effect on project costs could be 13% over what was initially calculated.

Figure 9 shows the relative importance of initial conditions on the cost of the project. This graph indicates that quality of design is the most important characteristic – by far – affecting project cost. Bidding bases and legal background are important conditions but quality of design has a paramount effect on costs.

5.3 *Analysis of combined conditions*

The mathematical model used for this analysis offers the attractive potentiality to evaluate the simultaneous effects in a individual way. We used this ability to evaluate the combined effects of the initial conditions with the conditions of interrelation.

Let's suppose the 'real' case of a contractor who has had a terrible experience in his relation with a particular owner but wants to decide if he will or will not bid on a project which shows initial conditions different to those of his earlier experience. Figure 10 shows the effects that various sets of initial conditions would have on quality of the project. The scale used for quality is relative and indicates the percentage of cases with quality inferior to that indicated, for which 50% represents the average quality. Even if the contractor counts on bad relations with the owner, if the project has started with good initial conditions, (see scenario 1), it is possible to achieve a completed project whose quality is superior to the average. An important

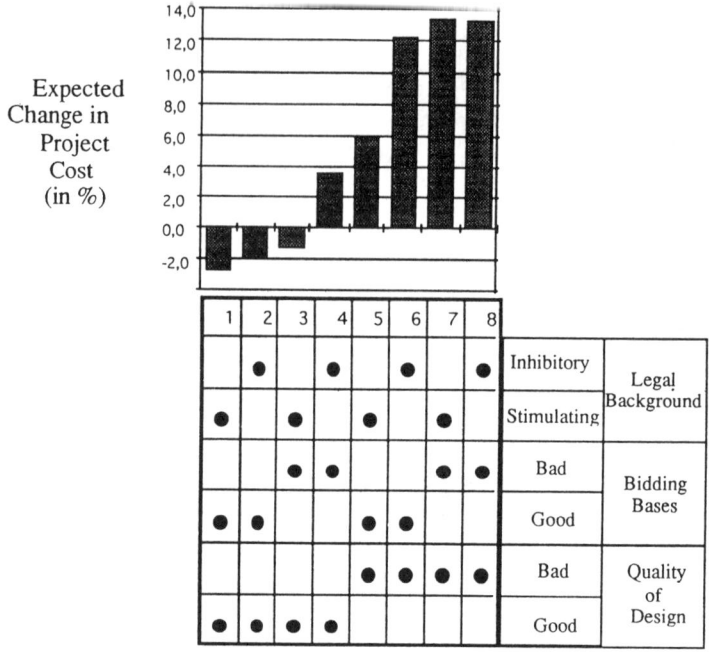

Figure 8. Impact of initial conditions on the project cost.

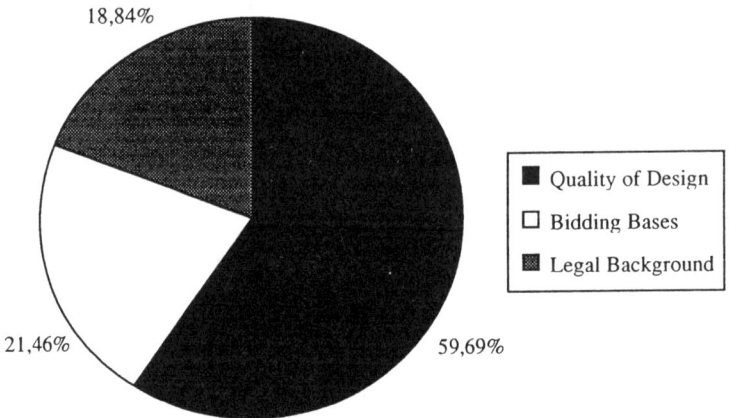

Figure 9. Relative importance of initial conditions on projects costs.

aspect of this scenario is that when the conditions of interrelation are negative, there is no additional negative effect on the quality by having negative initial conditions (see scenario 8).

Now suppose the case of a contractor who, in general, has had a good relation with an owner, but wants to decide if he will or will not bid on a project which may present initial conditions different to those of preceding projects with that owner. In Figure 11, the whitened zones of the bars represent cost reductions due to good conditions of the relation. The combined impact on project costs when there are very

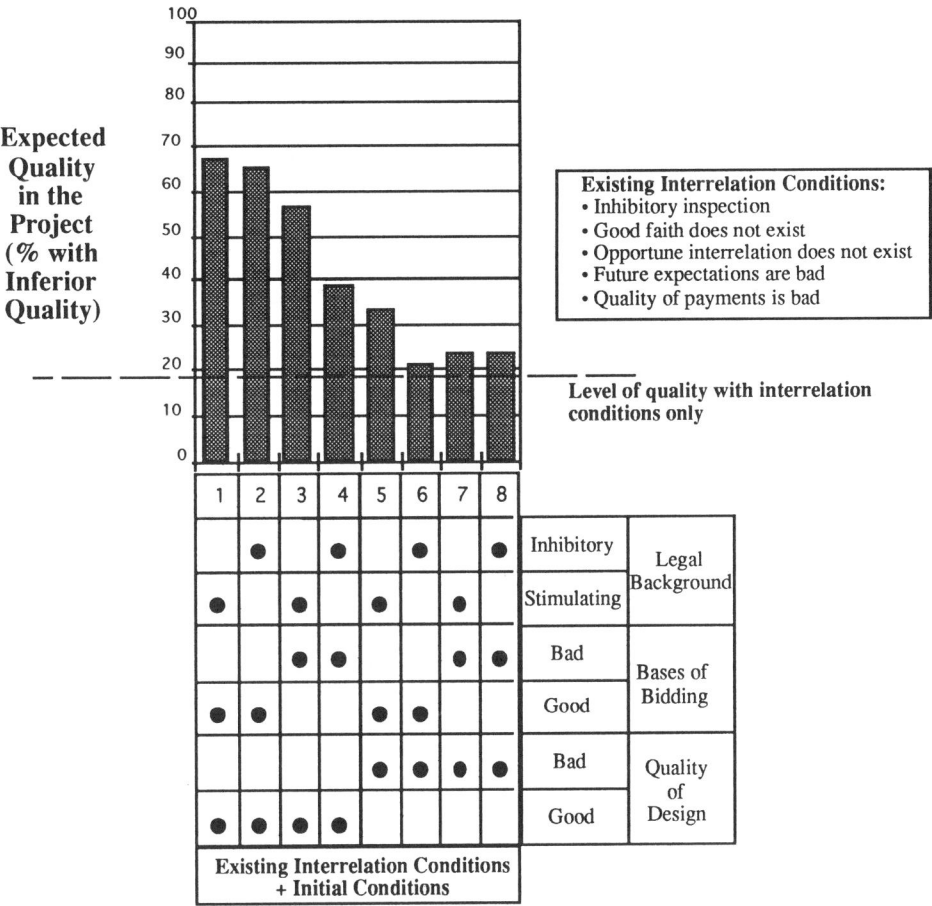

Figure 10. Combined impact on the quality project of the initial conditions with the worst-case scenario in conditions of interrelations.

positive initial conditions as in scenarios 1 and 2 is not detectable. However, cost climb to the extent that initial conditions worsen. Good conditions of interrelation account for, on average, 1% in scenarios 3 to 8. Probably, it is not possible to reverse the negative effects of bad initial conditions through a good relation.

6 ANALYSIS OF INFORMATION PROVIDED BY THE MODEL

Tables 1 and 2 summarize the relative importance that the variables studied have on the results of project, according to the perception of construction professionals who participated in analysis. Cost, schedule and the quality of contractor's management were the elements chosen by the representatives of the Chamber of Construction to measure the results.

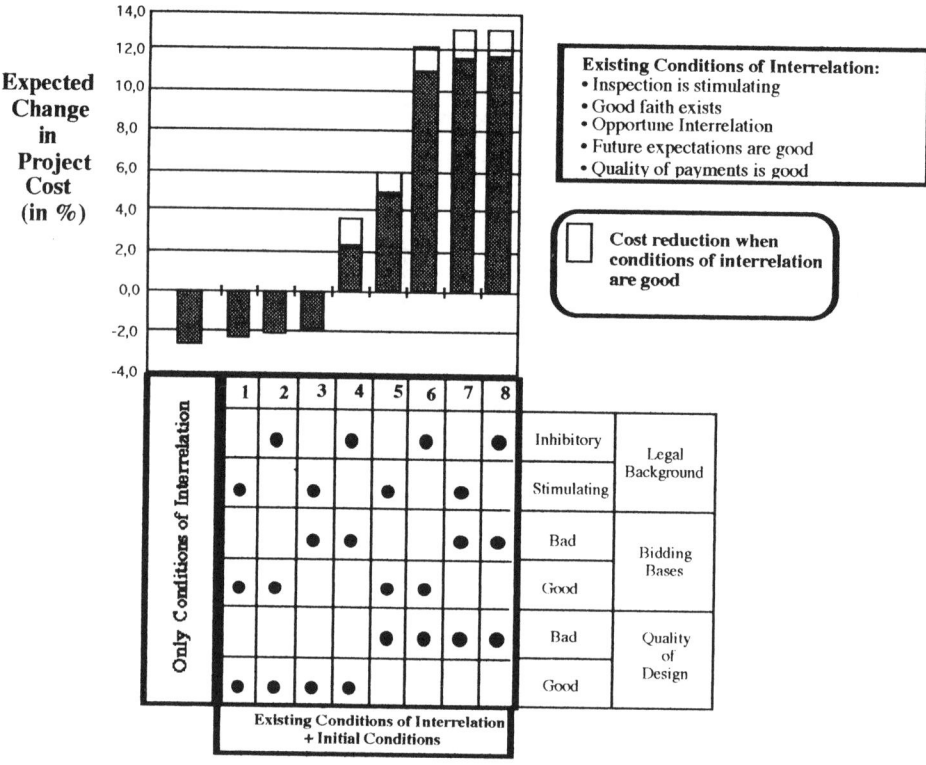

Figure 11. Combined impact on the project cost of initial conditions at a stage with conditions of interrelations.

Table 1. Relative influence of initial conditions on performance measure.

Initial conditions	Performance measure		
	Schedule	Cost	Quality
Quality of design	59%	60%	60%
Legal background	23%	19%	19%
Bases of bidding	18%	21%	21%

6.1 *Initial conditions*

Table 1 shows the influence of initial conditions on results. The enormous relative importance that quality of design has among them (60%) is remarkable; the other two factors, legal background and bases of bidding, are much less important at about 20% each.

No one will be surprised if an owner thinks that bad or good quality of design is of capital importance for the schedule and cost of a project. Designs thoroughly completed and based on reliable site information avoid modifications and favour good management by owners and contractors for the completion of project goals. However, it may be surprising that contractors feel as strongly about design as owners since, in general, the design is not their responsibility and design faults or modifications that affect their performance are not prejudicial – at least contractually – to them. However, the conclusions of the model respecting the importance of design, and linked to the lesser importance accorded factors representing the contractual basis lead us to suppose that:

– The contract does not always – or even usually – allow proper compensation for the costs that 'normal', significant or excessive changes of design mean for the contractor; and,

– For these reasons, contracts become excessively rigorous with respect to the quality and mutability of the design.

By contrast, owners and contractors agreed that their contracts should reflect the level of design, especially to provide major flexibility for designs based on uncertain site information, it is probable that construction companies would give more relative importance to the contractual factors which allow to them improve their economic results (costs) and management (schedules and quality), even when the designs suffer abnormal modifications, since they would be compensated for the expenses and difficulties that stem from aspects foreign to their own operations.

This hypothesis, which has yet to be demonstrated, prompts thinking about the usefulness of revising standard contracts, particularly those issued by governments, to improve their flexibility to the like of those contracts which include clauses of order of change and change of physical conditions. Such clauses and procedures that respect their intent create equitable conditions between owners and contractors and, therefore, promote good relations – and their benefits – between them.

Considering the importance of design, owners and contractors might be wise to collaborate to consider omissions and errors in drawings and their suitability to site conditions. They could reinforce these efforts with:

– More thorough site inspection to ensure constructibility within the parameters of cost, schedule and quality envisaged;

– Timely trouble-shorting by labor engineering employed by the contractor to request corrections, complements and materials required at the site; and

– The ongoing visits and cooperation of the designer, as well as technical meetings following inspections by the owner and the contractor.

6.2 *Conditions of interrelation*

Table 2 indicates the influence on the results assigned to the conditions of interrelation. The most significant are good faith and inspection, each with a similar relative importance of 40%. The other three factors weigh, together, only a 20%.

The opportune interrelation was defined as 'the speed of response in the communications between the parties when they need to exchange information about the project'. The small significance of this factor seems curious, particularly if we take into account the importance the results give to design and contemporary conditions wherein the designer is disengaged and off-site, unresponsive to the immediate demands of the construction process. Perhaps the author of this paper overrated opportune interrelation; but it could also be that in some way opportune interrelation is subsumed and implicit in good faith.

Table 2. Relative influence of conditions of interrelation on the performance measure.

Conditions of interrelation	Performance measure		
	Schedule	Cost	Quality
Good faith	41%	42%	44%
Opportune interrelation	5%	5%	6%
Inspection	41%	40%	35%
Future expectations	6%	8%	10%
Quality of payments	7%	5%	5%

Factors such as future expectations and quality of payment depend upon the owner, so the contractor can scarcely modify them; fortunately they have little importance.

Good faith has been defined as 'the confidence the parties to the project have in each other'. Confidence requires the participation of both parties. It does not depend sole and exclusively upon the attitude or will of contractor.

We have described inspection as either stimulating or inhibitory. The positive or neutral disposition of inspection does not depend exclusively upon the attitude or will of the contractor. Nevertheless, it is fortunate that the contractor has an important role in the two most significant factors of the conditions of interrelation. With his own attitude he may:
– Gain the confidence of the owner;
– Collaborate to get or to maintain stimulating inspection; and/or,
– Indirectly achieve a opportune interrelation with the owner.

Even when exception exist, the norm will be that this effort gives goods results, since the contractor's good management favours the objectives of the owner, so his total collaboration can be expected. In this context, it is absolutely advisable to seek excellent communications and collaboration between the parties (Campero 1992).

7 SUMMARY AND CONCLUSIONS

This chapter reports on a modelling methodology which forecasts the effects that certain characteristics of the owner-contractor relation would have on the results of a particular construction project in the area of public works. These results have been measured in cost, schedule and the quality of the project.

We developed a conceptual model as the formal representation of the perceptions of the modelling participants of how risk characteristics in the owner-contractor relation affect variables of the project and how these effects spread within a construction projects. The risk characteristics in the owner-contractor relation were split into two groups. One group, 'initial conditions', represents characteristics that create a formal framework for the project and are generally objective to the owner-contractor relation. Among these characteristics, quality of design given to the contractor as the basis of his work is clearly the most important. The second group, 'conditions of interrelation', presents less formal and more subjective elements than initial conditions.

In this group, the kind of inspection with which work is supervised and good faith in the owner-contractor relation are of major importance.

The methodology used to develop this model offers enormous advantages in a construction company's administration and planning functions. Construction managers, are frequently required to make decisions in situations where they can not count on precise and complete information and, therefore, are without the tools needed to make a reasoned and rigorous evaluation. This methodology offers a means for structuring a systematic discussion about factors and elements in the situation at hand.

The model is carefully designed as to be theoretically valid and easily implemented. Its transparency fosters consistency and verification of the information supplied at each stage of the modelling. The model does not replace the decision-maker's judgement but rather structures, enhances and validates it.

The application of the model to the owner-contractor relation focuses analysis and problem-solving efforts in those areas perceived to have a major influence on the results the contractor may expect from his management. In this case study, identification of significant factors points to the usefulness of efforts: 1) to ensure that contract flexibility reflects the quality of the given design; and 2) to increase confidence in the relations between, the owner and the contractor to form a work team with effective inspection.

ACKNOWLEDGEMENTS

The authors sincerely appreciate the collaboration of the Committee of Public Works of the Chilean Chamber of Construction, for the valuable support provided to this research. We thank also Fondecyt (project 1930665) and the Construction Training Corporation for their continuous support to this work.

REFERENCES

Alarcón Cárdenas, L.F. & Ashley, D.B. 1992. Project Performance Modelling: A Methodology for Evaluating Project Execution Strategies. *A Report to the Construction Industry Institute,* Source Document 80, The University of Texas at Austin, EEUU.

Ashley, D.B. & Alarcón Cárdenas, L.F. 1990. A Pilot Study on Construction Incentives. *Report to the CII Project Team Risk/Reward Task Force*, University of California, Berkeley.

Campero, M. 1992. Prevención y Manejo de Reclamos en Contratos de Obra Civil. *Revista de Ingeniería de Construcción*, Departamento de Ingeniería y Gestión de la Construcción, Pontificia Universidad Católica de Chile, N° 13, pp. 1-22. July-December 1992.

Gordon, T. & Hayward, H. 1986. An Initial Experiment with the Cross-Impact Method of Forecasting. *Futures* 1(2): 100-116.

Fast or concurrent: The art of getting construction improved*

PEKKA HUOVILA, LAURI KOSKELA & MIKA LAUTANALA
VTT Building Technology, Espoo, Finland

ABSTRACT: Fast tracking and concurrent engineering are approaches aiming at a shorter project duration. Fast tracking is a method already practised in construction projects, while concurrent engineering comes from other industries' product development projects. Both have been emerged as an alternative for the sequential approach of project realisation.

The purpose of this paper is to outline the essential features of fast tracking and compare them to those of concurrent Engineering, and to study the applicability of concurrent engineering principles to construction. The work is based on a literature study.

The main conclusions are:

1. Fast tracking is a practically oriented approach, without solid conceptual or theoretical basis. The essence of fast tracking is overlapping of design and construction, which does not always lead to an optimal design solution.

2. Concurrent engineering aims principally at reducing the duration of engineering time, increasing the value of the product and reducing the costs. Theoretically, this is achieved by reducing the share of those activities which do not directly contribute to the conversion of requirements to the final design, and by assuring that value is maximally added by those activities contributing to this conversion.

3. Concurrent engineering emphasising at customer satisfaction, team approach, concurrent process for design of the product and planning of production, strategic relations with suppliers and continuous improvement is fully relevant for construction.

1 INTRODUCTION

In recent times the main competitive factors in construction have been cost and quality. Cost competition has in the traditional construction process meant minimisation of costs in each phase in sequential construction process. Managing of quality has been understood as supplying of the products (building components) in conformity with the design that specifies the product. The described procedure has left little space for the product development and no incentives for renewal of the traditional construction process.

*Presented on the 2nd workshop on lean construction, Santiago, 1994

Today, the rapid supply of buildings that meet the life cycle needs of building owners and individual users is a new competitive factor in construction. One method that has been introduced for reducing the construction project time has been fast tracking: overlapping of design and construction activities. That has led in many cases to a shorter project duration.

In other industries it has been earlier understood that the essential factor in product development is rapid introduction of products that have features fulfilling the changing customer needs in each market segment. One of the methods that has been successfully applied in product development is concurrent engineering: an approach for integrated, concurrent life cycle design of products and their related processes aimed at shortening the lead time, reducing costs and giving added value to the customers.

This chapter presents first the definitions and concepts of both fast tracking and concurrent engineering as found from literature. The theoretical foundations of concurrent engineering are studied to be followed by a comparison of fast tracking and concurrent engineering. Finally it is discussed how the principles of concurrent engineering could be applied in construction.

2 FAST TRACKING

In literature, fast tracking has most often been understood simply as overlapping of activities: 'initial construction activities are begun even before the facility design is finalised' (Hendrickson & Au 1989). The Project Management Body of Knowledge defines fast track as follows (Project Management Institute 1987):

The starting or implementation of a project by overlapping activities, commonly entailing the overlapping of design and construction (manufacturing) activities.

Similar short definitions may be found in other sources (Willis 1986; Bennett 1991). Only rarely are more detailed recommendations for fast tracking presented. The study of Kwakye (1991) and the CII (1988) study are the only broader overviews on the problems and techniques of fast tracking.

Views on fast tracking vary. Part of writers emphasise the problems caused by fast tracking. In one Canadian study, the conclusion is that fast tracking often results in unexpected costs and does not necessarily lead to a shorter project duration (Fazio et al. 1988). This is confirmed in another study on American construction (Laufer & Cohenca 1990), which found that very low percentages of design completion prior to the start of construction may result in considerable construction delays. Tighe (1991) argues that fast tracking leads to less than optimum design and more costly construction.

On the other hand, there are many success stories of fast tracking. Often, such techniques as partnering, teaming etc. are mentioned as success factors (High 1992; Bakker et al. 1994).

Thus the conclusion is that fast tracking is a practically oriented approach, without solid conceptual or theoretical basis. The essence of fast tracking is overlapping of design and construction.

3 CONCURRENT ENGINEERING

3.1 *Background*

In literature, little or nothing was written about concurrent engineering before 1980. In the 1980's, companies began to feel the effects of three major influences on their product development: newer and innovative technologies, increased product complexities and larger organisations. Companies were forced to look for new product development methods. One of the most significant events in the concurrent engineering time line took place in 1982, when the Defence Advanced Research Projects Agency began a study to look for ways to improve concurrence in the design process. Five years later, when the results of that study were released, they proved to be an important foundation on which other groups would base further study (Carter & Baker 1992).

Ziemke & Spann (1993) state that concurrent engineering, as well as its subsets, design for manufacture (DFM) and design for assembly (DFA), are not recent concepts in both manufacturing and engineering, but rather parts of engineering design philosophy for many decades. They claim that in the present rush to emphasise concurrent engineering, DFM, and DFA, there has been a strong tendency to reinvent practices that were common during World War II and earlier.

Nihtilä (1993) says that already in the very early days of manufacturing new products and their manufacturing processes were designed simultaneously. This was a natural way of working since the organisations involved with the process were substantially fewer and smaller than those of today's manufacturing. Nihtilä calls simultaneous engineering what is meant by concurrent engineering in this paper. Another expression for the same thing is parallel engineering. Clausing (1994) rather talks about world class concurrent engineering differentiating it from the traditional concept that he calls basic concurrent engineering. Other expressions that can be found in the literature are collaborative engineering, synchronous engineering, life-cycle engineering, design for excellence etc. (Trygg 1993). It may well depend on the context of those buzzwords whether they should be regarded synonymous for concurrent engineering.

Concurrent engineering, as it is understood in the nineties, has definitions of its context, concepts of its essential features as well as guidelines for its implementation. These issues are described in detail in the following chapters.

3.2 *Definitions*

In the summer of 1986, the Institute for Defence Analyse Report coined the term concurrent engineering to explain the systematic method of concurrently designing both the product and its downstream production and support processes. That report provided the first definition of concurrent engineering as follows (Carter & Baker 1992):

Concurrent engineering is a systematic approach to the integrated, concurrent design of products and their related processes, including manufacturing and support. This approach is intended to cause the developers, from the outset, to consider all elements of the product life cycle from concept through disposal, including quality, cost, schedule, and user requirements.

This definition is now widely accepted. Carter & Sullivan (1994) have expressed it also in a shorter form:

Concurrent engineering is a systematic approach that brings product development and processes together for the entire company.

Kusiak (1993) defines concurrent engineering in a following way:

Concurrent engineering refers to practice of incorporating various values of a product into the design at its early stages of development. These values of address the entire life cycle of the product and include not only its primary functionality but also producibility, assemblability, testability, serviceability, and even recyclability.

Nihtilä (1993) uses the following simultaneous engineering definition:

Simultaneous engineering is the simultaneous development of the product, manufacturing and assembly equipment and processes, quality control and marketing. (Hartley & Mortimer 1990).

Numerous other definitions can be found from literature. Often they highlight some parts of concurrent engineering, such as team approach and improving of communication or tools for the customer interface or tools for automation of engineering activities.

3.3 *Concepts*

Smith & Reinertsen (1991) think of the task of managing the development of a new product as balancing effort towards four key objectives: development speed, product cost, product performance and development program expense. They say that the art of managing product development depends on making good trade-offs between these four possible objectives as presented in Figure 1 and suggest an economic model as a tool for a development project to make these trade-offs in a businesslike manner.

Noble (1993) reminds that among other benefits the motivation for concurrent engineering is primarily the economic one. As Figure 2 shows the best time to impact the economics of the product is in the conceptual/design phase. Therefore, it is critical that tools are developed to integrate economic issues within a concurrent engineering approach.

Selected concurrent engineering concepts are presented in Table 1 as picked up from literature. Clausing (1994) emphasises improved process and its clarity together with closer cooperation bringing improved unity. Schrage (1993) uses a ten-step approach with the focus on strong interface with customer and teamwork. Murmann (1994) mentions clear project objectives, responsibility of the project manager and cross-functional knowledge among important aspects. Carter & Baker (1992) point out organisational and communicational issues along the requirements and product development itself. Nihtilä (1993) concentrates on team approach and concurrent engineering tools.

Carter & Baker (1992) suggest the use of a company assessment questionnaire to characterise the present development environment of the company to be followed by determining the state where the company should be. The questionnaire can be plotted for a graphic view (Fig. 3) of where the company is and where it should be. The organisational issues include the understanding of individual employees and teams of

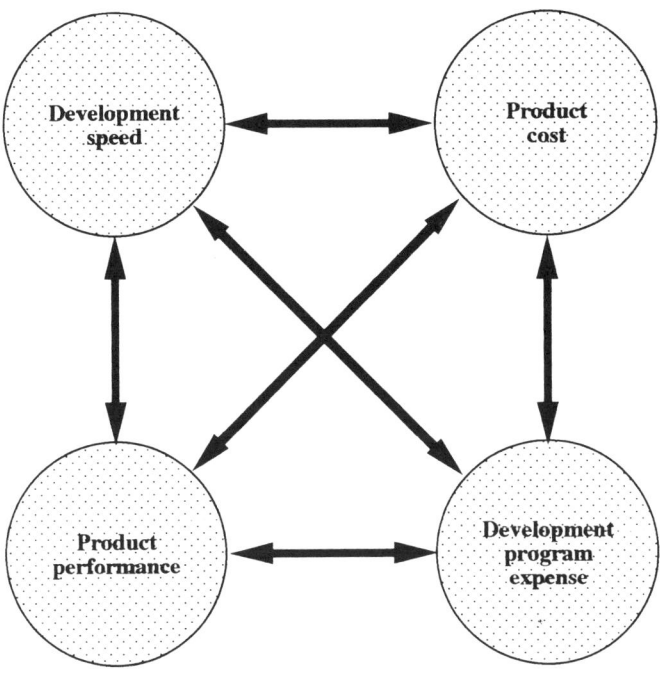

Figure 1. The six trade-offs between product development objectives (Smith & Reinertsen 1991).

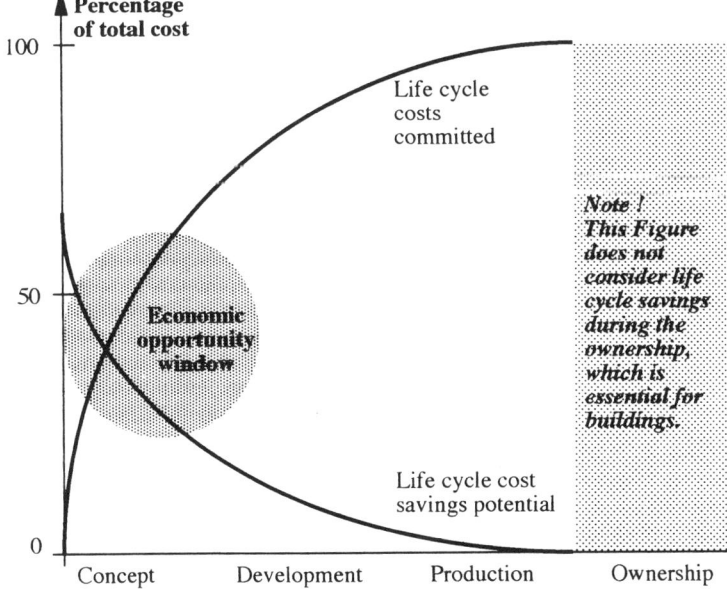

Figure 2. Window of economic opportunity for concurrent engineering (Noble 1993).

Table 1. Concurrent engineering concepts.

Basic concurrent engineering (Clausing 1994)	Concurrent engineering (Schrage 1993)	Project improvement activities towards time reduction (Murmann 1994)	Concurrent engineering (Carter & Baker 1992)	Simultaneous engineering (Nihtilä 1993)
Improved process (better game plan), improved clarity: 1. Concurrent process; 2. Focus on quality, cost and delivery; 3. Emphasis on customer satisfaction; 4. Emphasis on competitive benchmarking. Closer cooperation (better teamwork), improved unity: 1. Integrated organisation; 2. Employee involvement, empowerment; 3. Strategic relations with suppliers.	1. A top-down design approach based on a comprehensive systems engineering process; 2. Strong interface with customer; 3. Multifunctional and multidisciplinary teams; 4. Continuity of the teams; 5. Practical engineering optimisation of product and process characteristics; 6. Design benchmarking and soft prototyping through creation of a digital product model; 7. Simulation of product performance and manufacturing and support process; 8. Experiments to confirm/change high risk predictions found through simulation; 9. Early involvement of subcontractors and vendors; 10. Corporate focus on continuous improvement and lessons learned.	1. Define clearer project objectives; 2. Concentrate the resources on fewer development projects; 3. Use predevelopment to reduce technical uncertainty; 4. Improve project planning. 5. Improve paralleling and overlapping of development tasks; 6. Increase the competence and success responsibility of the project manager; 7. Improve expert and cross-functional knowledge; 8. Insure early manufacturability of the design concept; 9. Improve communication and communication behaviour of employees; 10. Intensify time and cost controlling in development.	Organisation: 1. Team integration; 2. Empowerment; 3. Training and education' 4. Automation support. Communication infrastructure: 1. Product management; 2. Product data; 3. Feedback. Communication infrastructure: 1. Product management; 2. Product data; 3. Feedback. Requirements: 1. Requirements definition; 2. Planning methodology; 3. Planning perspective; 4. Validation; 5. Standards. Product development: 1. Component engineering; 2. Design process; 3. Optimisation.	1. Team approach and simultaneous development of products and processes; 2. Project management; 3. Quality function deployment; 4. Design for manufacture and assembly; 5. Supplier partnership; 6. Continuous improvement; 7. Technical applications.

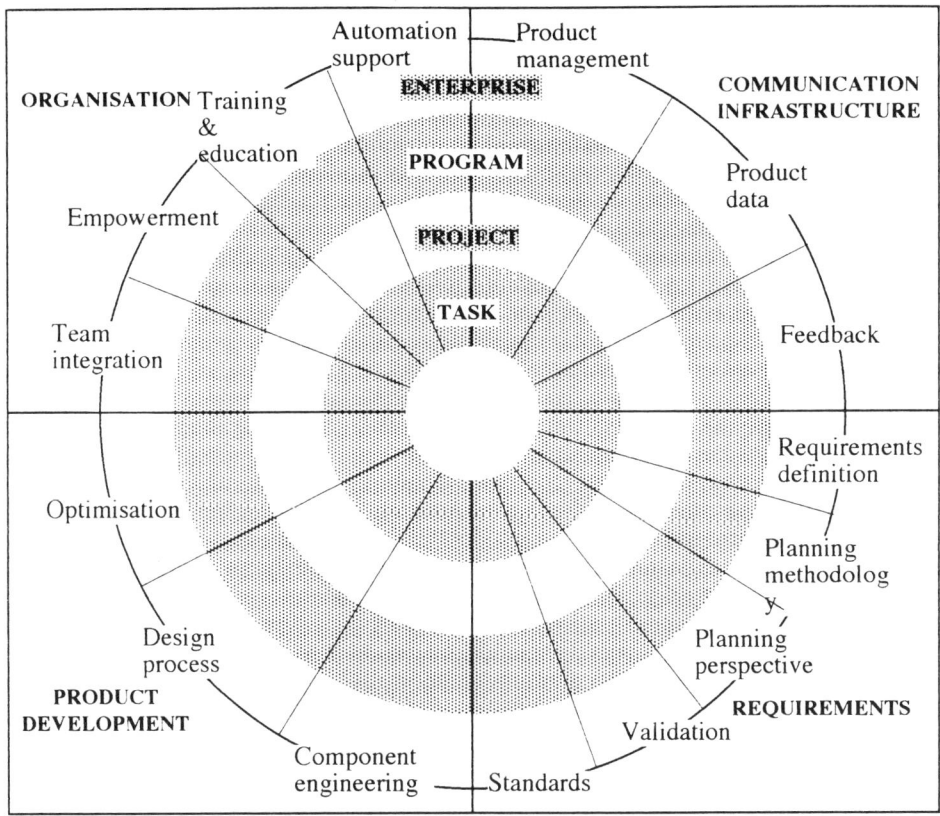

Figure 3. Concurrent engineering map (Carter & Baker 1992).

their roles and tasks in the context of the overall product development process, the responsibilities, training and education, and integrated tools. The communication infrastructure gives effective communication paths to product management, complete and accurate product data and feedback to handle customer expectations, product specifications, industry standards, and other requirements.

The requirement definition starts from converting customer needs upfront to the definitions, specifications, and design of the product so that at every state in the development process, individuals, teams, and managers can check that the requirements, specifications and designs are meeting customer needs. Product planning, evaluation, and design methods can happen both the bottom-up and top-down. The methodologies and validations for the design process are documented and measured. Managers must respond to the continuing evolution of technology.

The important concurrent engineering issues as presented here were picked from the definitions and discussion in the literature without studying thoroughly their background and conceptual basis. Those matters are analysed in Section 4.

4 THEORETICAL APPROACH TO CONCURRENT ENGINEERING

4.1 *Concurrent engineering as a conceptual and theoretical innovation*

Our basic argument is that concurrent engineering is based on a new conceptualisation of engineering operations. The traditional conceptual basis, that sees design and engineering as conversion, is restrictive, and in itself causes inefficiency. In order to eliminate these problems, we have to augment the conceptual basis of design.

In our opinion, design can effectively be conceptualised with the lean production paradigm (Koskela & Sharpe 1994; Koskela 1992), seeing any design[1] process simultaneously as:
– Conversion;
– Flow;
– Value generation.

The methods of concurrent engineering have been developed on basis of intuitive understanding of this wider conceptual framework. For advancement and diffusion of concurrent engineering, it has to be made explicit.

4.2 *Design is a conversion*

This traditional view holds design as activity (or task), where requirements are transformed into design fulfilling those requirements (Fig. 4). This big activity is then divided into subactivities, which are carried out by specialists. The perspective is on what people do in design. Thus, the focus may be on decisions, with the premise that the principal role of the designer is to make decisions Mistree et al. (1993), or on problem solving Murmann (1994).

A typical definition of designing is as follows Mistree et al. (1993):
Designing is a process of converting information that characterise the needs and requirements for a product into knowledge about a product.

In this perspective, development of design equals to making design activities more effective and efficient. In practice, this is realised by design tools (CAD, calculation models, simulation models, decision support tools) and by project management principles, methods and tools (called also systems engineering).

It is difficult to exaggerate the proliferation of this conceptual view. Traditional (and still most contemporary) methods of organisation, management and control are based on it. Also popular description techniques in information system development, like IDEF and SADT have this view as foundation. Project planning and management tools, like CPM, are another example.

However, in our view there are serious conceptual oversights in this view:

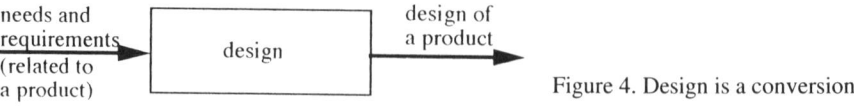

Figure 4. Design is a conversion.

---

[1] Design should be understood in this context equivalent to engineering.

– There are also activities in design that do not contribute to the conversion: for example, information is inspected, stored and communicated; these are not explicitly represented;

– For each design and engineering processes, there are customers, who have needs and requirements concerning this process; this is not explicitly represented.

The problem is, that these oversights are replicated and amplified in the various organisational, control, and communication structures that are based on this conversion view. Thus, the improvement potential related to these missing elements (non-conversion activities, fulfilment of customer needs) can not orderly be taken into account, even if participating individuals well perceive them.

In consequence, this view has contributed directly and indirectly to the many persisting problems in engineering projects:

– All requirements are not caught at the start;
– Design errors are detected in later phases, leading to costly rework;
– Long or no iterations for improving the design;
– Waiting for approval, instructions and information takes the major share of designers time;
– In general: long duration and inflated costs of projects, mediocre or low quality of the design of a product.

4.3 *Design is a flow*

This perspective, which originates from industrial engineering, focuses on what happens to information in design: 'things are made through the flow of information' Sekine & Arai (1994). The unit of analysis is the total flow of information[2]. In principle, there are four different possibilities: conversion, waiting, moving, and inspection (Fig. 5). In fact, only conversion can be design proper, other activities are basically not needed (and called therefore waste), and should be eliminated, rather than made more efficient. But a part of conversion, namely rework due to errors, omissions and uncertainty etc. is also waste.

In this perspective, improvement of design equals to eliminating waste. This means[3] (Fig. 6):

– Rework due to errors, lack of information, changes in scope etc.;
– Transfer of information;
– Waiting of information for the next step;
– Inspection and other non value adding activities.

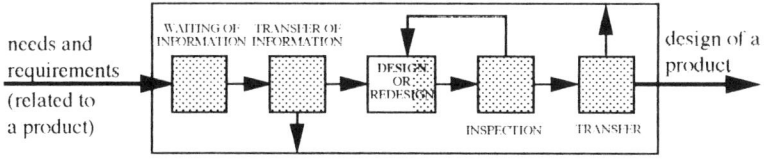

Figure 5. Design is a flow.

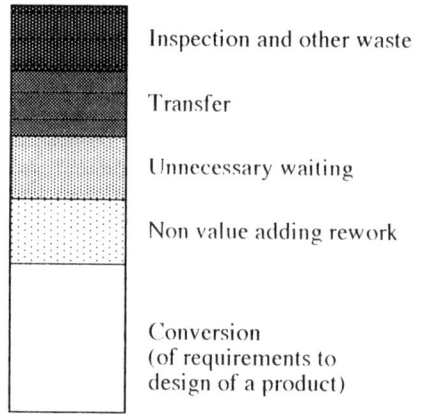

Inspection and other waste

Transfer

Unnecessary waiting

Non value adding rework

Conversion
(of requirements to
design of a product)

Costs of engineering   Figure 6. The formation of waste in engineering.

The significance of this view is caused by the fact, that the amount of waste is large in any complex operations, like engineering. When information flows are analysed in more detail, it is typically found that the share of conversion from the total flow time is very little. Cooper (1993) estimates that in design of large construction projects, there are typically from one-half to two and one-half rework cycles[4]. Thus, reduction of this waste provides a very worthwhile improvement potential.

The principles and methods of waste elimination are related to the root cause of each waste category;

– The major general cause for rework is uncertainty. Thus, it is paramount to reduce aggressively uncertainty especially in the early phases of the engineering project (Fig. 7) Bowen (1992). This requires:

1. The scope definition is done orderly, for avoiding scope changes;

2. All life cycle phases are considered simultaneously from the conceptual stage for avoiding iterations due to constraints in subsequent phases;

3. Prototyping, simulation etc. can be used to decrease technological uncertainty;

4. In later phases of the project, the design solution is practically frozen;

5. Design errors are reduced through quality management.

– The time and effort needed for the all necessary transfer of information can be reduced through team approach, especially when the team is co-located. In a team, much information can be transferred informally and orally, without paper and communication devices.

– The primary reason for long waiting times of information is that output from each phase is transferred to the following in large batches. Thus, decomposition of design tasks, intense informal communication and concurrency provide a solution to this.

– The causes of other non value adding activities vary, and so do corresponding methods. In general, waste can be attacked by the methods and techniques of business process re-engineering Davenport (1993). Non-compatibility of design tools

---

[4] However, Cooper does not distinguish between avoidable rework caused by errors etc. and iterations needed for improvement of the design solution.

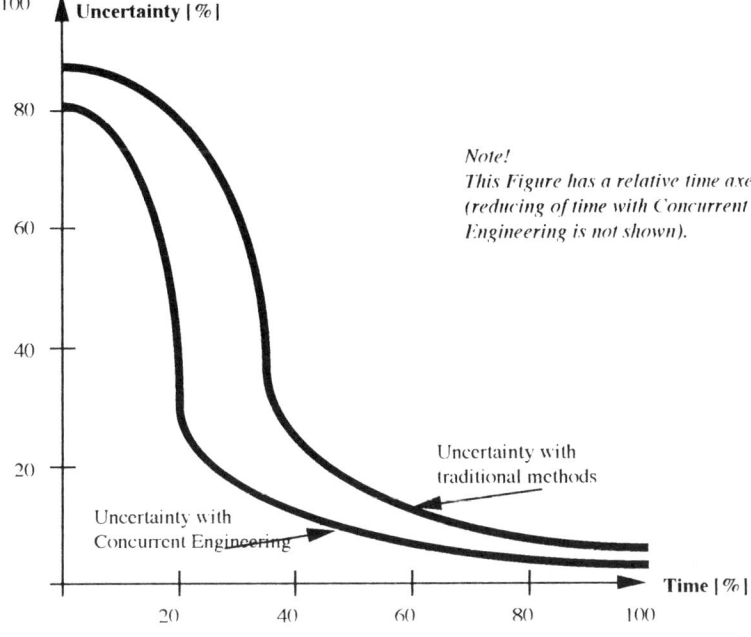

Figure 7. Development of uncertainty in traditional and concurrent engineering.

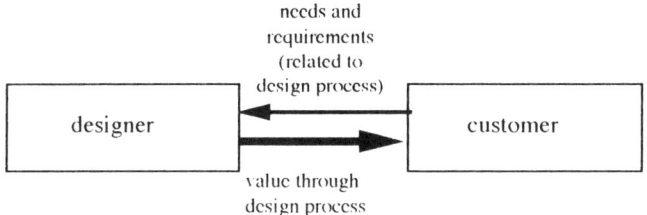

Figure 8. Design is generating of value.

cause one type of waste: manual data conversion. The term computer integration refers to standardisation of information structures so that various software and computers can be interfaced easily. Organisational borders cause non value adding activities to interorganisational flows; partnering may be used in order to reduce them.

4.4 *Design is value generation*

This perspective, which originates from quality management, focuses on value generated by design to its customer(s). Value is generated through fulfilment of customer needs and requirements (Fig. 8). Value consists of two components: Product performance and freedom from defects. Value has to be evaluated from the perspective of the next customer(s) and the final customer. It is difficult, often impossible to measure the absolute value. However, for practical application, measuring the relative value often suffices; for example the value loss in relation to the best practice value or theoretically best value.

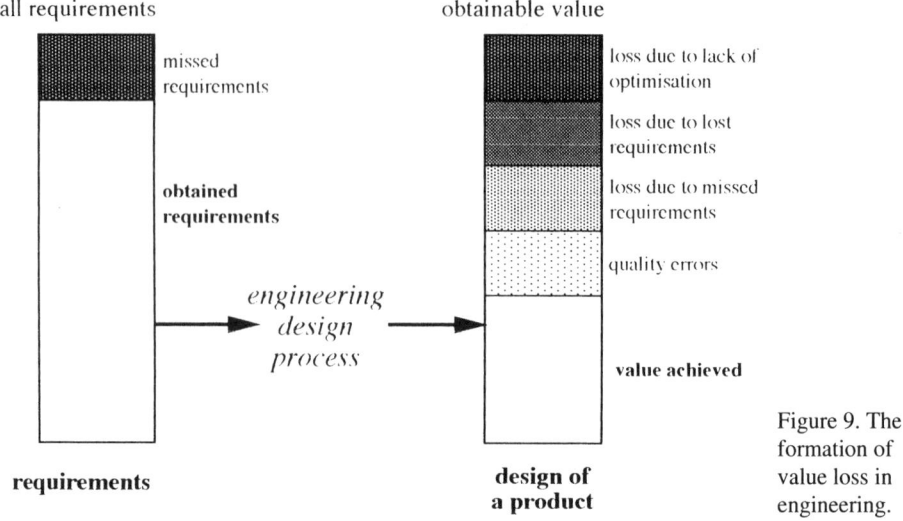

all requirements

missed
requirements

obtained
requirements

requirements

*engineering
design
process*

obtainable value

loss due to lack of
optimisation

loss due to lost
requirements

loss due to missed
requirements

quality errors

value achieved

design of
a product

Figure 9. The
formation of
value loss in
engineering.

In this perspective, improvement of design equals to reducing loss of value. Thus, the first question is, why does loss of value occur? There are four main possibilities (Fig. 9):

– Part of requirements are missed at the outset;

– Part of requirements are lost during the design process (for example, the design intention of a designer is not communicated to later steps, and may be spoiled by decisions in these);

– There is too little improvement and optimisation of design solutions (for example, constraints or opportunities of subsequent phases not taken into account);

– There are sheer errors in the final product design.

The corresponding solutions are:

– Rigorous needs and requirements analysis at the outset in tight cooperation with the customer;

– Systematised management of requirements, like the application of quality function deployment;

– For ensuring necessary improvement and optimisation, it is requisite to organise rapid iterations in major design issues; thus, all life cycle phases are considered simultaneously from the conceptual stage (Fig. 10). Also more formal tools like Taguchi methods, design for manufacturability or design for assembly may be used.

– Regarding errors, the quality management principles mentioned in the previous discussion on waste reduction can be used.

4.5 *Summary of essential features of concurrent engineering*

Some essential results of concurrent engineering are summarised in Figure 11. The major objectives of concurrent engineering are to reduce the duration of engineering, to increase the value of the product and to reduce its costs. Theoretically, this is achieved by reducing the share of those activities which do not directly contribute to

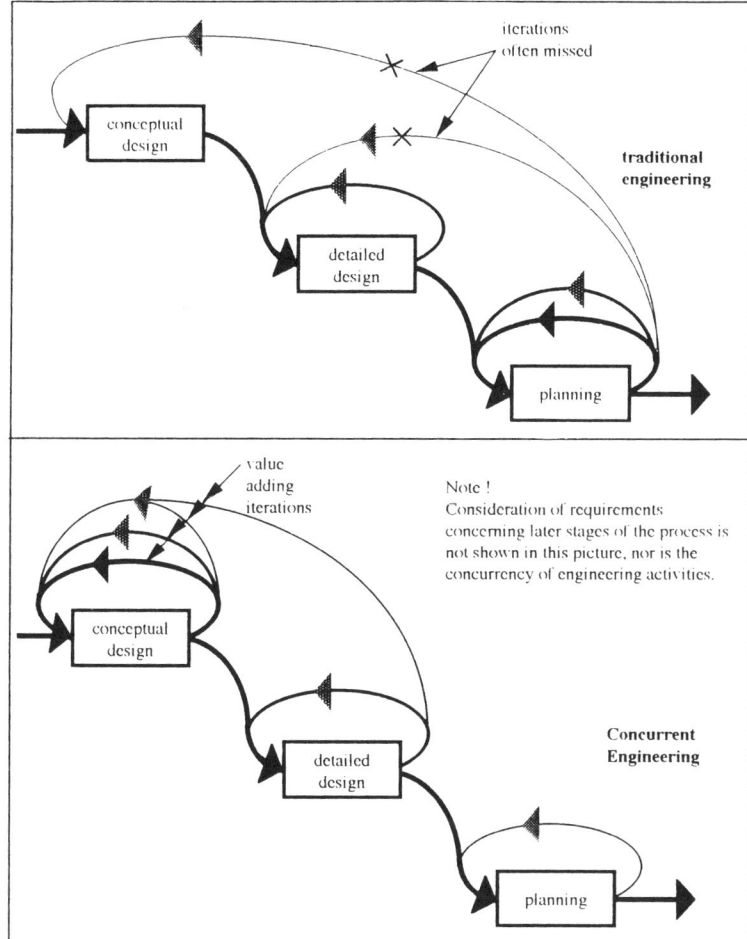

Figure 10. Iterations are carried out as early as possible in concurrent engineering.

the conversion of requirements to the final design, and by assuring that value is maximally added by those activities contributing to this conversion.

The rapid diminishing of uncertainty is a major distinguishing feature of concurrent engineering (Fig. 12). This is achieved though rapid iterations in the early stages of engineering, aimed at reducing uncertainty and/or improving the evolving design concept, and covering the requirements of all life cycle phases. Note that those iterations (or rather rework), which are caused by avoidable errors etc. are rather suppressed in concurrent engineering.

5 CONCLUSIONS

5.1 *Fast tracking and concurrent engineering*

In its original form, fast tracking is a brute force approach, based on the traditional (conversion) conceptualisation. The major characteristic of fast tracking, understood

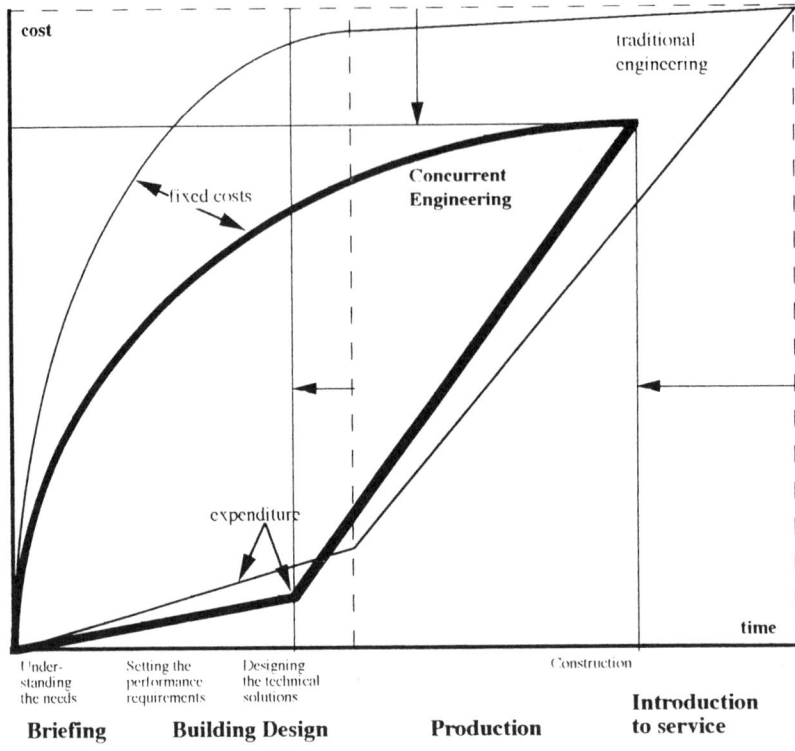

Figure 11. Essential results of concurrent engineering.

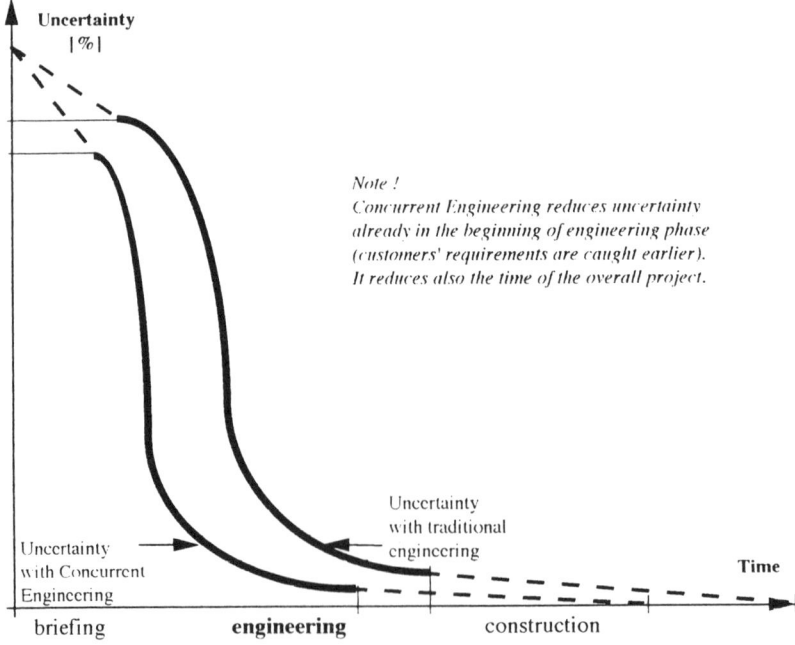

Figure 12. Development of uncertainty in construction, based on Figure 7 (Bowen 1992).

as mere overlapping of activities, is that uncertainty is increased in comparison to conventional (sequential) method. In consequence, often the total construction costs increase and the value of the end product decreases. Thus, other objectives are more or less sacrificed for speeding.

As explained above, uncertainty reduction is a major feature in concurrent engineering. It is also characteristic that improvements regarding all major objectives are pursued simultaneously.

Thus, fast tracking and concurrent engineering are clearly different, partly even opposing approaches. However, it has to be acknowledged that recently, several methods and techniques originating from concurrent engineering have been implemented in fast tracking projects, and fast tracking has thus started to integrate into concurrent engineering.

5.2 *Concurrent engineering from construction point of view*

Even if concurrent engineering has originated in other fields, it is in our opinion fully relevant for construction. It seems to give considerable improvement opportunities to many current bottlenecks in construction.

Concurrent engineering is conceptually and methodically more developed than the corresponding approach originating from construction: fast tracking. Thus, it should be very worthwhile to transfer the methods and techniques of concurrent engineering to construction context and to experiment with them in pilot projects. To some extent, this is already happening.

From construction point of view important aspects of concurrent engineering are customer satisfaction, team approach, concurrent process, strategic relations with suppliers and continuous improvement. These issues are not always emphasised within the traditional construction process.

Customer satisfaction in concurrent engineering means finding out also the unexpressed desires of the client for the whole life cycle of the buildings already in the beginning of the design process. That includes managing the facility, even recycling of the building components. There are certain techniques for finding out the customers' needs, such as quality function deployment (QFD), that may help in setting up the requirements that are not always presented by the client.

Team approach is suggested to facilitate with communication between different project participants. In many cases complicated construction projects are carried out with project participants that don't know each other beforehand. The communication issues bring often undesired waiting and errors, such as delays from transferring of information and rework due to implementing information that is not updated. Some tools, like QFD, force the participants to teamwork enabling direct communication and reduce extra costs and delays due to lack of needed information.

Concurrent process integrates the design of the products and their related processes. In fast tracking, the overlapping of activities leads to parallel design and construction. It does not necessarily include the constructability requirements that are focused in concurrent design and planning (design of product and planning of production activities), concurrent engineering.

Strategic relations with suppliers are often the weak points of construction compared to other industries. Just-in-time production as practised in the automotive industry requires a strong logistic network that doesn't exist in construction industry where building sites are messy with materials that are in risk of damage, theft and other loss while waiting for their installation.

Continuous improvement and benchmarking have been recently introduced to construction through quality related development projects. The traditional construction process is too often characterised by repeating of errors and copying of conventional solutions that are far from optimal ones.

REFERENCES

Bakker, Mosson & Cassells.1994. Fast Track Factory for Motorola. *Architects' Journal* 196 No 11, pp. 37-49.

Bennett, J. 1991. *International Construction Project Management: General Theory and Practice.* Butterworth-Heinemann, Oxford, 387 p.

Bowen, H. 1992. Implementation Projects: Decisions and Expenditures. In: Heim, J. & Compton, W. (ed.), *Manufacturing Systems: Foundations of World-Class Practice*, National Academy Press, Washington DC. 273 p.

Carter, D. & Baker, B. 1992. *CE Concurrent Engineering: The Product Development Environment for the 1990s*. Addison-Wesley, 175 p.

Carter, D. & Sullivan, T. 1994. *CE Concurrent Engineering: Best Practices for Global Success.* Mentor Graphics Corporation, 162 p.

Clausing, D. 1994. Total Quality Development. A Step-By-Step Guide to World-Class Concurrent Engineering. Asme Press, New York, 506 p.

Construction Industry Institute 1988. *Concepts and Methods of Schedule Compression.* Publication 6-7, 28 p.

Cooper, K. 1993. The Rework Cycle: Benchmarks for the Project Manager. *Project Management Journal*, Vol. XXIV, March, pp. 17-21.

Davenport, T. 1993. *Process Innovation – Re-Engineering Work through Information Technology.* Harvard Business School Press, Boston, 336 p.

Fazio, P., Moselhi, O., Théberge, P. & Revay, S. 1988. Design Impact of Construction Fast-Track. *Construction Management and Economics*, 1988, 5, pp. 195-208.

Hartley, J. & Mortimer, J. 1990. *Simultaneous Engineering: The Management Guide.* Industrial Newsletter Ltd, 1st edition, London, 132 p.

Hendrickson, C. & Au, T. 1989. *Project Management for Construction: Fundamental Concepts for Owners, Engineers, Architects, and Builders.* Prentice-Hall, Englewood Cliffs, 537 p.

Hirsh, B. 1992. A New Fast Track for Public Works. Civil Engineering *ASCE* 62 No 2, pp. 45-47.

Koskela, L. 1992. Application of the New Production Philosophy to Construction. *Technical Report* No. 72. Center for Integrated Facility Engineering, Department of Civil Engineering, Stanford University, 75 p.

Koskela, L. & Sharpe, R. 1994. Flow Process Analysis in Construction. *The 11th International Symposium on Automation and Robotics in Construction (ISARC), Brighton, UK*, 24-26 May, Elsevier, pp. 281-287.

Kusiak, A. 1993 (ed.). *Concurrent Engineering: Automation, Tools, and Techniques.* John Wiley & Sons.

Kwakye, A. 1991. Fast Track Construction. The Chartered Institute of Building, Occasional Paper No. 46, 36 p.

Laufer, A. & Cohenca, D. 1990. Factors Affecting Construction-Planning Outcomes. *Journal of Construction Engineering and Management*, Vol. 116, No. 1, March, pp. 135-156.

Mistree, F., Smith, W. & Bras, B. 1993. A Decision-Based Approach to Concurrent Design. In: Parsaei & Sullivan, *Concurrent Engineering: Contemporary Issues and Modern Design Tools*, Chapman & Hall, pp. 127-158.

Murmann, P. 1994. Expected Development Time Reductions in the German Mechanical Engineering Industry. *Journal of Product Innovation Management* 11, pp. 236-252.

Nihtilä, J. 1993. Simultaneous Engineering in Project Oriented Production. *MET Technical Report* 14/93, Helsinki, 84 p.

Noble, J. 1993. Economic Design in Concurrent Engineering. In: Parsaei & Sullivan, *Concurrent Engineering: Contemporary Issues and Modern Design Tools*, Chapman & Hall, pp. 352-371.

Project Management Institute 1987. Project Management Body of Knowledge of the Project Management Institute, Drexell Hill.

Schrage, D. 1993. Concurrent Design: A Case Study, in Concurrent Engineering. In: Andrew Kusiak (ed.), *Concurrent Engineering: Automation, Tools, and Techniques*, John Wiley & Sons, pp. 535-580.

Sekine, K. & Arai, K. 1994. Design Team Revolution: How to Cut Lead Times in Half and Double Your Productivity. Productivity Press, 305 p.

Smith, P. & Reinertsen, D. 1991. Developing Products in Half the Time. Van Nostrand Reinhold, 296 p.

Tighe, J. 1991. Benefits of Fast tracking are a Myth. *International Journal of Project Management*, Vol. 9, No. 1, pp. 49-51.

Trygg, L. 1993. Concurrent Engineering Practices in Selected Swedish Companies: a Movement or an Activity of the Few?. *Journal of Product Innovation Management*, 10, pp. 403-415.

Willis, E. 1986. *Scheduling Construction Projects*. John Wiley & Sons, New York, 462 p.

Ziemke, C. & Spann, M. 1993. Concurrent Engineering's Roots in the World War II Era. In: Parsaei & Sullivan, *Concurrent Engineering: Contemporary Issues and Modern Design Tools*, Chapman & Hall, pp 24-41.

Factors affecting project success in the piping function*

GREGORY HOWELL
University of New Mexico, Albuquerque, USA

GLENN BALLARD
Department of Civil Engineering, University of California, Berkeley, USA

The project manager promised to never be burned again by late pipe deliveries. On the next job, if there was a next job, early need dates for engineering and fabrication would be set. This would assure enough pipe to maintain continuous work with a full sized crew.

Later, on the next job, a lead pipe fitter explained to a visitor why he was installing the particular spool just rigged. He replied that it was the longest they could find. Pressed to explain the decision he said, ' Look – I'm getting beat up for productivity and long pieces make me look good'.

1 INTRODUCTION

The Piping Function Task Force (PFTF) of the Construction Industry Institute (CII) was challenged to find opportunities for step changes in performance. At first, finding improvements in the 20% plus range appeared impossible. No manager interviewed thought they could find such significant improvements in their operations.

The research was conducted in the context of emerging production management theory and in construction and manufacturing. This new approach has many names; TQM, world-class manufacturing, JIT, lean production, re-engineering etc. Some researchers believe the recent development of partnering and the increased utilization of design build approaches are really all partial implementations of a deeper change in construction thinking (Koskela 1992). By whatever names, these approaches enrich our ability to describe the construction process and provide the basis for better ways to manage projects which are ever more complex, uncertain and quick.

Against this backdrop, the opportunity for dramatic improvement hidden in current practice slowly became apparent. The results of specific hypothesis tested in this research are offered. These findings raise new questions about the piping process. The paper closes with two examples of how current control and management practices contribute to the problem of managing in the piping function.

1.1 *Toward a new perspective*

Project success on piping intensive projects is usually related to the success of the

*Presented on the 3rd workshop on lean construction, Albuquerque, 1995

piping function. The heart-line of the piping process is a long supply chain which moves information and then material from project definition through engineering, fabrication, installation to operations and maintenance. But efficient installation also requires prerequisite work, structural supports and equipment be in place. These are supplied through chains running parallel to the heart-line. This means that all advantage gained by improvements in the piping function may be lost if progress and quality is not maintained along the parallel paths. Rapid and efficient installation can only occur when the flows in all paths are coordinated to assure the coincident and timely availability of civil, structural and equipment requirements.

The PFTF, aware of the need to manage these flows, focused specifically on four key activities along the heart line of the piping process: Piping and Instrument Diagram (P&ID) preparation, vendor data, pipe fabrication and installation. These were chosen by a consensus of PFTF members early in the research as areas offering the greatest potential for improvement. This portion of the research examines the interfaces between fabrication and installation.

Previous CII research by the Project Organizations Task Force had broken new ground by revealing the nature and extent of uncertainty (Howell 1990; Carroll et al. 1991). It is not an oversimplification to say that under current management techniques, designed more for contract management than production management, uncertainty is equated to be risk.

Risk is to be avoided, shifted or accounted for by price. Risk is usually considered to exist or be applied at the contract or project level. As suggested by Koskela in this traditional approach, production activities are conceived as more or less independent sets of operations or functions which convert inputs to outputs. These activities are estimated, planned and controlled operation by operation for least costs and duration. They are improved periodically mostly by implementing new technology.

In a sense, production is managed by 'contracts' either with subs or internal divisions. The primary tools are the commercial contract, detailed estimate or budget and the CPM schedule. Planning is a matter of determining what should be done and how. Control is based on outputs, measuring actual progress and costs against planned (Laufer & Tucker 1987).

Under the emerging thinking supported by this research, production is reconceived in terms of material and information flows which can be tightly controlled for minimum variability and cycle time. Production is continuously improved in regards to waste and value by improving reliability of flows and periodically in regard to efficiency by implementing new technology. In this paradigm, planning is both a matter of determining what is to be done and how, and a matter of reducing uncertainty (Laufer & Howell 1993). One important function of the planning system is to shield workers at each station from that uncertainty management has been unable to eliminate (Ballard 1994; Ballard & Howell 1994; Ballard et al. 1994). Processes, particularly planning, are the focus of control (Plossl 1991; Ohno 1987; Shingo 1981). Control becomes a matter of assuring uncertainty is screened from work assignments and not reinjected into the flow of work by the selection and completion rules in use at each work station.

Reliable production planning is the immediate aim of this new approach. Primary tools are the detailed estimate, the CPM schedule, project status and resource availability data, and information on planning system performance.

Piping success requires minimizing the extent and effects of uncertainty during fabrication and installation. At present, uncertainty in the timing of deliveries of intermediate products from one continuing activity to another defines the production planning and management problem. For example, consider the uncertainty associated with the delivery of completed ISOs from the engineer to the fabricator and the delivery of fabricated pipe to the constructor. Lacking tools to minimize the uncertainty in these flows, managers strive for flexibility so that the project can proceed in the face of erratic deliveries and unexpected problems. On piping intensive projects, they rely on buffers to assure progress despite variations in the timing, sequence and quality of resources from upstream suppliers. Buffers dampen the effects of variations in the flow or resources and allow flexibility in the choice of work. Senior Managers have learned through time to size buffers to assure production continues despite uncertain deliveries.

Large buffers mask the sources of uncertainty while small buffers immediately make unexpected variation apparent. In the latter case, production planners who prepare assignments for workers have no choice but to use resources as they become available if they are to maintain production rates (albeit at increased cost). Conversely, the availability of backlog allows work to be selected to satisfy a variety of criteria such as 'first in, first out', easiest first, highest ratio of earned to expended effort first, or complete runs only. Thus buffers alone do not prevent uncertainty in the flow of resources. Production planners, those who make assignments to crews, can inject uncertainty when they select work to be done in the next period just to maintain production or when they fail to identify reliably the work to be completed in the next period.

Certainty can be maintained when the production planning system is reliable, that is, when people can do what they say they are going to do. The performance of a planning system can be measured in terms of reliability when production plans meet specific quality criteria. A certain flow of work reduces the need for downstream stations to maintain large buffers. Imagine that the plans of any station are reliable. The next station can then include work on their plan simply because upstream plans are highly certain. Planning effort is reduced and time made available to better match labor to work flow and to smoothing operational details of the planned work method.

2 HYPOTHESIS DEVELOPMENT

The Piping Function Task Force began working to understand the piping process by reviewing a diagram of the piping process portrayed as a critical path schedule. Meetings were devoted to the discussion of logical groupings of activities. Members visited various organizations with primary responsibilities for groups of activities. There was early and wide agreement that the interfaces between activities offered the major opportunity in the fabrication and installation phases.

A map of the 'heart-line' of the piping process from engineering through installation was developed to better understand the flow of resources and the interfaces. Draft maps were reviewed by 3 fabricators and task force members. The resulting map records the activities and planning processes found on piping projects. It provides a visual reference for beginning to understand the factors which produce uncer-

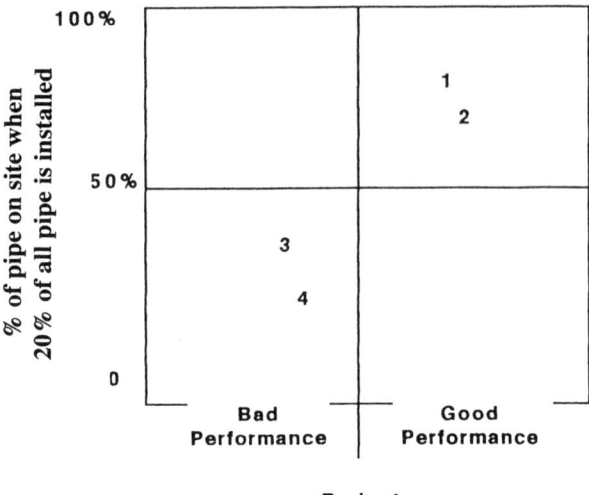

Figure 1. First data relating performance and piping buffers.

tainty within the piping heart-line, and the strategies used to cope with the uncertainty along the way.

While the map was being developed, data was being collected through site visits. Early results shown in Figure 1, suggested a relationship between the extent of piping backlog available on site early in the installation phase and project performance.

Projects 1 and 2 were as dissimilar as projects likely to be considered in this study could be. One was a large grass roots plant built with intense constructor involvement, extensive use of CAD, constructability, multiple problem solving teams, and a non-union workforce. Project 2, a much smaller job, was built with the classic separation of functions, a union workforce in the midst of an existing operating plant. Projects 3 and 4 were typical mid-sized projects struggling through a hand to mouth agony of drawing and material shortages.

The task force and the researchers were intrigued and surprised at the extent of the backlogs on Projects 1 and 2. It did not seem plausible that projects could tolerate the delay in starting installation needed to build the backlogs. Some felt Projects 1 and 2 were built under little schedule pressure or that the piping was not on the critical path. There was a serious question raised by members challenging project management claims that these jobs were constructed under great schedule pressure. Checking back with the project teams provided assurance that schedules were tight and piping critical. This surprise proved to be a turning point in the research and encouraged the task force and research team.

The pattern shown by the four points suggested finding projects with high performance and low backlogs would be one way to benchmark management performance. Buffer sizes might indicate the extent of uncertainty managers expect to experience. Perhaps some set of common circumstances, strategies, practices, and/or technology could be found on projects which fell in the bottom right quadrant of Figure 2. The idea was to find those projects which were best able to manage with a minimum of time and expense devoted to the development of large backlogs. Small

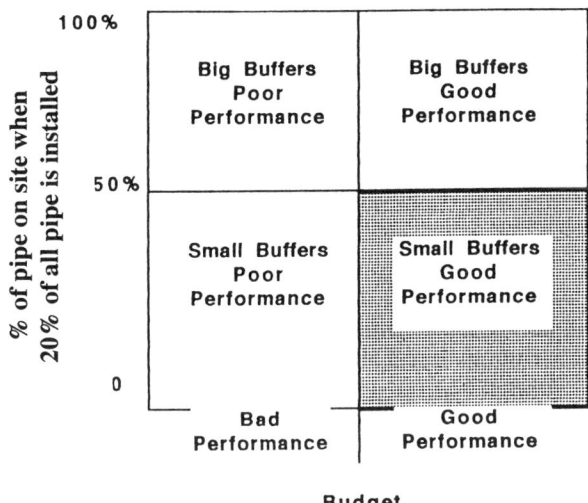

Figure 2. Proposed definition of benchmarks.

buffers could mean the management team had reduced uncertainty or was using some form of coping mechanism to minimize its effects.

At this point, the task force believed finding projects in the lower right corner would be straight forward. A draft protocol was devised which attempted to isolate specific project characteristics in order to determine when larger or smaller buffers might be appropriate. The resulting data request form was monumental and rejected. After some discussion the decision was made to use a simple protocol which gathered objective data on backlogs and performance and relied on a more subjective determination of project circumstances.

Thus the early site visits, interviews and task force discussions led to a series of questions. The answers to these questions provide the foundation for a new way to think about the piping process:

– What does the piping process look like in terms of flows and buffers from engineering through installation?

– Is there a relationship between buffers of pipe on site early in construction and piping performance?

– Does buffer size vary with the extent to which the project is complex uncertain or quick?

– Does project performance vary with the extent to which the project is complex uncertain or quick?

A final data request was prepared reflecting these questions. Commodity curve data was requested for the planned and actual provision of drawings to the fabricator, the planned and actual fabrication of pipe, and the planned and actual installation. Measures of piping performance, project performance and rework were recorded. Finally estimates of the project complexity, uncertainty and relative duration were requested. Data adequate for analysis was received for 24 projects.

The last data source measures the performance of the planning systems. Using the data collected at the 'last planner' level, the extent and causes of unreliable work flow can be identified. Early samples of this data led to more questions:

– Can the reliability of production plans be improved?
– Does increasing the reliability of planning increase productivity?

3 DATA

3.1 *Workmaps*

Understanding workmaps
Workapping was developed by Jerry Talley (1982) and Glenn Ballard. It is a standardized language used to document the way work moves through an organization. At first it may appear similar to a flow chart or an $IDEF_0$ chart, and in fact work mapping borrows from both. But workmapping makes critical distinctions necessary to understand the flow of resources such as materials and the development of plans related to a specific work activity.

In Workmapping, operations which converts inputs to outputs are represented by rectangles. Operations have two distinct types of inputs:

1. Resources, information or 'stuff' consumed in the conversion, enter the operation from the left and are typically shown as a trapezoid to suggest a bucket or buffer of resources. If no buffer is intended then outputs may be shown as a circle;

2 Directives, assignments or rules for doing work, enter from the top.

Note that the directives themselves are the product of a planning process. Inputs to this process are not shown in on Figure 3. Outputs are to the right. Thus, outputs of a process may be either resources for another operation or directives. The arrows on a work map can be read as 'goes to'. This allows the map to be read as if it were a language with subjects, verbs and objects. For example, the map below reads 'resources go to an operation which converts them into a product that goes to outputs. Planning converts some unidentified inputs into directives which direct the selection of input resources, and constrain the operation and outputs'. Figure 3 is a simple map of the planning and work process completed by the crew.

Directives are in the form of a weekly work plan (an assignment) prepared during a planning process. The output of planning is a directive to the crew while the output of resources supply systems provide inputs to crews. The output from the crew provides input to downstream crews. Control at this level is represented by a diamond which compares outputs to assignments. In reality, the diamond is a process itself which may result in changes to the resources, directives, or operation. For ease of

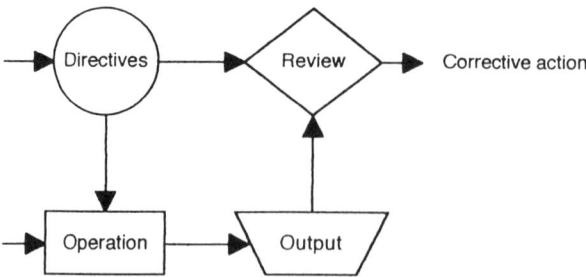

Figure 3. Workmap basics.

reading, these corrective action possibilities usually are not mapped. (It is worth noting that project controls compare the outputs with higher level directives such as budgets or schedules and not directly against assignments.)

On the map of the piping process from engineering through fabrication described in the following section, directives include 'work selection and completion rules' to indicate that the directives determine how work is selected from the buffer, and the rules for completion before release to the next station.

Directives are important because they are the basis for what work will be done and how. Thus the workmaps of the planning system may look the same but the work itself carried out under very different rules. One contractor may have and enforce a policy of completing all work when a spool is installed while in another the foreman or workers may select easy work to show immediate performance or may leave pipe only partially connected in response to pressure for increased production. The difference between these two sets of directives to do the same work will produce very different results. Thus the criteria for selecting specific items on production plans is a critical issue in planning at every step.

3.2 *Piping process workmap*

The workmapping language was used to develop Figure 4, a chart of the piping process from engineering through installation. This map records the 'way-it-really-is' on the majority of large process piping projects, for example, the map would be different if modules were prepared offsite. The map proceeds from engineering on the left to installation. The bold arrows represent the 'heart-line' activities. Parallel paths are simply identified as 'structural activities' and 'equipment activities'. These activities, provide resources necessary for the installation of pipe. While they are not detailed or studied here, the three paths make up a sort of continuing three horse race where the jockeys are all trying to tie. Each of the operations along the heart-line are linked to planning by directives, a set of work selection and completion criteria for each operation. These directives are the result of some planning process. Likewise all inputs to planning are not identified as they vary from project to project and company to company.

The preparation of Figure 4, the engineering to installation map, followed from the flow chart used by the piping function task force. A rough map was prepared for visits to sites and served to organize the discussions and interviews. Data collection points related to the flow of resources identified on the map and the data request used to determine the buffer sizes prepared. At this point the map was used in a series of conversations with fabricators. These conversations led to a more detailed map which identified the factors effecting the flow of piping through the fabrication. This map is a consensus document of common practice using an off-site fabricator. In addition to reviewing the map, fabricators commented on production management issues raised by the map. The results of these discussions is reported in the following section. A map for a project with on-site fabrication might look slightly different. The number of storage buffers between fabrication and installation might be smaller, the problems associated with the allocation of capacity might be simpler, and the planning system for the fabrication and installation phases might be more integrated.

Data in the next section will show the uncertainty related to the delivery against

Fabrication thru Erection 2

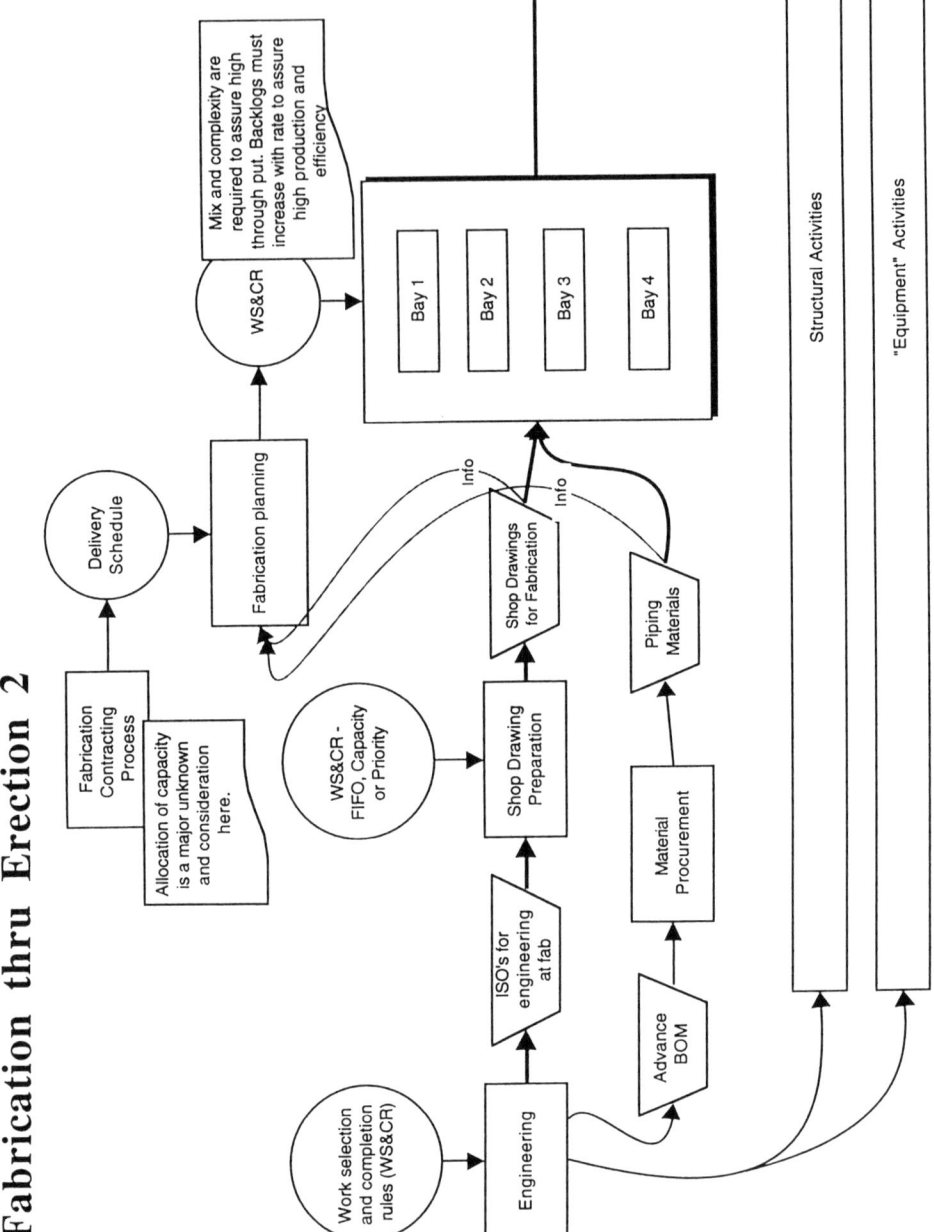

Figure 4. Workmap of the piping process.

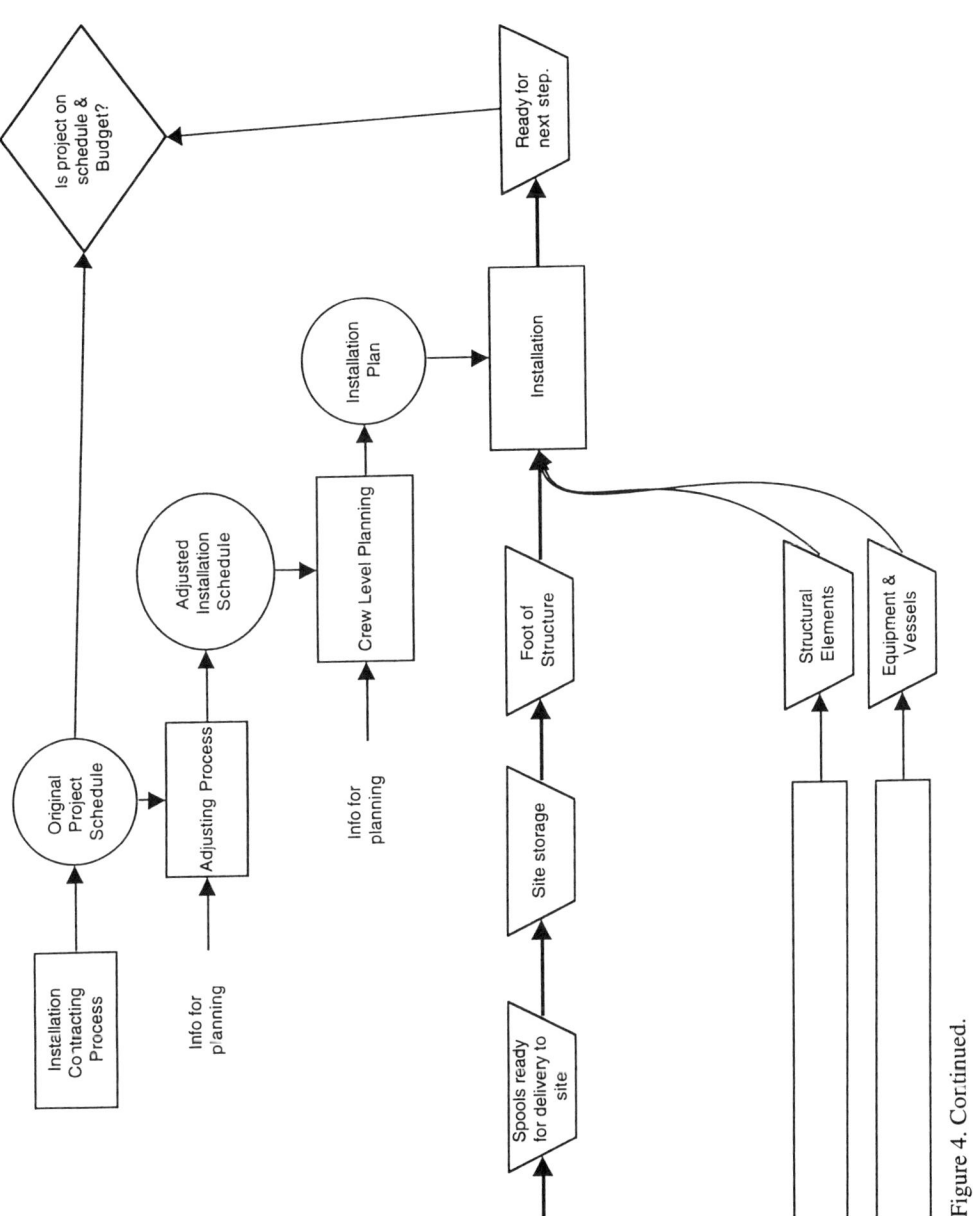

Figure 4. Continued.

plan along the heart-line. In reality the output of engineering often falls behind schedule. The next step is the preparation of shop drawings. Fabricators say the major problem is finding and organizing the specific information relevant to each spool. The information required for shop drawings is often spread through the design drawings, specifications and references. The electronic transfer of drawings is reported to be helpful as it reduces the time required to prepare shop drawings but may not reduce the variation in delivery caused by the need to collect and verify the detailed instructions of the engineer. Fabricators report is takes from one to two weeks to complete the shop drawings. They usually complete drawings in a first in first out (FIFO) sequence unless they have some specific schedule requirements or special requests. On occasion, they select drawings out of FIFO in order to keep shop fabrication lines fully occupied.

A short parallel path is shown on the map from engineering to a material procurement function. In some projects this occurs only with the arrival of drawings at the fabricator but a more advanced form of procurement is typical.

Once shop drawings are complete and materials available, fabrication can begin. Fabricators report it takes between 3 and 14 days to move a spool through the actual shop phase of fabrication depending on the complexity, size, painting and testing requirements of the spool. Fabricators say it rarely takes more than 5 days to complete the assembly of a spool started in the shop. In some cases the fabrication duration may be extended by either unusual painting or testing requirements or lack of capacity in these functions.

Conversations with fabricators (and then with constructors) often returned to the issue of capacity reservation and allocation. Fabrication is different from other activities in that shop capacity is relatively fixed. Since the fabrication business is highly competitive, full and continuous utilization of a shop is vital to the fabricator – just as the ability to quickly install pipe on site is vital to the installation contractor. Unfortunately, the frequent late release of drawings compared to plan makes full use of fabrication capacity difficult.

Fabricators admit to accepting more work than is possible in their facility because they count on late deliveries. Likewise project managers admit to setting early need dates to secure capacity. In many cases this two party game works but occasionally a fabricator is caught short. They then face wrenching dilemmas about which client to serve first. As a result, fabricators report doing everything they can to pin down when drawings will arrive. For example one fabricator reports that the teams of engineering and contractor representatives which were formed to work out quality problems also provided the unexpected benefit of intelligence on the arrival of drawings. (Here is the first example of many reports about the importance of leading information on the future flow of resources. This function appears to be a critical but often overlooked aspect of the procurement process. Unfortunately, it appears that engineers, fabricators and constructors bound by current forms of contract and lacking the trust to share the production information needed for integrated planning.)

The map shows the information on drawings and materials used by those doing detailed fabrication planning. Shop managers report they prefer a mix of sizes and complexity of pieces to keep all lines and all stations in their shops fully engaged. They report that the size of the backlog just in front of the shop must increase to assure high through-put and efficiency. Typical numbers are between 125 and 200% of weekly capacity.

The order that spools come out of fabrication is not necessarily the same as they went in. Differences between the input and output sequences may be due to simpler spools passing more complex ones on the line, spools loosing place in line because of quality failure recycle, and the difference in demand for, and rate of production by, lines dedicated to certain size ranges. Thus the buffer at 'ready for delivery to site' allow pipe to be resequenced for delivery.

Fabricators report a variety of practices for delivery of pipe to the site. In rare cases delivery is detailed for installation off trailers. At the other extreme are projects with no requirements for delivery packaging, timing or sequence. Fabricators uniformly reported that in many cases the first time they receive priority information is after the project has begun to experience delays.

Once delivered to site, pipe is typically stored in a lay-down yard and then moved to the foot of the structure just prior to installation. The lead time required to pull resources from the lay down yard varies depending on how it is organized. For example, on one site the time from order to delivery ran about 2½ weeks. On another spools were carefully checked on arrival, stored and pre-loaded onto farm wagons in reverse installation order. An order for delivery from the on-site yard to the structure was sent just prior to the installation of the last spool of the previous delivery. The next wagon load of pipe arrived in less than an hour. Obviously, as the uncertainty of delivery increases, it is in the interest of the pipe department to move more inventory to the foot of the structure. In several cases the only way to discover if materials were available was to requisition them for delivery and then to find matching sets of pipe and related materials.

The map serves as a reference for data developed in following sections. It is worth reflecting that the map demonstrates that variations in timing, rate and sequence of the production is affected by: 1) The timing, rate and sequence of the delivery of inputs to that station; 2) the directives, i.e. the work selection and completion rules at each station, and from variations which may arise during the operation. At the same time, the map suggests that uncertainty inherited from upstream flows can be removed from the flow and certainty reinjected if the directives do not allow passing work to the next station in an out of sequence or partially complete state.

3.3 *Commodity data*

Piping backlogs
Data on piping performance and the magnitude of piping backlogs was collected to test the original hypothesis that they were related. The data was plotted as shown in Figure 5 and the researchers were surprised that their initial assumption was not born out. Jon van den Bosch on CII staff suggested that other points be tested.

The researchers then explored the data searching for a pattern. Figure 6 shows the relationship between pipe on hand when 20% of the pipe has been installed and performance against budget. Successful projects have at least 60% of all pipe on hand when 20% has been installed. Other points were tried but those earlier taken at less than 20% installed were similar to Figure 5 and those taken later 30% showed the

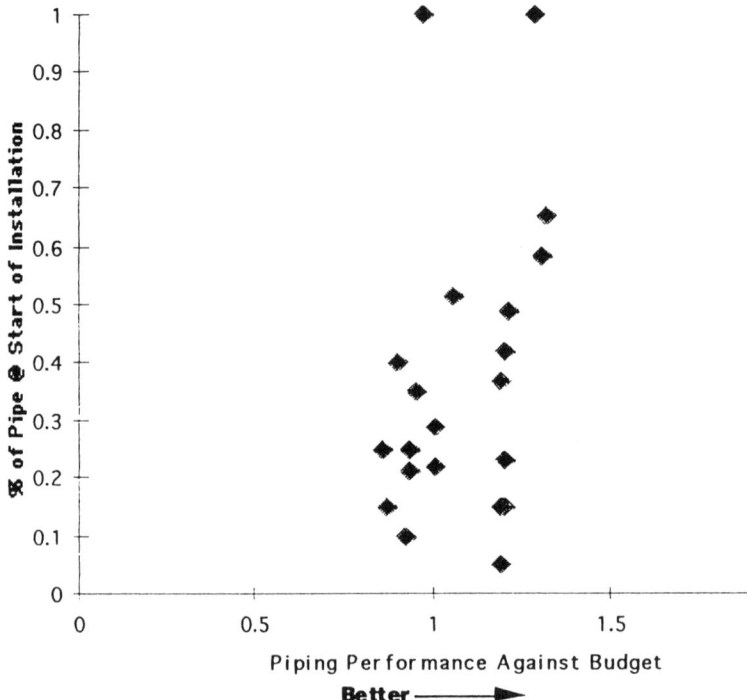

Figure 5. Piping buffer at the start of pipe installation.

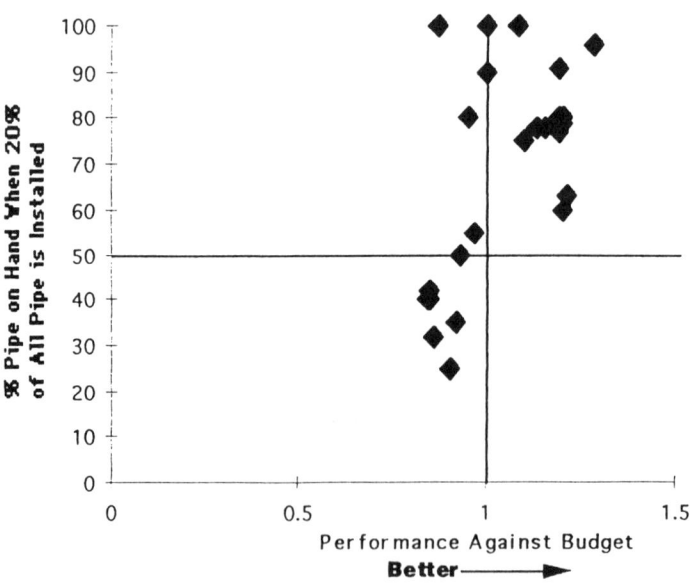

Figure 6. Piping buffer at 20% installed.

pattern less clearly. This suggests that the pipe on hand at the start of installation is less important than the ability of the piping supply chain to provide pipe to the site faster than it was being installed in the early phase.

PFTF members suggested a number of explanations for the difference in Figures 5 and 6. They said the of reasons for starting piping varies. On some jobs, a small amount must be installed well before production installation begins due to project logic constraints or the political requirement to show piping has started per schedule. Whatever the reason, the members strongly supported the view that a rapid build up of piping allowed the site supervision the greatest flexibility to maintain progress despite late deliveries of other commodities. In effect, piping progress could be maintained even if some equipment or supports were delayed. Projects which attempt to maintain production with small backlogs at the 20% stage clearly run the risk of low performance. All projects which met or improved on the budget show they had at least 60% of the total pipe on site when 20% was installed.

Two poor performing projects challenge this reasoning and conclusion. In one case a project has 100% on site at 20%. In fact it had 20% on site at the start of construction. Representatives of the company explained that the problems on the site were not related to materials, rather the extremely fast project was plagued with weather and labor problems. No explanation was available for the causes of the relatively small overrun on the project with about 80% of pipe on hand.

PFTF members and project participant explanations for why small backlogs were related to poor performance were consistent with the need for flexibility arguments. They agreed workers on projects with inadequate supplies of the right equipment, pipe and supports often would be forced to install partial runs or work on in a less effective sequence to make progress. These explanations were supported by the discussions with managers who worked on poor performing projects. On a typical poor performing project engineering was late and fabrication slipped in proportion to engineering. The project meets its target completion date but suffers serious overruns in piping. Participants explain the overruns by noting extreme problems in parallel paths of equipment and structures. It appears the job was paced by the late delivery of elements. The result was workers 'discovered' each day what work could be done as trucks came in the gate.

These graphs support the strong beliefs of task force members that prudent project management will wait to mobilize until engineering is complete enough to sustain a steady fabrication rate and that there is little advantage to be gained by trying to work with incomplete documents or supplies of pipe.

A typical commodity curves for successful projects is offered in Figure 7. Here construction progress is maintained even though there is a dip in the rate of engineering completion. This was a project rated as having an extremely short duration when compared to similar projects constructed by the company. On this project the fabrication does begin to slip behind schedule. The project manager explained that some drawings had been released with discrepancies and these delayed their fabrication.

The time required to build backlogs on successful projects varied. The average time to build backlogs was about 20 weeks on projects which were rated as having the shortest relative duration. Since few projects were available attempts at statistical analysis of the time required compared to other project variable showed no particular relationship.

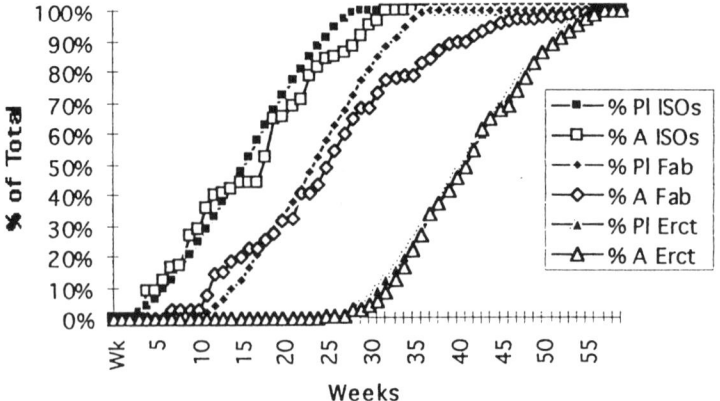

Figure 7. Planned actual commodity curves.

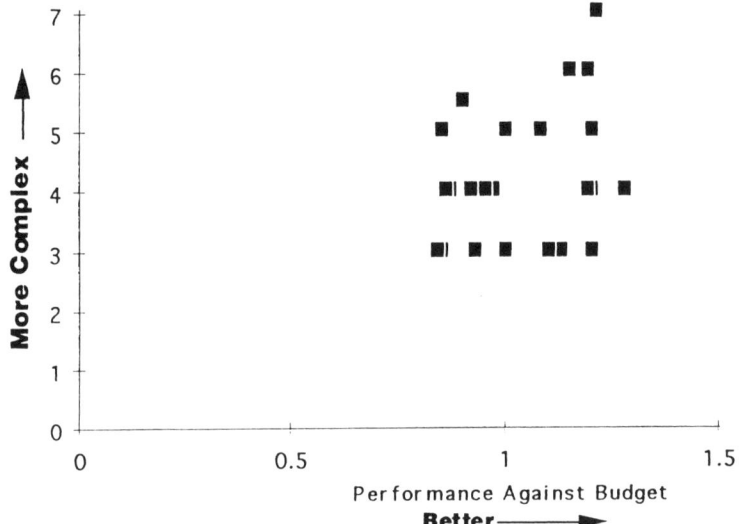

Figure 8. Project performance versus complexity.

Testing commodity data for other explanations of performance
As mentioned before, measures of project complexity, uncertainty and quickness were collected. Both project performance against budget and backlog data was plotted against this data. Schedule performance was not plotted as the data was not requested which distinguished original from final schedule.

Figures 8-10 show that the extent to which a project is complex uncertain and/or quick (CUQ) does not affect performance against budget. Projects at each level may be below or above budget. It appears that these factors are taken into account during the estimating process. Likewise there is no evidence that tougher projects are more likely to show poor performance. This can be seen in Figure 11 where measures of CUQ are added together and called 'toughness' It makes sense that performance against budget would not related to toughness because contractors must do a good

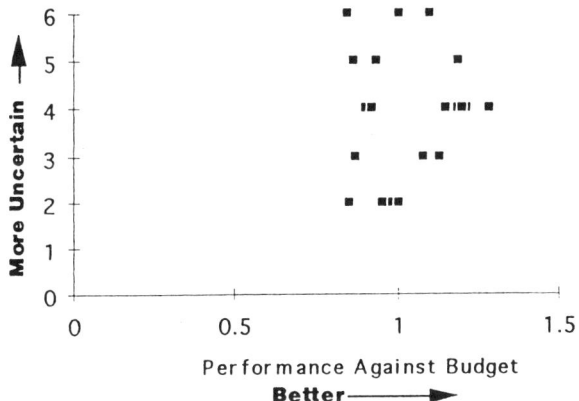

Figure 9. Project performance versus uncertainty.

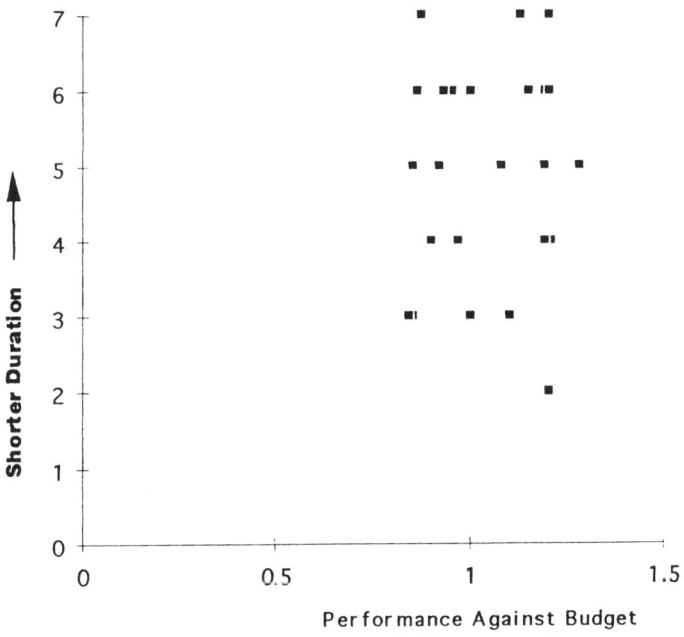

Figure 10. Project performance versus quickness.

job of adjusting their estimates to project circumstances. Contractors appear to make and loose money on both tough and easy projects.

Note that the average degree to which a project was CUQ is nearly the same for high and low performing projects. Both average complexity and duration pressure was higher on high performing projects and the average uncertainty was slightly higher on low performing projects. The differences in the averages are clearly much smaller than the wide variation of CUQ shown for either high or low performing projects.

The conclusion must be that the extent to which a project is CUQ is not related to project success.

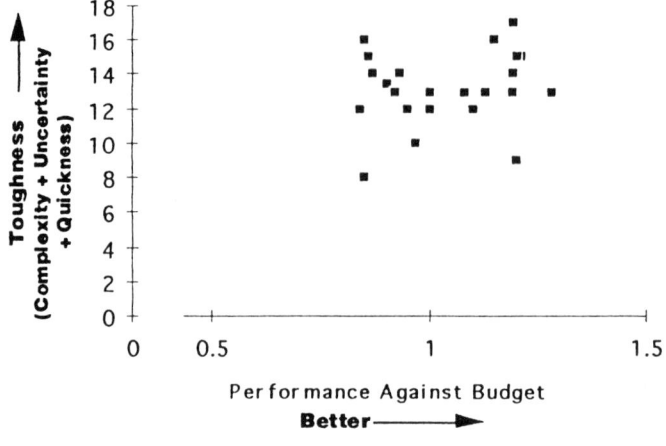

Figure 11. Project performance versus toughness.

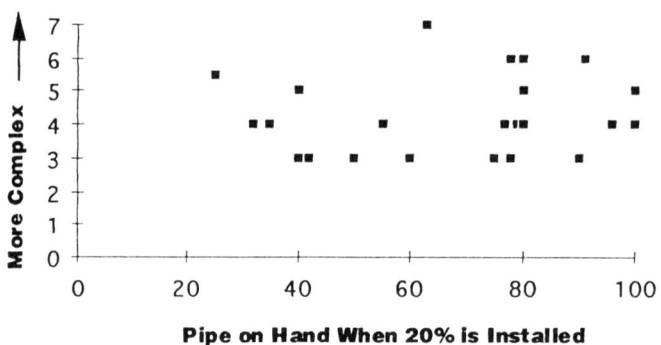

Figure 12. Backlogs of pipe at 20% versus complexity.

CUQ and piping backlogs

Next the data was reviewed to see if there was any pattern in the relationship between project circumstance and the extent to which piping backlogs were developed. It seemed plausible that managers might choose to have larger backlogs when faced with more uncertainty or smaller backlogs on very quick projects.

Surprisingly, the data shows no such patterns. Reports of the extent to which a project was CUQ, or 'tough' show no relationship to piping inventory strategy. This means that about the same level of piping backlogs are found across a wide range of project circumstances. There appears to be no difference in the backlog for across the range of complexity (Fig. 12) or uncertainty (Fig. 13).

The extent of buffers on relatively certain projects – those rated for uncertainty at 2 out of 7 on Figure 13 – range from 40% to almost 100%. The project with 40% project is one of the lowest performing sites reported.

There is a slight suggestion in Figure 14 that extremely quick projects actually have slightly larger backlogs than less rushed projects. This is opposite the predicted pattern. No ready explanations come to hand except that speed requires ready access to resources.

A review of the commodity curves on one project with the most aggressive

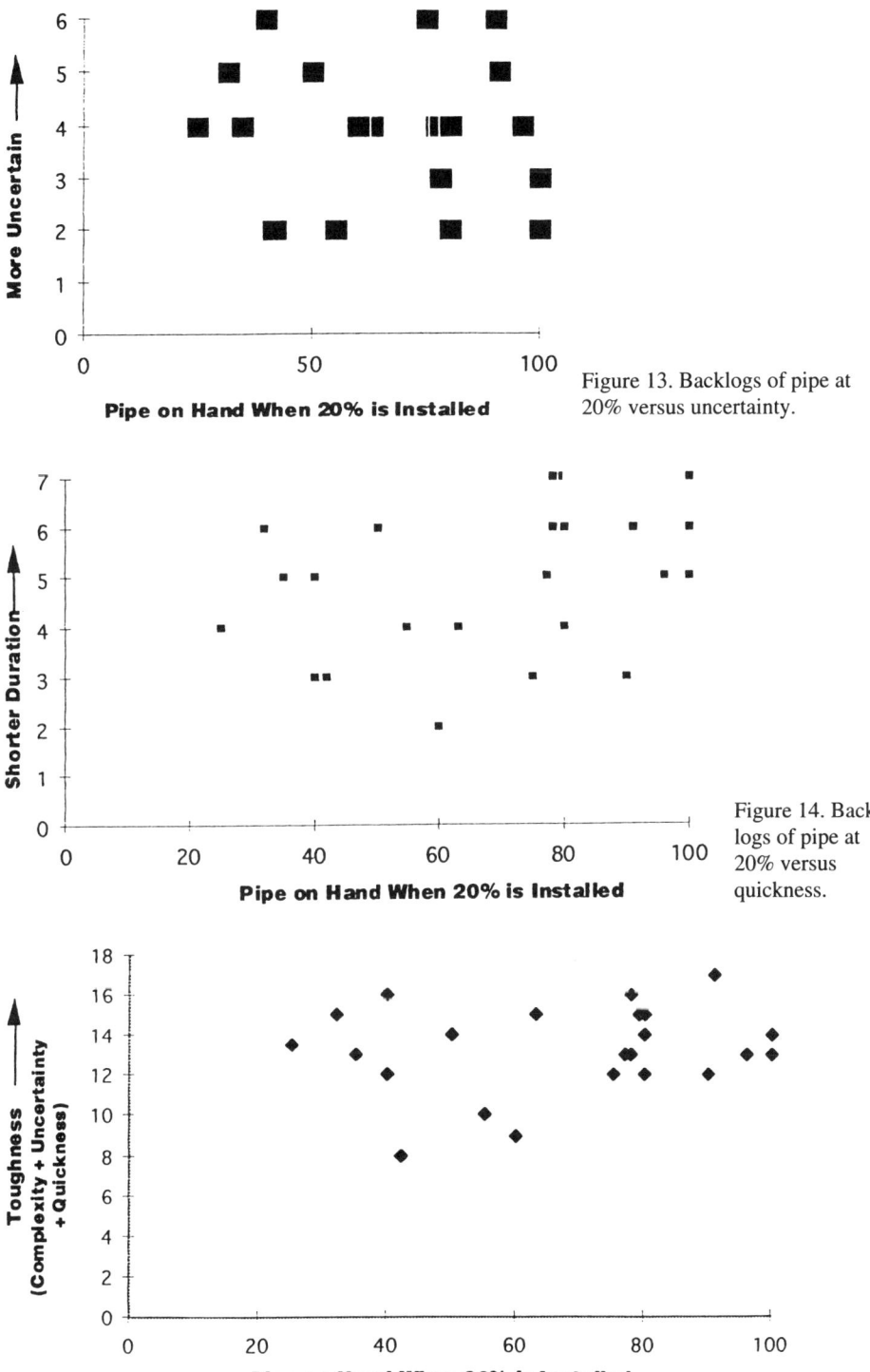

Figure 13. Backlogs of pipe at 20% versus uncertainty.

Figure 14. Backlogs of pipe at 20% versus quickness.

Figure 15. Backlogs of pipe at 20% versus toughness.

schedule ratings shows that the typical piece of pipe on project 'B' was in the lay-down yard on site for 22 weeks out of a 56 week period for fabrication and construction. Engineering required about 24 months and construction was completed in 21 months.

Finally, there appears to be only a slight suggestion that backlogs are smaller on less tough projects – at least they were on a few.

3.4 *Conclusions from the backlog data*

The original questions can now be stated as hypothesis with conclusions:
- The amount of pipe on site at the start of installation is related to
  project success   False
- Large buffers are necessary for project success   True
- Large buffers are sufficient for project success   False
- Buffers are sized to project toughness   False
- Project 'toughness' is associated with project success   False

It appears that the locations and size of buffers is not logically explained by the extent to which a project is complex uncertain or quick. This is in line the often expressed desire of project managers to maintain flexibility to cope with uncertainty. The rule in place appears to be 'the more (and the sooner) the better – no matter what the circumstance.

But as a stable flow of work requires reliable planning; buffers alone do not assure a stable flow. How reliable is planning?

3.5 *Last planner data*

The construction industry devotes tremendous energy and resources to planning projects and developing the schedules, budgets and other requirements that collectively tell project personnel what they *should* do. The process of translating these directives into work is shown in Figure 16. In the current form, project management

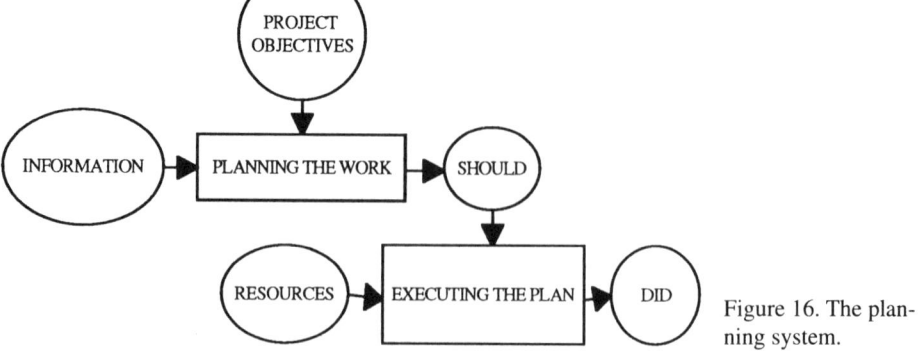

Figure 16. The planning system.

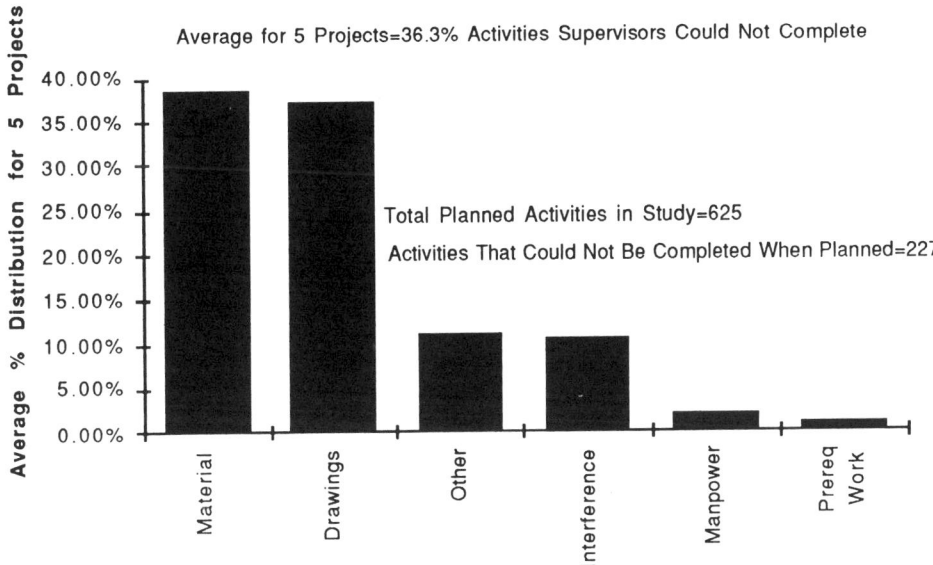

Figure 17. Match of 'should' and 'did'.

thereafter monitors and enforces conformance of *did* to *should*. Planning at the beginning of the project is replaced by control during project execution.

If this approach to project planning and management worked perfectly, *did* would always match *should*. Actual measurement reveals that what actually gets done differs from what is supposed to be done roughly 1/3 of the time. Figure 17 shows data from 5 piping intensive construction projects. 625 activities were scheduled on the short term look-ahead schedule during the study period, of these 227 were not completed as scheduled. The percentage of planned activities completed was 64%; i.e. the percentage of planned activities not completed was 36%. This shows that the lack of fit between what *should* be done and what *can* be done is substantial and systemic.

Note that materials and drawings account for almost 80% of the reasons why assigned work was not completed. Improving the match between should and can requires that only material sound plans are assembled. This is the key to reducing flow uncertainty, improving the coordination capability of plans and reducing the requirement for either large intermediate buffers or extensive crew level delays. This match is unlikely to improve without measurement and corrective action. In fact under pressure for production, the match erodes as workers select the easy work from that which can be done. Thus management supports increasing flow uncertainty in the name of making progress.

Last planner defined
Decisions regarding what work to do in what sequence over what durations using what resources and methods are made at every level of the organization, and occur throughout the life of the project. Ultimately, the 'last planner' produces assignments that direct physical production as in Figure 18.

This 'last planner' is last in the chain because their planning process output is an

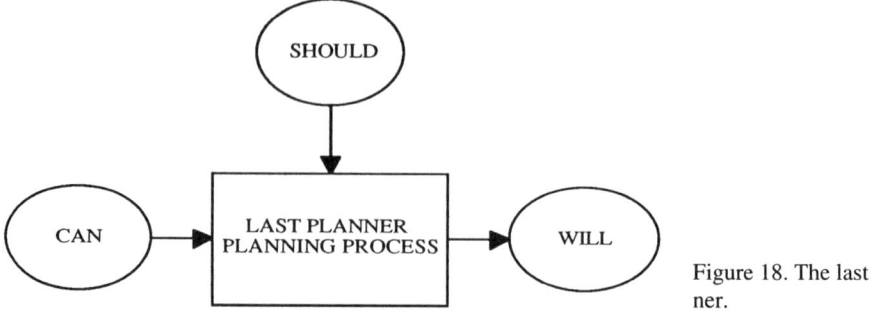

Figure 18. The last planner.

assignment to production workers and is not a directive for a lower level planning process. The work selection rules actually used by the last planner are important because the last planner retains significant authority over the details of the work processes and in many cases the specific pieces of work to install.

Last planner data collection methodology
More than 30 contractors from 16 different projects have participated in the Last Planner research. The 7 of those contractors principally or solely engaged in piping are included in this report.

Participation has occurred typically in two stages. In both stages, contractors were asked to have their foremen or supervisors write down what work they planned to do the following week, then report what work was completed and reasons for noncompletion. Initial results were taken as measurements of the degree to which *did* matched *should* (see Fig. 17). If the contractor initiated a screening process for forming a backlog of workable assignments from which to select weekly work plans, subsequent measurements reflected the degree to which *did* matched *will*.

Last planner data: Match of 'did' with 'should'
Data on the match between should and did was collected for seven contractors directly involved with the installation of pipe on petrochemical or large process plants. Contracts ranged from $3 million to $30 million. Data was recorded for several weeks and is presented as an average in Figure 19.

Some contractors reacted to the mismatch of *did* with *should* by implementing processes to select, size and screen potential assignments before placing them on weekly work plans. This made possible the identification and elimination of causes

Case #1	Contractor 1	33%
Case #2	Contractor 2	52%
Case #3	Contractor 3	61%
Case #4	Contractor 4	70%
Case #5	Contractor 5	64%
Case #6	Contractor 6	57%
Case #7	Contractor 7	45%
Average		54%

Figure 19. Last planner data: Match of 'did' with 'will'.

of non-performance. Here planning system performance is measured by Percent-Plan-Complete (PPC). This is a simple calculation of the percentage of activities listed on the weekly work plan which were in fact accomplished. This is not a direct measure of progress as some items may be larger than others. The idea is to expose the extent to which plans are reliable indicators of the work which will be done in the coming period.

Case 1: PPC increased from 33% to 70% over a period of several months, primarily from sizing crew assignments to target productivity and screening for material soundness.

Case 2: This contractor received materials from a general contractor, and continued to experience substantive materials-related delays. The last planner system allowed work assignments to be screened for these shortages, and PPC steadily increased from 52% to 75% by removing such constraints as access, scaffolding and lifting equipment.

Case 3: This contractor was able to implement the last planner approach from the outset of the project, and that proved to be an advantage. PPC increased from 61% to 90%, with the lack of accurate materials information from the general contractor the only recurrent cause of non-completion.

PPC and productivity
Increasing PPC should also increase productivity. First, as a consequence of screening for constraints and matching the amount of work assigned to the number of worker hours to be expended, delays and idle time should decline, thus reducing non-productive labor time and increasing productivity. Data support this assumption (Fig. 20). But more important, once crews are producing what their plans say they are going to produce, downstream managers should have a more stable work environment and be better able to match their labor to work flow while optimizing the allocation of shared resources. Data on this effect is not yet available and will be more difficult to collect.

Case 1: Extensive design-caused rework and late delivery of spools prevented this contractor from achieving good performance relative to worker hour budgets. How-

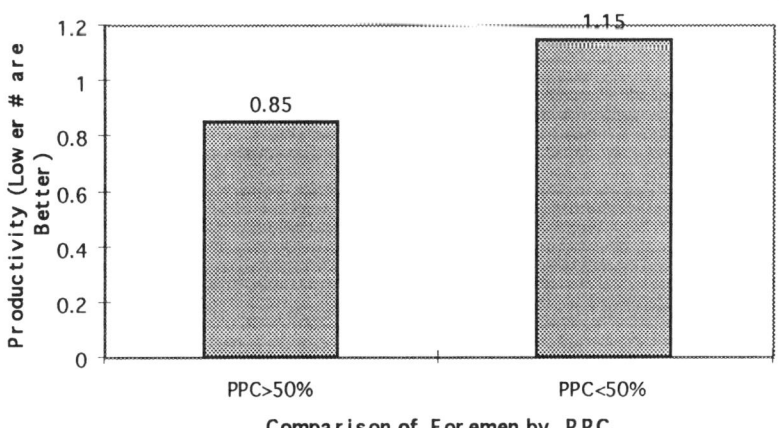

Figure 20. PPC and productivity.

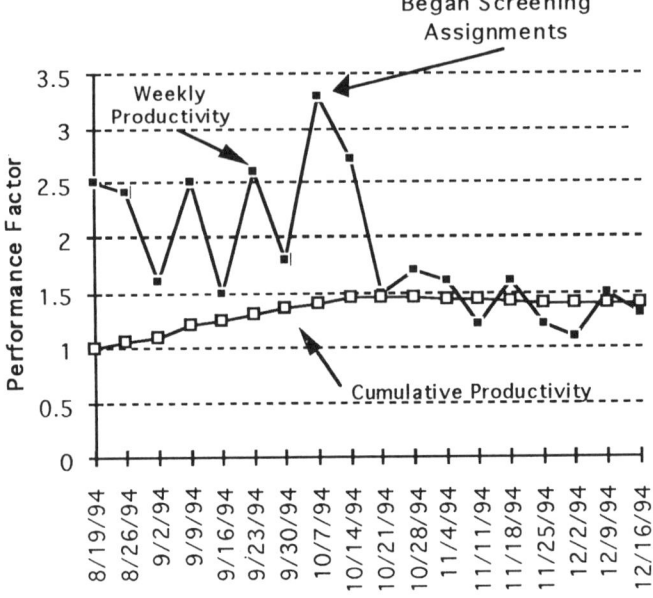

Figure 21. Effect of screening on performance.

ever, the beneficial impact of increasing PPC was evident in the superior productivity achieved by crews with PPC > 50% as compared to crews with PPC < 50% (Note: This contractor measures performance factor (PF) such that lower is better).

Case 2: This contractor productivity was deteriorating until the work assignments were screened to assure they were in fact practical, that is all resources were in control. The screening process was introduced in mid-October 1994 (Fig. 21). Subsequently, weekly variations in productivity declined dramatically, and cumulative PF levelled off. If the negative trend had continued, the contractor would have overrun its labor budget by 50%.

Case 3: This contractor underran its labor budget by 20% on a project on which most contractors overran by 50%.

Conclusions from last planner data
1. Current project management practice does not result in a good match between what should be done and what is done. Control techniques in use are unable to identify and correct mismatches.

2. Planning at the last planner level is unreliable. It is probably impossible to form meaningful detailed plans or to effectively coordinate work by relying on the plans of upstream activities when forming weekly work plans.

3. Planning reliability can be improved. Detailed crew level planning is necessary in order to shield production from uncertainty and variation in the flow of directives and resources. Systematic selection, sizing and screening of potential assignments improves the percentage of planned activities completed (PPC).

4. Improving PPC improves productivity in the immediate crew and sets the stage for improved performance downstream.

4 REFLECTIONS ON THE DATA

Remarkably, managers are not sizing backlogs to account for project complexity, uncertainty or quickness. No projects were identified which demonstrated both high performance and small backlogs. Large buffers almost seem a requirement for success even though they eat up valuable time and money.

Equally surprising is the failure of current construction work planning systems to produce reliable plans. While this research did not examine planning reliability in the engineering phase, it appears unlikely that it is any more reliable than field level planning. By contrast several of the fabrication plants involved in this research do exhibit highly reliable work planning.

Large buffers reduce the need for reliable planning as they allow some work to get done despite uncertain flows. Large buffers allow flexibility and mask the extent of uncertainty passed along because current planning systems fail to control work flow. A sort of circular reasoning is apparent – we need large buffers because planning is unreliable and unreliable planning makes large buffers necessary. Is it possible to significantly improve project performance without improving planning?

The causes of unreliable planning are simple to understand. The planning system is not designed to produce reliable production plans and its planning system performance is not measured. Planning systems rest on the assumption that resources necessary for planned activities will be available and resource supply systems fail to provide accurate forward information on availability.

Worse, the application of current control systems adds to the problem by pressuring people to do the wrong thing so the reports will look good. This is demonstrated by Figure 22 (Senge 1994). It shows how pressure for improved performance re-

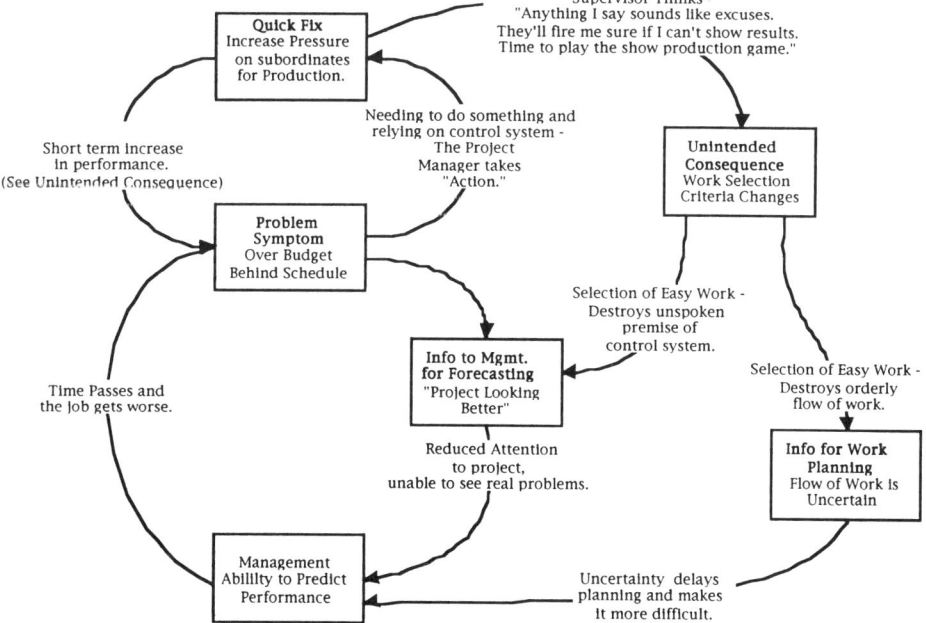

Figure 22. Quick fix - more trouble later.

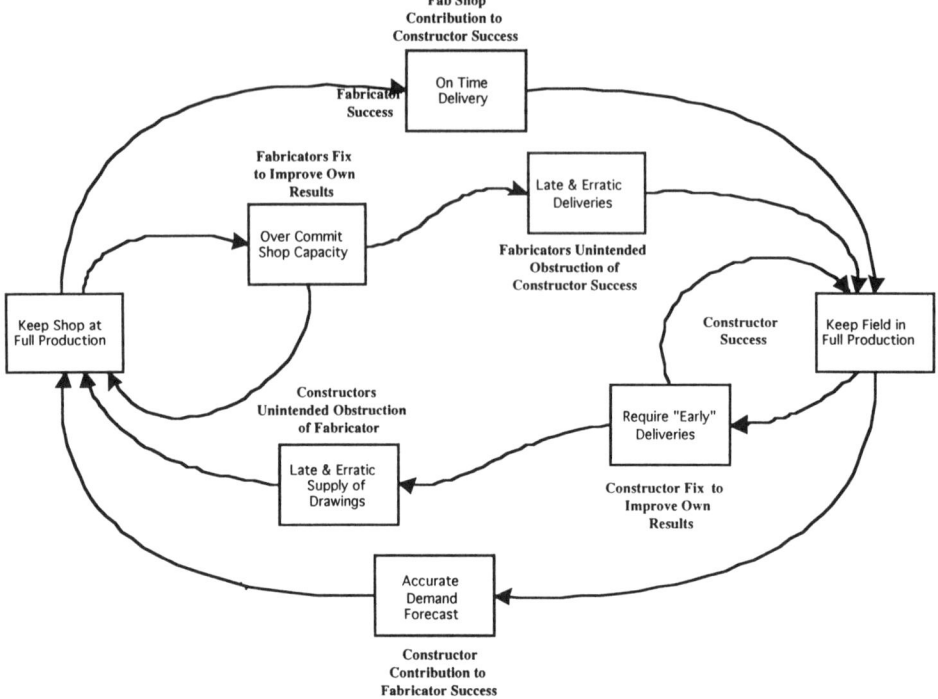

Figure 23. The 'need date dance'.

duces the orderly flow of work. As performance deteriorates, managers put pressure on subordinates to improve performance. Work selection rules are altered to do the easiest work first and the extent of the problem is masked for a short time. The possibility for learning or real improvement on a project is destroyed by the system designed to tell management the status of the work. The only solution is to install a planning system which can produce reliable assignments and then to measure its performance.

And why are large buffers located between fabrication and installation? Why not have the buffer ahead of fabrication since it is more reliable?

The answer appears related to a sort of game played by engineers, fabricators and constructors. The problem centres on the determination of need dates and the allocation of fab shop capacity. This is illustrated in Figure 23. Here the steps taken by both parties to assure their own success have a negative effect on the other party. Burned on each project, one learns to fudge on need dates and the other on available capacity. There are a number of possible solutions. The EPC constructor or owner could buy shop capacity. This would shift the risk and cost of late engineering on to engineering. Another idea would be to implement an integrated planning system from engineering through fabrication and installation. Buffers could be reduced and their location shifted by establishing common work boundaries and priorities, and by monitoring planning system reliability.

Current practices lead last planners to choose easy work in order to look good, and more senior managers to play the capacity-need-date game. Both were widely identified by participants in this research but they are difficult to document. Few are willing to confess their part alone and the design of planning systems masks the extent of misuse. But together these two system dynamics suggest that the extent of backlogs on projects may result more from the way people play the larger construction system and less from the explicit consideration of the minimum size of backlogs required to maintain progress given the actual state of the job.

REFERENCES

Ballard, G. 1994. *The Last Planner*. Northern California Construction Institute, Monterey, California.

Ballard, G. & Howell, G. 1994. *Implementing Lean Construction Stabilizing Work Flow*. Conference on Lean Construction, Santiago, Chile, September 1994.

Ballard, G., Howell, G. & Kartam, S. 1994. Redesigning Job Site Planning Systems. *Proceedings of the American Society of Civil Engineers Conference on Computing on Construction, Washington, D.C. June 1994.*

Carroll, J. et al. 1991. *Organizing for Project Success*. Construction Industry Institute Publication 12-2.

Howell, G. 1990. *How Owners and Contractors Organize Project Teams*. Construction Industry Institute Publication Source Document 53.

Koskela, L. 1992. Application of the New Production Theory to Construction. *Technical Report* No. 72, Centre for Integrated Facilities Engineering, Stanford University.

Laufer, A. & Tucker, R.L. 1987. Is construction project planning really doing its job? A critical examination of focus, role and process. *Construction Management and Economics* 5(3): 243-266.

Laufer, A., & Howell, G. 1993. Construction Planning: Towards a New Paradigm. *Project Management Journal.*

Ohno, T. 1987. *Toyota Production System*. Productivity Press.

Plossl, G. 1991. *Managing in the New World of Manufacturing: How Companies Can Improve Operations to Compete Globally*. Englewood Cliffs, N.J. Prentice-Hall.

Senge, P. 1994. *Fifth Discipline Field Book*. Doubleday, N.Y.

Shingo, S. 1981. *Study of the Toyota Production System*. Japan Management Association.

Talley, J. 1982. President, Organizational Diagnostics, 164 Main St, Los Altos CA 94022. Phone 415-941-1495.

Construction supply-chains: Case study, integrated cost and performance analysis*

WILLIAM J. O'BRIEN
Department of Civil Engineering, Stanford University, California, USA

ABSTRACT: This chapter provides a case study of production, inventory, and trans-portation costs and performance in a construction supply-chain. Building theory from the case evidence, trade-offs between production, inventory and transportation costs are explored and placed in a framework which can be used to assess these trade-offs with regard to costs for the supply-chain as a whole as well as for individual actors. The impact of uncertainty on cost performance is detailed together with a discussion of the implications of uncertainty for organizational arrangements in the supply-chain.

1 INTRODUCTION

Construction industry personnel and construction researchers have been showing an increasing interest in the application of supply-chain management ideas to project production. Supply-chain management is a set of concepts and methods that have been developed in the manufacturing industry in recent years. Application of these techniques has saved manufacturing industry hundreds of millions of dollars; a recent paper describing supply-chain efforts at Digital Equipment Corporation (DEC) claims a savings of one hundred million dollars (US) over a one year period Arntzen et al. 1995). Concurrent with these manufacturing efforts, construction personnel have become increasingly aware of the importance of materials supply to project performance. A Norwegian study shows an increase in the value of materials as a percent of project cost from 59% in 1970 to 67% in 1992 (Albriktsen et al. 1993). Similarly, a study of industrial plant construction reported workers spending 6% of their time waiting for materials (Bell & Stukhart 1987). The combination of productivity impact on-site and a growing recognition that materials delivered to construction sites often undergo complex production processes off-site has led construction personnel to look to supply-chain management for application to construction.

Despite this growing interest, little is known about the factors affecting performance in construction supply-chains. In the literature, major starting points linking off-site production to site production are the investigation of site materials management by Bell & Stukhart (1987) and the inclusion of materials lead times into a CPM formulation by Shtub (1988). Gray & Flanagan (1989) review the role of subcontractors

---

*Presented on the 3rd workshop on lean construction, Albuquerque, 1995

in building construction. A case study by O'Brien (1993) investigates relationships between suppliers, subcontractors and contractor in the case of a major delay on a building site, while a concurrent conference paper (O'Brien & Fischer 1993) proposes a high level framework to approach the analysis of supply-chains. A recent paper by Wegelius-Lehtonen (1995) reports on a stream of research in Finland that has focused on the logistics (inventory and materials handling) costs in supply-chains, but does not include an analysis of production costs at either end of the chain. While these efforts provide a starting point, there has been little theoretical or empirical investigation of integrated production and inventory decisions in construction supply-chains. It is the goal of this chapter to contribute to our understanding of construction supply-chains from an integrated perspective as one of the lessons of manufacturing supply-chain analysis is that such a perspective is necessary to achieve global rather than local optimization. A further important goal of this chapter is investigation of the role of uncertainty with regard to the timing of production in the supply-chain as uncertainty of this type is known to be detrimental to the performance of manufacturing supply-chains (Davis 1993).

Of the research performed in construction, the most extensive is that reported on by Wegelius-Lehtonen (1995). Using an analysis of five case studies of the logistics costs of plasterboard delivery and following the Just-In-Time production methodology of eliminating inventory (see Baudin 1990; Koskela 1992), she makes five prescriptions for materials management in the supply-chain: 1) plan and order deliveries 1-2 weeks before needed on-site; 2) order deliveries to come to the site just as needed; 3) divide the purchases into smaller deliveries and plan the time points and the contents of the deliveries accurately; 4) do not order extra pieces to the installation areas; and, finally, 5) move the material directly to the installation place. It is not clear from the Wegelius-Lehtonen paper under what conditions her prescriptions are applicable and some literature contradicts her prescriptions; a paper by Howell et al. (1993) demonstrates through case studies and analysis the need to hold inventory on-site to reduce the detrimental effects uncertainty in production rates has on overall site performance. The unit of analysis is different in the Howell and Wegelius-Lehtonen papers as the former focuses on site production and the latter on logistics costs. At some point these two perspectives must be reconciled to determine the trade-offs, if any, between production costs and logistics costs in the supply-chain. This in known to be a challenging problem in the manufacturing literature, but such an integrated production and inventory analysis has been outlined by Cohen & Lee (1988) and manufacturing firms have adopted supply-chain analysis to improve their internal operations and for negotiation of supply arrangements with supplier firms (Arntzen et al. 1995; Davis 1993).

To contribute to the literature, this chapter investigates the production and inventory decisions of multiple firms within the construction supply-chain to develop an understanding of supply-chain operations at a systems level. Basic questions to be answered concern: 1) the applicability and need for integrated supply-chain models in the analysis (and, hence, to provide a framework to reconcile the views of Howell et al. (1993) and Wegelius-Lehtonen (1995)); 2) the extent to which manufacturing methodologies can be used in the analysis of construction supply-chains; and 3) determination of factors affecting supply-chain performance. As this chapter is largely about increasing our understanding of supply-chains, additional areas of contribution are terminology to discuss and measures to analyze supply-chain performance. Furthermore, as supply-chains involve multiple firms and actors, this chapter also contributes to our understanding of inter-firm relationships. As Hinze & Tracey (1994) observe, relatively little is known about the contractor-subcontractor relationship, and it is safe to say that even less is known about relationships with firms off-site.

2 METHODOLOGY

Following the methodology of Yin (1989), an exploratory case study was performed. An exploratory case is justified given limited knowledge of construction-supply chains (particularly from an integrated production and inventory perspective). The goal of case research in this context is analytic generalization, and thus case research makes a contribution to theory but does not claim empirical validation of this theory. With this in mind, it was decided to study a single project to develop as rich a data set as possible to allow development of a theory along the areas and questions described in the introduction. Future research, building on this chapter and other research, may want to make more definite propositions and perform multiple cases using an experimental validation approach as outlined by Yin.

To identify a project for investigation, the following criteria were used: First, it should be a relatively small project (or be a project able to be divided into small segments) to allow the researcher to identify and understand the major influences on production. Second, the project should have a design that does not call for special construction techniques or fabrication procedures which call into question the ability to generalize beyond the case. Third, data about the supply-chain should be accessible to the researcher. Following these criteria, the Buchhaugen building project located in Trondheim, Norway was identified. This project meets all the criteria above and has the added benefit of being a project where the actors took an interest in supply-chain performance with the introduction of a novel delivery system inspired by Just-In-Time models. This system closely follows the prescriptions of Wegelius-Lehtonen and as such provides an excellent basis for analysis and comparison from a systems context. Further description of the delivery system is given in the case study section of this chapter, immediately following this section.

The study methodology is as follows: The case study investigates questions as described in the introduction, but as an exploratory study, research was not limited to these questions. Part of the study's goal is development of a context for further research in this area, and the data collected must be analyzed with a view to refining the questions asked. Also following this approach to exploratory research, the only definite proposition of the study is that it is possible to view firms as single entities, following the analytic approach of (Davis 1993) to aggregate micro-units of the supply-chain into larger components such as factories. Each factory in the Davis chapter corresponds to the relevant production/inventory portion of a firm in our analysis. Following this proposition, units of analysis for the study are the project, the firm, and in the context of the supply-system used, delivery of 'units' as described in the following section. Data was gathered about each unit of analysis. Project data largely comprised general project information and description of the inter-relationship between actors. Firm level data was collected in a series of structured interviews with representatives from the general contractor and from supplier firms who were both inside and outside the special project delivery system. Data about deliveries and supply system performance was collected from project records.

Care was taken to assure the validity of the research data and findings. Yin (1989) identifies four areas of importance: Construct validity, internal validity, external validity, and reliability. Construct validity involves the creation of correct operational measures for the concepts being studied. As an exploratory study, part of the investigation is to identify such measures and it is difficult to apply this criteria. However, multiple sources of evidence were used to develop constructs, and case study participants reviewed relevant sections of the report and the analysis of delivery data was discussed to determine if case descriptions and constructs were realistic and relevant to the participants. Internal validity concerns itself with causal relationships and is not generally applicable to exploratory case research which seeks to report. External validity determines the ability of the theory generated by the research to be generalized beyond the bounds of the case study. Here, steps were taken in project selection to make the theory more amenable to generalization, and the theoretical propositions of this study were made as specific and explicit as possible to allow future testing. Finally, reliability of the case, that is, removal of errors and biases within the case and ensuring the repeatability of the study, was sought by developing and following an interview protocol to ask project participants similar questions about case facts and their views of the supply-system. Answers were evaluated by checking the internal consistency of the information gathered from each participant and by checking for factual consistency across the data provided by each participant. Participants also reviewed the case report for factual errors.

Following the case study methodology of data collection followed by analysis and analytic generalization, the remainder of this chapter presents the case study, analyzes data within the case study, and follows with several theory building sections. Contributions are summarized at the end of the chapter.

3 CASE STUDY

3.1 *Project description*

Located in Trondheim, Norway, the Buchhaugen project is a 25 million NOK (excluding MVA, the Norwegian value-added tax) green-field, wood built condominium construction site with a floor area of approximately 3100 m^2. Buchhaugen is both developed and built by Veidekke A.S. (Divisjon Nord), Norway's second largest general contractor. Buchhaugen project was designed completely before construction began, although customers buying apartments had the option to specify finishing details such as heating control systems.

Buchhaugen project is unusual in its supply system. Based on an experimental Danish project built from 1991 to 1993 (Bertelsen 1993), the Buchhaugen project was to be built under a unit-based construction and delivery system. This system divides the project into units or small packages of material, each of which is a part of an apartment (e.g. a unit may be: interior walls, second floor, apartment 35) or block of apartments (e.g. roofing tile, apartments 35-43). Units are to be delivered to site in the location they are needed, just as they are needed on-site. Originally inspired by the 'Just-In-Time' delivery system pioneered in Toyota assembly plants, the rationale for this system is a decrease in materials storage and handling on-site, with a commensurate decrease in construction costs. Additionally, materials waste is to decrease through increased use of pre-cut materials and less breakage on-site (due to less materials handling). In return for timely delivery of units, suppliers get a rolling three week look-ahead for units needed by the site. The original Danish project estimated savings of 10% in site construction costs due to increased productivity and decreased materials waste (costs to suppliers are unknown).

As Buchhaugen is an experimental project, Veidekke A.S. entered an informal partnership with E.A. Smith Bygg A.S., a firm specializing in the supply of building materials (company descriptions follow in subsequent sections, including Veidekke, E.A. Smith, and the supplier firms Langmoen and Johs. Rasmussen). Veidekke held a contract with E.A. Smith to supply building materials with the exception of the foundation and the heating, plumbing, ventilation and electrical systems, as well as doors, windows, and kitchen installation. E.A. Smith held the contracts with its suppliers; total value of these contracts was 3 million NOK (this excludes the value of site installation). In all cases, contracts were based on unit prices (here, unit describes square meter or other appropriate measure, not unit as defined by the special delivery system). Veidekke wished for competitive market pricing for materials costs and did not wish to pay more for materials delivered under this system.

The unit based, three week order system works as follows: In week zero, order the units needed the following three weeks (weeks one, two, and three). In week one, order units that are needed in weeks two, three, and four. This system gives a three week look-ahead for all the units ordered. Additionally, the day of the week the unit is to be delivered on-site is also specified, although later in the project the three week look-ahead did not specify the day of delivery due to uncertainty about site conditions that far in advance. In theory, each unit should be ordered three times, once for the three week look-ahead, once for the two-week, and once for the one-week. In practice, units were sometimes ordered with less than three weeks notice or were ordered and then the delivery date was pushed back. A meeting was schedule each Monday morning to order units on the three-week look ahead; this was held with the site staff, a representative from E.A. Smith, and with input from the workers. (This Monday meeting follows a Friday afternoon meeting held with workers to assess site progress). In practice, holding the meeting on Monday meant that the effective order schedule predicted four weeks in advance, from the Monday order to the Friday of the third week. As the job progressed, it was found convenient to move the meeting to Tuesday mornings.

As general contractor, Veidekke hired carpenters and installed materials delivered by E.A. Smith. Other subcontractors installed their own materials. These subcontractors did not participate in the special unit delivery system, although building systems were installed concurrently with Veidekke directed work. No problems were observed with interaction between Veidekke work and work performed by subcontractors, and the special delivery system neither disturbed or was disturbed by the subcontractor work.

Production by carpenters drove the overall pace of production; workers were not held waiting for materials. The original schedule called for production to begin in

Table 1. Summary schedule statistics for Buchhaugen project.

Schedule	Project duration (weeks worked)	Activities	Made	Comments
Original	30	58	Nov 1994	
Rev 1	31	107	Feb 1995	First resource loaded schedule
Rev 2	31	107	April 1995	Revised resource loading, doubling of carpenters on-site
Rev 3	41	126	June 1995	Final schedule

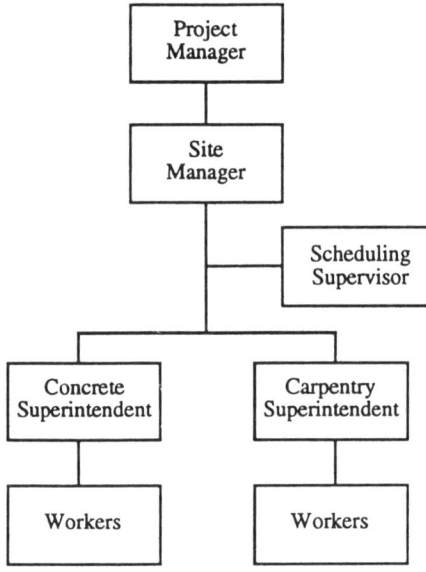

Figure 1. Veidekke site supervisions.

November 1994 with completion in June 1995. The schedule underwent four revisions as shown in Table 1.

Final completion of the project was in mid-September 1995. The project fell behind the original schedule due to worse than anticipated weather conditions in the winter and production slower than originally estimated. To meet hand over dates for the apartments sold, Veidekke doubled the number of carpenters on-site (from 15 to 30) from late April through June 1995. For improved project control, it was found necessary to increase the number of schedule activities used to describe the project.

Veidekke site supervision is shown in Figure 1. The project manager is not present on-site and has responsibility for multiple projects. The site manager is responsible for the site. Reporting to the site manager are two discipline superintendents who are in a staff position, as well as a scheduling supervisor. Carpenters were originally organized as a single group, although this proved difficult to manage and information feedback for predicted progress (and hence, unit delivery) became unreliable. Eventually the 30 carpenters were broken into 5 teams. Each team is paid an incentive wage based on amount of work completed, although they are guaranteed a high base wage. Construction on the various apartments was done individually or in small groups, so it was easy to assign and track work completed by each carpenter (although there was some confusion before the workers were split into smaller teams).

Units for the Buchhaugen project were designed by the carpentry superintendent with input from an E.A. Smith representative. The basic rationale was to determine a logical construction sequence. Units were then formed around this sequence, with units coming from single suppliers or from multiple suppliers (in this case, the unit was packaged at the E.A. Smith warehouse prior to site delivery). Each unit listed each piece needed for construction; wood was pre-cut except for mouldings, and other pieces (such as gypsum board) were ordered from stock sizes closest to the size needed on-site to reduce materials waste and time spent on-site cutting to size. Estimates of duration to complete an activity were not used in the formation of units, nor was there a link to the schedule, although part of a unit designation is where it is to be delivered on-site. Planning and development of the units was found to be very labor intensive, although listing all of materials at the start of the project did free time for increased site supervision during the project. Units were delivered at the designated time, although there were some problems in delivering units to the correct location as space at the designated site was not always available.

3.2 *Veidekke A.S.*

Veidekke A.S. is a general contracting firm undertaking a diverse range of construction activities principally in building, heavy construction, and asphalt. Additionally, Veidekke operates a small property division. Veidekke Nord, the division building the Buchhaugen project, has yearly revenues of 200 million NOK and a staff of 60. Veidekke acted as both owner and contractor on the Buchhaugen project, although this is an unusual arrangement for Veidekke as in recent years it has chosen to concentrate on its core contracting business. Originally scheduled to be developed in the 1980's, construction was delayed due to a downturn in the housing market.

The Buchhaugen project was chosen as the site for Veidekke's experimentation with the unit delivery system as it had several roughly equivalent apartments, which seemed to be a good match for repetitive delivery of units. Expectations going into the project were that site productivity would be increased due to less materials handling and more time spent on installation. Materials waste was expected to decrease. Total inventory (off-site and on-site) was also expected to decrease following the 'Just-In-Time' inspiration of the unit delivery system. Veidekke realized that initial planning needed to create the units would be more than in the traditional system, but expected administration to be less during construction. With feedback from the carpenters, there was an expectation that the unit delivery system would effectively drive itself. Furthermore, much of the work that would go into delineating the composition of the units would have to be done at some point as materials must be ordered whatever the delivery system.

In all, Veidekke's experience with the unit delivery system is positive. Compared to historical data, materials waste for wood is 10% less on the Buchhaugen site than on similar sites run under traditional delivery systems, while total waste decreased by about 7 to 9%. It is felt that site productivity increased under the system, but it difficult to give a definite figure as there were delays on-site due to inclement weather and because the initial estimate of site-productivity was overoptimistic. Units were generally delivered at or close to the designated site location, although there were some problems with conflicting storage space or storage space unavailable due to ex-cavated fill that was not removed. Veidekke surveyed the carpenters and found that 24 of 27 preferred the unit system, feeling that the unit system made it easier to know what materials to use and install, gave a cleaner building site, and that site manage-ment had more time to interact with the workers. Although there was a learning curve for the carpenters to adjust to the new system, it was preferred to the traditional delivery system where items are delivered in bulk. An example of this is the delivery of windows and doors, which were delivered in large lots and had to be stored on-site in an old house, requiring considerable on-site materials handling.

Although the unit delivery system worked well, delivering supplies when needed, Veidekke staff believes that some improvements can be made. Translating the design into units proved to be a more time-consuming task than originally thought. More traditional sites make less use of precut materials; site staff found that the extra step of listing materials for precut took too much time to calculate and cross check. Cross checking was important as in the unit delivery system there is no room for miscalcu-lation of component dimensions. When asked, staff thought that a CAD system which would automatically give pieces and measurements would be a great aid in the formation of units.

Beyond the time-consuming task of creating the units, there was agreement that there were too many units to track and control effectively. In some cases, a unit was ordered multiple times for different apartments, but in other cases different units had nearly identical manifests. This caused some confusion in tracking and location of the unit when delivered to site. Small differences in the design of the apartments caused the creation of these highly similar units; it would be better to designate a single unit for these apartments and do the necessary customization work on-site. Site staff also felt that the size of the units was often too small (contributing to the problem of too many units), although it is difficult to define the 'correct' size of a unit. Ideally a unit should have the properties of: 1) convenient size for transporta-tion; 2) not being nearly identical to other units; 3) contain items from the same supplier used at the same time; and 4) allow efficient use of cranes for positioning (trucks often have a small crane which allows this).

Site staff also had the expectation that the unit delivery system would effectively run itself. This was found to be a false expectation; the unit delivery system required continuous attention by site management. There were a number of factors which contributed to this, including the need for locating units when the originally delivery location was unavailable and the problem of tracking and predicting site progress. There were many changes in projected delivery date due to difficulty predicting site progress three weeks ahead. According to the site staff, the one week notification means delivery is probable, the two week notification means likely, and the three week notification is made with less confidence, although site staff characterized it as more than a guess. Actual day of the week delivery is needed is difficult to predict and highly subject to change. Furthermore, holding meetings on Monday of the or-dering week (with input from workers made the previous Friday), means there is a four week period from order to the end of the three week order period. Predicting site progress this far into the future is difficult and ordering as late as possible in the order week would give greater certainty to the order process.

To give greater ability to correctly forecast the flow of units needed by the site, increased site control was needed. This can be seen in the evolution of the schedule (see Table 1), where three revisions were needed to adequately predict site progress. The third revision, which is fully resource loaded and correctly estimates from site experience the amount of time needed to complete an activity, links each worker or team to units to be delivered and installed. Until this level of planning was performed, it was difficult to be certain about future site progress. Problems in forecasting productivity were exacerbated by the design variations from apartment to apartment which required different amounts of work time. Highly repetitive construction, such as may be found in block housing units, would be better suited to the unit system as once the first few blocks are built actual production rates are known.

3.3 *E.A. Smith Bygg A.S.*

E.A. Smith Bygg A.S. is a supplier of a full range of construction materials to projects and to individuals. Sales offices serve Norway's regions; each typically has a business office for interaction with project clients, a store for display of products for sale to individuals, and a warehouse supporting each of these business operations. The Trondheim sales office has a staff of 55 people, with total revenues of 330 million NOK per year. Sixty percent of these revenues come from sales to projects. Projects supplied by E.A. Smith vary in size from very small projects to larger projects of 3 to 5 million NOK. Most projects are small, ranging from 25 to 150 thousand NOK.

E.A. Smith is a supplier of products, not a producer. It maintains relationships with various producers and sells their products to individuals and may act to represent producers to project clients. As a supplier to projects, E.A. Smith supplies directly from its stocks and acts as a coordinator for materials supply from producers. In this role, E.A. Smith generally holds the contracts with the suppliers and passes materials orders from the site to suppliers. Often, this is done hurriedly, with sites giving only a short lead time before the materials are needed on-site.

E.A. Smith's competence as a supply coordinator led it to join the Buchhaugen project in this role. Expectations were that the unit system with three weeks look ahead would remove the crisis orientation of more traditional delivery systems, making improvements for all parties involved in the construction process by decreasing inventory and smoothing production flow. E.A. Smith's experience on the Buchhaugen project matches many of the expectations. The rolling three week system did reduce the crisis aspects of the system and made it easier to manage. Suppliers also liked the system as the three week order cycle generally increased lead time and allowed more time for them to plan their production.

However, there were some problems with the operation of the system. While the order system was less hurried, overall paperwork and administration costs increased. This is due to the increased number of orders and the extra attention needed for transportation and materials handling arrangements. One unanticipated problem with the unit delivery system is that most units are too small to fill a truck; this caused great problems in making shipment efficient. To match delivery volume with transportation capacity, two strategies were employed: First, supplier firms were often able to combine orders from different sites that were in the same region as the Buchhaugen project. This is possible mainly in seasons of high construction volume. Second, a supplier can combine multiple units in a truck and allow E.A. Smith to hold them in its warehouse and deliver as needed. These strategies of accommodating supplier transportation needs increases the inventory in the E.A. Smith warehouse to levels above that in more traditional delivery systems. Increased inventory levels in the warehouse were exacerbated by delays in projected delivery times; often, materials would be delivered to the E.A. Smith warehouse and held in stock for some time as the delivery date to site was pushed back. This is costly for E.A. Smith as it increased material handling costs and carrying costs as E.A. Smith pays its suppliers for delivery when supplies were in E.A. Smith hands, but Veidekke pays when materials are delivered to site. In some cases where very rapid delivery of materials was needed and suppliers were unable to accommodate the schedule, E.A. Smith would deliver supplies directly from its stocks that it uses for its store. In this case, E.A. Smith charges a premium over previously agreed upon unit costs for delivery from suppliers.

In general, E.A. Smith is happy with the operation of the delivery system and hopes to use it on future projects. However, a number of improvements can be made to the system from E.A. Smith's perspective: First, a more integrated information system to reduce paperwork is needed. Compatible computer systems with direct modem communication would be preferred. Second, delivery to the site now occurs every day of the week. To allow suppliers to efficiently combine orders and reduce the materials inventory at E.A. Smith, it would be desirable to deliver to site on just two, pre-determined days per week. Third, the increased demands for inventory storage in this system are a concern to E.A. Smith. Although able to accommodate the demands of the Buchhaugen project with existing resources, many concurrent projects operated with the unit system may require a larger warehouse and more workers. In this regard, E.A. Smith would like to be compensated for those cases where site-delivery is pushed back and E.A. Smith is left holding inventory for some time.

3.4 *Norske Skog – Langmoen A.S.*

Langmoen A.S. is a division of Norske Skog, a 7.3 billion NOK per year business with 5000 employees. Langmoen is part of the Norske Skog group of building materials companies (Norske Skog Bygg), which combined offer a full range of wood-based building materials. Focusing on lumber products, Langmoen A.S. operates three mills located in eastern-central Norway, processing 475 thousand cubic meters of wood per year. Revenues are 915 million NOK per year, with 840 employees. Each mill is operated independently with some specialization; the mill in Brumunddal that supplied the Buchhaugen project is the only mill of the three that cuts profiles and does specialty work. Langmoen customers are building projects such as the Buchhaugen project and firms such as E.A. Smith A.S. that stock lumber products for resale to individuals.

As a supplier of lumber products, Langmoen A.S. offers a large range of standard board sizes (e.g. length, thickness, breadth) as well as standard profiles and shapes. The bulk of orders are for delivery of standard sizes, although more and more orders contain at least some precut work. Contracts are generally based on unit prices, with precut work priced anywhere from 13-20% (averaging 15%) higher than the price for delivery of standard sizes. In addition, for orders that contain less than 25 precut pieces of the same dimension, Langmoen adds 10% to the price for that dimension. However, the market is very competitive, and seldom is this 10% markup possible. Often, Langmoen will make agreements with repeat customers such as E.A. Smith to restrict price markups to no more than 15%.

Mill policy for production of standard sizes and shapes is to make long production runs to economize on machine set-up time. Inventory is then held and individual orders are filled from stock. Packing of orders is done in two locations at the mill; workers can be rotated in an out of the packing job as needed. Traditionally, packing is done in a rush as building sites give a short lead time for supply of materials. The unit delivery system for the Buchhaugen project is appreciated for the extra lead time it gives; packing under this system is done early to avoid the rush.

The Langmoen Brumunddal mill maintains an exclusive relationship with a trucking firm located in Brumunddal, and is also located along a railway which is useful for shipping large orders. Transportation capacity has never been a problem; it is possible to get a truck with just a few hours notice. However, efficiency in transportation is important as truck rental between Brumunddal and Trondheim costs 150 NOK per square meter of floor space (full or empty). Where possible, Langmoen attempts to ship a full truckload of materials. Buchhaugen deliveries caused problems as units did not fill a truck. At times, Langmoen was able to combine the Buchhaugen orders with deliveries to other sites, but this was not always possible (especially in winter). Langmoen found E.A. Smith a great help as E.A. Smith was willing to hold units in inventory in its warehouse, allowing the filling of trucks.

Although most Buchhaugen orders were a mix of stock and precut work, the need to precut materials posed little problems for the mill. Precut work consists of taking stock pieces and cutting them to size. This can be done very efficiently with the use of a finger jointing machine which allows stock pieces to be combined and then cut to the designated precut length. Efficient planning on the finger jointing machine eliminates materials waste in the precut stage of manufacture. Without this planning, materials waste ranges from 10 to 15%. Effective use of the precut machine requires two to three weeks of lead time; this time is used for production planning and combination of orders from different projects to maximize machine efficiency and to reduce setup time. Here, production efficiency is sensitive to the mix of materials needed by other projects the mill is working on concurrently. Typically, there are 2-3 precut-projects a day worked on by the mill, although each order can be seen as a project by itself.

Where possible, mill policy is to make all pieces of the same size needed for an order at the same time unless there is a very long lead time and materials would be held in stock for several months. It is possible for the mill to accommodate short lead times of one to two weeks, but this leaves less time for production planning and leads to inefficiencies in production. Other than knowledge of materials wastes the mill does not estimate the cost of short lead times, but production inefficiencies stem from increased man-hours due to the rush nature of the job and increased internal transportation of materials, as well as increased setup time for the machines.

In general, Langmoen A.S. is happy with the performance of the Buchhaugen project. The three-week rolling schedule allowed improved performance with precut and packing work. The three week notification and listing of parts needed in units also gave comfortable certainty to the order with no sudden changes in materials needed by the project. Langmoen also found that the orders and unit descriptions supplied by E.A. Smith were adequate for its own use and little additional paperwork was generated; for example, the list of materials for a unit generated by Veidekke and E.A. Smith was released directly to the shop floor with no alteration by Langmoen staff. There is some concern about transportation efficiency within the Buchhaugen delivery system, although Langmoen is happy with the support provided by E.A. Smith. A final worry is certainty of lead time performance demanded by the site; if units are demanded with less than the three week lead time in this system, then costs for Langmoen will increase.

3.5 *Johs. Rasmussen A.S.*

Johs. Rasmussen A.S. (commonly known through the brand name Nor-Dan) is a large supplier of windows and doors to the Norwegian construction market, having a 22% share of the market in 1994 with expectations that this will expand in 1995 through the acquisition of a new factory. Revenues in 1994 totalled 420 million NOK (90 million NOK from export to Europe) with budgeted sales of 520 million NOK in 1995. Johs. Rasmussen operates four factories located across the southern part of Norway, with regional sales offices serving all of Norway. Staff size, including sales and manufacturing, is approximately 700.

The Nor-Dan product line consists of a large number of standard shapes which can be combined to create custom windows and doors suited customer requirements (size, shape, thermal properties, colour, etc.). Approximately 80% of production is for special orders and 20% of production is make to stock. Make-to-stock production consists of a number of standard windows and doors which are primarily sold to makers of prefabricated houses. Special orders go to construction sites. The large number of standard shapes and needs of the building sites make each special order unique.

Each of the four factories operated by Johs. Rasmussen A.S. specializes in one type of production (windows or doors) with one factory capable of handling highly specialized orders. Total production across the four factories averages 1000 windows per day and 110 doors per day. (These numbers vary depending upon the complexity and size of the order in question). Basic factory policy is to operate at full capacity. Production of stock windows and doors is used to smooth overall production. There are seasonal variations in the 80%-20% split between custom and stock production, with relatively more stock production in the winter period due to low construction volume.

Lead time for special orders varies from four to six weeks, depending upon the degree of customization and prefabrication (e.g. painting in the factory rather than on-site requires extra-lead time for drying). The first one to two weeks of lead time are used for production planning and confirmation of the order specifications with the customer; items requiring specialty stocks from suppliers are also ordered over this period. The remaining time is used for production; this seldom takes more than three weeks. Over this time period, many orders are processed simultaneously. For

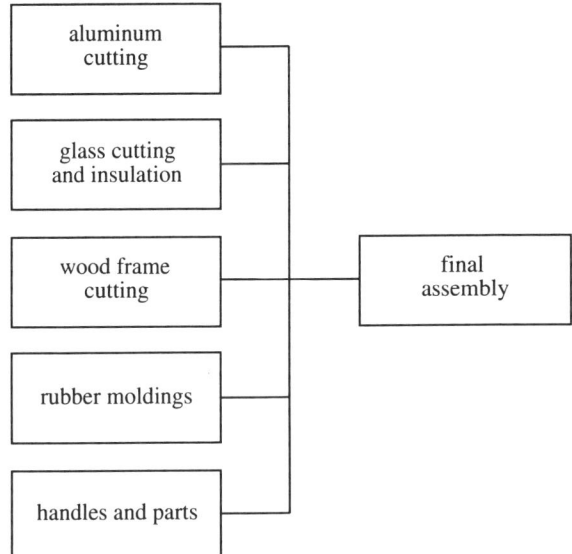

Figure 2. Arrangement of work
centres in a typical window factory.

any given window or door, actual time spent in production is small compared to total time in the factory. Most of the time is spent waiting in batches for orders to be processed through the various work centres in the factory. Figure 2 shows the typical organization of work centres in a window factory. Suppliers deliver raw materials to each of the five different work centres which make the constituent parts of the window. These centres process orders in batches. Where possible, different orders with similar parts are processed together to reduce setup time and increase total throughput. Nonetheless, there are economies of scale that come with increased order size. Windows are sold on a unit price basis with cost dependent on the number of windows ordered. An order of 1-3 windows (of a single type) is priced 25% more than and order of 8 or more windows; an order of 4-7 windows is priced 10% more than an order of 8 or more windows. If many more than 8 windows are ordered, a quantity discount may be given. This non-linear price schedule roughly represents the costs of factory production.

In general, the factory maintains stocks of raw materials. Suppliers give discounts for large orders and factory management has found it economically beneficial to maintain stocks of materials that are large relative to most customer orders. This inventory buffer effectively decouples factory production from supplier production. Only large orders of three million NOK or more require special consideration of materials supply, particularly when adjusting production to meet changes in customer delivery dates. Rasmussen generally contracts with one supplier for each material type; suppliers tend to be large firms compared to Rasmussen. Suppliers are evaluated on quality, price, and capability to deliver reliably (closely related to the size of the firm). Contracts are generally made on a unit price basis. Little exploration has been done to develop a more closely connected logistics relationship, although sales channels may work together (e.g. to educate the customer about the various thermal properties of glass products of the supplier).

Where possible, the factory produces goods for an order in large batches to obtain scale economies and ships them directly to the customer. This reduces the amount of finished goods inventory on hand, reducing the need for storage space and materials rehandling. This policy also speeds payment from customers. Here, transport considerations are important. Shipment is made to fill trucks or railroad cars as there is a fixed cost per truck and car used for delivery.

In consideration of shipping costs and supply capability, Rasmussen decided not to participate in the special delivery system on the Buchhaugen project (for which it delivered the bulk of windows and doors). Total contract value was approximately 600 thousand NOK, consisting of 400 windows and doors. Production was in two different factories (windows and doors) located in separate parts of Norway (Otta and Egersund). Production for the windows was made in two series due to colour variation. Three main deliveries were made to site, two by trailer truck of approximately 200 thousand NOK each and one by train (to Trondheim and then by truck to site) of 100 thousand NOK. A fourth delivery of 100 thousand NOK was made to the regional sales office located 20 kilometres south of Trondheim city centre. Three deliveries were made from the sales office to the Buchhaugen site.

The regional sales office has a 900 m^2 warehouse. This warehouse is a new acquisition for Rasmussen, starting operation over the course of the Buchhaugen project. If the warehouse was present during the entire Buchhaugen project, Rasmussen would have been able to participate in the system by shipping all goods from the factory to the warehouse and then delivering from the warehouse to the site as needed. Such a delivery system is more expensive than delivering large batches directly to site. Added costs include delivery by small truck (estimated to be 1200 NOK per delivery for the Buchhaugen site given driving time from the warehouse), extra warehouse costs, costs of carrying inventory (at least 5% interest cost) without payment, and extra materials handling costs. These costs must be evaluated on a case-by-case basis. Implementation of such a delivery system will not affect factory production schedule; production will still be scheduled in large batches. Implementation of the Buchhaugen delivery system would also be possible in regional districts without a warehouse provided a surrogate warehouse is provided by an actor such as E.A. Smith.

4 DELIVERY DATA AND ANALYSIS

A historical record of deliveries is available through the unit order sheets for each week. As units are delivered just as they are needed on-site and each unit is relatively small, this delivery record closely matches site production. It is also worthwhile to note that the site was not kept waiting for materials; rather delays in site production held up deliveries. Thus uncertainty on-site influenced performance in the off-site supply-chain rather than uncertainty in the supply-chain affecting on-site performance.

Data from the unit order sheets from the start of the project in November 1994 through June 1995 (representing the majority of all units delivered) was assembled in a spreadsheet to create a history for each unit, detailing the first order date, projected week of delivery for each week ordered, and final delivery date. This data set represents a near complete record of deliveries (there were very few deliveries made after June 1995) made to the project, and, through a record of changes made in the projected delivery date for each order for each unit, a history of the uncertainty of site production. Beyond comparing scheduled versus actual progress on-site such as the schedule performance index (SPI) outlined by Barrie & Paulson (1992), construction research literature does not contain any measures or recommendations to analyze such a data set, so a series of measures were created for this analysis.

In the unit order system, each unit should have a three week lead time from first order to delivery, and be ordered once on the three-week look-ahead, once on the two-week look-ahead, and once on the one-week look-ahead. Lead time distribution for all the units ordered is shown in Figure 3, where each bar represents one unit. Units are sorted chronologically by week of delivery, but weeks are not delineated. Some weeks have more deliveries than other weeks. As can be seen in Figure 3, there is considerable uncertainty in the delivery system with many units being ordered with less than three weeks lead time and others ordered with much more than three weeks lead time. If the system had worked perfectly, the graph would show a three week lead time for all units (it is important to note that due to the week-long holidays around Easter and Christmas, some units would have a four week lead time).

While Figure 3 gives a sense of uncertainty in project production, showing both over optimistic and under optimistic delivery times throughout the project, Figures 4, 5 and 6, describe the uncertainty for each unit. As for Fig. 3, each bar represents a unit and units are sorted chronologically by delivery date. Figure 4 shows the number of times one-week notification was given for delivery for each unit. The norm is for the one-week notification to be given once for each unit. As shown in Figure 4, most units were given the one-week notification only once, with a few units given one-week notification two or three times. Thus very few units due to be delivered were delayed at the last minute and project production can be seen to be quite certain within a one-week time frame.

Figure 5 shows the number of two-week notifications given for each unit; as for one-week notifications, there should be one two-week notification given for each

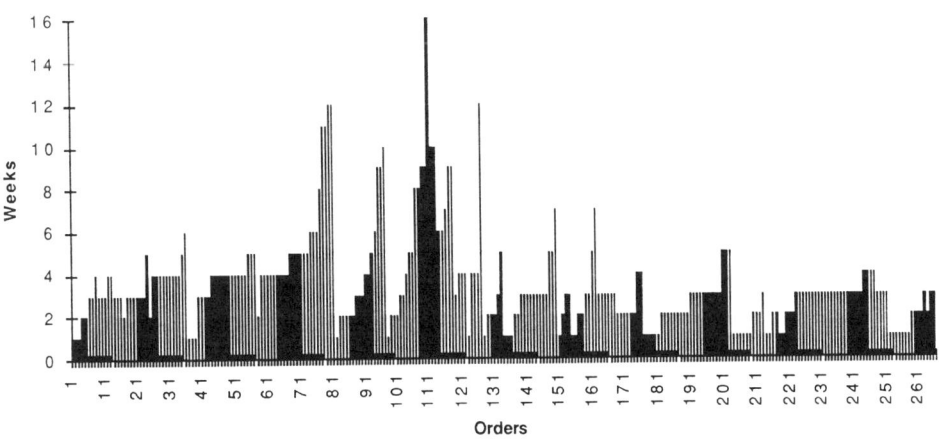

Figure 3. Lead time distribution of all units.

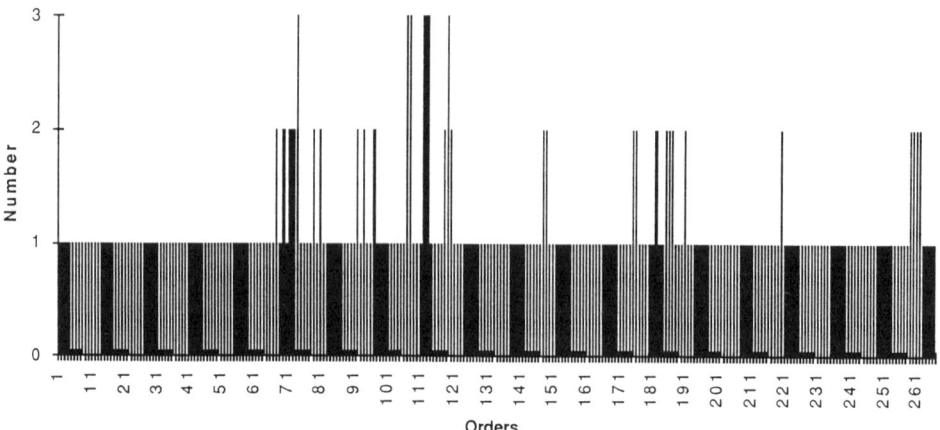

Figure 4. One-week notifications.

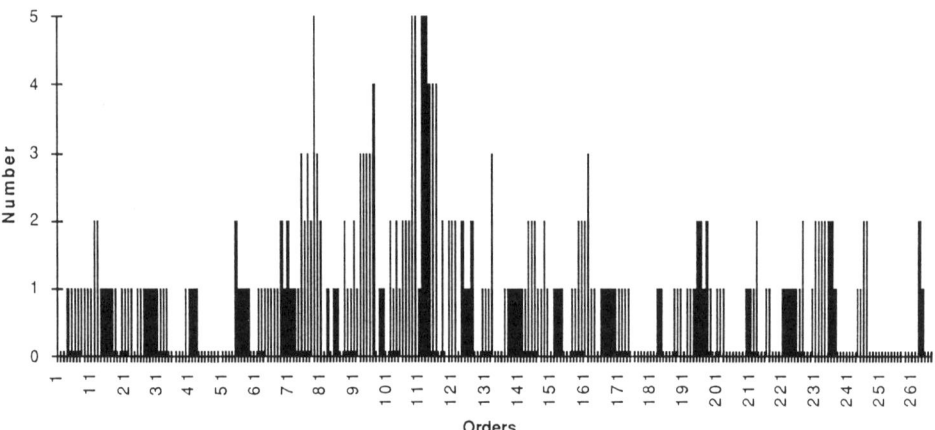

Figure 5. Two-week notifications.

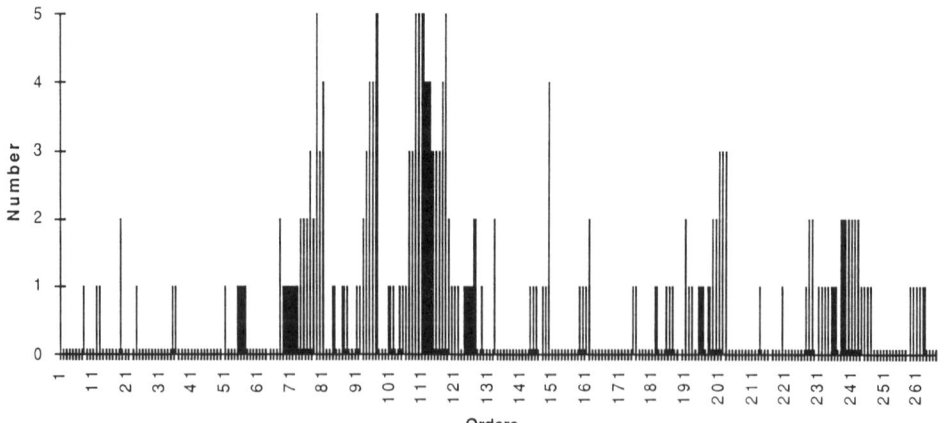

Figure 6. Number of changes in projected delivery date.

unit. Figure 5 shows great deviation with this norm, with many units being ordered with no two-week notification and several others having delivery pushed back (represented by multiple two-week notifications). Project production is much less certain with a two-week forecast than with a one-week forecast.

Figure 6 gives the number of changes in projected delivery date for each unit. With the unit order system, there should be no changes in projected delivery date as each week on the order cycle should indicate the same delivery date. It can be seen that many units had a change in projected delivery date, which can signify either a delay or an acceleration. Furthermore, many of the units have more than one change in delivery date, signifying that uncertainty about unit delivery dates can remain unresolved for several weeks.

Critical statistics from Figures 3-6 are as follows:
- Percent of deliveries with less than three weeks lead time: 33%
- Percent of deliveries with more than three weeks lead time: 36%
  (This includes deliveries around Easter and Christmas which have a
  four week lead time).
- Percent of deliveries with no two week notification: 34%
  (This includes deliveries with more than two weeks lead time but
  have no two week notice due to changes in projected delivery date).
- Percent of deliveries with only one week notification: 14%
- Percent of deliveries with a change in delivery date: 43%
- Percent of deliveries having a change in delivery date with more
  than one change in delivery date: 40%

The data indicate that one-third of all deliveries have less than three weeks lead time, and almost one-half of these have but one week lead time. On the other hand, many units have more than three weeks lead time, and if a unit is delayed once, there is a good chance that it will be delayed again. In all, the data indicate that certainty about site production needs decreases greatly past the one-week look-ahead.

The weekly unit order system suggests a week based analysis of project performance. Although sorted in chronological order by delivery date, Figures 3-6 give little data about the flow of materials to the project and relative uncertainty from week to week. A measure of uncertainty that allows comparison is needed. As mentioned above, construction literature provides little guidance about how to construct such a measure. Two general approaches suggest themselves: First, it is possible to investigate the uncertainty leading up to any given week. A second approach is to look forward from any given week and measure the future uncertainty. A hybrid measurement combining forward and backward looking measures is also possible. Of these approaches, a forward looking measure is appropriate as orders are made each week for future weeks.

A simple but inclusive measure of forward looking uncertainty was constructed for each week as follows: First, for each week, determine all the units that have been ordered but not yet delivered – these units are said to be *in process*. For these units, count all the future changes in projected delivery date, ignoring changes in projected delivery date made in previous weeks. (A backward looking measure would look only at changes made up to the week in question, while a hybrid measure might look at all changes in delivery date for units in process). For each week, sum the changes for all units in process and divide by the number of units in process. Division by the

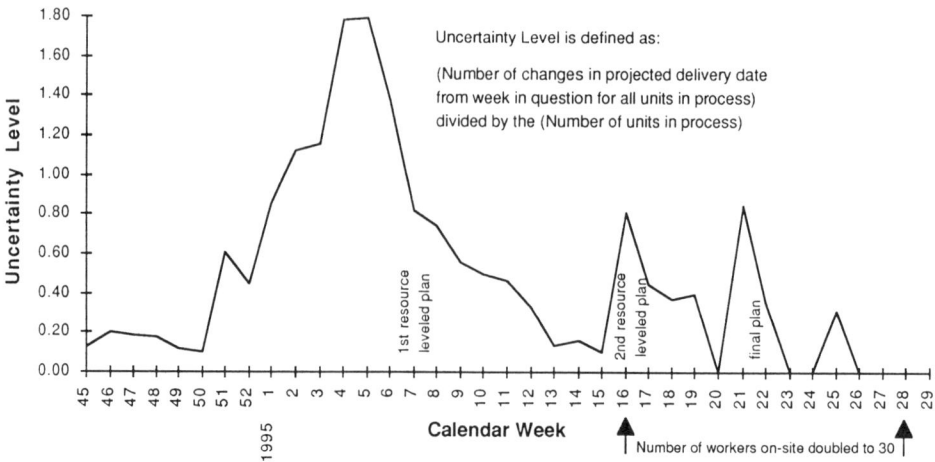

Figure 7. Weekly uncertainty level.

number of units acts as a norm allowing comparison from week to week; thus a week with five changes and five units in process will have a higher uncertainty level than a week with five changes and ten units in process.

Uncertainty level is shown for each week of the data set in Figure 7. Weeks are shown as calendar weeks, and each schedule revision is indicated on the figure. Initially, the project had a small level of uncertainty which grew in late December and peaked in early February. Uncertainty dropped dramatically after creating a resource levelled plan, although this plan still underestimated activity duration. The early part of 1995 also was subject to unseasonably bad weather, increasing uncertainty on site. Uncertainty increased somewhat after increasing the number of carpenters on-site with the second schedule revision, although some weeks later when the final revision was made, uncertainty almost disappeared. (Those units delivered after week 29 are reported to have been subject to very little uncertainty in delivery schedule).

Figures 8, 9 and 10 compare the uncertainty level with data about the flow of units to the site. Figure 8 plots the uncertainty level per week together with the number of deliveries made each week. (Uncertainty level in Figs 8-10 has been multiplied by ten to make visual comparison easier). Figure 9 plots uncertainty level per week together with the number of new orders for units placed each week. Figure 10 plots the number of units in process (ordered but not delivered) and the uncertainty level.

As is visually apparent, the only correlation is between uncertainty level and units in process as shown in Figure 10, and even this correlation is weak. Units in process plotted against uncertainty level in a scatter graph (Fig. 11) shows some clustering. For any given week, when there are fewer than 20 units in process, the uncertainty level is relatively low. When there are more than 20 units in process, the uncertainty level is likely to be much higher. This corresponds to site staff observations that it was difficult to track and control the large number of units. The more units in process, the more relative certainty decreases. This makes intuitive sense given the limits of human rationality.

Figures 10 and 11, together with the uncertainty level and chronology of site events (Fig. 7), suggests that uncertainty on-site is subject to a number of influences

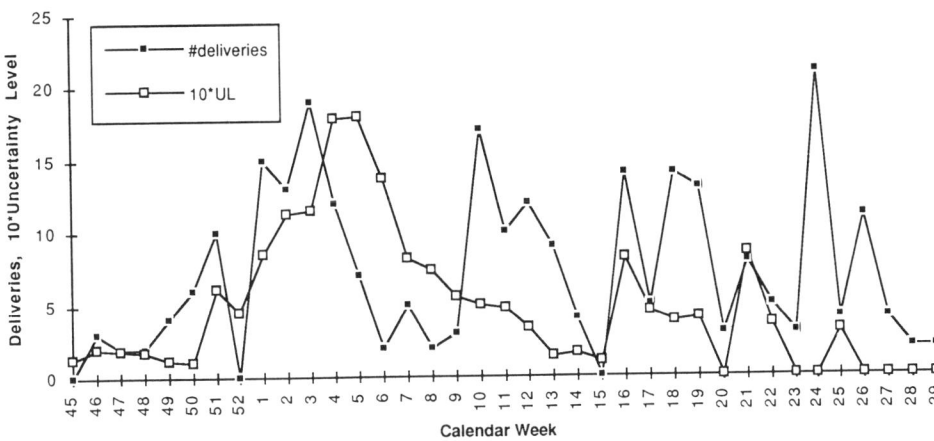

Figure 8. Uncertainty level and number of deliveries per week.

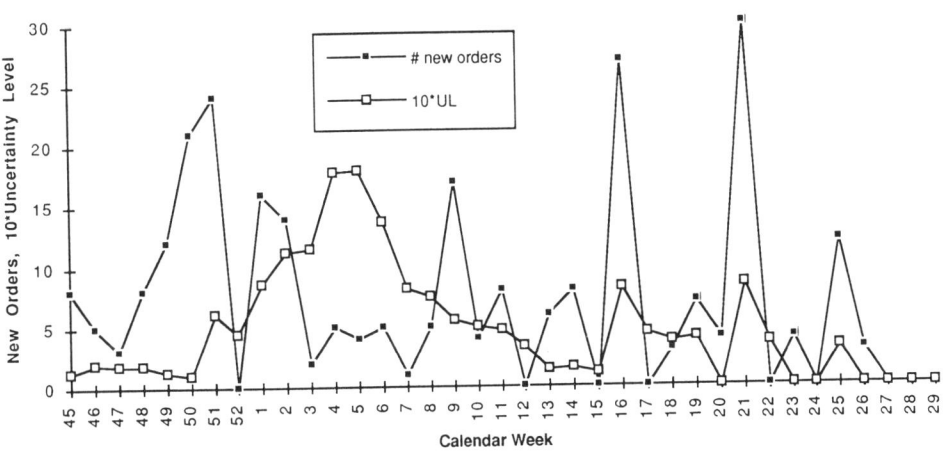

Figure 9. Uncertainty level and the number of new orders per week.

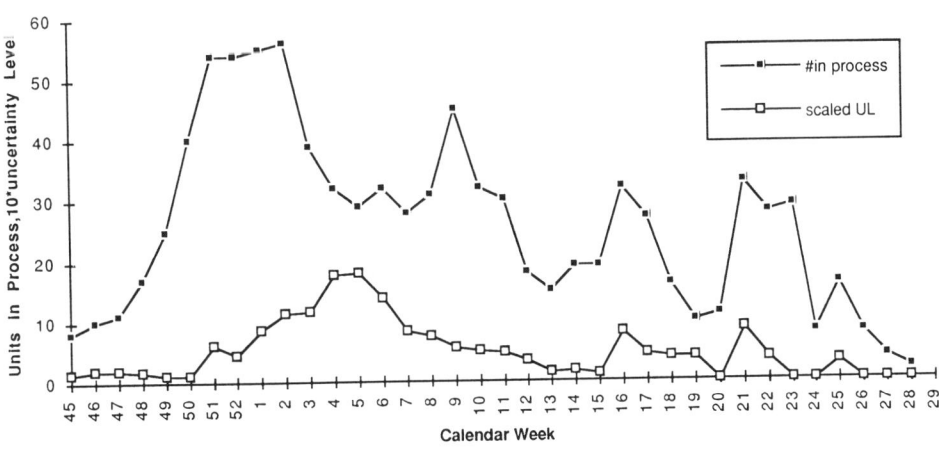

Figure 10. Units in process and uncertainty level.

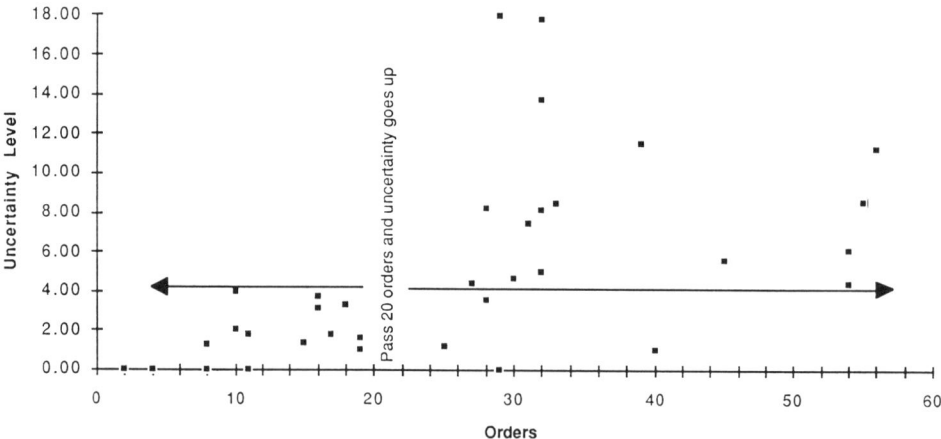

Figure 11. Units in process versus uncertainty level.

which have different magnitudes. Identifiable influences on the certainty of site production are weather conditions, accuracy of productivity estimates and schedules, the number of units in process, feedback from the workers, and the level of detail of the schedules (together with the assignment of workers to scheduled activities). Comparing uncertainty levels over the course of the project, it appears that weather and general accuracy of schedules and estimates have a greater impact on certainty than do schedule detail, volume of units in process, and worker feedback. We conclude this by observing that uncertainty was greatest during the period of worst weather conditions and least accurate schedules. Furthermore, uncertainty continued throughout the project until schedules were fully detailed and resource loaded (Revision 3), with careful assignment of workers to scheduled activities, and good feedback mechanisms were established. This suggests that planning and control necessary to accurately forecast project production requires information as complete as possible and that partial measures to improve certainty levels give only limited improvement.

5 UNCERTAINTY AND PRODUCTION COSTS

Uncertainty in site production clearly affects the cost and performance capabilities of the supply-chain. These costs manifest themselves in three ways: efficiency in transportation, efficiency in production, and efficiency in inventory handling. Transportation efficiency comes from filling trucks to capacity and this can require sharing truck space across projects. Effective management of this process requires certainty as to the week and day units are needed. Efficiency in production comes from the ability to plan production effectively in the context of other orders that need to be processed. Uncertainty can disrupt these plans and increase costs. Efficiency in inventory handling comes from reducing the total stock of inventory and reducing the amount of internal transportation that goes on at any location. Uncertainty in the delivery cycle can increase inventory held in the chain by pushing the delivery date back after production of the unit has occurred.

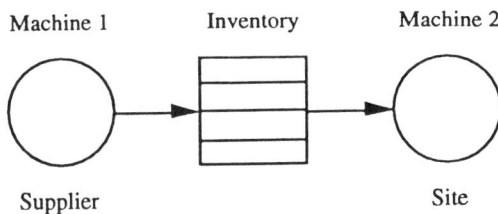

Figure 12. The suppliers site relationship as a two machine system with inventory buffer.

In this section we explore the tradeoffs between inventory, transportation, and production costs in construction supply-chains. A number of simple examples are given to demonstrate that tradeoffs between these costs occur even in the simple, deterministic case. Analysis of these tradeoffs becomes more elaborate in the presence of uncertainty, and a manufacturing two-machine production-inventory model (Fig. 12) is used as a basis for this analysis. This model is used to develop a typology of supplier classes which are affected by (and manifest) uncertainty differently. In the development of this typology, some critical distinctions between project production and manufacturing assumptions are made.

It is always good to reduce inventory if all else can be held constant. The unit system accomplishes this on the Buchhaugen site, allowing increased on-site production together with decreased materials waste and inventory costs. However, it is less clear that this is the case for Buchhaugen suppliers. Consider truck transport which costs 5000 NOK per trip. A firm makes units and each unit fills one-half a truck. A unit is needed on-site in week one and in week two. The cost of holding a unit in inventory on-site is 1000 NOK per week, and the cost of holding a unit in inventory in a warehouse off-site but close to site (as E.A. Smith has done) is 250 NOK per week with transportation cost from the warehouse to site of 500 NOK per delivery. There are three possibilities: Deliver a unit from the supplier in week one and in week two; total cost of transport and inventory storage is 10,000 NOK. The second possibility is to deliver both units directly to the warehouse near the site in the first week and deliver each to site as needed; here total cost of transport and delivery is 6250 NOK. The third possibility is to deliver both units directly to site on the first week, with a total cost of 6000 NOK. Clearly, there are tradeoffs between inventory storage and transportation costs and total costs may be less if inventory is held somewhere in the supply-chain. This is true even in cases of complete certainty about when units are to be delivered. With uncertainty, costs increase as orders are often made on short notice (recall that 33% of deliveries were made with less than three weeks lead time), eliminating the ability to use transportation capacity effectively. Delaying delivery to site can have the same effect, as a unit that was to share space with other units on a truck is now not to be delivered and the truck goes partially empty (alternatively, the unit can be delivered to site or the warehouse close to site, increasing inventory costs).

The example in the preceding paragraph is deliberately structured to show an example where storage on-site is cheaper than storage through an intermediary such as E.A. Smith. It is entirely plausible that on-site storage can be a more cost-effective solution than storage anywhere else in the supply-chain. However, storage off-site has many advantages. There is less of a chance of breakage, and productivity on-site will not be lowered due to excess materials handling. Furthermore, a dedicated storage site may be more efficient at handling materials than on-site storage. Nevertheless, inventory costs occur wherever inventory must be stored in the supply-chain, and it is important to recognize the case for on-site storage if inventory must be held.

The tradeoff between storing inventory and transportation costs is shown above, demonstrating a case for holding inventory even under complete certainty about site production. A similar example can be created for off-site production in the presence of set-up costs (present both for Langmoen A.S. and Johs. Rasmussen A.S.). Set-up costs occur when a machine must be taken off-line in order to prepare it for production. It is desirable to make as few set-ups as possible to maximize production on the machine, and thus parts of same type are generally made in large batches. It is generally more efficient for a supplier to produce many units in advance and hold them in inventory than to produce small lots just before shipping. This is a common practice in manufacturing operation, and although there is a trend to reduce set-up time and cost, it is very difficult to eliminate set-up times completely in a production environment requiring some degree of customization. Indeed, it has been estimated that 75% of production is done in customized lots, and these are generally produced in a job-shop environment where inventory is held to assure high machine utilization (Askin & Standridge 1993). Thus even in an environment of complete certainty about demand and production there is a case for holding inventory.

Interaction between production and inventory is more complex under conditions of uncertain demand and production. Abstractly, we can represent the site and supplier as a two machine system with an inventory buffer as shown in Figure 12. The supplier (Machine 1) produces parts which are held in inventory until needed by the site (Machine 2). This basic layout is a subject of extensive study in operations management literature, and it has been clearly demonstrated that the presence of an inventory buffer between machines is necessary to increase throughput (by avoiding starvation of Machine 2) under conditions of uncertainty or variability in production times (Baker et al. 1990; Conway et al. 1988). Rather, the important question is how big the buffer should be. This, of course, is directly in contrast to the Just-In-Time (JIT) inspiration of zero inventory, but JIT is meant to operate in conditions of low variability both in production and in demand (see Baudin (1990), Karmarkar (1989), and Zipkin (1991) for commentary on the applicability of JIT to various production environments). Clearly demand variability is present on the Buchhaugen site, where 43% of all deliveries have at least one change in delivery date. Thus not only do delays cause inventory to be held because of unit production being performed before it is needed, it is also useful to hold inventory to avoid excess demands on the supplier when there are rush orders (remember that 14% of all Buchhaugen deliveries had only one-week lead time).

Building on the two machine model of Figure 12, it is possible to develop a typology of supplier types. The first type is one where the inventory buffer is very large, effectively decoupling supplier production from site production. This is the case for commodity items which are produced and held in stock and for customized items produced in large batches. Langmoen stock supplies are of this type as is the production of Rasmussen, which produced all the windows for Buchhaugen in two long production runs. In this case, the relevant costs of the unit system are logistics costs (transportation, storage and materials handling) and costs of site production.

Here the tradeoff is between the costs of more deliveries (together with benefits to site production) and the costs of increased inventory. In the case of windows, costs for delivery in units as opposed to large batches are estimated to be in excess of 1200 NOK per delivery. This cost must be offset by increased production on-site. Of course, this delivery cost is quoted for delivery of windows from a Rasmussen warehouse, and may be reduced through sharing truck space with other batches.

Suppliers who produce custom items with long lead times form the second class. Pre-cast concrete is a good example of this type of production, where extensive production work must take place off-site to fabricate the forms and cure the concrete. Production often involves multiple stages with design review. Risk due to uncertainty entails completion of items before needed on-site. This involves two components: the internal risk due to variability in the completion times for each stage, and external risk due to shifting capacity requirements. Production generally requires a large portion of available supplier capacity over the relevant production period. Figure 13 depicts the interaction between capacity, cost and time for the supplier (O'Brien et al. 1995). Bars on the capacity-time plane show the firm's allocation of capacity to different projects over time. The firm must decide when in time to allocate capacity to projects. As a goal, the firm tries to minimize its total expenditures (subject to project constraints), which is the area under the expenditure path traced along the cost surface. Costs are an increasing function of capacity utilization for systems with variability due to congestion effects (Banker et al. 1988). Interaction between projects, represented by placement of bars on the capacity-time plane, can cause costs to increase or decrease and place conflicting demands on available capacity. When site uncertainty requires the supplier to change schedule, capacity constraints may be encroached. This has cost implications for the supplier and for the site as the supplier may not be able to accommodate the change in schedule. Thus analysis of the interaction between site and supplier is primarily an assessment of the buffer period which should be assigned between the end of production and start of installation on-site (in essence, the time over which inventory is planned to be held). Delivery costs in a unit system versus traditional delivery methods can then be assessed if appropriate.

A third type of supplier has high production volume compared to site demand, set-up costs are low, and production makespan (the time from start to completion of fabrication) is short compared to the order period (a week in the case of Buchhaugen). Here, the supplier has the capacity to make and supply the orders as needed and re-

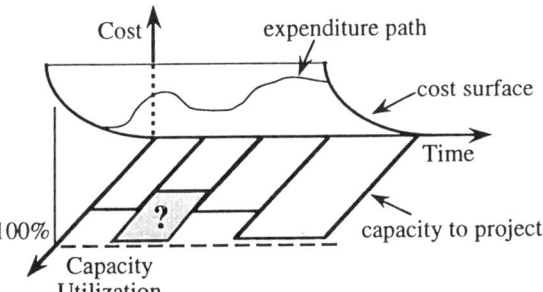

Figure 13. A supplier's cost-time-capacity relationship.

quires little lead time for production. The supplier is relatively unaffected by uncertainty in site production, able to produce ahead of schedule or delay production with little effect on production costs. Inventory in the system can be minimized relative to transportation costs. This is an ideal kind of supplier for the unit order system, but such a supplier must by definition produce a range of highly standardized units in volume, with only limited customization (see Askin & Standridge (1993) for a discussion of customization ability under different production systems).

Many suppliers may make use of standardized parts and customize them before delivery; in this case, it is the customization step which drives production time. This leads us to the fourth class of suppliers, those with positive set-up costs, a makespan of the same order of magnitude as the order period, and site demand a significant portion of production capacity. This is an important class of suppliers; Langmoen precut is an example. Here, lead time is important for production smoothing. As a generic example, assume the factory plans production on a weekly basis. Orders arrive with a one week lead time. Then the factory must fill all orders as they arrive, and capacity utilization is dependent on the total volume of orders each week. In some weeks, there may not be enough capacity to fill all orders. Now suppose that orders are made three weeks in advance, as in the unit order system. In this case the factory has the ability to smooth production. As described in a previous example, set-up costs may be reduced by producing larger batches of items (enough to meet several weeks of demand). Clearly costs are decreased in this system as no matter what the conditions, the supplier always has the option of doing what it was going to do if it had just one week lead time.

If a supplier is of type four and usually processes orders in small batches, then the three-week unit order system reduces costs for the supplier. Langmoen precut is a good example of this as they have the ability to efficiently plan production on their machines given an ability to find common board sizes among various orders and run them through their machines at the same time, reducing setup costs and materials waste. Maximization of machine efficiency takes both time to plan and knowledge of production demand over several order periods – Langmoen needs 2-3 weeks lead time to obtain this cost-effective production. Here, we observe that costs go down although finished goods inventory increases.

Problems arrive for suppliers of type four when there is uncertainty in the lead time given for orders. Clearly, rush orders destroy the ability of a supplier such as Langmoen to efficiently utilize its production capacity. Delays may cause less of a production problem when trying to combine orders, but in either case there are capacity problems. Referring to Figure 13, the capacity problems of suppliers with long lead times are also relevant here as changes in the time orders are needed can require production to be shifted and capacity constraints violated. This is most obvious in the case of a rush order, but delays also cause problems. Assume there is a delay and the supplier decided it is best to postpone production in favour of other production. In a traditional manufacturing analysis with assumptions of the independence of orders (Gershwin 1994), no further impact of delays would be assessed. However, the finite nature of project production alters the conditions as if an order is delayed, it is likely that site production is delayed. This implies that at some point site production is likely to be accelerated and there may be a flood of orders for the supplier to process which will conflict with capacity constraints. The history and context of site production is relevant to decision making on the part of the factory.

Costs for the factory are heavily dependent on the certainty of site production; the difficulty of the Buchhaugen site to accurately forecast more than one week in advance does not match well with supplier needs for multi-week forecasts. Assessing supplier costs and production policies under uncertainty is a difficult task, however. There are costs that come from capacity constraints and from the inability to adequately plan production. Here, suppliers of class four may be divided into two subclasses: Those with production synergy among orders and those without. Langmoen precut is an example of a supplier with production synergy as it combines production from different orders to economize on set-up costs and materials waste. Another example of production synergy comes from set-up times that are dependent on the previous production run (see Gershwin (1994) for a detailed discussion and approach to modelling production under set-up costs). An example is painting, where it is quicker to switch from a lighter colour to a darker colour than from a darker colour to a lighter colour as less cleaning is needed. The optimal solution to the set-up problem is to cycle from lighter to darker colours, switching back to the lightest colour only after the darkest colour is applied. Uncertainty in demand can hurt production efficiency by removing the ability to pool orders for set-up efficiency or by interrupting the optimal cycle of setups, as well as by imposing capacity related costs as described in the preceding section.

Type four suppliers with little production synergy among orders are primarily affected by the capacity problems caused by variability in demand. Even this is a difficult problem to assess. Figure 14 is a simulation diagram drawn using Stella II, a systems simulation software package. This is a model of a simplified factory production with no synergy between orders. Rectangles are reservoirs which hold stocks,

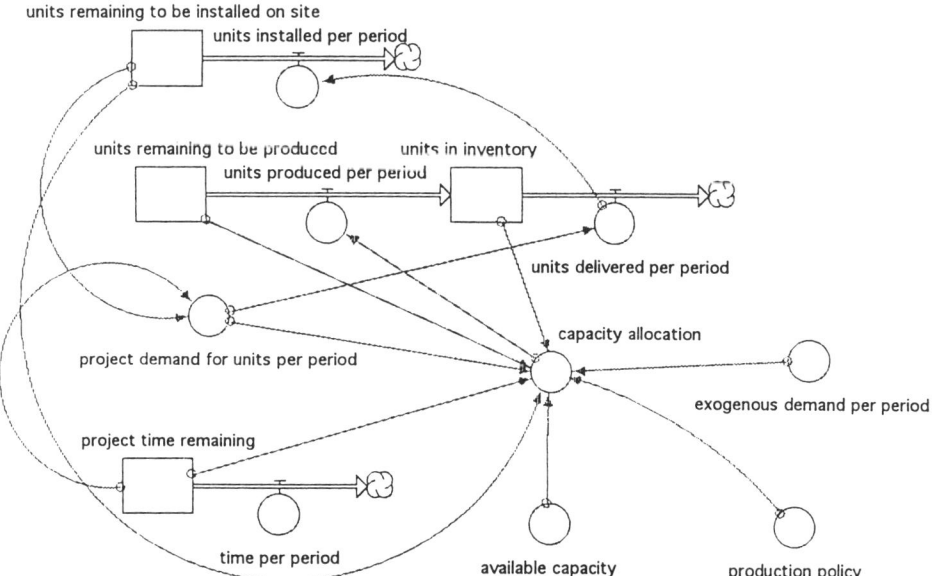

Figure 14. Simulation diagram of supplier production with influences.

circles are indicators, and circles with valves attached to twin bars regulate the flow of stock between reservoirs (clouds consume flows and are used when no stocks are needed for measurement). This model is made with production planning in mind, but can also be used as a basis for assessing the costs of uncertain events. Each production period, the factory must decide how to allocate available capacity meeting the current demand for units and demand from other projects (exogenous production). The factory can supply from inventory or produce directly for site demands each period. Additionally, the factory can produce units ahead of demand and hold them in inventory. Project demand for units is influenced by the units remaining to be installed on-site and the time remaining until project completion (or activity completion). Capacity allocation each period is determined by the project demand for units and exogenous demand, together with a number of influencing factors: First, the number of units held in inventory can act to increase or decrease unit production per period depending on the size of the desired buffer between factory and site. The size of this buffer is influenced by the uncertainty in demand and expectations for future demand; this expectation is influenced by the second and third factors, time remaining for project (activity) completion and the units that remain to be on-site. The number of units that remain to be produced are also a factor in the production decision as this can influence lot-sizing decisions. Available capacity has been separated from the capacity allocation decision to show that this is a finite amount which may be subject to adjustment through worker overtime, etc. (Truly complex models would include the possibility of machine failures when determining available capacity). Finally, production policy has been separated from the capacity allocation. Models of this complexity are difficult to optimize and thus comparison between different production policies is one of the tasks to be performed in analysis. A simple and logical production policy would be to produce units to inventory if there is extra capacity during the production period. This provides a buffer for periods of high demand and allows reduction of set-up costs. This is not necessarily an optimal policy should inventory costs be high or if uncertainty in demand is low. Clearly, supplier production under uncertainty involves a complex set of decisions and much research remains to be done in this area. However, the model of Figure 14 together with the approaches summarized in Gershwin (1994) provide a starting point for case by case analysis and more general research.

The analysis above has largely focused on the impact of site uncertainty on supplier costs. This is motivated by the nature of the Buchhaugen deliveries, where site production drove supply-chain production. However, the typology of supplier classes and capabilities of each class are relevant for assessment of the impact of supplier uncertainty on on-site production. Suppliers of type one and three are unlikely to affect site production given the nature of their production, while suppliers of type two and four will affect site production due to problems in capacity allocation. Here, rather than assessing the impact of site uncertainty on capacity allocation, the question is turned around and the impact on supplier production due to uncertainty in demand from other projects must be assessed. Visually, in Figure 13 this means moving project bars into the shaded bar specific to the project in question; in Figure 14 this means making exogenous demand a random variable.

The analysis for each class of suppliers is summarized in Table 2, together with a graphical mnemonic for each class. Suppliers of class one can be seen chiefly as

Table 2. Suppliers types and capabilities.

Class	Supplier production costs	Supply-chain inventory	Comments
One	Little interaction with site conditions	No effect at supplier; supply chain inventory dependent on transportation arrangementsand costs	Increased use of precut and prefabrication may tend to decrease the importance of this class
Two	Greatly influenced by uncertainty requiring adjustment of scheduled production; cost of adjustment determined by distribution of capacity demand over time	Holding inventory creates a buffer between site and supplier; the important decision is not how much to hold but how long to hold it	Generally outside the unit system, although if production can be delivered as units, costs and assessment methodologies are as in class one
Three	Little interaction with site condition	No effect at supplier; supply chain inventory dependent on transportation arrangements and costs	The JIT ideal; not amenable to high levels of customization
Four	Can be reduced if order lead time is increased over traditional systems, but uncertainty increases costs	Tendency to increase due to production efficiency; can decrease if traditional systems produce and order only in large batches	Unit system has the greatest potential to reduce total costs within this class but uncertainty diminishes efficiency gains

stocks of inventory which supply the site. Costs to be assessed here are largely the costs of different logistics systems, where the costs under the unit system are likely to increase due to more deliveries (but there are possibilities to reduce this cost by sharing transport with other firms). Class two suppliers are shown as a series of machines linked together (representing long, possibly multi-stage production) supplying the site via an inventory buffer. Costs here are primarily the costs stemming from uncertainty in time of production and interaction with capacity constraints on the part of the supplier. Logistics decisions are primarily those concerning how much time to use as a buffer between site and supplier production. Class three suppliers are shown as a large machine (representing high production volume) supplying the site through a small inventory buffer. As in class one, costs are primarily logistics costs. Suppliers of class four are depicted as a machine supplying two sites via two inventory buffers. Multiple sites represent the influence of capacity demand on production costs and capabilities. Costs here are a function both of logistics costs and of production costs and are strongly influenced by uncertainty in site demand.

6 SUPPLY-CHAIN PERFORMANCE

Using the assessment of the impact of uncertainty on supplier costs given in the previous section, it is possible to construct an appraisal of overall supply-chain performance. Figure 15 shows the supply-chain structure for the Buchhaugen site, with

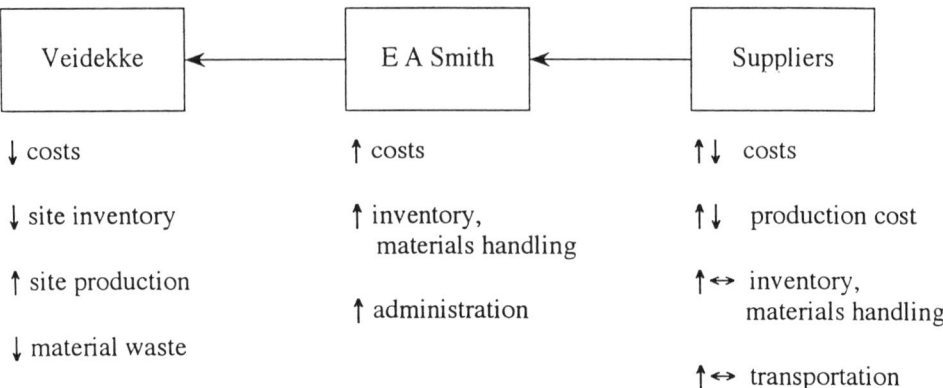

Figure 15. Costs in the supply-chain.

suppliers delivering to E.A. Smith and E.A. Smith supplying the site. Units were also delivered directly from supplier to site, bypassing E.A. Smith, but this relationship is not shown for simplicity. Total costs and relevant cost categories for each of the parties in the unit system are shown.

Clearly, total cost for Veidekke has decreased. Site production increases, materials waste on-site decreases, and site inventory decreases (with a decrease in associated carrying and handling costs). Materials wastes have decreased by 7-10% by comparison with historical averages. Labor productivity is thought to have increased, but the inaccuracy of the initial estimate and severe winter weather conditions make an accurate determination difficult. The original Danish project which served as a model for the Buchhaugen project claimed overall savings of 10% for site costs.

Costs for E.A. Smith are clearly increased in the unit system; E.A. Smith must spend more time on administration to process unit orders and to arrange efficient use of transportation capacity. Increased use of information technology may reduce these administration costs. Inventory and materials handling costs are also increased for E.A. Smith not just because of delays but also because of the need to store materials for short terms to economize on transportation costs. E.A. Smith had the ability to meet the extra demand of the unit order system with its existing level of capacity; if more sites use the order system, E.A. Smith may need to expand its warehouse and staff size. While this will increase total costs, the marginal costs per project will go down due to economies of scale. These scale economies come from efficiency of materials handling in the warehouse through specialization and from increased efficiency of transportation through a greater ability to share truck space across multiple projects. Within geographical constraints, sharing truck space among multiple suppliers is also a possibility. With this ability, inventory held by E.A. Smith will actually decrease as there is less need to hold inventory to economize on transportation cost. This economy of scale argument also justifies the maintenance of E.A. Smith as an entity separate from Veidekke. While Veidekke could assume most of the responsibilities of E.A. Smith for transportation of units to the Buchhaugen project, it could not do so for multiple projects in a region.

Supplier costs are the most difficult to assess and are dependent on the production technology employed by the supplier. For suppliers of type one and three, which have little interaction between site production and factory production, costs are dependent solely on transportation efficiency. Costs are likely to increase in the unit system due to an increased number of deliveries. At best, costs in the unit system will be equal to those in traditional systems where suppliers deliver in large batches. Suppliers with long lead times (Type 2) did not play an important role on the Buchhaugen site and comment is reserved until discussion of the extensibility of the unit system, given below. Suppliers of type four which have production decisions interacting with site demand are the most difficult case to analyze. Transportation costs are dependent on efficient utilization of truck capacity and can be assessed in same manner as for other supplier types. Production costs can decrease with the increased certainty of three weeks lead time, and it is this synergy between supplier costs and site demand which holds the most potential for savings in the supply-chain. However, uncertainty in site demand can destroy these efficiencies and increase costs. As the unit ordering system is novel, it is likely that type four supplier costs will have a tendency to decrease as other production will not necessarily have the certainty of the unit system. If a unit type system becomes more common and suppliers have greater certainty in production planning than presently, supplier costs will be more dependent on the uncertainty is site demand. In summary, costs for suppliers can go up or down; inventory costs will likely stay the same or increase for type four suppliers as they will hold more inventory to smooth production. Production costs can go up or down depending on uncertainty, while transportation costs will tend to increase due to more deliveries.

Total costs for the supply-chain can increase or decrease depending on the relative magnitudes of costs for Veidekke, E.A. Smith, and the suppliers. There does not appear to be a system which causes improvements for all parties in the supply-chain, and costs are clearly influenced by the magnitude of uncertainty in site demand. If the unit system spreads to more common usage, transportation costs will tend to decrease due to economies of scale, but supplier costs will become even more sensitive to site uncertainty. In this case Veidekke needs to take even greater steps to control and predict site production; as described in the data analysis section, this requires well estimated, highly detailed, resource loaded schedules with good feedback from the workers.

Within the context of production costs for suppliers, certainty of site production seems to be the major obstacle to application of the unit system to other sites. In this regard, sites with high repetition in production are natural sites for implementation of the unit system. However, uncertainty tends to increase with units in process, foreshadowing difficulties using the system on large sites. Extending the unit system to on-site subcontractors with a labor force not under the direct control of Veidekke also causes difficulties in information feedback as there is an extra organizational layer that information must filter through. Under the current system of contracts which penalize firms for delays on-site, it is natural for firms to be optimistic when forecasting production. This does not fit with the unit system need for accurate information. The Buchhaugen site does demonstrate that it is possible for the unit system to work alongside more traditional arrangements, and these arrangements can continue if steps are taken to isolate unit based site production from uncertainty in other systems. Finally, suppliers of type two (long lead times) are somewhat incongruous with the philosophy of the unit system. Clearly uncertainty in delivery times from type two suppliers could be very disruptive to the unit system, and an increase in the buffer time between supplier production and site demand may be required to effectively isolate the unit system from uncertainty. If this is impossible (e.g. if off-site production is on the critical path), then the unit system may not function well.

7 ORGANIZATIONAL ARRANGEMENTS IN THE SUPPLY-CHAIN

It is worthwhile to differentiate between costs for individual actors and costs for the supply-chain as a whole. Under the current fixed price contract system, costs of uncertainty in the supply-chain are not observed by Veidekke, and hence it has little incentive to improve demand certainty as long as suppliers are meeting delivery dates and stocks on-site are not increasing. The fixed price system, based on market pricing for traditional systems, does not allow exploration of potential savings in supplier costs. Thus while Veidekke could be missing some savings of the system, on the whole it is pushing risk for site uncertainty to its suppliers.

Supplier response to this risk when determining prices will determine the ultimate cost for Veidekke, however. Suppliers must recoup their costs; if the unit based systems become common suppliers will learn how to include the cost of uncertainty in their bids and Veidekke and other general contractors will pay for the price of site uncertainty. Thus one possible organizational arrangement of the supply-chain is to continue with fixed price contracts, where market prices will reflect the cost of unit delivery with an expectation of uncertainty. As these prices will not explicitly represent the cost of uncertainty, sites will have little incentive to improve uncertainty performance (in fact, they may be penalized for improved performance by spending more on control systems with no reduction in materials costs). Thus a fixed price system will lead to a greater level of uncertainty and higher aggregate costs than what may be achieved through the unit system.

The problem is to make the costs of uncertainty explicit to the actors who have control over uncertainty. This can be accomplished through giving a base price for units if there is no uncertainty and a price for units if there is uncertainty in the time of delivery. This is a form of non-linear pricing (Wilson 1993) which is used to induce more efficient behaviour on the part of consumers and producers. Let us suppose that there are identifiable costs of rush orders and that these costs stem primarily from production and transportation inefficiencies. Similarly, suppose that the primary cost of delays stem from increased inventory costs (implying that production policy is to produce and hold inventory in the case of a delay). Contracts would explicitly represent these prices, providing a base cost with no uncertainty, a cost for delay, and a cost for acceleration. With such a price schedule, Veidekke would directly observe the costs of uncertainty and have an explicit incentive to reduce uncertainty. Furthermore, Veidekke would evaluate bids based on the expected cost of the unit under uncertainty rather than on the base unit cost.

Such a non-linear price schedule could also be used to determine the costs of changes caused by suppliers that did not meet schedule; as such, the non-linear price schedule complements the traditional contractual system of assigning costs for changes to the parties responsible for those changes. Nonetheless, non-linear pricing represents a dramatic departure from traditional fixed price systems, not only in the mechanics of payment but also is the philosophy which drives the contractual system. As described above, fixed price contracts tend to push risk down the supply-chain (for which the contractor and owner may pay higher prices), while with a non-linear price schedule the general contractor explicitly assumes more of the risk in return for a lower base price. It is reasonable to question if such a non-linear price system can be maintained in a competitive construction market.

Wilson (1993) provides a comprehensive survey of non-linear pricing for firms and markets and states that non-linear pricing is feasible if four preconditions are met: 1) the seller has monopoly power; 2) resale markets are limited or absent; 3) the seller can monitor customers' purchases; and 4) the seller has desegregated demand data. These preconditions are largely determined for volume production and require some interpretation for customized, project production. Monopoly power determines the ability of a firm to set prices which are differentiated from the market. Substitution of products by competitors limits the ability of firms to set a non-linear price schedule if they trying to claim above average profits. Competition between suppliers does not prohibit an uncertainty based non-linear price schedule, but it does suggest that market discipline will cause suppliers to offer very similar schedules so none will be able to claim higher than average profits. Resale markets are not a problem for uncertainty based non-linear pricing as they do not exist for customized production. As stock production (supplier Types 1 and 3) is not influenced by demand uncertainty, non-linear pricing for these type of goods will be determined by transportation efficiency. A reseller of goods will likely be subject to the same costs of uncertainty as the supplier while having little opportunity for arbitrage by obtaining goods in volume at a lower price. Thus resale markets are not likely to cause a problem for non-linear pricing in a project environment. Nor will Wilson's third and fourth points pose problems for the implementation of uncertainty based non-linear price schedules. The ability to monitor customer purchases is necessary to assess costs and correctly bill the customer; such a monitoring capability is built into the unit order system. Desegregated demand data is used by the supplier to design a price schedule which is suited to customer demand; primarily, this data is used to set prices to increase profits in the presence of monopoly power. As noted above, an uncertainty based price schedule is different in inspiration as it is used to recoup costs rather than to obtain extra-normal profits, and the collection of desegregated demand data is a prerequisite only insofar as it is needed to assess production costs. Although subject to some interpretation for project environments, Wilson's preconditions for the applicability of non-linear pricing are met and such a system appears feasible.

An uncertainty based non-linear pricing schedule can be implemented in several forms. Given synergy in transportation by combining units, a contractual system where the general contractor holds contracts with each of the suppliers and has to somehow separate shipment costs from producer costs may be difficult to operate (in the least, it will ignore transportation savings). A better system is one following the Buchhaugen arrangement where Veidekke held a contract with E.A. Smith and E.A. Smith held contracts with the suppliers. This allows E.A. Smith to assess transportation synergy and the affect of uncertainty on these costs. E.A. Smith can then provide a nonlinear price schedule to Veidekke for each unit based on transportation costs and supplier production costs. Such a system would follow the description given above where rush orders are the primary influence on supplier production costs.

Wilson observes that although it is possible to construct very complicated non-linear price schedules, much of the benefit of such systems can be obtained with simplified structures. This is important commentary as it is easy to foresee that complicated, context dependent price schedules will be needed to adequately reflect the price of uncertainty. The price schedules described in the previous paragraph are based on simplified assumptions about the regularity of production. Put simply, a price schedule that gives the same cost of uncertainty for units no matter when they are produced does not take into account the sensitivity of production to aggregate demand, whether through synergy among orders or merely through competing demands for capacity. If these conditions are important to production cost, then the price schedule must take them into account to truly assess the cost of uncertainty. Thus rather than prefixing all prices, a contractual system would contain clear rules to determine the price of each unit given observed conditions of uncertainty and production. Such a system has the greatest potential benefit for exploring and minimizing costs in the supply-chain, but is very difficult to construction and operate. Furthermore, such a system requires a great deal of trust and joint risk sharing among project participants. When production costs are not strongly influenced by capacity and synergy considerations, or when there are a large number of units ordered and there is some expectation that costs can be averaged, Wilson's comments on ability of simple price schedules to obtain most of the benefits of non-linear pricing seem to be justified and a context sensitive pricing system would not be preferred over simpler systems.

Four alternative organizational arrangements have been proposed in this section; Table 3 summarizes the elements of each system. These organizational arrangements are based on contractual form; little attention has been given to the contractual and cultural arrangements which must accommodate these forms, although some discussion has been made as to the risk attitudes required. Thus while Wilson's preconditions for non-linear pricing have been met, much work and research is needed to

Table 3. Potential contractual forms for unit delivery.

Contractual form	Risk attitude	Comments
Fixed price	Risk shifting	No incentive for improved uncertainty performance
Non-linear, no provision for transportation synergy	Risk sharing	Incentives for improved uncertainty performance, lowered base prices with explicit representation of uncertainty costs
Non-linear, provision for transportation synergy	Risk sharing	Similar to above, but preferred as it allows transportation synergy to be accounted for and savings to be shared; also maintains a single point-of-contact between site and suppliers
Non-linear, case-by-case analysis of unit costs given prescribed rules for determination	Complete risk sharing	Most difficult to implement, but may be required to adequately capture supplier costs and minimize supply-chain costs

flesh out the details of such arrangements. Similarly, the relative benefits of each system to manage uncertainty and to reduce costs needs further investigation. If improved project controls cannot reduce uncertainty beyond a certain level, then non-linear pricing may have few benefits. However, multiple forms of pricing can be employed on a single site; the form most appropriate to supplier and site needs can be used where appropriate. Thus it is possible to foresee an arrangement whereby a single supplier and the site may have a very close relationship and assess production costs on a case by case basis, while other arrangements with suppliers are based on simpler, arm's-length contractual arrangements.

8 CONTRIBUTIONS

Beyond the general contribution of a case study investigating integrated production and inventory decisions in a construction supply-chain, this chapters makes several contributions to knowledge. First, the tradeoffs between transportation, inventory, and production costs are clearly demonstrated, showing the need for and applicability of integrated models. The potential for such systems to identify global savings has also been demonstrated (the simplest case being the implementation of a unit delivery system with the provision of extra demand certainty for type four suppliers), as well as the difficulty of finding global improvements which reduce the costs for all actors.

As a second contribution, this case study identifies the importance of uncertainty in timing on supply-chain costs and performance and develops some measures to analyze the nature and extent of this uncertainty. Closely following manufacturing theory, the need for and role of inventory in the supply-chain has been demonstrated. The link between uncertainty and inventory (and between uncertainty and production costs) in the supply-chain has been further explored through the development of a typology of four classes of supplier. Here, a critical distinction between manufacturing analysis and project specific analysis has been made by showing that supplier production decisions are sensitive to the history of production, violating traditional modelling assumptions. This opens new lines of research and allows much more specific questions to be asked about supplier production capabilities in the context of project production than was previously possible.

The firm based analytic framework, together with a typology of suppliers, provides a basis for examination of inventory, transportation, and production costs from a systems perspective and provides a context for evaluation of other supply-chain research. For example, the supply-chain research reported on by Wegelius-Lehtonen (1995) can be shown to be applicable to suppliers of class one and three which have production decisions detached from the rest of the supply-chain. This basic framework is extensible to include more types of suppliers and also provides a starting point for analysis of more complex behaviour. For example, structural steel has a long initial lead time for design and fabrication, but milling of shapes can be done on relatively short notice (while being subject to set-up synergy through combination of orders and to capacity constraints on the milling machine). Thus steel supply combines the elements of type two suppliers (long lead times) with type four suppliers, and such a combination may be used as a basis for graphical representation and analysis.

A final contribution of this research is toward development of an understanding of organizational arrangements in the supply-chain by proposing four variations in contractual structure and linking these variations to potential supply-chain performance and risk attitudes on the part of the project participants. The proposed non-linear pricing structure is novel to construction and differs from some of the assumptions detailed in Wilson (1993). In this sense, a critical differentiation between standard economic theory and project environments has been made, and research areas have been opened not just to distinguish the costs and benefits of different contractual structures on construction supply-chains, but also for the potential of non-linear pricing to induce efficiency in competitive markets.

ACKNOWLEDGEMENTS

The author wishes to thank the members of Veidekke, E.A. Smith, Langmoen, and Johs. Rasmussen for their time and input in the development of this report. Funding for this project was graciously provided by the Prosjektstyring 2000 program based at Norges Tekniske Høgskole, Trondheim, Norway. Members of the Institutt for Bygg-og Anleggsteknikk were particularly helpful and supportive of this research.

REFERENCES

Albriktsen, R.O., Bergan, R., Kjær, K.N., Larsen, K.A., Schreiner, P. & Skogstad, H.P. 1993. *Perspektivanalyse for bygg-og anlegg 1995-2005* (Report No. 32/93). ECON Senter for Økonomisk Analyse, Oslo, Norway.

Arntzen, B.C., Brown, G.G., Harrison, T.P. & Trafton, L.L. 1995. Global supply chain management at Digital Equipment Corporation. *Interfaces* 25(1): 69-93.

Askin, R.G. & Standridge, C.R. 1993. *Modelling and Analysis of Manufacturing Systems*. New York: John Wiley and Sons, Inc.

Baker, K.R., Powell, S.G. & Pyke, D.F. 1990. Buffered and unbuffered assembly systems with variable processing times. *Journal of Manufacturing and Operations Management* 3: 200-223.

Banker, R.D., Datar, S.M. & Kekre, S. 1988. Relevant costs, congestion and stochasticity in production environments. *Journal of Accounting and Economics* 10(3): 171-197.

Barrie, D.S. & Paulson, B.C. 1992. *Professional Construction Management* (3rd ed.). New York: McGraw-Hill.

Baudin, M. 1990. *Manufacturing Systems Analysis With Application to Production Scheduling*. Englewood Cliffs, New Jersey: Yourdon Press.

Bell, L.C. & Stukhart, G. 1987. Costs and benefits of materials management systems. *ASCE Journal of Construction Engineering and Management* 113(2): 222-234.

Bertelsen, S. 1993. *Byggelogistik I og II, materialstyring i byggeprosessen (Construction logistics I and II, materials-management in the construction process, in Danish)*. No. Boligministeriet, Bygge-og Boligstyrelsen (København).

Cohen, M.A. & Lee, H.L. 1988. Strategic analysis of integrated production-distribution systems: Models and methods. *Operations Research* 36(2): 216-228.

Conway, R., Maxwell, W., McClain, J.O. & Thomas, L.J. 1988. The role of work-in-process inventory in serial production lines. *Operations Research* 36(2): 229-241.

Davis, T. 1993, Summer. Effective supply chain management. *Sloan Management Review,* p. 35-46.

Gershwin, S.B. 1994. *Manufacturing Systems Engineering.* Englewood Cliffs, NJ: Prentice-Hall, Inc.

Gray, C. & Flanagan, R. 1989. *The Changing Role of Specialist and Trade Subcontractors.* Ascot: The Chartered Institute of Building.

Hinze, J. & Tracey, A. 1994. The contractor-subcontractor relationship: the subcontractor's view. *ASCE Journal of Construction Engineering and Management* 120(2): 274-287.

Howell, G., Laufer, A. & Ballard, G. 1993. Interaction between subcyles: one key to improved methods. *ASCE Journal of Construction Engineering and Management* 119(4): 714-728.

Karmarkar, U. 1989, September-October. Getting control of Just-in-Time. *Harvard Business Review,* p. 122-131.

Koskela, L. 1992. *Application of the New Production Philosophy to Construction.* (Technical Report No. 72). Centre for Integrated Facility Engineering, Stanford University.

O'Brien, W.J. 1993. Construction supply-chain management: An exploratory case study (unpubl. paper). Dept. of Civil Engineering, Stanford University.

O'Brien, W.J. & Fischer, M.A. 1993. Construction supply-chain management: A research framework. In: B.H.V. Topping & A.I. Kahn (eds), *CIVIL-COMP-'93,* Information Technology for Civil and Structural Engineers (pp. 61-64). Edinburgh, Scotland: CIVIL-COMP Press.

O'Brien, W.J., Fischer, M.A. & Jucker, J.V. 1995. Economic view of project coordination. *Construction Management and Economics, forthcoming.*

Shtub, A. 1988. The integration of CPM and material management in project management. *Construction Management and Economics* 6: 261-272.

Wegelius-Lehtonen, T. 1995. Measuring and re-engineering logistic chains in the construction industry. In: *International Federation for Information Processing Working Conference on Re-engineering the Enterprise.* Galway, Ireland:

Wilson, R.B. 1993. *Nonlinear Pricing.* New York: Oxford University Press.

Yin, R.K. 1989. *Case Study Research: Design and Methods* (Revised ed.). Newbury Park: SAGE Publications, Inc.

Zipkin, P.H. 1991, January-February. Does manufacturing need a JIT revolution? *Harvard Business Review,* p. 40-50.

APPENDIX: THIRD AND FINAL REVISION OF THE BUCHHAUGEN CONSTRUCTION SCHEDULE

Stray thoughts

Design aspects: Produce for very low lead time items. This reduces total inventory and increases flexibility.

Design units that can fill a truck, be put up easily.

Geographic location of suppliers: Collocation of suppliers who make products needed at same time.

Combine orders/suppliers problem: If every body uses this system, will it be tough to match delivery days (now can be flexible on when to deliver to traditional sites)?

Geographic location of sites: Order from suppliers who also supply sites nearby. (Do this if not there is some delivery buffery in time of day delivery; possible interaction with other sites – e.g. site 1 doesn't need it today so don't bother to deliver to site 2)

Design schedule around lead time requirements: Not just adequate lead time but structure installation sequence to reduce variation where it is costly to adjust.

Changes: First order of magnitude: Well estimated schedule, weather. 2nd oom: Productivity tracking, assignment, feedback.

The Veidekke observation of too many units corresponds with the data on changes.

Day of delivery is important for efficient use of transportation – what is the two week to one week variation in delivery day? Also in this vein, should there be a predetermined superunit to group deliveries efficiently for transport?

'The history and context of site production is relevant to decision making on the part of the factory' – this is a good tie to schedule certainty.

Natural extensions of supply system
1. Just contract with E.A. Smith and confine NL pricing arrangements to them. Let EAS assume costs and risks further down the chain.

2. Extend down all the way into supply chain. This allows case by case expense calculation. Choice between the two systems dependent on sensitivity of supplier production to other project needs, ability to average costs over a number of units as well as dependence on the stability of the mixture of other demand:

– Selection of pricing arrangements - how much uncertainty can be controlled;

– Path dependence – time to pick an org system now before it is too late;

– Pricing context – if there is a stable demand for other types of projects, shifting can be said to cause few problems in synergy and costs are primarily the setup time increase. capacity problems remain;

Incentive difficulties and price mark-ups; risk shifting under linear unit pricing. At the moment can get good prices, because of supplier unfamiliarity with the system. But is missing the possibility for even lower costs as costs are currently set by market. When the unit system or similar system is more mature, then costs will be set based on an assessment of uncertainty.

Rapid construction as a change driver in construction companies*

LAURI KOSKELA, PETRI LAURIKKA & MIKA LAUTANALA
VTT Building Technology, Espoo, Finland

ABSTRACT: In this chapter, time based management as an improvement method in a construction company is considered. The outcome of a continuous drive for construction time reduction is called rapid construction. The difference between fast tracking (where speed is a goal in itself) and rapid construction (where speed is also used as a means to achieve other objectives) is clarified. The theoretical and implementation issues of rapid construction are discussed. The underlying theory is reviewed, the relevant methods analyzed, and suitable measures presented. The impact of the economic environment on rapid construction is analyzed. Then initial experiences from a project initiated by VTT Building Technology, aiming at implementation guidelines of rapid construction, are presented.

1 INTRODUCTION

Time based management (TBM) has recently been successfully used in a variety of industries (Stalk & Hout 1989). However, there are rather few studies focusing on this area in construction (Hastak et al. 1993; Månsson 1994; Ireland 1995; Construction 1995).

It is understood that the best results of time based management would be accomplished by considering the whole building process from design to construction, as proposed in (Hastak et al. 1993; Ireland 1995). However, in this chapter, we consider time based management as an improvement method in construction companies. We define the resultant rapid construction as follows: the outcome of a continuous drive for shortening the construction time by a construction company and its partners.

Thus, the focus is on improvement action that can be launched by the construction company itself and its close allies. We justify this focus with following reasons:

– Traditional procurement methods, where construction companies do not have many possibilities to affect the design solution, are still practised, and the situation seems not to change rapidly.

– Construction companies need to improve their operations also in the case of new procurement methods.

– Success in improving internal operations may later lead to more comprehensive process improvement, initiated by construction companies.

*Presented on the 3rd workshop on lean construction, Albuquerque, 1995

– Most construction companies are small and cannot easily afford sophisticated co-operative arrangements, like partnering, concept designs, etc.

The underlying theory is reviewed, the relevant methods analyzed, and suitable measures presented. The chapter is based on a current study of the potential of time based management in construction companies, a survey on literature and related on-going research.

2   RAPID CONSTRUCTION IN COMPARISON TO OTHER CURRENT APPROACHES

2.1   *Rapid construction in comparison to fast tracking*

Fast tracking is an approach aiming at a shorter project duration, having already been practised in construction projects. In literature, fast tracking has most often been understood simply as overlapping of activities: 'initial construction activities are begun even before the facility design is finalized' (Hendrickson & Au 1989). The Project Management Body of Knowledge (1987) defines fast track as follows (similar short definitions may be found in other sources):

– *The starting or implementation of a project by overlapping activities, commonly entailing the overlapping of design and construction (manufacturing) activities.*

The study of Kwakye (1991) and the CII study (1990) are broader overviews on the problems and techniques of fast tracking. It is an accepted fact that fast tracking costs more (Kwakye 1991):

– *Fast tracking costs more because the accelerated production rate is above the optimum level of production (the level at which the marginal productivity becomes disproportionately expensive).*

The major difference between fast tracking and rapid construction[1] is the following: Fast tracking is achieved through operating above the optimum rate of production, whereas in rapid construction, the thrust is in transferring this optimum rate of production (Fig. 1). We could also formulate this as follows: Fast tracking leads to an increase of problems (as observed by Fazio et al. 1998), whereas rapid construction aims at decreasing the number of problems.

2.2   *Time based management in comparison to quality improvement*

In recent years, various quality improvement initiatives have become increasingly popular in construction companies. Because rapid construction, as an application of time based management in a construction company, can be seen as a competing approach to quality improvement, it is useful to clarify their differences.

Despite the popularity of quality improvement, its value has increasingly been questioned. Harari (1993) claims that only about one-fifth of all QI programs (across

---

[1] In a recent Construction Industry Institute study (1995), the terms schedule compression (corresponding to fast tracking) and schedule reduction (corresponding to rapid construction) are used in a similar comparison.

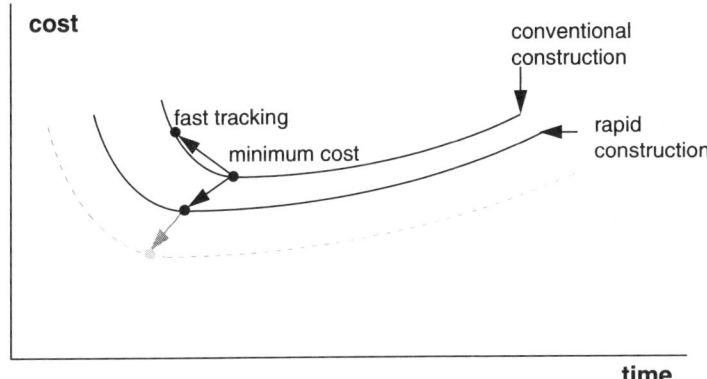

Figure 1. In rapid construction, the aim is to transfer incrementally the time-cost curve, whereas in fast tracking, a point on the existing curve is selected.

industries) achieve tangible results in terms of quality, productivity or financial improvements. One common explanation to this is that this failure is due to a lack of focus to results in quality programs (Atwater & Chakravorty 1995). Another, related explanation forwarded is that quality programs have primarily been internally focused, rather than customer focused (Howe et al. 1992).

Similar critical discussion is starting also in construction. It is argued that the ISO certificate does not guarantee the benefits of quality improvement (Hansen 1994), and that there are fundamental shortcomings in the International Standard ISO 9001 from the building industry viewpoint (Couwenbergh 1994).

In our view, TQM and other quality improvement approaches are partially overlapping with time based management. Quality improvement basically addresses defects and failures to fulfil customer requirements. There are proven methodologies and tools for quality improvement.

In contrary, time based management addresses all non-value adding activities as a source of improvement. The methodology of TBM is less structured. The strong point of TBM is that its features, like clear objectives, immediate monitoring and feedback, are effective vehicles for energizing a behaviour change. On the other side, it is a limitation of TBM that customer requirements, others than time-related, are not addressed.

Thus, quality improvement and time based management are best seen as complementary: TQM efforts can be enhanced through TBM features and vice versa. Especially, we assume that the enhanced focus and change energizing capability of TBM are crucial.

3 THEORETICAL BACKGROUND

3.1 *General theoretical background*

Time based management has been extensively used in manufacturing and other sectors since the 1980's (Stalk & Hout 1989). Although an explicit theory for time based management is lacking (Bartezzaghi et al. 1994), it is commonly agreed that the elimination of non-value adding time components (waste) in the processes in question is the primary rationale for time reduction. Also, shorter cycles make it also possible to make improvements more quickly. Time is seen as more useful a metric than cost and quality because it can be used to drive improvements in both. In addition to this forced elimination of wastes, time reduction provides a faster delivery time to the customer, a simplified management of production and other benefits (Schmenner 1988). A strong positive correlation between speed and efficiency in different manufacturing industries has also been proven statistically (Holmström 1995).

In manufacturing, the central concern is the lead time (the time from when a customer order is taken until it is shipped). There are a number of practical strategies for lead time reduction (Hopp & al. 1990), including elimination of work-in-progress, smoothing the work flow and elimination of variability.

3.2 *Theoretical background in construction*

The situation in construction differs in many ways from the typical situation in manufacturing. The production in construction is assembly-type, where different material flows are connected to the end product. Considering a simple example, the total construction time can be described as follows:

$$T = b(n - 1) + t$$

in which T = total time; n = number of tasks; b = time buffer between the start of consecutive tasks; t = duration of a task.

This is graphically illustrated in Figure 2. Note, that in practice tasks have different durations and the time buffers between tasks vary. The first task is usually planning and site preparation (not usually shown in schedules), the second one often excavation, and the next ones ordinary construction tasks. However, the principles discussed next are valid also in the more complex, practical situations.

Thus, there are three major ways of reducing the construction time (Fig. 3):

1. Increasing the speed of tasks. This can naturally be achieved through increased labor and more effective machinery. However, after the optimal rhythm has been reached, the costs will rise. A more interesting alternative is to eliminate non-value adding subactivities from each task. As it is well known, error correction, multiple

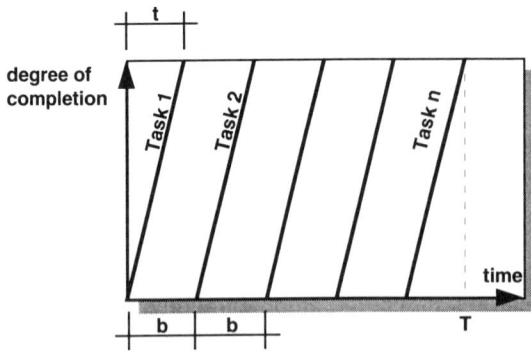

Figure 2. Formation of construction duration.

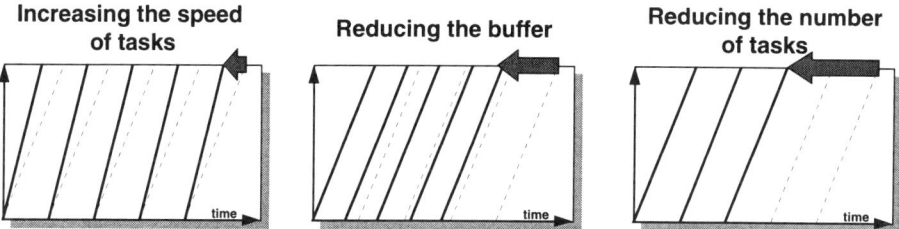

Figure 3. Theoretical alternatives for construction time reduction.

handling, waiting etc. make up a large part of the working time, and provide thus a good potential for improvement;

2. Reducing the buffer between consecutive tasks (increasing the overlapping of tasks). The buffers are planned between tasks in order to prevent waiting due to variability in task speed. Thus, reduction of variability is needed (Kaplinski 1993). Variability is caused by human factors, environmental impacts and disturbances flowing from earlier phases of the project. More detailed planning and monitoring is required. The reduction of disturbances necessitates long term activities in the construction company;

3. Reducing the number of tasks. The first possibility is the transfer of tasks off-site (off the critical path), to upstream phases of the respective material (or information) flow. This means increased precutting, pre-assembly and pre-fabrication: the possibilities vary from precut gypsum boards to prefabricated bathrooms' elements. A second possibility is provided by alternative, more constructable design solutions. Utilization of multi-skilled work gangs also decreases the number of distinct tasks.

In practice, the construction time is often also shortened by reducing the planning time; however, this might lead to a prolongement of the total construction time.

It has to be noted that there are generic problems encountered in construction time reduction:

– The increased level of subcontracting also means that a growing part of the site activities are not in effective control by the construction company;

– The one-of-a-kind character of construction means that disturbances, variations etc. are common; the sensitivity to disturbances will easily increase in time compression.

Like in manufacturing, there are evidently several benefits for a construction company from rapid construction:

– Added competitivity in relation to clients with time pressures;

– Enhanced cost and quality performance;

– Added scheduling flexibility: Even in projects with predetermined (by the client) normal schedule it becomes possible to choose the starting date more freely, depending on resource availability. Added time for design and planning might often be beneficial;

– Added capacity: With shortened project durations, the project throughput capacity of the company is increased.

4 IMPLEMENTATION ISSUES

4.1 *General considerations*

Experience in manufacturing suggests that there are four key issues that are required in the implementation of time reduction (Koskela 1992):

– Management commitment: Leadership is needed to realize a fundamental shift of philosophy, with the goal of improving every activity in the organization;

– Focus on measurable and actionable improvement, rather than just on developing capabilities;

– Involvement: Employee involvement happens naturally, when organizational hierarchies are dismantled, and the new organization is formed with self-directed teams, responsible for control and improvement of their process;

– Learning: Implementation requires a substantial amount of learning.

In fact, time based management is often explicitly used to achieve corporate change, where the corporate culture and individual behaviours are changed (Hart & Berger 1994). Time as an easily understood measure and target provides a catalyst suitable for these purposes.

In our view, these issues are equally valid in construction. However, regarding corporate renewal and mental changes, construction clearly differs from typical manufacturing settings. Interdependent tasks, a large number of participants, a unique coalition of project team members and time pressure all contribute to a need of stable institutionalized behaviour by each participant (Kadefors 1995). Thus, there is a strong industry culture, in addition to corporate culture, which controls behaviour. The industry culture is, of course, more difficult to change by the action of one particular company, as noted in (Ekstedt & Wirdenius 1994).

4.2 *Measurements*

Like in other industries, the conventional measures of construction, which most often focus on cost or productivity, fail to make waste visible and to stimulate continuous improvement. There are a number of characteristics to be required from a new set of measures (Maskell 1991):

– They must be directly related to improvement goals (and thus strategy for improving production);

– They primarily use nonfinancial measures;

– They are simple and easy to use;

– They provide fast feedback;

– They foster improvement (not just monitor it).

In our view, the major issues to be measured are related to waste (non value adding activities), value (for the customer) and improvement efforts. It is possible to find measures, which reflect indirectly or partially these issues (Table 1).

4.3 *Measurement of construction duration*

In the context of rapid construction, the continuous measurement of construction durations is a primary task. For this, a graphical tool as presented in Figure 4 can be

Table 1. Measurements for rapid construction.

Measures on waste	Measures on value	Measures on improvement efforts
Construction duration	Customer satisfaction	Training of employees
Total work hours	Defects at hand-over	Site order index
Conformance to schedule	Amount of warranty costs	Number of improvement initia-
Accident frequency	Project delivery accuracy	tives per employee
Defects found under construction		Number of co-development rela-
Disturbances under construction		tions with subcontractors and
Accuracy of material deliveries		material suppliers
Amount of (physical) waste to be		
disposed from site		

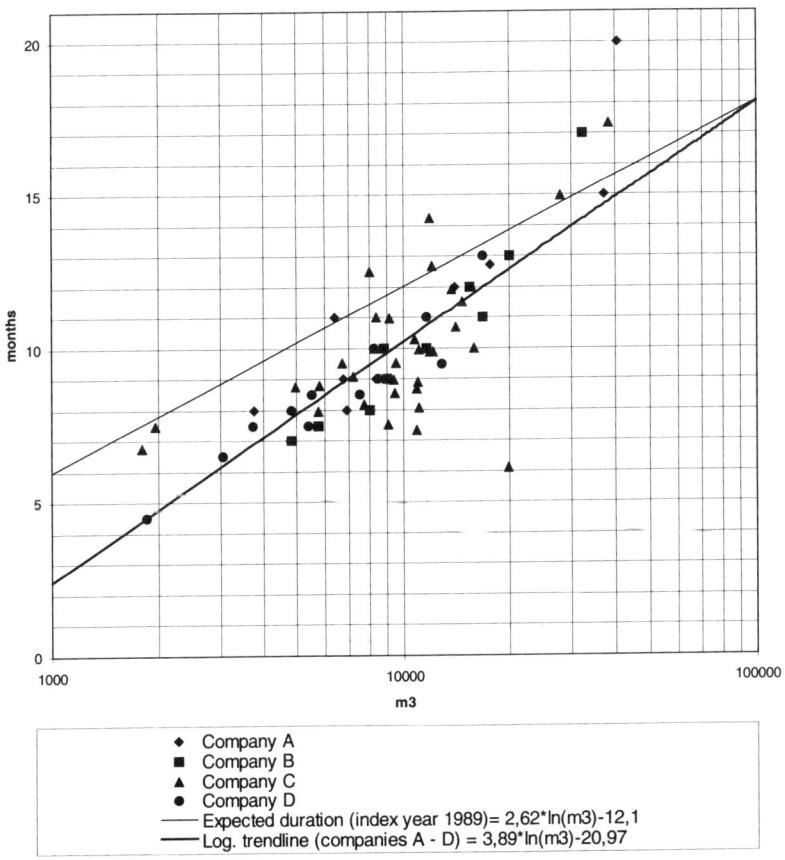

Figure 4. An example of the use of the construction duration analysis tool. It was used in VTT's rapid construction project to benchmark the site performance of the participating companies. Analysis shows that construction durations have improved about 20 % within the last five years.

utilized. The tool is developed to compare realized construction durations with expected, statistical construction times of apartment buildings. The formula used is based on empirical data (Nousiainen 1992) and uses only one variable, building volume (m^3). Similar tools that compare time-cost performance or time-floor area performance are developed also by Kumaraswamy (1995). NEDO (1988) developed a similar, simple tool for construction time estimation.

Statistical tools as presented above, have two ways of using. They can be used to evaluate the current plan or realized construction time of a single project against the statistically expected construction time. Unfortunately, in most cases, the formula appears to be too rough. It should consider more project characteristics and thus have more variables. Another field in which statistical tools can be used is in measuring continuous improvement at the company level. In Figure 4, for example, a trendline shows the current level of improvement.

5   IMPACT OF ECONOMIC ENVIRONMENT TO CONSTRUCTION DURATION

Preconditions for rapid construction depend on both the development intensity of construction companies and the business environment of the whole industry. Changes in the business environment, such as economic trends, new laws and regulations, etc., may effect on construction durations and set limitations for applicability of different rapid construction methods.

5.1 *Industrial agreements, legislation and other governmental control*

Legislation and industrial agreements have either direct or indirect impinge on construction durations and, especially, on the methodologies that are applied in rapid construction. Such controlled and regulated matters are:
  – Working hours, (overtime and shift work restrictions);
  – The use of unskilled, temporary workers;
  – Bad weather payment practice etc.
Also taxation practice has indirect effects on construction duration. For example, in Finland, the change in the VAT practice improved competitiveness of prefabricated products and thus got ground for shorter site times.

5.2 *Impact of business cycles to construction durations*

Economic environmental complexity and business cycles have an effect on construction durations (Walker 1995). Figure 5 shows changes in construction durations of one stairway apartment buildings in Finland between 1983 and 1993. It shows that during the boom construction durations become longer. It can also be seen that there is no remarkable long time improvement in durations. This is probably a consequence of the fact that clients and developers in housing do not usually have any major interests to speed up construction. Instead, in industrial and commercial building, the improvement might be more evident.

Business cycles impact upon supply of resources. During the boom many factors

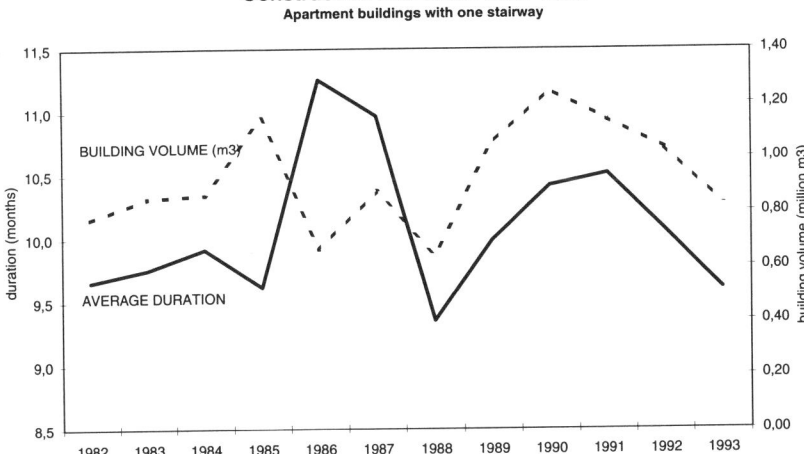

Figure 5. Business cycles have an effect on construction durations.

hinder the effective production process and stretch construction times. Such factors are, for example:

– Long material delivery times;
– Shortage of skilled workers;
– Stoppages (strikes and lockouts; in Fig. 5, the effect of heavy industrial disputes during 1986 can be noticed).

On contrary, during the recession, construction times are shortened by good availability of resources and skilled labor.

5.3 *Rapid construction and business cycles*

Especially during the boom clients favour short construction times. Therefore, in such a situation, rapid construction gives clear competitive advantage for construction companies to get the job. But to implement rapid construction during the boom is difficult.

During the recession the attitude of construction companies towards rapid construction changes. Now it aims at cost effective construction. In other words, it is used to improve the production performance to lower the production costs. On the other hand, during the recession it is also easier to implement rapid construction in practice.

6 HOW TO START?

How to get time reduction started is certainly a serious problem in construction companies. For approaching this, an evaluation of the present status of improvement activities in the four participating construction companies was made in VTT Building

'Technology's project on rapid construction. An evaluation matrix was prepared, where the advancement in critical dimensions is described. The companies were evaluated in structured interviews and through site visits.

The average results from this evaluation are shown in Figure 6. Improvement has focused primarily on preplanning of production, production control, management and organization as well as quality. Product development and production process development are areas that are least attended. There are natural reasons to this: subcontracting, the traditional procurement method and the attention to cost, propelled by a deep recession.

Although all companies were active in quality development (with one certification and more planned), there were little measurable results visible from these activities. The average construction duration has decreased app. 20 % in the last five years, but this seems primarily to be caused by the recession (the volume of construction has decreased by more than half since 1990).

An evaluation of the present situation in regard to the four important implementation issues (mentioned in Chapter 4) was also carried out:

– Management commitment: Only in one company was managerial commitment to time reduction discernible;

– Measurable and actionable improvement: Although there were, to varying degree, measurements in use, they were not actively used for targeting and monitoring;

– Involvement: Especially the involvement of workers was on low level;

– Learning: Company training is primarily directed to quality and production control. Training for workers is scarce.

Thus, to sum up, it seems that the potential of time based management is largely untapped in construction companies. On the other hand, the TQM activities already introduced provide a good foundation for TBM.

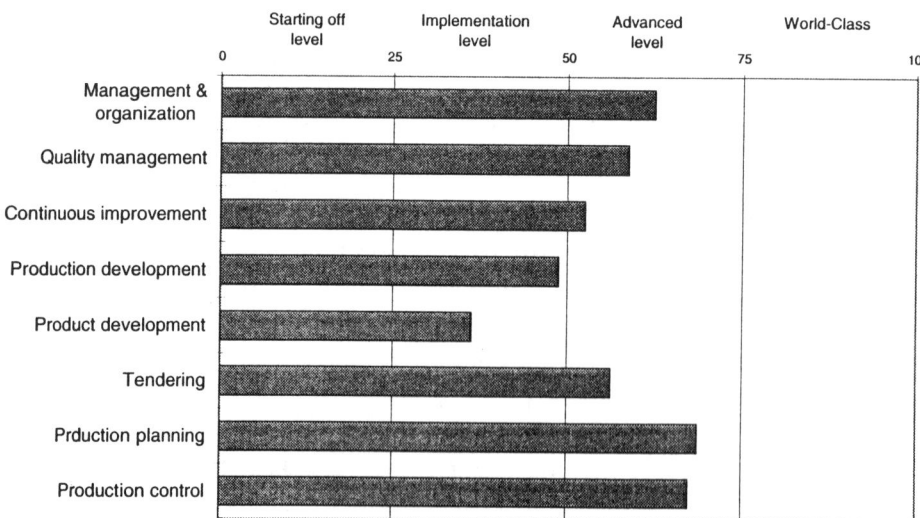

Figure 6. Analysis of the present status: average evaluated values.

In the light of these findings, following action is suggested for launching time based management in these companies:

1. Management should commit to time reduction and actively set targets for it;

2. Application of new measures, both time related and others, would provide a basis for target-setting and monitoring;

3. Work processes where time related decisions are made (especially tendering and production planning) should be systematized;

4. Methods and tools that promote involvement and change should be introduced (example: programme for disturbance reduction, where the whole company personnel participate).

After a successful take-off, further progress may be achieved through focusing on:

– Promotion of training and other learning;

– Product development (concept buildings, standardized details);

– Process development (supply chain development, improvement of site activities).

This suggested action programme is planned to be experimented in the participating companies.

7 CONCLUSIONS

Time based management seems to be an awaited addition to the armour of improvement methodologies in construction. It is seen to complement the now popular quality approaches in three respects:

1. Time related measures, especially construction duration, provides a powerful global measure for evaluating and monitoring the overall performance;

2. Its features, like clear objectives and immediate monitoring and feedback, are instrumental in energizing a behaviour change;

3. New improvement potential is opened: types of waste and slack unattended by, say, quality methodologies, are focused on.

However, there is still very little experiential knowledge from practical implementation of time reduction in construction. Methodologically, pioneering companies are rather much on their own. Nevertheless, if they succeed, the competitive benefits might be considerable.

Rapid construction provides for several interesting research themes. From strictly research point of view, the effects of time reduction in construction are still largely unproved. Construction time reduction highlights unsolved problems and poorly understood issues, like:

– Achieving cultural and behaviour change in construction;

– Designing project organizations and procurement forms that promote, rather than hinder, improvement.

REFERENCES

Atwater, J.B. & Chakravorty, S.S. 1995. Using the theory of constraints to guide the implementation of quality improvement projects in manufacturing operations. *Int. J. Prod. Res.* 33(6): 1737-1760.

Bartezzaghi, E., Spina, G. & Verganti, R. 1994. Lead-time Models of Business Processes. *International Journal of Operations & Production Management* 14(5): 5-20.

Concepts and Methods of Schedule Compression CII, SD55. 752 p.

Construction Industry Institute. 1995. Schedule Reduction. Publication 41-1, 33 p.

Construction Industry Institute 1988. Concepts and Methods of Schedule Compression. Publication 6-7, 28 p.

Couwenbergh, J.C.H. 1994. What is missing in the International Standard ISO 9001 regarding to the Building Industry. In: Odd Sjoholt (ed.), *Quality Management in Building and Construction.* Norwegian Building Research Institute. Pp. 508-513.

Ekstedt, E. & Wirdenius, H. 1994. Enterprise Renewal Efforts and Receiver Competence: The ABB T50 and the Skanska 3T Cases Compared. *Paper presented at the IRNOP Conference in Lycksele, Sweden,* March 22 - 25, 1994. 18 p.

Faster building for commerce. Nedo. 1988

Fazio, P., Moselhi, O., Théberge, P. & Revay, S. 1988. Design Impact of Construction Fast-Track. *Construction Management and Economics* 5: 195-208.

Hansen, R. 1994. Requirements on Quality Systems within the European Community – Certification and/or continuously improved ability? In: Odd Sjoholt (ed.), *Quality Management in Building and Construction.* Norwegian Building Research Institute. pp 481-486.

Harari, O. 1993. Ten reasons why TQM doesn't work. *Management Review,* January, 33-38.

Hart, H. & Berger, A. 1994. Using Time to Generate Corporate Renewal. International. *Journal of Operations & Production Management* 14(3): 24-45.

Hastak, M, Vanegas, J. & Puyana-Camargo, M. 1993. Time-Based Competition: Competitive Advantage Tool for A/E/C Firms. *Journal of Construction Engineering and Management* 119(4): 785-800.

Hendrickson, C. & Au, T. 1989. *Project Management for Construction: Fundamental Concepts for Owners, Engineers, Architects, and Builders.* Prentice-Hall, Englewood Cliffs. 537 p.

Hiolmström, J. 1995. Realizing the Productivity Potential of Speed. *Acta Polytechnica Scandinavica, Mathematics and Computing in Engineering Series* No. 73. Helsinki. 55 p.

Hopp, W.C., Spearman, M.L. & Woodruff, D.L. 1990. Practical Strategies for Lead Time Reduction. *Manufacturing Review* 3(2): 78-84.

Howe, R., Gaeddert, D. & Howe, M. 1992. *Quality on Trial.* West, St. Paul, MN. 155 p.

Ireland, V. 1995. The T40 Project: Process Re-engineering in Construction. *Australian Project Manager* 14(5): 31-37.

Kadefors, A. 1995. Institutions in building projects: implications for flexibility. *Scandinavian Journal of Management* 11(3).

Kaplinski, O. 1993. Diminishing non-uniformity of construction processes. *Construction Management and Economics* 11: 53-61.

Koskela, L. 1992. Application of the New Production Philosophy to Construction. Stanford University. Centre for Integrated Facility Engineering. *Technical Report* #72.

Kumaraswamy, M. & Chan, W. 1995. Determinants of construction duration. *Construction Management and Economics* 13: 209-217.

Kwakye, A. 1991. *Fast Track Construction.* The Chartered Institute of Building, Occasional Paper No. 46, 36 p.

Månsson, O. 1994. Time based management applied in Skanska, a construction company in Sweden. pp. 344-348.

Maskell, B. 1991. *Performance Measurement for World Class Manufacturing.* Productivity Press, Cambridge. 408 p.

Nousiainen, A. 1992. Shortening of Construction Period. Helsinki University of Technology. Faculty of Civil Engineering and Surveying. Master's thesis. 101 p. (in Finnish).

Project Management Institute 1987. Project Management Body of Knowledge of the Project Management Institute. Drexell Hill.

Schmenner, R.1988. The Merit of Making Things Fast. *Sloan Management Review,* pp. 11-17.

Shapira, A., Laufer, A. & Goren, I.N. a. Cutting Housing Construction Time: The Israeli Experience 1990-92. National Building Research Institute, Technion. Mimeo. 23 p.

Stalk, G. & Hout, T. 1989. Competing Against Time. Free Press, NY.

Walker, D. 1995. An investigation into construction time performance. *Construction Management and Economics* 13, 263-274.

A process approach to design for construction*

MIKA LAUTANALA
VTT Building Technology, Espoo, Finland

ABSTRACT: Production and construction requirements are often missed or misunderstood in design stage due to the separation of design and construction. This causes waste: the product specification is far from easiest possible to be build and it does not consider the possibilities which suppliers' capability offers. Traditionally construction requirement consideration is based on designers' personal experience from construction.

The chapter discusses requirements for consideration of production requirements in design phases. Then a process to catch construction requirements into the product specification is proposed. The process integrates company and project level. It implements principles of concurrent engineering and supports continuous improvement. Finally future research needs are discussed.

1 INTRODUCTION

1.1 *State of the art*

Production and construction requirements are often missed or misunderstood in design stage due to the separation of design and construction. This causes waste: the product specification is far from easiest possible to be build, and it does not take notice of the possibilities suppliers' capabilities offer. Traditionally constructability requirement consideration is based on designers' personal experience from construction.

Manufacturing industries have been actively developing design for X (manufacturing, assembly, cost, life-cycle,...) and concurrent engineering methodologies. Drastic improvements in productivity achieved through design for manufacturing and assembly (DFMA) have been reported. With some products assembly time has been reduced even up to 90% mainly through reducing the number of parts, simultaneously product development cycle has shortened and manufacturing cost decreased (Youssef 1994; Huthwaite 1990; Kirkland 1988).

Separated design and construction have possibly been the main reason for ignorance of constructability and design for construction research in Finland, but also incentives for R&D have been missing. Not until recently design-build approach has

*Presented on the 3rd workshop on lean construction, Albuquerque, 1995

237

become more common. We have national de-facto product standards to manage construction requirements, but these do not meet current needs within the industry. There are not formal models how to apply project and producer specific construction needs into the product specification.

In the manufacturing industry the main reason for design for manufacture and assembly is that up to 70-85% of a product's cost is already committed through decisions made during the product design stage, even though only 5% of the total development costs have been expended (Huthwaite 1990; Ong 1995).

1.2 *Benefits of constructability*

A survey performed by Russell et al. (1992a) report constructability programs to result in 5.7% savings in project cost. They have also analysed four case studies, which implied savings from 1.1 to 10.7% of total project cost (Russell et al. 1992b).

Also quality defect costs caused by design give us a low estimate for constructability cost. Josephson (1994) have surveyed various quality defect studies, which report quality defect cost to range from 0.4 to 25.9% of total construction cost. Out of this cost design and planning cause 14-78%.

Pesonen (1995) has analysed five completed projects and found their quality error costs to range from 1.8 to 3.7% of total construction cost. In these cases design and planning phases have caused 48-62.9% of the quality costs, resulting in 0.92-2.3% of total construction cost. These figures, however, include only rework caused by design, and poor efficiency in production has been ignored.

Following case study suggests 50% savings in production labour cost to be obtainable in precast concrete production through improved constructability. If there would not have been benchmarking of three units, and only one unit would have been analysed, the estimated saving potential would not have been this high. Analysing just one unit would have underestimated the potential benefits.

1.3 *Case study: Precast concrete delivery*

VTT was in 1993 involved in a study where three precast concrete units from three different countries were compared. The focus in the study was to compare design and project management and how these contribute to the performance of production. The major finding was that the more production and engineering were integrated the better was productivity both in production and in engineering. This suggests that improving constructability makes also the engineering more effective. This is possible using product standardisation: In the unit with best productivity, engineering and production have together developed product standards, which meet the production requirements. Simultaneously engineering receives templates for their work. It needs to be articulated that product standardisation does not mean standardisation from customer's point of view.

The engineering man hour expenditure in the best case was 1/4 of the highest one! All the credit cannot be given to the integration of engineering and production, nor can the persons be blamed. Cultural differences in organising construction project had significant impact on the effectiveness of engineering. Also the product was diff-

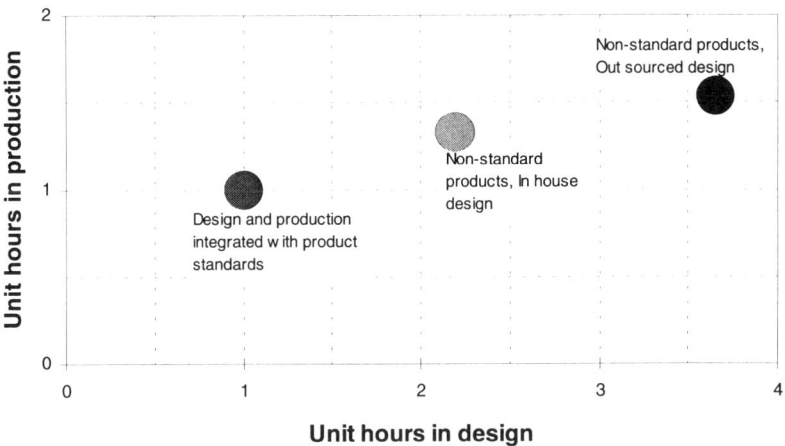

Figure 1. Comparison of the effectiveness of production and engineering (h/m^3, beams and columns) in three precast concrete units. The scale has been normalised.

erent from engineering point of view, even if for the customer there were not any difference.

1.4 *Concepts*

Construction industry Institute defines constructability as 'the optimum use of construction knowledge and experience in planning, design, procurements, and field operations to achieve overall project objectives' (Russell et al. 1992).

One of the most referred definitions of buildability is the one presented by CIRIA: 'The extent to which the design of the building facilitates ease of construction, subject to the overall requirements for the completed building' (Griffith 1986; Adams 1989; Hyde 1995).

Moore & Tunnicliffe (1994) criticise that CIRIAs definition can be seen to impose constraints on design creativity (and quality of the building) and this hinders it from being generally accepted. They propose a definition to buildability: 'That design philosophy which recognises and addresses the problems of the assembly process in achieving the construction of the designed product, both safely and without resort to standardisation or project level simplification'.

I would like to add a lean dimension to the definition. First I understand missing constructability as a waste; improving constructability is reducing all the unnecessary that is not required in the construction process to obtain the quality customer is asking for. The unnecessary can be rework, activities which do not add value to the customer or poor productivity. The second point is to relate the constructability to the quality of the building and customer satisfaction: Ease of construction which is achieved at the expense of reducing customer satisfaction, is not improving constructability. This is concluded in: 'DFC is a design philosophy which aims at reducing all the unnecessary in construction process which does not contribute to the quality of the building'.

More pragmatically constructability can be seen as a performance measure on what has been designed in respect with the contractors' resources and capabilities, and simultaneously taking into consideration customer needs.

Constructability is combination of:
– Manufacturability;
– Prefabricability;
– Transportability;
– Assemblability.

All the above remarks are well addressed in the definition of design for manufacturing by Youssef (1994): 'A design philosophy that promotes collective and integrated efforts of a number of teams involved in planning, organising, directing and controlling all activities related to products and processes from idea generation to a finished product or service such that:
– Available design, manufacturing and information technologies are efficiently utilised;
– Teamwork is emphasised;
– Redundancies and non-value-added activities are eliminated;
– Enterprise integration is promoted; and
– Customer requirement and quality are built in the design'.

1.5 *The problem*

There are two main problems to be solved. First we need to know what makes a building, or building product, easy to build (manufacture and assemble) and secondly how to design the product accordingly. Thus research has to develop tools and methods to identify which makes some alternative constructable and tools to support designers to design accordingly.

The chapter suggests a process oriented approach to DFC to manage the problem. First requirements on DFC process are surveyed and then a DFC process is depicted formally using IDEF0. Needs for future research are derived from the process model.

2 REQUIREMENTS FOR DESIGN FOR CONSTRUCTION

2.1 *Support team work*

Basically, the point in constructability is the co-operation of design and production and how the design team can keep production requirements in view. We can develop various tools to help designers to prepare constructable specifications, but still the team work approach is most efficient as Hovmark & Norell (1994) have stated. They have identified four different levels for application of design tools: guidelines, analysis of product features, product reviewing and team building. They conclude that design tools become more effective in the order given.

2.2 *Continuous improvement and organisational learning*

Continuous improvement is necessary to maintain competitiveness and to keep the company on the survival path. Design for construction system should also contribute to this.

It is enormous task to collect all relevant data on constructability. Currently this is managed by personal experience over a long period of time. People are running projects and learning by trial and error. Typically same mistakes are repeated by different persons. Know-how is not shared systematically within the organisation and organisational learning is thus hindered.

To provide organisational learning and continuous improvement five main components are required. First, we need to set goals. Second, we need to measure the performance to know where we are and are we improving and achieving our goals. Third, the results of the measurements need to be fed back and fourth, action has to be taken based on the measurement and feedback. And finally, gained experience need to be disseminated and shared among the organisation.

2.3 *Improve total performance of the project*

A construction project is composed of numerous deliveries which are often treated independently. As though the performance of each would not be dependent of the others. Still, the customer will choose the contractor based on the total performance of the delivery (price, quality, delivery time, etc.), and the total performance will (or should) decide, who will get the next project.

Major challenge in a DFC system is to support continuous improvement and learning over company borders and thereby to improve the performance of the delivery chain as a whole.

2.4 *Emphasise initial stages of design and engineering*

Emphasise initial stages of process, i.e. conceptual planning and early involvement of construction. The earlier construction is involved the better chances to impact the result of the design it has. Possible changes to the design are much more cheaper to do at the early phases of the design. Changes in the later phases will often lead to extensive rework in design.

A solution to support robustness of design process is modular product design. If the design can be divided into independent tasks, each task can proceed parallelly and changes in one task do not reflect into another task. Inevitably this requires tools to manage design process and it also sets requirements on product itself.

2.5 *Document history*

During a project, or from project to project, the environment may change. Hence, a decision which has been good once may be improper in the next situation. To be able to use experience data effectively and to be able to make project evaluation objectively, historical record of decisions made (what, why, who and when) need to be maintained. Not only decisions, but also the facts decisions are based on, should be documented. This supports also continuous improvement.

2.6 *Probabilistic approach*

Constructability, like all natural systems, has statistical nature. There is always variation, which furthermore causes risks. DFC should have probabilistic approach to be able to reduce variation and help in managing risks. Causal models would not able to cover this.

2.7 *Flexibility*

The environment where companies are operating is changing continuously. Accordingly the system needs to be able to adapt to continuous change.

The environment for a project is different from project to another. Competitiveness or goodness of different solutions vary along with the state of the business: price of labour compared with prefabrication varies, delivery time of some specific product changes etc. Solution, which is good today or in this project, may not be acceptable tomorrow or in the next project. System needs to be flexibly adaptable to new conditions. This requirement restricts us to rely just on checklists which list acceptable or non-satisfactory solutions.

2.8 *Reactiveness*

The faster feedback is the more effective it is. If the time between decision (action) and feedback is long, a lot of other decisions based on the fault decision have been made before the corrective action has been taken. Design is proceeding based on fault decisions and corrective action will lead to extensive redesign.

On the other hand, the faster feedback is the better it hits the problem. If feedback is very slow, it is difficult to identify the real reason for the problem. For example, if constructability problems are audited after the project has been completed, it may be difficult to trace the reasons from the early phases of the design, and latest design phases are rather been blamed. Reasons for problems become invisible.

Thus we can conclude that:
– Feedback from site is needed already during construction, not just after the completition (which would be too late for the very project);
– Design needs fast feedback right after each major decisions. Support should rather be proactive than reactive. Reactive DFC system leads rather to redesign than right-at-the-first-time design.

2.9 *Transparency in design*

The main target or measure, which is used in our development of DFC system, is to reduce construction time. Rzevski (1993) among others emphasise that, to reduce lead time, information should be easily available for all participants in a project. This is also necessary to understand relationships between different subsystems and to be able to focus on the total performance improvement.

3 DESIGN FOR CONSTRUCTION PROCESS

A framework for design for construction has been created and described using IDEF0 process modelling technique. The basic notation of IDEF0 is depicted in Figure 2. The model aims at meeting the requirements listed above. Implementation issues have not yet been considered.

Design for construction aims at improving both constructability of a project and constructability of products (Fig. 3). Improving constructability of a project also contributes to the product development.

The DFC process model is described detailed in two levels: Company (strategic, Fig. 4) and project (operational, Fig. 6) levels. The company level deals with setting constructability metrics, analysing project realisation data and taking required action.

Selected measures should depict the constructability of the product in respect with the available process, and help to focus required action. A measure should not either be too laborious to use. Realisation data (experiences from previous projects) will be analysed statistically helping to define new measures. Realisation data is thus depicted as an input.

The amount of different measures needs to be reasonable; there should be only handful of measures. Too numerous measures would lead to 'bureaucracy' and energy will be consumed on measuring while it should be spent on acting.

Constructability goals and metrics should support corporate strategy and the strat-

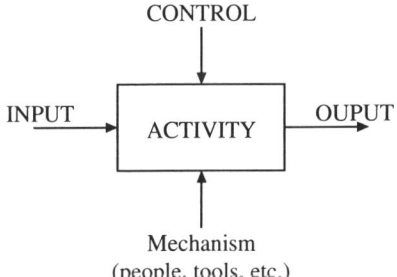

Figure 2. IDEF0 notation.

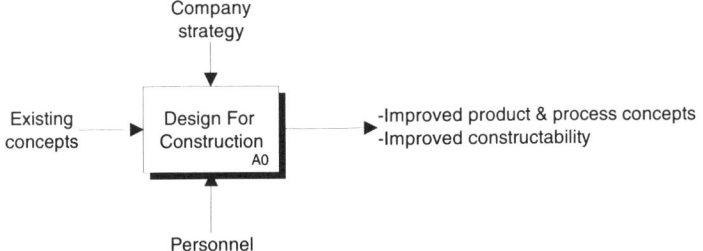

Figure 3. The design for construction process model.

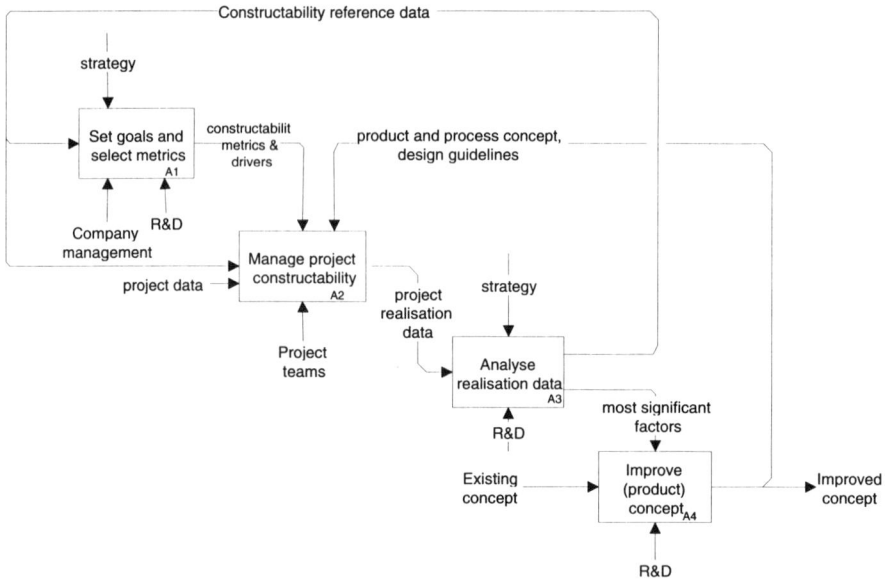

Figure 4. Node A0 Improve constructability. Design for construction-process on company level.

egy is thus a control to setting goals and metrics in terms of IDEF0 notation. The company level has strategic nature.

'Analysing realisation data' activity collects experience from past projects and formulates the data in a form where it will be reusable. A major task is also to identify which are the most significant factors impacting constructability, and as a result to control product and process development.

Company level DFC addresses the requirement of continuous improvement and organisation learning. It provides the company with four main components of organisation learning: set goals, measure, act and share. Continuously evolving product concepts are used for sharing (see Fig. 5). Also feedback is partially supported via product concepts, but also direct feedback is required at the project level.

The project level of the DFC process describes the constructability related information flow between the main activities of a construction process. The process starts with design (and engineering), which has user requirements as an input, and company level product (and process) concepts and design guidelines as an initial control. The purpose of the guidelines should not be to lead the architect and engineers to select the optimal design. Rather, these guidelines should enable the designers to be aware of the impact different alternatives have on construction (incurred substantial cost, caused problems, etc.).

Simultaneously with the design starts construction planning, although the intensity is lower in the beginning. Planning defines requirements and constraints on the product based on overall schedule and available resources. If the output of design is called 'product model', the output of planning represents 'process model'.

Constructability of the product is analysed by evaluating the compatibility of 'product model' and 'process model' and comparing them with the reference data.

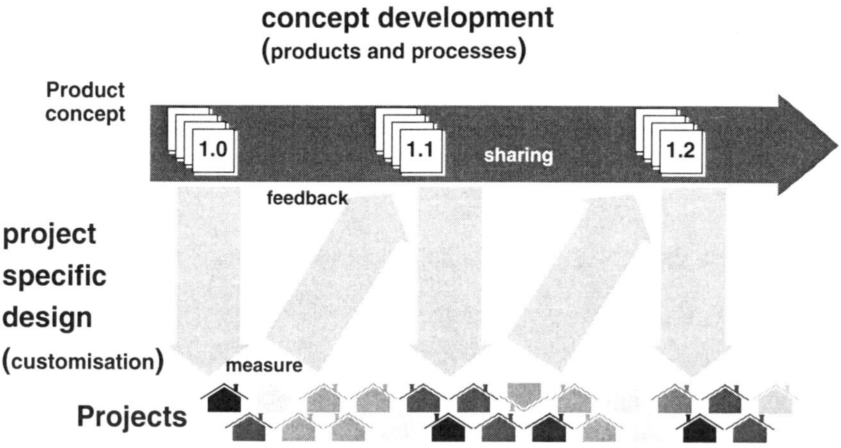

Figure 5. Continuous improvement and organisational learning through continuously evolving product and process concept. Modified from (Nihtilä 1993).

more detailed. Still the main benefit is expected from the early phases of the project and the main effort should be put on the initial stages of the project.

An important task within the constructability analysis is to record the history of the decisions: why a certain alternative have been chosen, etc. This is necessary for distinguishing project specific constraints. Still the system should not become bureaucratic.

Constructability analysis is the point where team approach occurs in the DFC process model. Planning provides the analysis with construction expertise and knowledge. Additionally company level constructability metrics are controlling the constructability analysis.

The purpose of DFC system is to improve interaction between design and construction. Planning is the interface between these activities. In the model constructability analysis is intentionally placed after the planning activity to emphasise that the purpose is to evaluate product specification in respect with available processes and that planning should be involved from the early phases of the project. Still I want to emphasise that IDEF0 charts are not representing a schedule. The activities are overlapping and they are iterative.

Various DFC tools for different needs are applied during the constructability analysis:

– Tools to evaluate constructability of a product;

– Tools to support co-operation of design and construction and to enhance team work. Simple tools which help team members to focus themselves on the product and process instead of persons, would be most useful in this purpose. One example of this kind of tool is quality function deployment (QFD);

– Tools to outline production requirements;

– Tools to measure constructability of a realised project (collect feedback);

– Tools to identify constructability factors.

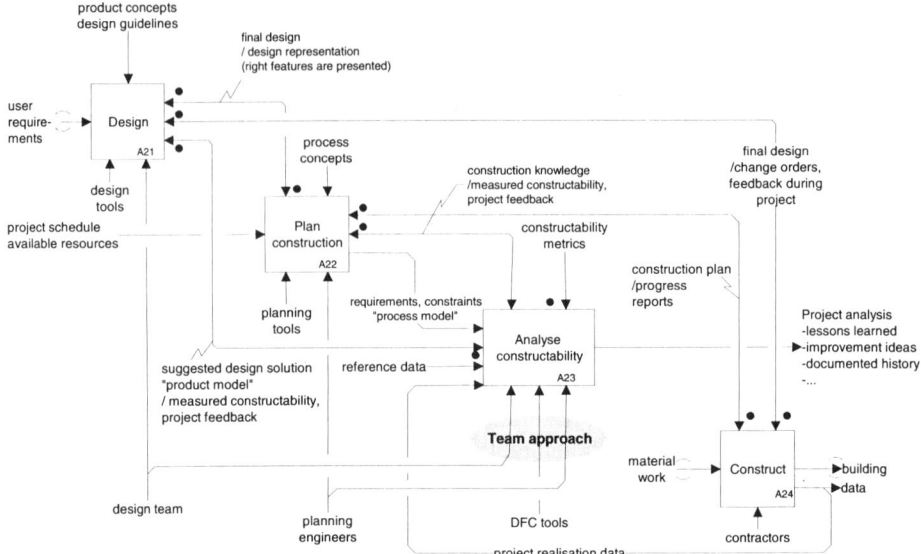

Figure 6. Node A2 Manage project constructability: Design for Construction-process on project level.

In constructability analysis-activity also project feedback from construction during and after the project is processed. On the other hand there needs to be direct and rapid feedback from construction to design and to other phases of the project prior to construction.

4 DISCUSSION

The presented design for construction process is a theoretical model, which is created to study the information flow that is needed to manage constructability on company and project levels.

The defined process is generic. It can be implemented in various kinds of organisations and the process could be focusing on many other aspects instead of constructability with minor or no changes on the process. Replacing the word 'constructability' with 'requirements' would give us a process model for design for requirements.

The model is also used to pinpoint needs for future research. To further strengthen this purpose the activity 'Analyse constructability' on the project level should be decomposed.

How to represent reference data
The presented model is heavily relying on reference data and team approach. Research needs to address the question, *how to represent reference data* so that it is efficiently reusable from project to project. If the data is unstructured and informal, it

will be practically impossible to use in the long run. Probabilistic approach is an essential element in reducing variation and furthermore reducing risk level.

The basis for the data should be the selected constructability metrics or any other data which is recorded for performance measurement. From research standpoint study on *generic constructability factors* would be the initial task to create structure for the data. Fisher (1991) has studied constructability factors of reinforced concrete structures. He concludes that the factors can be formalised and be represented in a knowledge base.

Manufacturing industry has been able to identify the generic factors affecting assemblability: number of parts, handling and insertion. For each of these more detailed guidelines exist (Boothroyd & Dewhurst 1991).

Modelling manufacturing information

As mentioned earlier, constructability is a performance measure on what has been designed in respect with the contractor capabilities. Analysing constructability will mean comparison of the product specification and contractor profile. There is also a need to represent the contractor (producer) profile systematically as well as constructability factors. A simple example of this is, what is the available crane capacity (moment capacity), which will have impact on the maximum weight of components depending on the layout of the building.

Tools

A DFC system will materialise in a set of tools which are applied in the construction process. A handful of tools which are easy to implement and use are necessary. First we need tools to help in building the reference data; tools to identify constructability factors. Other possible tools would be:
– Tools to evaluate constructability;
– Team work tools;
– Tools to support concurrent work of design, engineering and planning;
– Existing tools like QFD, DSM, SPC, Taguchi, etc.

In the first place the tools should encourage design construction collaboration and help persons to focus their discussion on the subject, the product, rather than on persons. The substance of the tool is secondary, though important.

Benefits

The described process offers a structured and systematic view on design for construction, which helps in increasing understanding on DFC. The process model can serve as a map where the components of DFC system and their interfaces can be defined and studied both in research and while building a corporate DFC system. Currently the model is not very detailed and it needs to be further defined for a specific purpose in industrial use. A company specific DFC process can also be used as part of a quality manual.

REFERENCES

Adams, S. 1989. *Practical Buildability*. London: Butterworth. 122 p.

Boothroyd, G. & Dewhurst, P. 1991. *Product Design for Assembly*. Wakefield, RI: Boothroyd Dewhurst Inc. 136 p.

Fisher, M.A. 1991. Constructability Input to Preliminary Design of Reinforced Concrete Structures. Stanford: Stanford University, Center for Integrated Facility Engineering, *Technical report* no 64. 104 p.

Griffith, A. 1986. Concept of Buildability. IABSE Workshop. Zuerich, 1986. Zuerich, in-house publishing. IABSE report 53. pp. 105-116.

Hovmark, S. & Norell, M. 1994. The GAPT Model: Four Approaches to the Applications of Design Tools. *Journal of Engineering Design* 5(3): 241-252.

Huthwaite, B. 1990. Checklist for DFM. *Machine Design* 62(3): 163-167.

Hyde, R. 1995. Buildability as a design concept for architects: a case study of laboratory buildings. *Engineering, Construction and Architectural Management* 2(1): 45-56.

Josephson, P.E. 1994. Causes of defects in building. Göteborg: Chalmers tekniska högskola. Report 40. 186 p. (in Swedish).

Kirkland, C. 1988. Meet two architects of design-integrated manufacturing. Plastics world. December 1988.

Moore, D.R. & Tunnicliffe, A. 1994. Development of an Automated Design Aid (ADA) for improved Buildability and Accelerated Learning. In: Chamberlain, D.A. (ed.), *Automation and Robotics in Construction XI*. Amsterdam: Elsevier Science B.V. pp. 163-170.

Nihtilä, J. 1993. Simultaneous Engineering in Project Oriented Production. Helsinki: Metalliteollisuuden Keskusliitto MET. 84 p. (Technical Report 14/93).

Ong, N.S. 1995. Manufacturing cost estimation for PCB assembly: An activity-based approach. *Int. J. Production Economics* 38 (1995) 2-3, pp. 159-172.

Pesonen, J. 1995. The profitableness of quality control in constructing projects. Tampere: Tampere University of Technology, master's thesis. 106 p. (in Finnish).

Russell, J.S. Radtke, M.W. & Gugel, J.G. 1992a. Project-level model and approaches to implement constructability. Austin, Texas: Construction Industry Institute. 224 p. (Source Document 82).

Russell, J.S. Gugel, J.G. & Radtke, M.W. 1992b. Benefits and costs of constructability: four case studies. Austin, Texas: Construction Industry Institute. 125 p. (Source Document 83).

Rzevski, G. 1993. The Integrated product development process: Issues and methods. In: Roozenburg, N.F.M., *Proceedings of ICED 93*, Volume 1. Zürich: Edition Heurista. pp. 493-498.

Tichem, M. 1993. Design for Manufacturing and Assembly; A Closed Loop Approach. In: Roozenburg, N.F.M., *Proceedings of ICED 93*, Volume 2. Zürich: Edition Heurista. pp. 1033-1040.

Youssef, M.A. 1994. Design for Manufacturability and Time-toMarket. Part 1: Theoretical foundations. *Int. Jrn of Operations & Production Management* 14(12): 6-21.

Continuous improvement in construction management and technologies: A practical case*

HERNÁN DE SOLMINIHAC T. & ROBERTO BASCUÑAN
School of Engineering, Catholic University of Chile, Santiago, Chile

LUIS GERMAN EDWARDS
ENACO Las Vertientes, Santiago, Chile

ABSTRACT: One of the most important factors within a continuos improvement policy is the incorporation of new technologies. The best way to increase productivity and optimize resources is by an improvement in training activities, in construction methods, administrative methods, and in the incorporation of new materials. 'Los Benedictinos' project, built by the construction company 'Enaco Las Vertientes' in Santiago Chile, is a real example in which an improvement is attained due to the company concern and the joint work with 'Dictuc' the consultant branch of the 'Catholic University of Chile'. The implementation of Dictuc recommendations together with the company's own innovations gives important results both in quality as in productivity. As an example of this the different methods adopted by the company in order to solve the problems found along the works, together with the impact of the new implementations such as mobile warehouses, concrete plant, changes in the organization policy and, the most important one, a stimulus for team working are shown. The productivity increase and the complaints decrease as a result of the new policy leads to the conclusion that initiatives like the ones adopted in the example must continue to be implemented in order to break the construction conservatism currently found in Chile.

1 INTRODUCTION

1.1 *General considerations*

The concern about improvement and progress of the Chilean construction industry, that have a great potential for growth, is an urgent matter. According to the Chilean Chamber of Construction, the investment will grow progressively until it reaches in 1999 almost six billion dollars. On the other side, the national construction industry faces very important challenges among which we can mention market competitivity, opening of the international markets, projects complexity, increased demands on quality and pressure to reduce the projects work schedules. The current strategies of the construction sector regarding technical projects, materials and methods are based on low cost and plentiful human resources. With growth of this activity, reaching almost full employment, we can appreciate less availability of qualified hand labor so

*Presented on the 2nd workshop on lean construction, Santiago, 1994

we must look for formulas to use it in the best possible way, this means that we must build more with the same resources. Due to the great construction industry current and future development, and due to the fact that the products generated by this industry directly affect the society performance and development, at the same time that are intensely used by the society members, we must look for new technologies in the design area, materials and construction methods, and also in personnel management.

It can be shown, with some examples, the true feasibility of improvement in the construction area. This is the case of productivity, which reaches quite lower indexes compared with the ones reached by other productive sectors in the country. Data given by the 'Productivity and Management Service (SPG)' of the 'Catholic University of Chile' indicate that related to the construction activity, an average of 1/3 of the journey is not made use of. On the other hand, if we study the building construction methods from 20 years back we notice that both, some construction systems and some materials used are almost the same as the ones currently in use. Lastly it can be said that most of the personnel management systems carried out nowadays in the country's construction industry are the result of experience in works management. Furthermore, this experience does not remain in the company, but only in the workers, and in view of the high hand labor rotation existent in this sector, a great loss of experience is produced.

This national overview has carried some construction companies to take true conscience on the topic, like it is the case of the 'ENACO Las Vertientes' company which, in the construction project 'Los Benedictinos', has demonstrated a real interest in developing better construction systems, especially in the quality and productivity areas.

The project 'Los Benedictinos' consists on a houses group for the middle high class located in 'Los Domínicos' sector at 'Las Condes' commune, in lands property of the company. The houses have between 135 m^2 and 140 m^2 built in sites between 350 to 500 m^2, at the present time its prices goes from US$ 148,000 to US$ 187,000 according to their size, location, and site. This project was begun in 1988 towards the end of the year and consider the construction of approximately 1200 housings, of which 750 have already being delivered to their owners, at a rhythm of 150 annual houses. This project will be developed in several stages, undertaking a term of around 12 years in total. Due to the magnitude and duration of the work, there was a concern in the company for improving the productivity, quality and security. The service of 'Dictuc' through the 'Productivity and Management Service' of the 'Catholic University of Chile' was contracted, among others, in November 1992. The interest of this work is to analyze this particular case giving a description of the different implementations through time and the respective aftermath in the works productivity, quality and cost.

1.2 *Objectives*

The main objective of this paper is to show, by means of the project 'Los Benedictinos' the importance of a constant concern in improving the current construction systems. We expect that it be useful as an example for others constructing companies, and thus achieving the optimization of the existing resources, bringing benefits for the companies and contributing to the country development.

1.3 *General methodology*

This analysis has been planned in four stages. The first is a description of the different construction methods that have been in use from the beginning of the work. The second stage is of diagnosis, where the actions that helps to overcome the problems of efficiency and quality are listed. The third stage will be the evaluation of construction work evolution by means of productivity, quality and costs curves. Lastly, the fulfilment of objectives and their respective conclusions will be analyzed.

The methodology employed in all these stages considered three studies carried out by the 'Catholic University of Chile'. The first of them, dated January 1993, is called 'study on the Operations and Construction Systems of the Work Los Benedictinos', and it describes the problems and recommendations for the work in its second construction stage in a detailed way. A second study, dated March 1993 and called 'Study on Motivation and Management of Personnel', shows the results of a survey to 70 company's workers and to 17 workers of subcontract activities. Lastly, in June of the same year, is carried out a study for the improvement of the concrete system named 'Cost study on concrete systems for the work Los Benedictinos'.

It will be also considered in the methodology, the reports delivered to Enaco Company by the SPG which renders its service to the company from November 1992. These monthly reports shows data related with the productivity of each activity and the results are compared with the national pattern.

Lastly, it is important to mention that in the elaboration of this paper we relied on the support and participation of Enaco Company professionals, whom delivered indispensable information for the development of this report. These data rendered by the construction company, plus the visits to the job, completed the necessary information needed to know the evolution of the work Los Benedictinos project.

2 WORK DESCRIPTION

The construction work 'Los Benedictinos' was divided in three stages within its history. The selection of these stages was based on important changes in the construction methods. These changes are:
   a) From confined masonry to reinforced masonry houses; and then
   b) From reinforced masonry to reinforced concrete houses.

2.1 *Confined masonry houses*

This method was the first to be implemented and it basically consisted in the construction of houses with hand made bricks which were structurally supported by reinforced concrete pillars and beams. This method did not added any novelty respect other houses built with the same construction system.

2.2 *Reinforced masonry houses*

The construction of traditional houses is changed to reinforced masonry, which do not need to be structurally supported by reinforced concrete beams since the reinforcing steel goes through the bricks. These bricks are hollow and factory made. These bricks with reinforcing steel inside are filled with the same mortar used to bond them together, all of which makes the wall structurally resistant. Another important change in the construction process was the decision of isolating the houses not only at the roof, but also at exterior vertical parameters. A system of thermal panel or house peripheral lining was also designed, forms by a 10 mm thick sheet of expanded polystyrene and a 15 mm thick plaster-cardboard sheet.

This thermal panel come to replace the interior stucco and for this reason the overcharge did not have much incidence. At the same time this solution had much acceptance among the buyers in spite of the disadvantage regarding the fact that they had solid walls but they would have inconvenient with the pictures fixation system or with the hollow sound produced by a knock.

The causes for the changes mentioned above basically were:
– Improve the quality of the offered product;
– Due to environmental reasons the offer of craft bricks diminished in Santiago and as the project needed a constant supply of this material, they opted for the pressed brick alternative with a much more sure supply;
– For the same scarcity of craft bricks reason, their cost increased to the level where the pressed bricks were a real alternative;
– Additionally, one of the work expensive activity, due to its complexity and slowness, was to form and lay the concrete pillars. For these reasons it was opted to eliminate it with the new method.

2.3 *Reinforced concrete houses*

By the middle of 1992, within the construction boom, a scarcity of pressed bricks occurred obligating to paralyze the work for approximately 2 month with the consequent damage. Additionally the increase in the materials and labor cost caused the need to look for construction solutions with more stable costs and that were not affected by possible scarcity problems. It was decided to design and construct houses with reinforced concrete structure with the consequent increase in quality.

With the implementation of this third method (Figs 3 and 4) the quantity of concrete used diminished in relation with the masonry construction. This fact could seem strange, but it is explained because the masonry houses need bigger foundations when compared with concrete houses. According to the company data they expect a saving, as a result of this new system, around 8% of the building structure. This new method is also used to industrialize a great portion of the building structure, that is the case of the forms, which were subcontracted with Doka, a system of great acceptance in Europe.

This industrialization brings hidden benefits, since besides increasing productivity, it facilitate an industrialization of the finishing activities, due mainly to the fact that certain elements can be brought prefabricated to the exact dimensions without the necessity of adjust them at the job. As an example of this, we could list the aluminium windows frames, doors, windows, etc. which generally differ, product of errors in the building structure, with the dimensions in the drawings. Due to the lack of experience in Chile with this type of forms, at the beginning there were some implementation and training problems, but they are being solved little by little and by now the prospective productivity has already been reached. All these systems must work with cranes and the company has in operation 6 cranes of 3 different types.

Currently the houses are being constructed with a peak technology for our country, the following are its most relevant characteristics:
 – 12 cm thickness reinforced concrete walls;
 – Reinforcing steel of the walls made of a double mesh Acma type;
 – 11 cm thickness reinforced concrete slabs;
 – Prefabricated roof structure with Gang Nail System;
 – Peripheral walls thermal isolation made of 40 mm expanded polystyrene and 15 mm 'volcanita'.

This modern system has been of great utility in the work, not only related to isolation, but it also simplify many houses construction activities, specially related to service tubing since now they go superimposed on the concrete wall and hidden by the thermal panel.

3 DIAGNOSIS

At this section we will try to give a small description of the work in all their stages, finding the problems that have been encountered, the actions carried out and all the innovations made by the company with the purpose of improvement.

This concern, on the side of the company, for the quality improvement of the houses has resulted in a great prestige earning in little time. This can be verified by the fact that more of the 50% of the sold houses, had been acquired by recommendations of the current owners, showing that way the good results of the system.

3.1 *Problems*

There had been several technical problems that the company had to face, but the most ordinary and relevant are the following two: shrinkage cracks, and condensation.

Shrinkage cracks
The confined masonry system use craft bricks, which without doubt produces a greater water absorption and because of it, shrinkage cracks in the walls. In order to eliminate this problem several solutions were tried like the use of stucco with a mixture of plaster and sand or the improvement of the mortar mix, but non of these solutions succeeded in eliminating the problem. This fact made the company responsible to the owners and had to paint and fix the cracks two years after the purchase. In the second stage, that of reinforced masonry, the problem improved enough. The machine made pressed brick absorbed less water than craftmade and therefore the shrinkage diminished considerably, but was not eliminated. Only by the middle of this stage of construction the problem was solved upon eliminating the interior stucco and replacing it by thermal panels. This modern system is based on covering the inside of the peripheral walls with a sheet 15 cm thick of plaster-cardboard in the case of houses of masonry and with 4 cm of polystyrene and a sheet of 'volcanita' afterwards in the concrete made houses, this due to the fact that concrete is a bad

thermal isolating material. Besides the great improvement in the acoustic and thermal isolation as a result of this system, the stucco activity and the costly shrinkage cracks repairs were eliminated obtaining with this new system a more economical alternative.

Condensation

Another big problem that has been had along the construction of the work is condensation, which, in spite of being a design problem, the company tried to solve in order to diminish its effect. In order to have condensation three simultaneous factors must be present, exterior low temperature, interior humidity and bad thermal isolation. If successful in diminishing whatever of these factors condensation will not be produced. With the low temperatures in this sector nothing can be done. To diminish the interior humidity the catalytic stoves, that produce humidity upon burning a high percentage of air, were changed for balanced throw stoves. On the other hand, as was mentioned, upon the change to a system with thermal panels the thermal isolation is increased. With these two factors the condensation in the interior of the houses was reduced with the consequent increase in the residents life quality, which is reflected in a decrease in the users complaints.

3.2 *Some implementations*

In the studies made by the 'Catholic University of Chile' a series of recommendations that would increase productivity and quality of the work were put forward. Some of these recommendations will be listed here, plus the company innovations that were really implemented with success.

Mobile warehouses

According to studies of the SPG 15% of the labor were busy in material transportation, i.e. $9 million (chilean pesos) of the payroll were spend in this item. On the other hand, a wooden warehouse, of 2,4 m width, 6 m long and 2,7 m high, mounted on a metallic chassis with wheels, cost around $1,6 million (chilean pesos) plus tax. The implementation of two of these warehouses reduced by a half the materials transportation producing a monthly saving of $4,5 million, which added to the warehouse keeper salary, paid for the investment in the first month and produced an important saving from the second month on. The location and the supplies rely on the advance state of the surrounding houses.

Bathroom

They put 6 chemical bath in strategic places in order to reduce the distance to these facility and to improve the worker's motivation. According to the reports of the SPG the percentage of labor occupied in journeys to the bathroom was 0.72% of the total. The implantation of these baths reduced at least to the half the journeys to the bathroom with which a considerable saving for this concept is had, fact that more than compensates the cost, since besides producing a saving, the company is proportioning a confort benefit to the workers, who before this had to travel long distances in order to satisfy their necessities.

Concrete plant

In a very detailed and complete study prepared by the 'Department of Engineering of the Catholic University of Chile', it was arrived to the conclusion of the necessity of implementing a concrete plant. The alternatives given in this study shows savings over 14%, where it has on the other hand a better controlled quality and as a result a more even quality related with the mobile concrete mixer used in the work. This concrete plant implementation was justified in full in accordance with the necessity of quality and productivity improvement in the concrete works.

Quality organization

In the diagnosis study a not very clear and definite policy on quality management was detected. For the work administration the prime objective was the schedule fulfilment leaving the quality required by the project as an element apparently secondary. Since the importance of this matter, the management defined and documented clearly their policy and objectives related to quality. In this way a quality management system was implemented redefining the work administration prime objective, which to day is focused in obtaining the project required quality given by the cost, time and use of the established resources. Also an organization devoted to solve all the existing problems in the houses already delivered was created. The user, in the moment he receive the house, fill a form with the different details and observations he could appreciate and the company take action to solve them.

Training

One of the surprises found by the survey made to the workers was their interest in training. This to the point that they were willing hours to attend to courses dictated by the company after the work hours. A series of in the job courses were dictated in 1993, highlighting the ones about human relationships were special attention was given to he importance of each person work. These courses were attended by supervisors and workers alike and had so much success that it was dictated four times.

Another important aspect that was approached that year was the one concerning to the illiteracy of a group of workers. After 6 weeks courses, 14 workers learned to read and write, staying in classes until after 8:30 p.m. 2 days a week. There is no need to say the attitude of them toward the company after discovering that reading and writing opened a new world for them. Lastly, but not less important, was a 7 month course attended by approximately 25 supervisor from different 'Enaco' projects about the topic 'Supervision techniques in quality and productivity'. This course was dictated by 'Procal' in the 'Extension Centre' of the 'Catholic University of Chile'. These supervisor learned about the basic tools to carry on with these projects and, with time, they have transformed themselves in leaders of their people, which is fundamental in these processes.

Revising the survey the company become aware that near 70% of their workers did not had completed basic education. To improve this a system was planned in which a group of workers prepared themselves to give exams valid to the Ministry of Education by the end of 1994. Another plan is a program of 14 quarterly courses that are being dictated through 'INACAP'. These courses have approached technical aspects of the construction and have reached much acceptance among the workers.

The future plans are directed toward a more cultural instead of technical education. For this they are preparing personal developing programs based on human quality. In other words, general culture programs approaching topics of international news, theatre, classical music, etc.

Communication between workers

In the above mentioned study, a degree of insatisfaction within the organization was also detected, aroused by communication and trust problems between the superior management and other levels. Insatisfaction due to the little information that the middle levels had about the decisions taken by the higher levels was also manifested. This is related to a low team work spirit. Another mentioned aspect was that of the lack of communication between field workers and administration, particularly concerning with an orderly outlining of field work necessities. This carries to a high pressure by the field managers to have their problems solved quickly (operation based on crisis). With the purpose of improving this situation more effective communication channels were created through minutes or short memos on decisions or actions that affect other people work within the organization and that them should know about. Periodic work meetings that include personnel related with the principal functions within the organization (acquisitions and warehouses, administration, field specialties, etc.) are carried out. Concerning the general interest information it was put on news boards in service accessible for the personnel in sectors of greater circulation. In this aspect a mailbox was also implemented in which the personnel deposit their opinions on the problems that concern them, and they generally give the solution also. As an example of this we can mention the chemical bathrooms, the same ones that in spite of increasing the productivity, were found lacking in hygiene by the users and give the placement of semi mobile bathrooms as a solution, with more sound sanitary facilities. This is already implemented. It is important to highlight that all the letters deposited in the mailbox, anonymous or not, are answered and placed in the news board as soon a possible.

Additionally, the job manager invites for breakfast in the offices a group of 7 workers every week with the only purpose of getting to know each other more and that way they are kept informed directly of the future plans. In these reunions the conversation topic is free. Until now the topics seldom has been related directly with the work at hand and these conversations has resulted very effective for the communications improvement between workers and management. Following the same rule, the head field professional has lunch periodically with the supervisors getting to know more about the work problems. In these lunches the work plans or job program are made and at the same time a more close communication with the people that is call to be job leaders is kept. Lastly the company develops, within the job organization, the team work concept, which for its great importance will be described in greater detail.

3.3 *Team work*

Promoting a team work mentality is probably the topic of greater importance among the implementations above mentioned. Apart from being an excellent personnel training way, it has a direct effect in improving the companionship, the work atmosphere and productivity. Team organization and making their work efficient is a complicated task and is mainly based in the fact that the people that works directly in the processes is the more able to solve the problems or to improve the efficiency of these processes. Next some of the measures that has been implemented in this direction since the beginning of the new policy will be mentioned.

Courses
At the beginning of 1993 the company makes all the workers participate, from supervisors down, in continuous improvement training courses dictated by 'Procal'. The results of these courses are in plain view at the job, the improvement spirit of the workers is highly beneficial. When asked about the job done is commonplace to hear answers like: 'It still can be done better', which is a surprise coming from a sector that usually is not identified with the company.

Filming
This new system, that began to operate by the middle of 1993, consist in having a person film the different work teams. The team gathers once a week to see the film, making themselves a review of their work performance. To these sessions outsiders that could criticize the workers do not attend. The idea is that they can see for themselves their own work. This system had have a double benefit in productivity, on one hand upon revising their work method they can improve upon detecting their errors. On the other hand, when they are working when the filming is taken place they feel under pressure and work better and faster.

Work methods
The work team methods, that are part of a series of continuous improvement implementations, are based in this case in two aspects. The first is the weekly report that all the work teams must present related to several problems and their respective solutions. On the other hand monthly meetings are carried out which are attended by a guide team composed by the field manager, the administrative manager and all the professionals in charge. In these meetings all the reports coming from the different work teams are read and a decision is taken on what problem must be dealt with first. The system tends to the personnel solving their own problems avoiding the loss of time of the administrative heads in solving these problems and on the other hand the personnel is satisfied because they feel themselves as being part of the implementations. When this methods was implemented was remarkable the seriousness with which the personnel took to this method, delivering perfectly composed typed reports, and were the contributions, in respect to problem detection and their solutions, were a great help.

Keeping up to date technologically
In the team work concept is was decided as very important to have a coordination between the different 'Enaco' projects at professional level. Currently there are 12 field professionals that meets for lunch twice a month to coordinate their needs. These meetings are held in the different works taking advantage of the visit to the work in order to interchange know-how and keeping up to date with the different projects. Occasionally other relevant persons from outside the organization are also invited to give technological lectures similar to the ones given by the Chilean Cham-

ber of Construction. It can be concluded that the company consider of first necessity to be as up to date as possible in the new technologies and this is achieved mainly listening to those that knows more about the specific topics.

Results reached up to now
The results of this new system were quite immediate since the different work teams propositions were very favourable. That is the case of the warehouse keepers team, who presented a budget, with exact figures for a new metallic warehouse, that would replace the 8 wooden warehouses. The advantages of this new warehouse, of which the management had not realize, were so evident that it construction was carried out. An other example that can be mentioned is that of the crane team that presented a system for night illumination. Their position was not that they did not wanted to work at night but a solution to be able to see the load in a surer form to lift it, this again reflect what brings the new team work mentality. Another example is that of the finishing and subcontracts team which made a mesh of finishing activities, from the aluminium placement to the final tide up. This mesh ended up as some king of Gantt Chart that was of great help because for the first time the critical finishing activities were determinated. Lastly, teams that had made other contributions can be listed, creating new scaffolds and concrete vessel systems which have facilitated different tasks.

Another important point that has been achieved is that the workmen have understood the concept of organization internal client, key factor for continuous improvement development. As an example of this concept it can be said that currently the concrete slab team knows the importance of the electric tubing, or simple things like the aggregate base team knowing the importance of delivering a well levelled ground to the concrete placement team that comes later.

4 EVALUATION

At this stage it will be shown the conclusions of the previously described implementations. The index variations are better represented in the productivity, cost and quality curves. That way the productivity can be appreciated according to the different construction methods (Fig. 1), relating the quantity of houses built every month versus the month from the work beginning.

4.1 *Productivity*

The labor productivity data is taken from the SPG from 'Dictuc' of the 'Catholic University of Chile', which had information from the end of 1992 on, date from which this services were contracted. The labor divided in three working categories can be seen in Figure 2:

– Productive Work (PW): Is the time a worker spends in productive actions of any of the construction units;

– Contributory Work (CW): Is the time devoted to the necessary support tasks to carry on with the productive actions. This category has been divided into that are: transportation, cleaning, training, measuring, others;

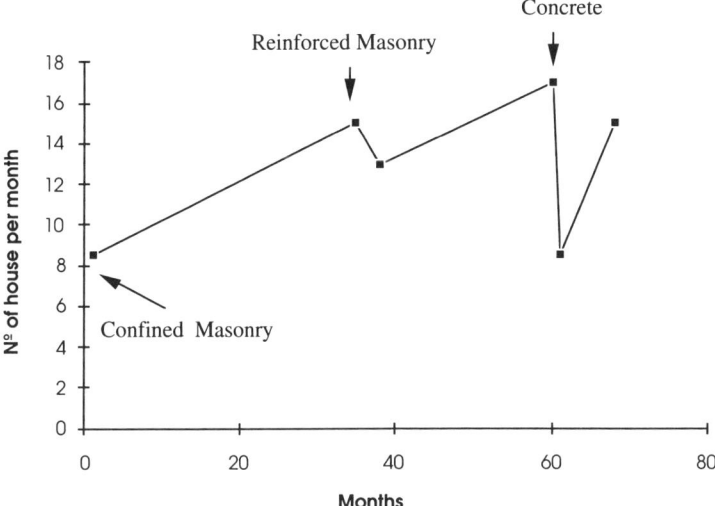

Figure 1. Number of houses per month as a function of the construction method.

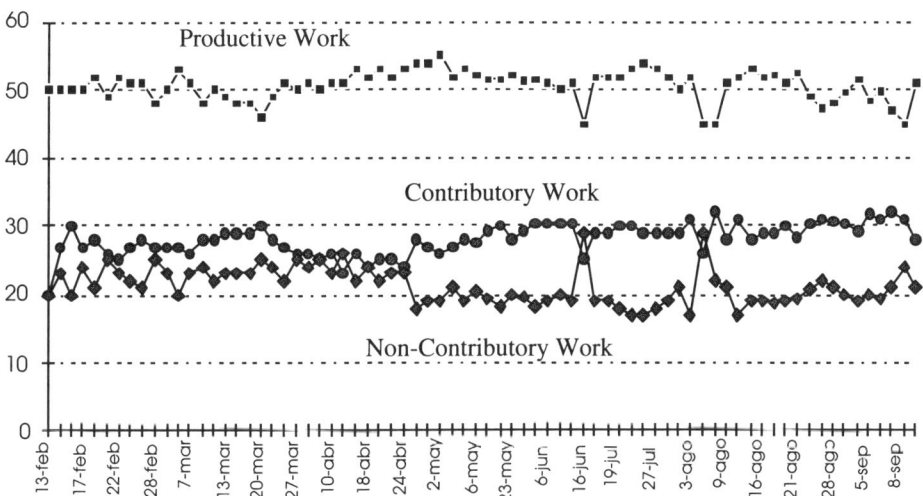

Figure 2. Productivity of the manpower.

– Non Contributory Work (NCW): Is any activity that do not correspond to the previous categories and that implies unproductive time like journeys, rests, idle moments, work sequence, physiologic necessities, etc.

These general indexes, compared with standard national values (Benchmarking). Table 1 can be transformed into an effective tool for the productivity control of the human resource, identifying in a short time horizon, which specialties and what sectors are having problems.

Table 1. Activity level on November 1992.

Activity level	Productive	Contributory	Non-contributory
Optimum	60	25	15
Normal	55	25	20
Average in Chile	45	23	32
Los Benedictinos	45	24	31

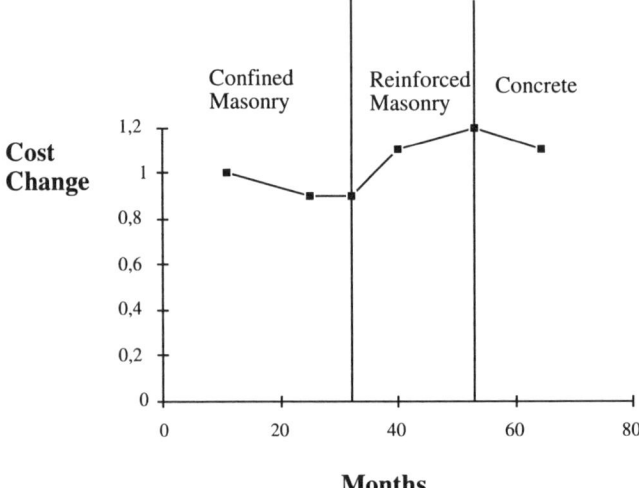

Figure 3. Unit cost per house over time.

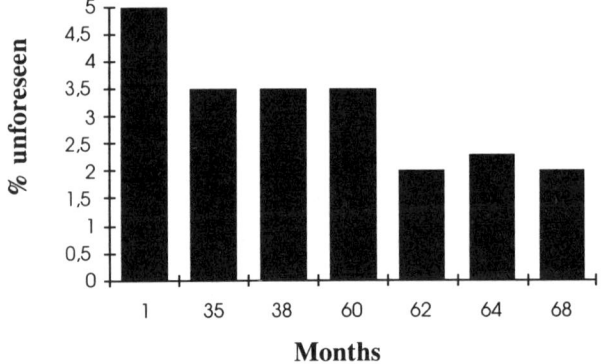

Figure 4. Change on unforeseen cost.

4.2 *Costs*

This index does not reflect on its own the variation on the houses costs since it consider many other factors independent of the constructive system. That is the case of the different negotiations the company carries out with suppliers for the payment terms and prices which would give an erroneous reading refereed to construction costs. (see Figs 3 and 4).

Table 2. Number of claims over time.

Months	% of people that claims	Number of claims	Claim index
1	90	15	13,5
35	90	15	13,5
38	90	10	9
60	70	5	3,5
64	70	5	3,5

4.3 *Quality*

It is very difficult to find a system that measures quality exactly therefore was opted for assessing the claims variation along the construction period, which represents a real alternative in order to measure the houses quality. Like was already said, there is a special organization in charge with solving the problems and claims from the users, and with this it was not difficult to determine the variation through time (see Table 2). The claim index is given by the multiplication between the percentage of people that reclaims and the quantity of claims, and it can be appreciated that there has been a progressive decrease along the work.

5 CONCLUSIONS

The professionals in charge of the job 'Los Benedictinos', introduce the importance of getting to know as soon as possible the construction method, even before the design. The search for new construction techniques and the improvement of personnel management systems tends to improve the work productivity and quality. On the other hand a cost reduction can be appreciated, and this makes these ideas a necessary and feasible implementation.

The great economical development we found in Chile can be stooped by the foreign market competition or by the lack of modernization in the construction sector. For these reasons mechanisms must be found to incorporate and develop new technologies in equipment, materials, construction methods, quality control, training and other concepts that has been achieved, in some degree, by the developed countries. The incorporation of these concepts to our idiosyncrasy is urgent since they have a direct influence in productivity increase.

Additionally, the construction sector strategies in reference to technical projects, materials and methods, are based on the existence of abundant and low cost human resources. With the increase of this activity, reaching almost full employment, a lesser availability of qualified labor can be appreciated, and this carries us to search for formulas to make use of the existing resources in the best possible way, i.e.. that we could construct more with the same quantity of resources.

Lean manufacturing of construction components*

LAURI KOSKELA
VTT Building Technology, Espoo, Finland

JUKKA LEIKAS
Mecrastor Corp., Espoo, Finland

ABSTRACT: This chapter is based on a research project undertaken by a consultancy company, a research institute, and six construction component manufacturing firms. The goal was to draw up a methodology for initiating lean production activities in construction component manufacturing, and to verify the potential of lean production in this industry through practical experiments.

The resultant conceptual and methodological framework is outlined, and the implementation procedure is presented. Process improvement and redesign initiatives implemented by participating firms, along with related results and benefits, are analyzed. Finally, the feasibility and significance of lean production in construction component manufacturing is discussed.

1 INTRODUCTION

Since 1991, the Finnish economy has experienced a deep recession. The volume of building construction in 1994 is less than half in comparison to 1990. Practically all organizations in the construction industry have been forced to downsize. Also the construction component industry has been severely hit by the recession. Thus, there are great pressures to raise profitability and competitiveness, but rather little money to invest for these purposes. 'Lean production', promising significant benefits in short term and with modest funding, seems like an ideal solution in this situation.

This was the background of the research project 'Lean construction component manufacturing', undertaken by Mecrastor Corp. (consultancy company), VTT Building Technology, and six construction component manufacturing firms . The objective was to draw up a methodology for initiating lean production activities in construction component manufacturing, and to verify the potential of lean production in this industry through practical experiments.

Among the firms, there was both small and medium sized firms as well as larger companies. The participating firms and the product lines selected for experimentation were as follows:
 – Gyproc Ltd: Gypsum boards;
 – Lujabetoni Ltd: Prefabricated concrete facade elements;
 – Metsäpuu Ltd: Kitchen fixtures;

*Presented on the 2nd workshop on lean construction, Santiago, 1994

– Novart Ltd: Kitchen fixtures;
– Parma Ltd: Bathroom modules;
– Rakennusbetoni-ja elementti Ltd: Lightweight concrete elements for partitioning walls.

The division of work was as follows. The theoretical and conceptual framework of lean production was created by VTT Building Technology. The methodology was set up by Mecrastor and VTT Building Technology. The consultants of Mecrastor provided methodological training and facilitated the pilot projects in companies.

2 LEAN PRODUCTION

The conceptual framework of lean production has been outlined in earlier work (Koskela 1992; Koskela 1993; Koskela & Sharpe 1994). In the following, we summarize the main features. The underlying idea was to simplify and unify the conceptual and methodical framework so, that it can be used without difficulty even in small and medium sized firms.

2.1 *Conceptual framework of lean production*

The conceptual framework of lean production combines three different views of production and operations:
– Material or information is converted (traditional view);
– Material or information flows (just-in-time view);
– Value is generated through fulfilment of customer requirements (quality view).

Thus, production is seen as flow processes, which are composed of:
– Conversion activities;
– Flow activities: moving, waiting and inspection;
– Customers, for which value is generated.

The intrinsic flow process goals are to decrease process cost and duration and to increase value for the customers. The value consists of two components: product performance and freedom from defects (conformance to specification). Value has to be evaluated from the perspective of the next customer(s) and the final customer. In opposition to cost and duration, it is difficult, often impossible to measure the absolute value. However, for practical application, measuring the relative value often suffices; for example the value loss in relation to the best practice value or theoretically best value.

An important distinction is based on the insight that not all activities generate value. In flow processes, we distinguish value-adding and non value-adding activities:
– Value-adding activity: Activity that converts material and/or information towards that which is required by the customer;
– Non value-adding activity (also called waste): Activity that takes time, resources or space but does not add value.

Note that conversion activities are usually value adding, but not all. Similarly, flow activities are usually, but not always, non value adding.

The improvement of non value adding activities should be focused on their reduction or elimination, whereas value adding activities have to be made more efficient.

Now, the rationale of lean production may be presented as in Figure 1. We have three options for improving production:
– Reducing the costs (and duration) of value adding activities through increased efficiency;
– Reducing the costs (and duration) of non value adding activities (waste), through elimination of these activities;
– Reducing the value loss.

The potential of lean production is embodied in the two latter options; the first one has been customarily used.

2.2 *Principles and methods of lean production*

Given that we have to reduce waste and value loss, what are the principles and methods to do this? A number of principles and methods exist for controlling, designing and improving flow processes (Koskela 1992). They are closely related to the respective cause of waste and value loss. Regarding waste, its causes and corresponding principles are summarized in Table 1.

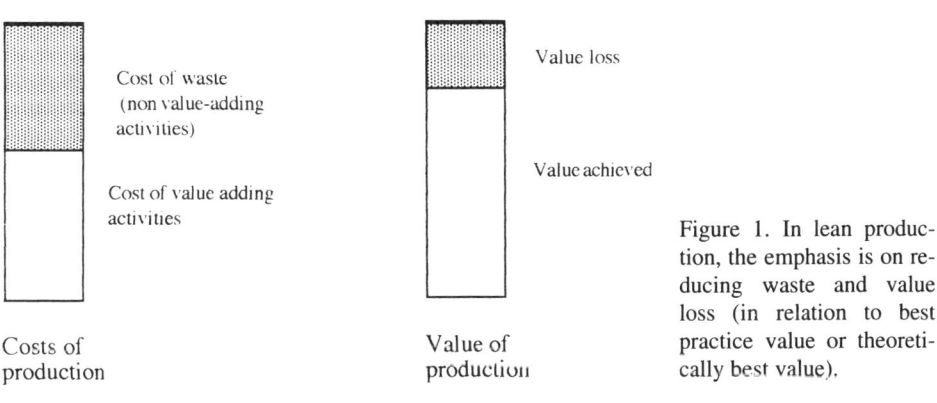

Figure 1. In lean production, the emphasis is on reducing waste and value loss (in relation to best practice value or theoretically best value).

Table 1. Causes of waste and corresponding lean principles or methods.

Cause of waste	Lean principle or method
Hierarchical organization	Process oriented, team based, flat organization
Process not in control (excessive variability)	Reduction of variability
Waste not recognized, not measured	Process charting for identification of non value adding activities; compression of cycle times
Long and complicated information and material flows	Simplification
Process rigidities	Increased flexibility
Suboptimization	Focus on whole processes
Confusion and disorder	Visual management

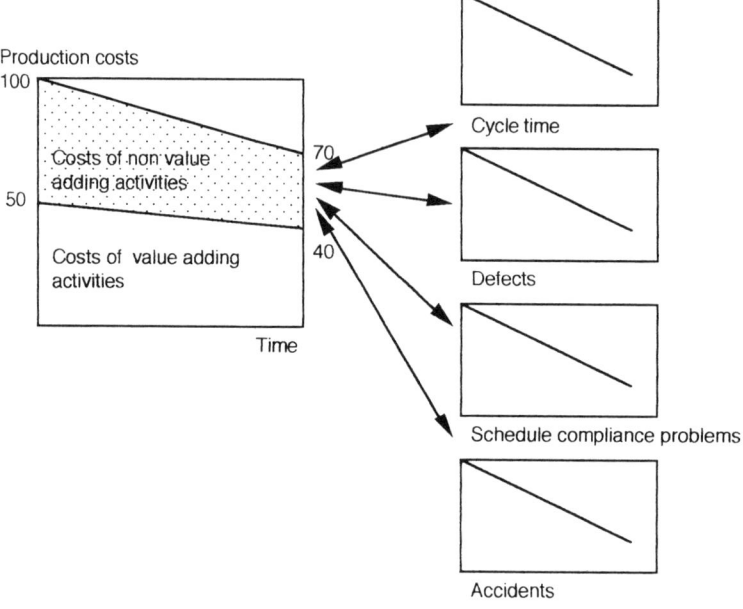

Figure 2. It is not practical to measure the total costs of non value-adding activities, however, indirect or partial measures may be used.

2.3 *Measurements in lean production*

Given that the implementation of lean principles is underway, how do we know that we are making progress? Even if it is not practical to measure the total waste in the process, it is possible to use indirect or partial measures of waste (Fig. 2). The same applies for loss of value.

3 METHODS OF IMPLEMENTATION

Each firm organized a team to carry out the project. In most firms, there were worker participation in the team. In all cases, there were support from the top management. The work was directed and carried out by the team and facilitated by the consultant. The progress in each firm was reported and discussed in joint meetings where all involved organizations participated. The experimental implementation in firms was carried out from October 1993 to June 1994.

The experimentation was positioned by the firms in different ways, according to their situation. Some firms used the project for learning and trailing the lean methodology, some for augmenting TQM activities. However, in some firms the project was considered as the main thrust regarding the improvement of the product line in question.

The implementation started through selection of major macro-processes as subjects for experimentation. These were further divided into micro-processes, which could be analyzed separately. The processes were then charted and measured: especially the waste (rework, material waste, idle time due to waiting etc.) and value loss was analyzed and their causes were investigated. Improvement goals were stated and corresponding action was planned, aiming at a redesigned process or improvement of the existing process.

4 FINDINGS

4.1 *Concepts and methods of lean production*

In general, the conceptual and methodological approach worked well. A number of difficulties were encountered:

– The distinction value adding activity/non value adding activity was perceived as too simplistic or harsh. Some people were offended when a part of their tasks was labelled as waste. Thus, a third category was introduced: activities without value added. These are activities that are essentially wasteful but necessary under current operating procedures (Monden 1994).

– Methods taught, analysis forms prepared etc. were not followed in detail, but rather selectively. However, this can not be considered as a major problem, because internalization of the main concepts of the approach proved to be much more important than strict observance to step-by-step guidelines.

Making waste visible (by flow charting or measurements) turned out to be the most important single feature of the conceptual and methodological approach. It motivated all, both workers and management, to think creatively about new solutions and to accept changes to time-honoured practices.

4.2 *Analysis of the current situation*

Most firms focused on waste, rather than value loss, when analyzing the current situation. In all cases waste, that earlier had existed unobserved, was detected and made visible.

Typically, only 20% of all steps in order processing or production were evaluated as value adding. In order processing, a major share of working time, 25-40%, was allocated to finding missing information, rework or other non value adding activity. Similar figures were found in production.

The major internal causes for waste were as follows:

– The traditional organizational model, leading to fragmented and long work flows, and narrow tasks;

– Layout problems: tasks or functions requiring communication or material transfer are located so that the distance is long.

A part of waste was caused by the specific features of the customer industry, construction. Some examples:

– There is a tendency to place construction component orders with missing information, obviously due to incompleteness of design; the amount of order changes is also relatively high;

– Due to the generally poor dimensional accuracy in construction, one firm has to make measurements on site and produces then the components according to these, instead of relying on measures as presented in drawings;

– Traditionally, design and production of buildings are carried out by separate organizations; designers lack knowledge on manufacturability and constructability.

4.3 *Improvement initiatives*

An overview on the improvement initiatives is given in Table 2. Besides the actions presented in Table 2, there were numerous instances where waste could be eliminated through minor, immediate changes.

Table 2. Overview on process redesign/improvement implemented and related results and benefits.

Firm	Process analyzed	Process redesign and improvement	Results and benefits
Gyproc Ltd	Aftertreatment of boards (cutting, handling, storing, packaging)	Layout changes and other operational changes aiming at streamlining of material flows	Implementation still underway
Lujabetoni Ltd	Sales-order-production-delivery process	Team based, flat organization has been implemented. Information and material flows have been streamlined	Productivity, quality, profitability and worker participation have clearly increased
Metsäpuu Ltd	Sales-order-delivery process	Technical changes: Collocation of key persons, EDI. Operational changes: objectives and incentives for teams. Improvement of standard routines and related training	Identification of improvement potential is evaluated by the firm as the most important result
Novart Ltd	Order-delivery process	Various operational changes implemented: parametric cost estimation for prebids, streamlining of the flow of order information to production planning. Decision made on redesign of the total information flow, including the elimination of invoicing as a separate task	Improvement of customer service, reduction of errors related to order information, increased productivity
Parma Ltd	Order-production-delivery process	Team based production organization has been implemented. The role of designers has been changed to that of project leader (each order is a project)	Production cycle time has decreased, productivity increased, workers participate actively in improvement
Rakennus-betoni-ja elementti Ltd	Order-production-delivery-installation process	The product offered to the customer is being changed from elements to ready partitioning walls; various standard routines for the order-delivery-installation process were developed	Sales process will change; implementation still underway. Better service to customer and reduction of waste are anticipated to be the most important results

The most extensive changes were carried out by Lujabetoni and Parma. These are described in more detail.

At Lujabetoni, the pilot project stimulated the following changes:
– Two organizational levels were abolished;
– Production was organized as self-directed teams of 6-8 persons;
– Purchasing was simplified: the teams order directly materials from suppliers;
– The role of foremen was changed to that of facilitator and trainer;
– Sales strategy was changed: sales and internal design department work as a team at project offering stage;
– Several changes, required by users, to information systems were realized;
– Continuous measurement of sales and production was started: on each Monday, the profit of the previous week is disclosed (to all) and compared with the target;
– The practice of having 2-3 men making repairs to facade elements on site – all over the country – was abolished (instead, the quality of elements delivered from the factory was increased).

At Parma, the changes were as follows:
– Production was organized as self-directed teams of 4-6 persons. The team takes care of the majority of assembling operations of a bathroom. It also handles short term planning. A member of the team goes to the site to carry out the hand-over inspection together with the contractor.
– The role of the designer was extended to that of a project leader (order = project). The project leader takes the overall responsibility for the smooth progress and timely delivery of the order. He becomes involved as early as possible, already when customer requirements are charted. He is in regular, direct contact to each team that is producing bathrooms for his project.

It is clear that most initiatives addressed internal causes of waste; it is more difficult to influence external causes, especially those originated by customers. However, one firm introduced a policy of not forwarding an order to production planning before all necessary specifications are delivered by the customer (the complete kit concept (Ronen 1992)). This proved to be effective.

4.4 *Results and benefits*

In all firms, the results of the pilot project were evaluated as good or excellent. This is reflected in the comments by firms on results presented in Table 2. In the following, more detailed information is presented for the two firms analyzed above.

Lujabetoni Ltd. reported the following results and benefits:
– The capacity (average production volume) has increased 20% due to lean initiatives;
– The costs of goods sold (COGS) have decreased 5-10%;
– Claims and quality costs in general have decreased essentially;
– The atmosphere in the factory is excellent.

At Parma Ltd., results and benefits are as follows:
– Production is organized into 19 tasks; in the earlier situation, there were 36 tasks. It is estimated that this alone will result in saving 5% of the working time (consumed earlier in waiting);
– The number of different drawings made for production was halved;

– Production cycle time has decreased;
– Workers are motivated and participate actively in improvement.

In both of these firms, the lean principles are being transferred to other production lines, and the improvement of the piloted production line is continuing.

All in all, the results exceeded the expectations of the participants of the project.

5 CONCLUSIONS

This experimental implementation of lean production clearly shows, that this approach can provide significant and rapid benefits in construction component manufacturing. It is also understood that the potential of waste (and value loss) elimination was by no means exhausted in these experiments: there is ample room for continuous improvement.

Thus, lean production is applicable and worthwhile in construction component manufacturing. In this industry, the same internal causes for waste, as in other manufacturing industries, exists. However, beyond that, the erratic and undisciplined nature of the customer industry, construction, provides an additional source of waste.

For manufacturing related causes of waste, the methods and techniques developed in other industries are applicable also in construction component manufacturing. However, regarding construction related causes of wastes, new solutions are needed. Experimentation, development and research is thus required.

The improvement potential detected gives support to the argument that in industrialization of construction, poorly controlled design, fabrication and site processes have often consumed the theoretical benefits to be gained from industrialization (Koskela 1992). It is not enough to change construction to look like (traditional) manufacturing, rather the total design-fabrication-erection processes should be designed and improved so that real and significant benefits emerge.

It is customary to view information technology and automation as the major means for improving efficiency in construction component manufacturing. These results rather support the thesis (Koskela 1992): 'In work flows in construction, it is more profitable to initiate process improvement activities than to automate parts of the present work flow'. Thus, both lean production and information technology/automation should be seen as major improvement approaches, which complement each other.

ACKNOWLEDGEMENTS

The described project was funded by Technology Development Centre, Mecrastor Corp. and the six participating companies. The project was initiated and organized in cooperation with the Finnish Association of Construction Product Industries and VTT Building Technology.

REFERENCES

Koskela, L. 1992. Application of the New Production Philosophy to Construction. *Technical Report* No. 72. Centre for Integrated Facility Engineering. Department of Civil Engineering. Stanford University. 75 p.

Koskela, L. 1992. *Process Improvement and Automation in Construction: Opposing or Complementing Approaches?* The 9th International Symposium on Automation and Robotics in Construction, 3-5 June 1992, Tokyo. Proceedings. pp. 105-112.

Koskela, L. 1993. *Lean Production in Construction.* The 10th International Symposium on Automation and Robotics in Construction (ISARC), Houston, Texas, USA, 24-26 May, 1993. Elsevier. pp. 47-54.

Koskela, L. & Sharpe, R. 1994. *Flow process analysis in construction.* The 11th International Symposium on Automation and Robotics in Construction (ISARC), Brighton, UK, 24-26 May, 1994. Elsevier. pp. 281-287.

Monden, Y. 1994. *Toyota Production System.* Second Edition. Chapman & Hall, London. 423 p.

Ronen, B. 1992. The complete kit concept. *Int. J. Prod. Res.* 30(10): 2457-2466.

Lean productivity and the small private practice*

DAVID EATON
Department of Surveying, University of Salford, UK

ABSTRACT: This chapter describes a lean productivity strategy for a firm that is not currently the overall cost leader in a mature market segment. It uses the analysis of the Chartered Quantity Surveying Practice (CQSP), a service provider in the United Kingdom construction industry as the focus of such analysis. A CQSP that is not currently the cost leader can, using lean productivity philosophy, reconfigure the value chain to find new cost curves which may actually be lower than the overall market low cost producer for certain clients and certain services. A viable strategy can thus be created for a firm that could not otherwise compete.

1 INTRODUCTION

Definition of CQSP

The CQSP's output and UK construction output in general is very slow to completion. This means that individual contracts are a significant proportion of annual production for all firms. This necessitates an approach to business that is risk averse. The service provision is largely tailored to individual clients requirements and thus cannot benefit from learning curve productivity improvements as fixed manufacturing can.

There is also a trend towards the globalisation of construction and therefore the globalisation of professional services for construction. There is also a new emphasis on the provision of service; the provision of a package suited to an individual clients needs rather than a standard package offered to all clients.

The CQSP's services are therefore highly fragmented, highly specialised, highly autonomous and extremely competitive, not ideal conditions for fundamental improvements.

The small CQSP firm also has the disadvantage of being unable to offer the range of services nor the scale of services of a larger competitor.

The advantages that the small CQSP firm possess must be used effectively and efficiently to compete with the larger firm.

Typically, the smaller CQSP firm claims that lower overheads are possible because of the reduction in the integrative and differentiation activities, this in turn

---

*Presented on the 2nd workshop on lean construction, Santiago, 1994

leads to the possibility of lower total bids for work. This aspect will be examined in more detail later.

2 THE ORIGINS OF LEAN PRODUCTIVITY

Lean productivity is defined as a management philosophy that emphasises that all activities consist of conversions and/or flows. Conversion activities are those actions that add value to the transformation of material or information into a final product. Flow activities are non-value adding activities which exist merely to link conversion activities together. typically, flows are activities such as inspection, testing, waiting, moving, delivery, etc.

Thus all activities within the small CQSP firm can be classified and examined and improved independently of each and every other conversion or flow.

This new production philosophy undoubtedly had its origins in Japan of the 1950's. However the philosophy has such a combination of attributes that a more definitive origin is difficult to achieve.

A prominent element of this philosophy was the Toyota Just-in-Time (JIT) production system (Womack et al. 1990). This system consisted of the elimination of inventories, reduced set-up time, small lot production capability and the elimination of production peaks and troughs (collectively referred to as Heijunka 'production smoothing').

At the same time, quality issues where being addressed in Japan, and indeed America, by a workforce-wide involvement in incremental production improvements (Kaizen) and importantly the refinement and consolidation of service groupings with inter-related holdings and strong 'relational' (Kay 1993) contracts, commonly referred to as Keiretsu.

These three elements, Heijunka, Kaizen and Keiretsu are the basis of a production management philosophy that emphasises that all activities consist of conversions and/or flows. This new production philosophy is termed lean productivity.

3 THE AIMS OF LEAN PRODUCTIVITY FOR THE SMALL FIRM

All the service activities of the CQSP can be controlled and improved if a methodology is created that allows rational analysis. Typical quantity surveying service provision is fragmented and segmented to such a degree that current analytical frameworks (such as management hierarchy diagrams and job-description procedures) fail to recognise the inherent heterogeneity. The lean production analysis of activities into flows and/or conversions can be used as an analytical tool that may be create the opportunity for an alternative cost curve for the small firm.

As previously mentioned, the smaller firm may not offer the range or scope of services of the larger competitor. Rather than seeing this as a disadvantage it can be turned into a significant advantage. Because of the reduction in range and scope, the small CQSP firm will have considerably fewer flows and conversions within the organisation. More attention can therefore be given to each flow and conversion activity. Because of the small size of the firm it may be possible for certain activities to be defined as a Leibnizian 'monadal closed system' (Hamlyn 1990). Each monad (closed sub-system element) reflects the process from its own point of view, but has no effect upon, and is not affected by, any other monad.

In the small CQSP a monad may be an individual (or small team of workers) that performs a large proportion of added-value conversion activities with a disproportionately small amount of flow activity. (An example of this would be a small bill of quantities preparation group who receive the architects design drawings and convert these into a fully quantified construction activity tender document).

This monadal approach allows continuous incremental improvements to be made in isolation, without reflection upon the impacts that such changes will make on other elements. This is certainly not applicable to the larger firm where the interactions of integrative and differentiating activities are more complicated.

Figure 1 shows a schematic transformation system for a focused firm. In the simplified form shown the small firm is modelled by only four Leibnizian 'monadal closed systems'. The four monadal sub-systems are namely: commission, develop, produce and deliver. They are linked to each other only by flow activities, represented by the shaded flow envelope. Surrounding the flow envelope is the environmental supra-system which places a constraint on the operation of the entire CQSP system. No details of this supra-system are shown but they can be modelled. Thus within each monad conversion improvements can be made that have no effect what-so-ever on the rest of the system. Any flow activity within a closed monadal group (intra-monadal flow) can also be altered without effecting the rest of the system. The only cause for management concern are the inter-monadal flows (those flows that move the process between monads).

Figure 2 models the lean productivity improvements of the small firm over time.

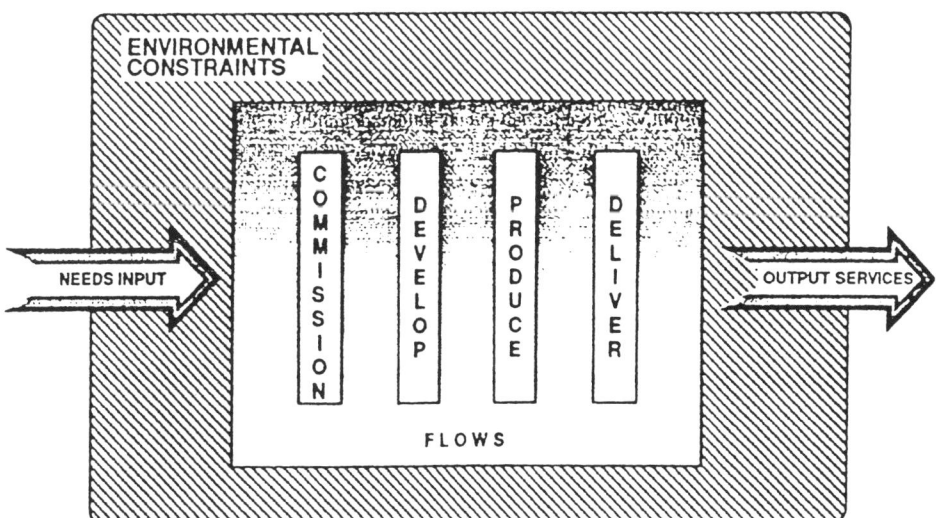

"Conversion activities performed by indicative collateral groups, named: Commission, Develop, Produce and Deliver"

Figure 1. Schematic transformation system for lean productivity.

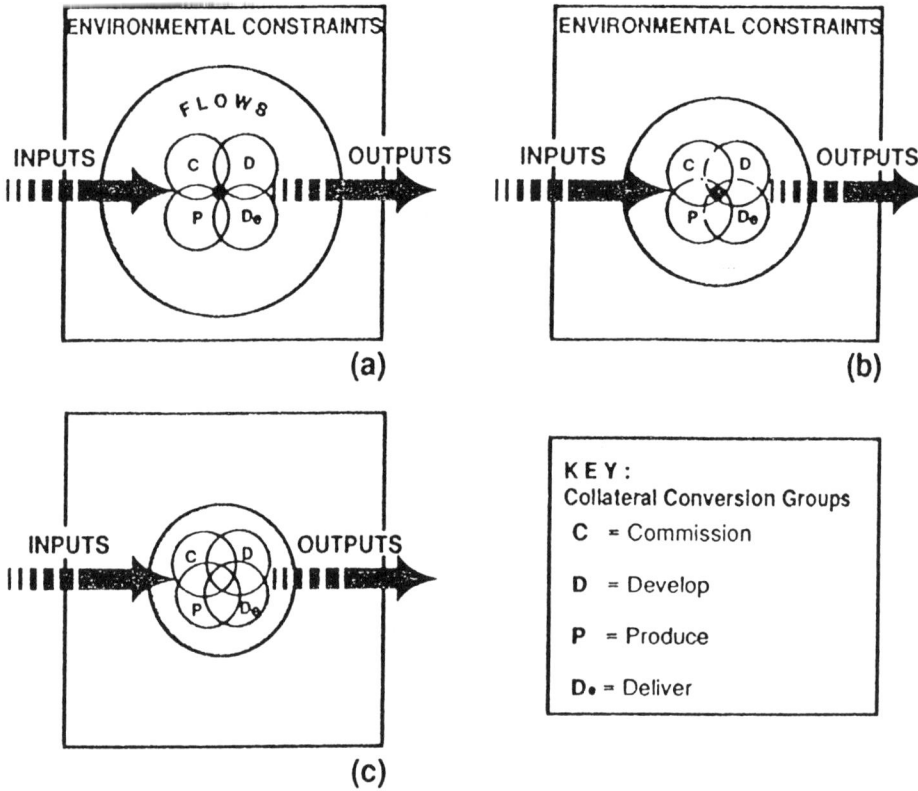

Figure 2. Lean production congruence improvements.

By concentrating on the monadal conversion activity attention is directed towards reducing the time and/or effort required for each activity. This productivity improvement is represented by the monad shrinking in size as time passes from (a) through (b) and ending at (c). The intra-monadal flows are examined so that flows within a monad are modified so that they converge on a focal point at the axis of the monad, this also represents a productivity improvement, and is likewise represented by the shrinking size of the individual monad.

These intra-monadal flow activities are also examined and reduced or eliminated so that the flows shrink until they match the outlines of the individual monads.

Intra-monadal flows are examined to either eliminate or reduce such non-value adding activities. The productivity improvements shown in this way are depicted by the monads moving closer together creating a smaller and more closely knit system. (i.e. overlaying each other) and converging on a central axial point for the small firm. This is indicated by the central black dot in 2a and 2b. Movement towards this central axial point creates system congruence (Eaton 1993), yet another recognised production productivity improvement technique.

The extent of the total lean productivity improvement is shown by the decrease in area of the small firm system. (compare the size of the system boundary circle in 2a with 2c).

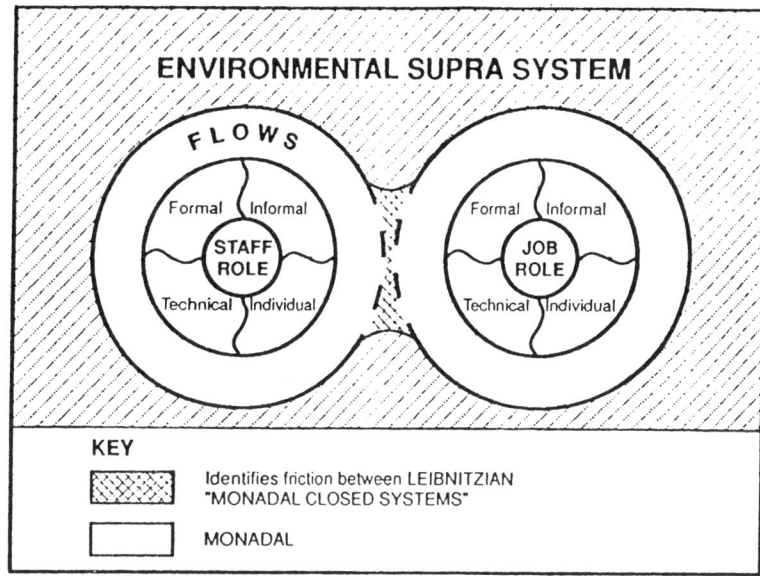

Figure 3. A multi-systems lean productivity analysis.

Figure 3 shows that the organisational sub-system analysis framework is independent of the lean productivity analysis by using a multi-systems to represent a conversion and flow analysis model (Eaton 1993). The cross hatched areas show the additional flows that are necessary because of system and sub-system incompatibilities. This area would be a prime candidate for targeted high impact lean productivity improvements.

4 ACHIEVING LEAN PRODUCTIVITY FOR THE SMALL FIRM

The implementation of a lean productivity strategy for the small firm may be considered as a four stage process (Betts et al. 1994). An initial or preliminary stage is created that allows the analysis and separation of service provision, (as provided by the individual professional) into flows and conversions. Stage 1 then allows the conversions to be examined and improved by current quality control techniques, whilst the flows can be simplified, or if possible, eliminated. Stage 2 allows the conversions to be improved by typical quality assurance techniques, whilst the flows are improved by further simplification and automation. Stage 3 sees the conversions improved by total quality management techniques, whilst the flows that remain are fully automated.

5 CONCLUSION

I have demonstrated that the adoption of the lean productivity production philosophy has great potential for improving the performance of the focused firm that is not the market leader.

This can be done by adopting a number of lean productivity improvement techniques (Eaton 1994).

The validation of these contentions is awaited, however, there are problems associated with lean productivity techniques that need to be addressed immediately. The strategy implicit in lean productivity is that of extreme centralisation of policy and isolated decentralisation of service provision. This will require more talented production groups capable of managing themselves. This has inherent implications for management styles, systemisation of service provision and mobility of workforce and management. Such implications need to be evaluated.

REFERENCES

Betts, M., Fischer, M. & Koskela, L. 1994. The purpose and definition of integration. *ASCE* (Submitted for publication).

Eaton, D. 1993. The development of a conceptual model for the improvement of technical quality within the quantity surveying practices of the United Kingdom construction industry. *CIB W-65 Symposium proceedings: The University of The West Indies. Trinidad & Tobago.*

Eaton, D. 1994. *Lean Productivity Improvements for Construction Professions.* Forthcoming paper in 13th International Congress of Cost Engineers, Association of Cost Engineers. London.

Hamlyn, D.W. 1990. *The Penguin History of Western Philosophy,* p. 158, Penguin.

Kay, J. 1993. *Foundations of Corporate Success,* pp. 55-60, OUP.

Womack, J.P., Jones, D.T. & Roos, D. 1990. *The machine that changed the world.* Rawson Associates.

Lean production productivity improvements for construction professions*

DAVID EATON

Department of Surveying, University of Salford, UK

ABSTRACT: Lean productivity is seen to be a possible methodology for addressing many of the fundamental faults in traditional service provision for the construction industry professions of the United Kingdom. The increasing globalisation of construction activity will require all the construction service professions to change and adapt or they will wither and die.

This chapter will explore the origins of lean productivity as exemplified by manufacturing industries world-wide. It will trace the potential development of lean productivity into the construction industry professions. It will then explain why and how lean production can be of benefit and offer predictions for its exploitation within the professional services associated with the industry.

1 INTRODUCTION

The chronic problems associated with the UK construction professions are well known: Low productivity, insufficient quality, poor co-ordination, high cost, etc. A number of solutions have been proposed to address parts of these problems. These solutions tend to be versions of procedures adopted by the manufacturing industries and modified to suit the conditions of the construction industry (Koskela 1992). For example, quality assurance (QA) has been advocated as a remedy for poor quality (BSI 1987); computerised integration of design and procurement as a remedy for low productivity (Betts et al. 1994) and electronic data interchange for poor coordination (Dym & Levitt 1991).

It is apparent that whilst there is much similarity between construction and manufacturing, (and therefore innovations from manufacturing are typically cited as suitable for introduction into construction) the results of the introduction of such procedures, have, however, failed to match expectations, and frequently fall short of the improvements achieved in other industries. It is apparent from the introduction of these procedures as attempted remedies that the problems addressed are not definitive for construction but are merely symptomatic of a more deeply rooted and fundamental problem for the construction industry. The root of the problem is defined by the nature of the construction industry itself.

Whilst it has no characteristics that are unique to construction, sharing features

---

*Presented on the 2nd workshop on lean construction, Santiago, 1994

with many other industries, it does has a combination of these characteristics that is unique (Hillebrandt 1984). The construction industry is subject to exceptional national and international cyclicality in total demand, fragmented and sporadic client led changes in specific work type demands which are extremely volatile and sensitive to economic and environmental changes.

The construction industry professions' output and construction output in general is very slow to completion, this means that individual contracts are a significant proportion of annual production for all firms. This necessitates an approach to business that is risk averse. The service provision from the professions is largely tailored to individual clients requirements and thus cannot benefit from learning curve productivity improvements as fixed manufacturing can.

There is also a trend towards the globalisation of construction and therefore the globalisation of professional services for construction. There is also a new emphasis on the provision of service; the provision of a package suited to an individual clients needs rather than a standard package offered to all clients.

The construction professions are therefore highly fragmented, highly specialised, highly autonomous, typically under-remunerated in comparison to other professions and extremely competitive, not ideal conditions for fundamental improvements.

It would not be remiss to also mention that there is a distinct element of 'me-too' within the professions. There is some evidence that developments are widely adopted for fear of being left behind rather than any fundamental belief in the efficacy of the development.

However a new production philosophy, rather than a new production development may well change this. This new production philosophy is termed *lean productivity*.

2 THE ORIGINS OF LEAN PRODUCTIVITY

The new philosophy undoubtedly has its origins in Japan of the 1950's. However the philosophy has such a combination of attributes that a more definitive origin is difficult to achieve.

A prominent element of this philosophy was the Toyota (JIT, Just-in-Time) production system, (Womack et al. 1990). This system consisted of the elimination of inventories, reduced set-up time, small lot production capability and the elimination of production peaks and troughs (collectively referred to as Heijunka 'production smoothing').

At the same time quality issues where being addressed in Japan, and indeed America, by a workforce wide involvement in incremental production improvements (Kaizen) and importantly the refinement and consolidation of service groupings with inter-related holdings and strong 'relational' contracts, commonly referred to as Keiretsu.

These three elements, Heijunka, Kaizen and Keiretsu are the basis of the lean production management philosophy that emphasises that all activities consist of conversions and/or flows.

Conversions are activities that add value to the transformation of material or information into a final product.

Flows are non-value adding activities which exist merely to link conversion activities together. Typically flows are activities such as inspection, testing, waiting, moving, delivery etc.

3 THE AIMS OF LEAN PRODUCTIVITY FOR THE PROFESSIONS

All service activities can be controlled and improved if a methodology is created that allows rational analysis. Typical service provision is fragmented and segmented to such a degree that current analytical frameworks fail to recognise the inherent heterogeneity. Figure 1 illustrates the philosophy of lean productivity for the construction professions.

The essential elements of the lean production philosophy can be examined separately under distinct headings, however, the benefits can only be achieved by the holistic application of all the elements. The elements are described below:

Process transparency: The essence of this element is that the process cycle should become more transparent. The lean productivity techniques provide a framework for a detailed critical analysis. Each part of the process is analysed and then catergorised as either a flow or a conversion activity. Having thus separated each activity it then apparent as to how the activity can be improved. For flow activity (which adds cost but not value) the aim is to eliminate the activity in its entirety. If elimination is not possible then a secondary aim would be to reduce the duration of such flow activities. For a conversion activity the emphasis is on managerial support for improvements in the effectiveness and efficiency of conversion activity. These conversion activity improvements create an opportunity for increases in productivity. The essential elements of the managerial support for conversion activity relate to unstructured or semi-structured problem solving, coordination and communication improvements and the integration of the managerial role into the domain of worker activity. Thus worker self-control and self-management becomes the focus of the entire process. Responsibility moves from the supervisor to the worker, this reduction in power-distance indexing (Hofstede 1980) creates cultural conditions under which the entire process is subject to worker review and examination rather than managerial examination of individual elements. By passing control from supervisor to worker a balance of flow and conversion improvements can be attained, rather than a lop-sided concentration on conversion activity only, which tends to be the typical managerial approach.

Service Activities are considered as material and information flow and conversion processes which are

- controlled for minimal variability and cycle time

- improved continuously with respect to waste and added value

- improved periodically with respect to efficiency and effectiveness by implementing new and/or changing the conversion process

- improved periodically with respect to efficiency and effectiveness by eliminating or reducing the flow process

Figure 1. Lean productivity philosophy for the professions.

Conversion improvements: The conversion activities are examined by the worker and incremental alterations are made to the conversion process which either; reduces the variability of output, and therefore improves the quality of the typical product; or improves the effectiveness of the conversion process thus reducing the conversion cycle time and consequently improving productivity. To these tangible improvements must also be added the intangible improvements that are related to motivational aspects of the worker, such as the improvement in the quality of working life, experiential learning benefits and empowerment improvements.

Flow improvements: The flow activities are examined and improvements achieved by reducing the share of non-value adding activities as part of the whole process. This can be done by incremental improvements in the flow process to improve productivity or occasional process redesign which eliminates elements of the flow process therefore simplifying the entire process cycle by minimising the number of steps and linkages and thereby increasing productivity. Thus flow improvements can achieve cost reductions for the whole process and / or time reductions for the whole cycle.

Kaizen improvement conditions: By advocating and incorporating incremental and continuous improvements into the process the workforce has a lean productivity ethic inculcated. This ethic creates the culture and conditions necessary for such incremental and process redesign. The workforce has proprietary responsibility for the entire process and thus the drive for improvement becomes workforce 'pushed' rather than management 'pulled'.

Keiretsu created improvements: Regional clustering of 'relational' contractors can establish the 'insider' thus avoiding regional cyclicality and product diversity. This avoidance of regional diversity allows the development of a geographical focus strategy which can then establish the 'image' of a local service, which in turn aids the creation of a regional 'localised' mass demand. This regional keiretsu can develop a sophisticated demand management system because of the cultural and environmental 'relationships' between the parties.

This regionalising keiretsu can also prevent regional harvesting by a multi-regional or international group because of the local integrity of the markets.

Added value improvements: Can be achieved by a systematic consideration of customer requirements. This is achieved by value chain analysis of both the supply chain and the client demand chain. A focus of this analysis is the ability; to have an in-depth analysis of; and close contact to; the regional markets as an 'insider'. A 'local' image can be created which can be used to create a regional niche segment of a national market, this regionalisation can then be transposed into a 'mass' local market. The additional cost saving potential attributable to the benefits accrueing from scale efficiencies can then be reaped by the appropriation of supra-profits or by the reduction of service price and the subsequent increase in market share and consequential increase in normal profits. Repeated iterations of this 'local' market cycle effect can be used to fine tune the added-value of entire services or parts of activities.

A mass 'local' market will create a demand for a service product line that can be provided to other 'local' markets as a specialised focus or niche service, with the added benefit to the supplier of the ability to appropriate more of the total added value from the service.

Output flexibility improvements: The improvements in both flow and conversion activities can increase total productivity but, perhaps more importantly, can allow the introduction of conversion activity options at various stages in the process that improves the overall responsiveness to output requirements and creates a rapid and flexible variety of provisions that can be suited to individual client needs. Rather than Henry Fords' 'You can have any colour you want – so long as it's black!' you can have the Burger King 'You want it – You got it!'

Productivity and performance measures can thus become benchmarks rather than fundamental targets. They can become motivators rather than demotivators.

4 STRATEGIC IMPLEMENTATION OF LEAN PRODUCTIVITY

The corporate management of a company wishing to implement lean productivity should adopt a proactive approach and should follow a five stage plan together with the associated feedback mechanisms.

Mission statement and company objectives definition: The senior management of the organisation should be asking and answering the questions, What is our business and What will it be? A carefully structured response will create a vision of the needs of the organisation over the medium to long term planning horizon. This sets forth the organisations intent to achieve a particular business position. Having established the mission statement this is then converted into specific performance targets. The establishment of these targets challenges the organisation to close the gap between the desired business position and the actual business position. A results-oriented climate emerges from this process which moves the organisation in the intended direction. Two types of targets are derived from this exercise: financial and strategic objectives. Financial objectives are derived to preserve the critical financial viability of the organisation, whilst strategic objectives relate to the companies competitive advantage vis a vie comparable competitors, in such areas as product quality, customer service, reliability and company image and reputation. It is these strategic objectives that can be improved by the implementation of lean productivity philosophy and techniques.

Identification of processes and activities for examination: The next stage is to analyse all company processes and services and to separate these into individual activity chains. Each element of the activity chain is identified as either a conversion or a flow. The activity chain elements can then be examined for potential improvements using the philosophies of: process transparency, conversion improvements, flow improvements, Kaizen improvement conditions, keiretsu created improvements, added value improvements and output flexibility improvements.

Examination and measurement of existing activity practises: The activity chain analysis conducted to identify the processes and activities can now be used to evaluate current performance in each element. A first step prior to quantification can be the identification of typical failure points, such as waiting, waste and abortive effort in the activity chain. These points can then be measured and targets for improvements can be defined.

Identification of new improvement levers: These points then serve as a focus for initial examination for improvement potential using the lean productivity philosophies. It is then necessary to examine the whole array of lean productivity philosophies to create a strategic plan for the implementation of the productivity improvement.

Test prototype improvement: As with all strategic plans the proposed productivity improvement should be tested for effectiveness and efficiency. This will involve the entire workforce, since the holistic view is necessary. Typical activity eliminations or reductions can relate to transactional, geographical, automational, analytical, informational, sequential, knowledge management, tracking and direct connection improvements.

Feedback: As with all systems it is necessary to incorporate feedback mechanisms. With lean productivity this process assumes a major importance since many of the improvement possibilities are incremental in nature and require repeated iterations of the activity cycle to validate and evaluate productivity improvements.

5 ACHIEVING LEAN PRODUCTIVITY FOR THE PROFESSIONS

The implementation of lean productivity for the professions may be considered as a four stage process. An initial or preliminary stage is created that allows the analysis and separation of service provision,(as provided by the individual professional) into flows and conversions. Stage 1 then allows the conversions to be examined and improved by current quality control techniques, whilst the flows can be simplified, or if possible, eliminated. Stage 2 allows the conversions to be improved by typical quality assurance techniques, whilst the flows are improved by further simplification and automation. Stage 3 sees the conversions improved by total quality management techniques, whilst the flows that remain are fully automated. This process is illustrated in Figure 2.

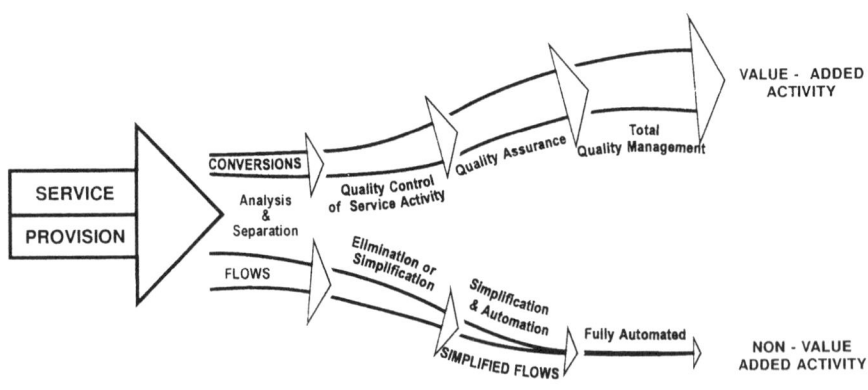

		PRELIMINARY STAGE	STAGE 1	STAGE 2	STAGE 3
Productivity Improvement Techique	Increase Process Efficiency		Reduce cost of non-conformance and increase process efficiency		Reduce or eliminate non-value added flow and increase efficiency of value added conversions

Figure 2. The development and implementation of lean productivity.

It should be emphasised that the lean productivity philosophy can be achieved by two approaches, both follow the same pattern but have a differing emphasis. The first approach is that of exhaustive examination. All activity chains are examined and all potential improvements are considered. However, this approach can involve a combinatorially explosive number of options which will entail solution filtering and pruning. A problem that is yet to be satisfactorily addressed (Eaton 1994). The second approach is that of high impact analysis. The organisation identifies a range of critical success factors (CSF's) that are vital to the attainment of the financial targets set for the company objectives contained within the mission statement. These CSF's are then analysed for activity chain flow and conversion elements. The process of lean productivity improvement is then applied as discussed.

6 AN ILLUSTRATIVE EXAMPLE

Figure 3 illustrates the lean productivity approach for a typical quantity surveying service, the production of measured bills of quantities for competitive tendering purposes. This single process is analysed, and an activity chain is constructed, (Fig. 3a) typical failure points are identified as outliers to the main flow. The individual activities are then allocated to either flow or conversion processes (Fig. 3b). This analysis then identifies that the activity chain could be improved by splitting into two sections, the first would require all flow activities to be achieved by electronic data interchange, whilst the second section would be achieved by the use of a computer package (Fig. 3c). The process illustrated would be reiterated over time and eventually the conclusion of the lean production philosophy would be the creation of a meta-system whereby the bill production process would be subsumed as a flow activity in the design and construction of a building. The entire bill production process has been reduced to a flow activity (Fig. 3d).

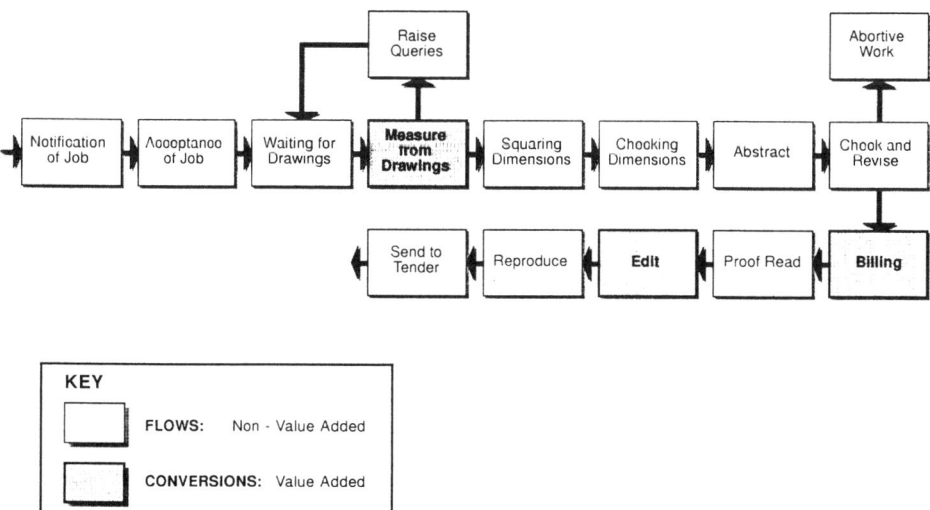

Figure 3. BQ production. As a flow and conversion process: A simplistic illustration. a) Quality control conversion and simplify flows.

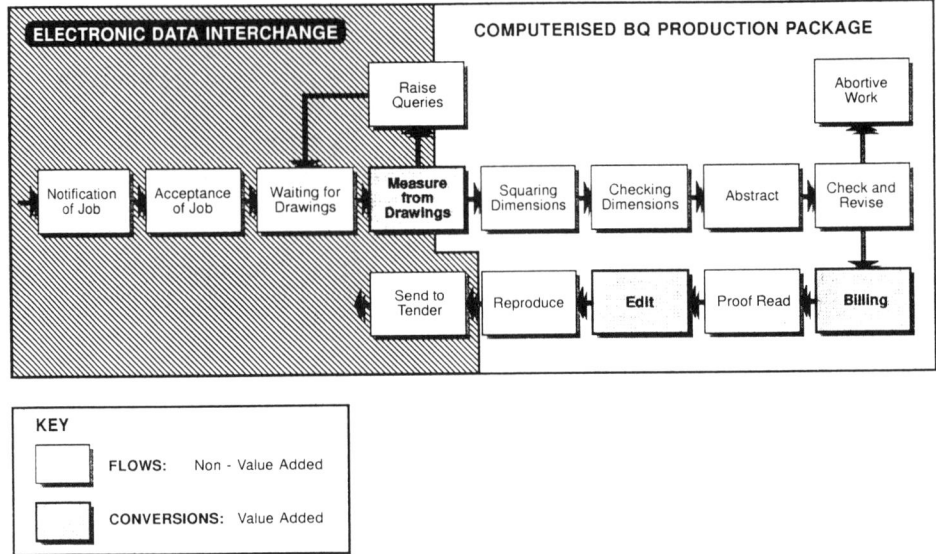

Figure 3 Continued. b) Quality assured conversion and simplified.

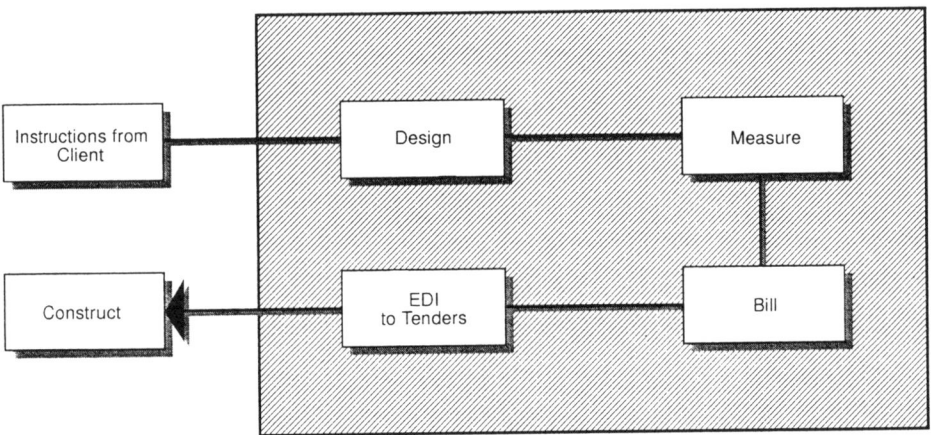

Figure 3 Continued. c) Total quality management and fully automated.

7 LEAN PRODUCTIVITY RESULTS

Figure 4 illustrates the potential productivity improvements that are possible by adopting the lean productivity philosophy in the analysis of service provision.

Cost and time reductions and hence, productivity improvements can be achieved in a number of circumstances. The lean productivity philosophy incorporates the benefits of *competitive advantage, competencies and capabilities theories* and addresses these with a number of distinct techniques.

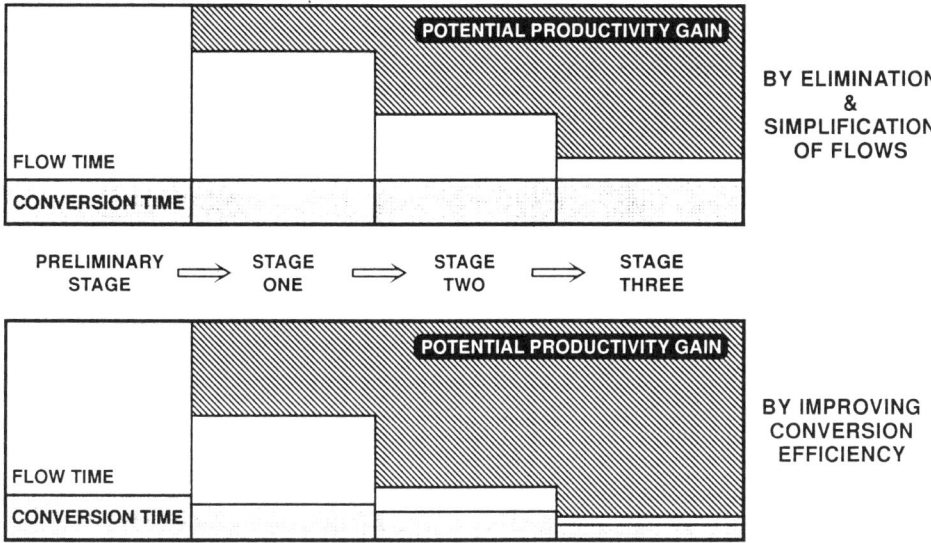

Figure 4. The lean productivity improvement potential.

8 LEAN PRODUCTIVITY DIFFICULTIES

There are, however, problems associated with lean productivity techniques that need to be addressed immediately. The strategy implicit in lean productivity is that of extreme centralisation of policy and isolated decentralisation of service provision. This will require more talented production groups capable of managing themselves. This has inherent implications for management styles, organisational structures, new skill requirements, systemisation of service provision and mobility of workforce and management. Such implications need to be evaluated.

9 CONCLUSIONS

The managerial emphasis on corporate and strategic issues rather than technical and process issues means that senior management within the organisation have an explicit commitment to the attainment of strategic targets. This then involves an implicit commitment to the productivity improvements identified under the lean productivity philosophy. This keenness will be transmitted to the workforce via the lean production techniques.

The techniques can be combined to create a culture that values and seeks better consistency in output. This can only improve the quality of service provision and the reputation of the service provider. The holistic attainment of strategic objectives which have been inculcated into the work ethic of the work force should create better service provision and improved client satisfaction.

Lean productivity can provide means of improving the speed of delivery of professional services by eliminating or minimising waste in flow activity and by the improvement in conversion activity. In some instances, activities that are currently perceived as conversion activity may be altered by technological innovation into a flow activity, that could be subsequently eliminated or minimised.

The techniques also provide the potential for process flexibility, the activity chain analysis provides an explicit framework for identification and evaluation of the contribution of each element to the overall process. This means that a basic service could be tailored to the requirements of individual clients by a reconfiguration of the service activity chain. Prior to this analysis, the impact of tailoring a service was difficult to forecast and thus had a high level of uncertainty associated with the cost implications of such alterations. The new activity chain analysis provides a mechanism for service definition and therefore a mechanism for risk identification and apportionment.

The flexibility of service provision means that potential innovative service implementation is possible. Changes can be evaluated and realistic cost comparisons can be made using the activity chain analysis. For example, basic service provision in one geographical area could be introduced in another area as a

differentiated high added value service with the associated appropriation of higher fees.

The lean productivity philosophy and the associated techniques offer a number of potential productivity improvements for the organisation providing services to industries. These potential improvements are equally applicable to the construction industry professions.

I have shown that the adoption of the lean productivity philosophy has great potential for improving the quality and performance of professional services to clients who currently are extremely dissatisfied.

What is needed is a strategy for implementing lean productivity within the construction services professions. I offer the following, in increasing order of difficulty and increasing order of effectiveness, as a tentative approach:

 – Reflection;
 – Recruiting;
 – Rewarding;
 – Re-orientating; and
 – Restructuring.

Reflecting means providing the worker with a periodic opportunity to reflect on the possibilities for flow and conversion improvements. A specific allocation of time for debriefing after each project is an essential element of the lean productivity process. A time for reflection on what went well together with the problems incurred is necessary to provide feedback for future improvements. I suspect (but as yet have no proof) that the crucial factor in workers determining future changes to enable productivity improvements is the setting of priorities by senior management. Senior management must make reflection and feedback the priority rather than rushing into the next project and repeating the same mistakes.

Recruiting of personnel is obviously a key factor. Personnel who are capable of identifying potential changes are obviously essential. This may entail training and further support. What is perhaps less obvious is that the manager of such lean productivity improvement schemes must also be selected carefully. Management becomes support to workers and not control of workers. The wrong managerial approach will lead to worker alienation and avoidance activities rather than improvement activities. Thus the managers of such a process must have the knowledge and recent experience of the workers activity. This presents a conflict with the approach of some firms in using the process as a training opportunity for junior managers or as a safe haven for otherwise redundant senior managers.

Rewarding is a crucial aspect of the lean productivity improvement process. The formal reward structure often acts as a disincentive. Why should the worker make suggestions for improvements if the entire benefit goes to the firm. The manager should also be rewarded for providing the type of management that evokes improvements from the workers.

Re-orientating the managerial structure so that managers face downwards within the hierarchy rather than the more typical upwards. Managers should expect their rewards to be generated by the performance attainment of those they supervise, rather than in recognition of the service that the manager provides to his immediate superior. A support relationship is necessary from the top downwards within the managerial hierarchy. This leads to:

Restructuring, the hardest way to achieve lean productivity improvements. It is however, probably, the most effective way. If the firm treats every managerial role as a coordinative requirement, therefore implying that it is a flow rather than a conversion, it can be examined to see if it can be eliminated completely. Such flows are costs. The managerial role is to coordinate the monads identified in the previous chapter. For each monad the organisation should ask how independent can it be made of others? What is the minimum number of intra-monadal flows and what are the essential mondal connections of these flows? Coordination then only applies when it is justified by monadal requirements or supra-system coordinative requirements.

The organisation which follows these principles will be decentralised, flat and lean.

REFERENCES

BSI 1987. *BS 5750 Part 0.* British Standards Institute.
Betts, M., Fischer, M. & Koskela, L. 1994. *The purpose and definition of integration.* ASCE (Submitted for publication).
Dym, C.L. & Levitt, R.E. 1991. *Knowledge based systems in engineering,* pp. 206. McGraw-Hill.
Eaton, D. 1994a. *Interpretative and modelling problems of risk and uncertainty in bidding techniques.* 1st Congress on computing in civil engineering, American Society of Civil Engineers, Washington, June 1994.
Eaton, D. 1994b. *Lean productivity and the small private practice.* 2nd International Workshop on Lean Construction Edifica Construction Fair, Santiago, Chile, September 1994.
Hillebrandt, P. 1984. *Economic theory and the construction industry,* p. 8. Macmillan.
Hofstede, G. 1980. *Culture's consequences,* pp. 65-109. Sage.
Koskela, L. 1992. *Application of the new production philosophy to construction.* CIFG Technical Report 72. Department of Civil Engineering, Stanford University.
Womack, J.P., Jones, D.T. & Roos, D. 1990. *The machine that changed the world.* Rawson Associates.

Toward construction JIT*

GLENN BALLARD
Department of Civil Engineering, University of California, Berkeley, USA

GREGORY HOWELL
Department of Civil Engineering, University of New Mexico, Albuquerque, USA

1 INTRODUCTION

The acronym JIT has been highly visible since the late 1980's, as manufacturing attempted to meet competitive challenges by adopting newly emerging management theories and techniques, referred to by some as lean production (Womack et al. 1991). What is JIT? What is its relevance for the development and implementation of lean construction theory and practice (Koskela 1992)?

Manufacturing JIT is a method of pulling work forward from one process to the next 'just-in-time', i.e. when the successor process needs it, ultimately producing throughput. One benefit of manufacturing JIT is reducing work-in-process inventory, and thus working capital. An even greater benefit is reducing production cycle times, since materials spend less time sitting in queues waiting to be processed. However, the greatest benefit of manufacturing JIT is forcing reduction in flow variation, thus contributing to continuous, ongoing improvement. Can this approach be applied to construction? What is 'construction JIT'?

2 CONSTRUCTION JIT VS MANUFACTURING JIT

JIT is a technique developed by Taichi Ohno and his fellow workers at Toyota (Ohno 1987). Ohno's fundamental purpose was to change production's directives from estimates of demand to actual demand – a purpose originally rooted in the absence of a mass market and the need to produce small lots of many product varieties.

In assembly line production systems managed by lean production concepts, the directives for production are provided by means of kanban from downstream processes. This system insures that whatever is produced is throughput, i.e. is needed for the production of an order. Kanban works as a near-term adjusting mechanism within a system of production scheduling that strives for firm and stable aggregate output quantities, and provides all suppliers in the extended process progressively more specific production targets as the plan period approaches, resulting ultimately in a firm 2-6 week production schedule. This system provides sufficient flexibility to adjust to actual demand, while assuring that all resources are applied to the production of throughput.

*Presented on the 3rd workshop on lean construction, Albuquerque, 1995

In manufacturing, the need for flexibility comes from a potential difference between forecast and actual demand. Many products are being produced, so it is important to minimize the time required to produce any specific type of product demanded. In construction, there is only one product produced once. And in the case of industrial construction, that product is the facility for producing manufacturing's products. It is consequently important to reduce the time needed to produce the facility, not necessarily the time to produce any component. (Note: This fact often conflicts with the different interests of the various organizations involved in a project). Further, changes arise from progressive definition of customer wants, so flexibility is needed in order to respond to late-breaking changes.

The application of JIT to construction differs substantially from its application to manufacturing because construction and manufacturing are different types of production, and because of the greater complexity and uncertainty of construction.

The extent and significance of uncertainty in construction has been adequately addressed in earlier papers (Howell et al. 1993; Howell & Ballard 1994; Howell & Ballard, publ. pending), but a moment's reflection supports the view that construction is complex. The number of parts, relative lack of standardization, and the multiple participants and constraining factors easily make the construction of an automobile factory more difficult than the production of an automobile in that factory. When this complexity is joined with economic pressures to minimize time and cost, that uncertainty results is not surprising. But is construction really a different type of production than manufacturing, or simply a more complex and uncertain version of manufacturing itself?

2.1 *What kind of production is construction?*

Construction is the final component in manufacturing's product development process. Construction is complete before manufacturing's production begins. Consequently, it is misleading to conceive construction as analogous to factory production (although some aspects of construction fit better in that analogy; i.e. fabrication). Construction is best conceived as a product development process, extending from product design through process design to facility (the manufacturing process tool) construction, the end result of which is readiness for manufacturing.

Admittedly, this is a best fit in the case of industrial construction, and becomes less plausible as we move toward the cookie cutter end of the industry spectrum, e.g. manufactured housing. There seems to be a gray zone between manufacturing and construction, where the work looks like construction because final assembly is done where the facility is to be used, but looks like manufacturing because all that remains of the process is to match production output with sales. This gray zone is obviously ripe for industrialization and mechanization, which ultimately pushes it over into the camp of manufacturing. The proper business of construction is completing product and process design. Once that is done, it is but a matter of time before wit and invention capture mere assembly for manufacturing.

Uncertainty is a necessary component in construction conceived as a product development process. The very purpose of the process is to surface and resolve trade-offs between means and ends, all the way from product design through facility construction. The management of projects so conceived is the proper terrain for lean construction concepts and techniques.

So, construction is a different type of production than manufacturing, and has greater uncertainty and flow variation. Is there an application for JIT in construction?

2.2 *Using JIT to reduce variation and waste: Manufacturing versus construction*

By minimizing inventories between processes, Ohno removed the safety stock that allowed a downstream process to continue working when a feeder process failed. He also required that operators stop the production line when they were unable to fix problems. Consequently, it became necessary to solve problems rather than simply passing bad product down the line. Problems also became highly visible since they could result in line stoppages. Forced confrontation with problems together with analysis to root causes produced a progressively more streamlined and smoother running production process, with fewer end-of-the-line defects and higher throughput.

How might this work in construction? Construction is schedule-driven. Given a well-structured schedule, if everyone stays on their part of the schedule, the work flows smoothly and maximum performance is achieved. However, as we all know, it is rare that projects perform precisely to their original schedule. Business conditions change, deliveries slip, a design requires correction, etc. If a schedule has sufficient slack in the impacted activities, changes may not impact end dates. When there is little or no slack, players are pressured to make it up in accelerated production. Where in this scenario is the opportunity or impetus for reducing variation and waste? Where are the buffers that conceal variation and waste?

3 TYPES OF CONSTRUCTION BUFFERS

There are two types of inventories that can serve the function of buffering downstream construction processes from flow variation (Fig. 1). The most familiar type is piles of stuff; materials, tools, equipment, manpower, etc. These piles of stuff may

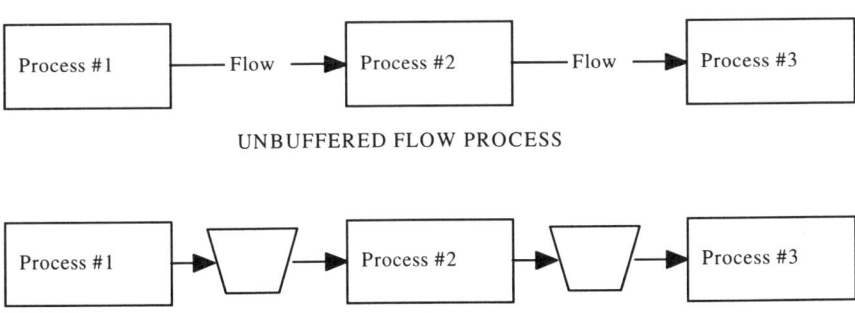

UNBUFFERED FLOW PROCESS

BUFFERED FLOW PROCESS

Figure 1. Use of buffers in flow processes.

originate in decisions to insert certain time intervals between scheduled activities, e.g. between fabrication and installation of pipe spools. Consequently, while they take the form of stuff, they often also represent time added to project duration, so I call these 'schedule buffers' (Howell & Ballard, publ. pending).

Less familiar are inventories of workable assignments, produced by planning processes that make work ready for downstream production (Ballard & Howell 1994). These buffer by enabling a reliable, predictable flow of output from each process. They need not imply the existence of piles of stuff, depending upon the predictability of flow between supplier and customer processes. I will call these inventories of workable assignments 'plan buffers'.

3.1 *Functions of schedule buffers*

In the construction of process plants (petroleum, chemical, food processing, pulp and paper, etc.), projects are frequently fast track; i.e. construction begins before design is completed (Huovila et al. 1994). Late delivery of drawings and materials has led construction contractors to demand earlier delivery, reducing the time available for engineering to complete design, resulting in more delivery problems and demands for even earlier deliveries. This is clearly a vicious circle.

Large schedule buffers between suppliers and construction may shield the contractor from the impact of late deliveries (Howell & Ballard, publ. pending), but does nothing to address the root causes of variation. Further, the shielding is expensive, both in time and money. There is a better way.

A suggested rule: Place schedule buffers just after processes with variable output. For example, that suggests placing schedule buffers between engineering and fabrication, rather than between fabrication and installation. The fabrication and delivery processes are highly predictable, unless drawings are incorrect or incomplete, or drawings are pulled out of fabrication to be revised. A schedule buffer in front of fabrication would provide more time for engineering to complete its work and do it correctly. It would also provide the fabricator an opportunity to select and bundle work to meet his needs for production efficiency and the contractor's needs for quantities and sequence.

Another suggested rule: Size schedule buffers to the degree of uncertainty and variation to be managed. Research (Howell & Ballard, publ. pending) has shown that schedule buffers are sized without regard to the toughness of projects; i.e. their level of uncertainty. This amounts to wasting time and money accumulating piles of stuff not all of which is needed.

3.2 *Functions of plan buffers*

Schedule buffers do not replace plan buffers. Plan buffers are necessary even when schedule buffers are in place because having a pile of pipe does not provide a piping crew with workable assignments. Pipe spools must match with valves, controls, hangers, gaskets, bolts, welders, lifting equipment, etc. Structures for supporting the pipe must be in place. Preferably, the spools that can be installed are those that should come next in an optimum constructability sequence. Assembling physical components, reserving shared resources, determining optimum sequencing, and siz-

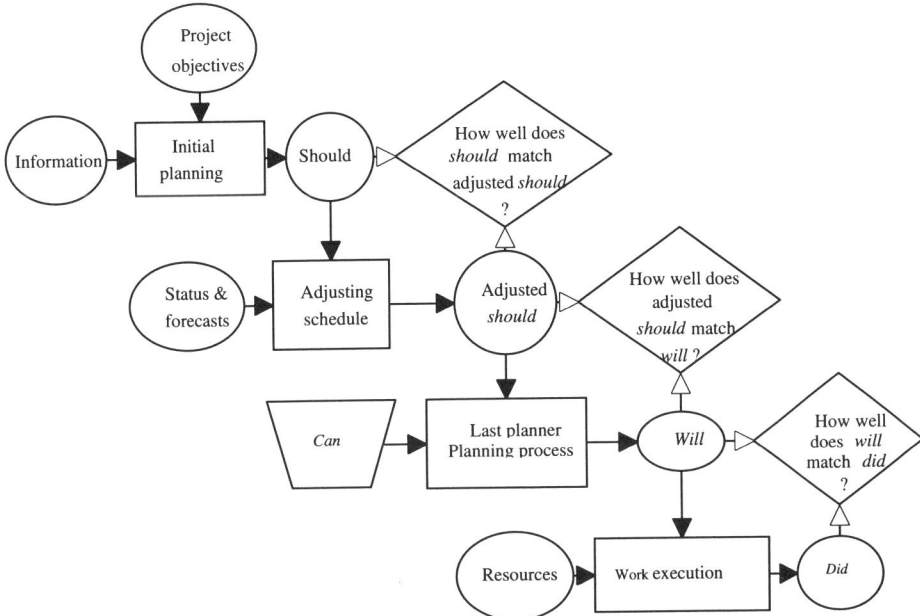

Figure 2. The construction planning process.

ing assignments to absorb the productive capacity of the crew is best done prior to making assignments and committing to what work will be done in the plan period, usually one week.

Plan buffers, sometimes called backlogs of workable assignments (Ballard & Howell 1994), are the outputs of make ready processes. They determine what *can* be done as distinct from what *should* be done (Fig. 2). Obviously, commitment to what *will* be done next week can only come from *can*, regardless of the pressure for production and the need to make up schedule slippages. The common practice of pressuring for production regardless of *can* is rooted in a theory of construction project management that disregards capability and management of flows in favour of schedule push and management of contracts (Howell & Ballard 1994).

By monitoring the match of *did* with *will* using the measurement of PPC, the percentage of planned activities completed, and acting on the root causes of non-completions, we can learn how to produce better plans and how to do what we plan to do. The implications for work flow, project durations and productivity are enormous.

Think of the complete construction process, from engineering through installation and start-up, as a complex of work processes, with work flowing from one to the next. When a downstream process attempts to plan its work and determine the resources it will need, it may have shielded itself from unreliable inflow using piles of stuff or schedule spacing. However, it only needs those piles of stuff if supplier processes cannot reliably do what they say they are going to do. If supplier processes consistently achieve PPCs near 100%, customer processes can plan their work and match resources to it. Reduction of schedule buffers and better matching of resources to work flow both contribute to reduction of project time and cost.

3.3 *Plan-pull versus schedule-push*

Make ready processes produce inventories of workable assignments by 'pulling' forward resources needed to do that work that will best contribute to throughput at each point in time. Resources were procured and distributed in accordance with schedules; i.e. the work was driven by schedule-push. Now the driving mechanism becomes plan-pull.

Plan-pull can include reference to successor readiness. For example, the decision to install this structure or that may be determined by the predicted delivery of equipment and piping for each of those structures. Generally, *should* is continuously tested against *can*, and *will* is selected from the best of the available alternatives. While plan-pull mechanisms are common, industry thinking has not recognized their role and importance.

4 A STRATEGY FOR CONSTRUCTION JIT

The JIT ideal is elimination of physical buffers (materials or time) *between* production processes, and the achievement of one piece flow (Howell & Ballard, publ. pending) *within* processes, i.e. batch sizes of one. Ohno was able to virtually eliminate such in-process inventories because production scheduling provided sufficiently stable coordination of flows.

Construction scheduling does not provide such stabilization. Consequently, it is not appropriate to simply eliminate physical buffers without first attacking the causes of variation and uncertainty. Even though manufacturing and construction share the same ultimate objective of reducing variation and waste, their strategies for achieving that objective must be different.

I propose as a strategy for construction JIT:
1. Better location and sizing of schedule buffers;
2. Immediate implementation of plan buffers and make ready processes in front of production processes; and
3. Progressive replacement of schedule buffers by plan buffers.

4.1 *Better location and sizing of schedule buffers*

It will require developing better assessments of project uncertainty and determining the quantitative relationship between buffers and the uncertainty they are intended to buffer. It will also require experimentation with relocating schedule buffers, to test the principle of locating buffers just behind processes that are the source of flow variation.

4.2 *Place plan buffers and make ready processes ahead of each production process*

The last planner (LP) initiative, has been described in some detail in previous chapters (Ballard & Howell 1994). Although it has been experimentally tested in both the United States and South America (Venezuela), it may be helpful to consider it as a research hypothesis.

Hypothesis: Production can be shielded from upstream uncertainty through planning.

– Benefits of the research: The last planner method of detailed production planning shields production from upstream uncertainty thus improving productivity, revealing sources of uncertainty and variation, releasing resources for further improving performance 'behind the shield', and providing a highly predictable near-term work flow to downstream processes.

– Methodology: a) Solicit engineering and construction projects from industry; b) Evaluate the crew/squad level planning systems of each participant; c) Help participants conform their systems to the last planner model; d) Develop measurements of comparative productivity: Before and after LP, Between LP and non-LP; e) Collect measurement data; i.e. percent planned assignments completed, planned productivity, and actual productivity (Fig. 3); f) Analyze measurement data and test hypothesis.

– Characteristics of the last planner method: a) Written weekly work plans for each front line supervisor and work group; b) Assignments drawn from a backlog of workable assignments created by screening for constraints and by acquiring necessary resources; c) Assignments expressed at the level of detail necessary for screening constraints and for statusing completion; d) Weekly work plans sized to target productivity; e) Front line supervisors participate in the selection and sizing of assignments, provide reasons why planned work was not done, and share responsibility with other craft or discipline supervisors for acting on reasons within their power to avoid repetitive failures; f) Craft superintendents/discipline supervisors see that others act on reasons beyond the reach of the craft or discipline.

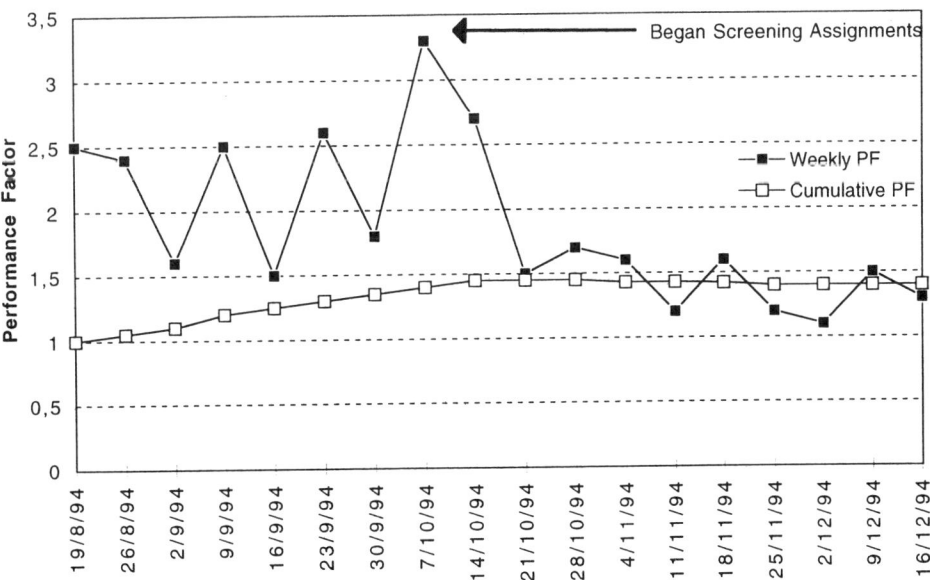

Figure 3. Screening and productivity.

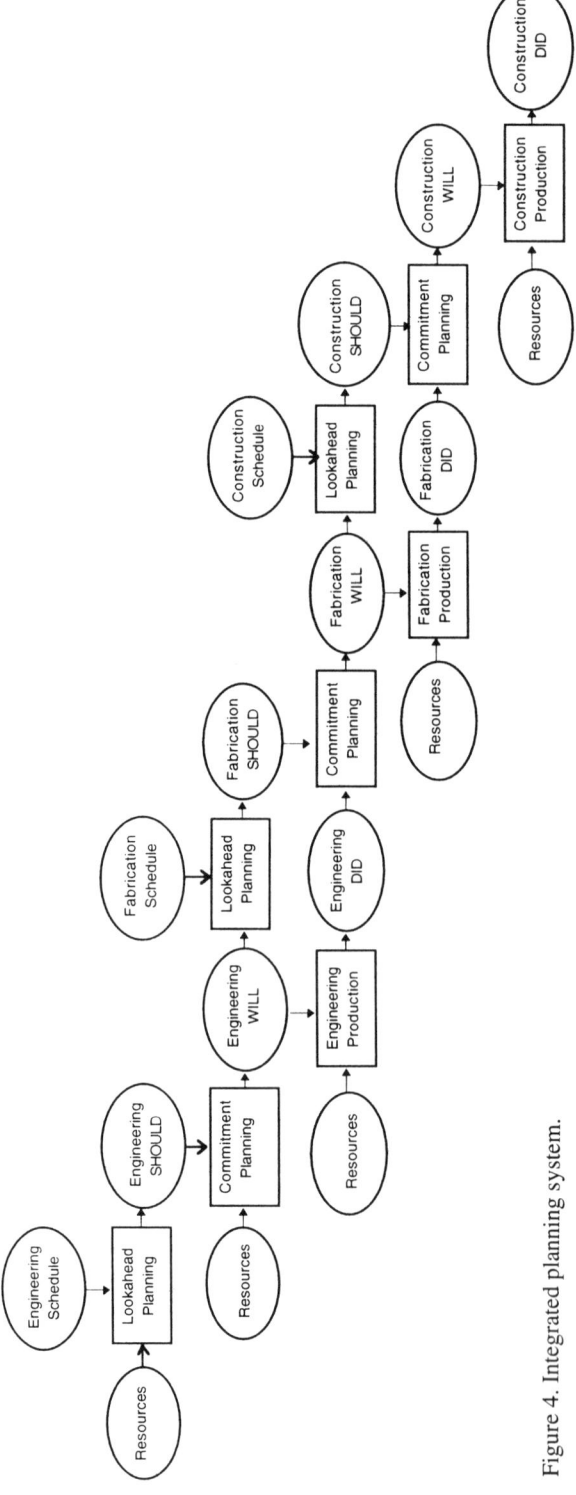

Figure 4. Integrated planning system.

4.3 *Progressively replace schedule buffers with plan buffers*

The long term goal of replacing schedule buffers with plan buffers. How will the construction industry achieve that level of predictability of work flow that will eliminate the need for piles of stuff or time between production processes?

Hypothesis: Work flow variation can be reduced:

– Benefits of the research: 1) Project duration can be reduced by reducing the buffers between EPC functions, and buffer sizes can be reduced if work flow variation can be reduced. 2) If work flow can be made more predictable, labor and other resources can be better matched to work flow, thus improving productivity;

– Methodology: Phase I: Identify and analyze examples of successful efforts (tools and techniques) to increase the predictability of work flow; and Phase II: Test tools and techniques in experiments sponsored by industry members.

– Examples of tools and techniques: a) Developing more accurate assessments of project uncertainty; b) Adjusting schedules using work packages and milestone screening. Strategy: Extend look ahead and make ready processes progressively further back into the future, from each customer station to each supplier station, until limits of predictability are met. Act on constraints to push back limits; c) Buying information to extend the accuracy and range of forecast deliveries; d) Producing more advance warning of changes in design, scope or quantities; e) Integrating supplier and customer schedules at the item (e.g. isometric) level of detail (Fig. 4). Redesign processes for determining need dates for delivery of vendor data to design and materials to construction.

5 CONCLUSION

Construction JIT will be advanced by implementing demonstrated techniques and industry research to test theories and develop new tools and techniques. Research topics have been proposed that constitute a strategy for implementing construction JIT.

Construction and manufacturing are different types of production, nonetheless a form of JIT is applicable to construction, in which physical buffers may ultimately be replaced by better managing uncertainty and eliminating the causes of flow variation. As the implementation of plan buffers propagates certainty throughout projects, productivity will improve from better matching labor to work flow, and project durations will shorten as physical buffers shrink with the flow variation they are designed to absorb.

A new way of conceiving the tasks and tools of construction project management has been proposed. Instead of relying simply on schedule-push, managers are advised to systematically employ plan-pull as a means of adjusting to uncertainty and insuring that resources are employed to maximum advantage at each point in time. Instead of concentrating management attention and effort on managing contracts and enforcing obligations, managers are advised to manage the flow of work across production processes and the various specialty organizations brought into a project to execute those processes.

REFERENCES

Ballard, G. & Howell, G. 1994. Implementing Lean Construction: Stabilizing Work Flow. *Conference on Lean Construction,* Santiago, Chile.

Howell, G. & Ballard, G. 1994. Lean Production Theory: Moving Beyond 'Can-Do'. *Conference on Lean Construction,* Santiago, Chile.

Howell, G. & Ballard, G.(publ. pending). Managing Uncertainty in the Piping Process. Source Document (number to be applied). Construction Industry Institute, Austin, Texas.

Howell, G., Laufer, A. & Ballard, G. 1993. Uncertainty and Project Objectives. In: *Project Appraisal.* Guildford, UK.

Huovila, P., Koskela, L. & Lautanala, M. 1994. Fast or Concurrent – The Art of Getting Construction Improved. *Conference on Lean Construction,* Santiago, Chile.

Koskela, L. 1992. Application of the New Production Theory to Construction. *Technical Report* No. 72, Centre for Integrated Facilities Engineering, Stanford University.

Ohno, T. 1987. *Toyota Production System.* Productivity Press.

Shingo, S. 1981. *Study of the Toyota Production System.* Japan Management Association.

Womack, J.P., Jones, D.T. & Roos, D. 1991. *The Machine That Changed The World: The Story Of Lean Production.* New York. 1st Harper Perennial Edition.

Building as never before*

HÅKAN BIRKE, JAN ERIC JONSSON & PETER TOLF
Arcona Ltd., Nacka Strand, Sweden

1 ARCONA

Arcona is a Swedish company listed on the Stockholm stock exchange. Consolidated annual sales are slightly over 1.5 billion Swedish kronor (eq. to over $200,000,000) and the total number of employees is currently in the range of 1,000. Arcona is organized into three business areas: Automobile, real estate and construction.

Arcona's core business is construction. Most of our assignments are turnkey projects (design-build contracts) conducted in close cooperation with the customer, and with full responsibility for costs, quality and time schedules. All construction work on-site is carried out by subcontractors under Arcona's supervision.

Our construction business is organized into three companies that work closely together. 'Arcona Project' is our project management organization with 80 employees including joint staff for computer services and technical development. The subsidiaries 'BSK architects' and 'Energo' (electrical, water and HVAC engineering) with a total of 120 employees are responsible for planning and design. 20-25% of their operation is related to turnkey projects within Arcona, the remainder are projects for outside customers.

Arcona's primary market is the Stockholm region. Our sales are in the range of 500 Million Swedish Kronor, equivalent to 70 Million US dollars. 80% of sales are related to design-build contracts, 20% to consulting activities. Profitability is higher than the industry average.

2 THE CONSTRUCTION INDUSTRY IN SWEDEN

Generally, throughout Sweden's construction industry, the education level as well as investments in R&D are lower than in other Swedish industries. The development of productivity is most unsatisfactory. During the last 20 years the time needed for final assembly of a square-foot residential house, the industry's largest product category, has increased by 20%. At the same time, production costs in real, inflation-adjusted currency more than doubled. Profitability in the industry has always been strongly dependent on the domestic market situation. The average profitability is rather low.

In our industry, very little attention is paid to the idea of 'process improvement'.

*Presented on the 3rd workshop on lean construction, Albuquerque, 1995

Productivity seems to be of very low interest and is hardly monitored. The industry looks upon itself as working on 'projects' and projects are by definition unique. It is generally thought that application of the 'lean production' concept to the construction industry is impossible or at least difficult. Furthermore, it is taken for granted that lean construction offers only marginal savings potential. A reason often referred to, is that there is not enough repetitive work in the construction business.

The behaviour of the customers – when buying from the industry – reflects this thinking. There is a strong tendency to sub-optimize, for instance by making *independence* a more important issue than competence and efficiency. As a consequence, the relationship between a customer and his consultants or contractors is often a win-lose or even a lose-lose relationship.

3 ARCONA'S GOALS

In Arcona's opinion, it is more feasible to apply the lean production philosophy to construction. Furthermore, the potential for increased productivity by doing so is substantial.

Since the industrial processes to which we refer are basically driven by individual customer's requirements, we realize that our construction projects can be allowed to be unique. The process and technique we develop must be able to handle individual demands on the final product as well as different decision-making processes from the customer's side.

Five-year goals were set for Arcona in 1994: 50% reduction of assembly time and 30-40% reduction of fixed costs depending on project type.

4 ARCONA'S METHOD

The following explains some important elements of our building method and some other activities we undertake to achieve our goals.

Time as the driving force. By setting tough deadlines on the final assembly time, we are forced to change the way we run projects. We are forced to develop our process and technique, we are forced to remove traditional professional barriers, we are forced to increase prefabrication and we are forced to dramatically raise the overall precision level in order to eliminate unexpected, unnecessary adjustment work.

Altogether this will lead to reduced consumption of all resources. Not only time for assembly, but also manpower, materials, transports and capital. This, in turn, leads to our primary goals, the reduction of costs as well as increased product quality.

Relations. Our relationship to our customers should be based on complete openness and supported by mutual incentives – a win-win relationship. We try to establish a long term relationship that makes process development and goal-setting important issues of both parties.

Precision. The high precision level we strive for in our work process is based on a detailed 'program' (job specification), an accurate plan for decision-making, and well-defined strategies and goals for the product and process, mutually established by the customer and ourselves.

To ensure that the established precision level will be maintained consistently through the process, we determine prefabrication units, assembly sequences and border lines between different contractors in the earliest stages of designing the product. That is one of the reasons why our strategic suppliers must be involved from the very beginning.

Planning. The schedules are very detailed and produced with high accuracy. The more we can rely on the prerequisites of the schedules and their accuracy, the more we can plan without slack time and the less we hesitate to 'assemble everything as late as possible'.

Continuous improvement of the process. Non value-added activities are continuously reduced or eliminated.

Continuous improvement of the technique. This is done in close cooperation with our suppliers. The goal is to reduce time used for final assembly, to a large extent by using fewer components with fewer and simpler fasteners. On top of that, contractors should always be allowed to complete their work on site in *one* sequence and without interruptions.

The JIT principle. Our goal is that no component in the entire supply chain should be picked from a storage shelf but manufactured on the customer's order and delivered in the exact amount and quality *just in time* for the assembly.

5 COOPERATION WITH STRATEGIC SUPPLIERS

One of the key issues of Arcona's method is the selection of, and long term cooperation with strategic suppliers.

The most strategic suppliers are the Designers and that is why we have proprietary resources in this area, mainly from our two subsidiaries, BSK Architects and Energo. The second most important group of vendors are the Electrical, Water and HVAC Installers. Since one and a half year ago, Arcona, BSK Architects, Energo and three outside installation contractors have formed a fixed team operating together on all turnkey projects. This team works towards jointly approved overall objectives, supported by a series of operative stage-by-stage goals, which are continuously monitored. Openness and mutual incentives are as important within this 'inner circle' as they are in our relationship with the customer. To keep up the pace and the motivation, it is also important that the team is kept together from project to project.

Continuous exchange of know-how takes place in a number of working groups, currently covering the following fields:
 – Planning;
 – Organization;
 – Problem definition (obstacles to efficiency in assembly work);
 – Process;
 – Technique.

6 MEASURING AND BENCHMARKING

It is important that we always know that we are heading in the right direction and moving forward fast enough. For that purpose, we have defined a set of supplementary goals in terms of numbers indicating the most important work method improvements we want to accomplish. These key figures are measured continuously and new, bolder goals are set for every new project.

Benchmarking is another important tool for speeding development. So far, we think that we have picked up much more inspiration and good examples from other industries than we have from our own.

Two of our three strategic suppliers are members of the ABB Group, the global Swiss-Swedish engineering company. Since 1991, all ABB companies have run a customer-oriented, time-based improvement project called T 50. Our cooperation with the two ABB companies has made valuable experience from the T 50 project available to Arcona.

7 TODAY'S SITUATION

As of today, we have recorded considerable progress in terms of shortened assembly times, increased prefabrication content, cleaner sites, more accurate planning and control, improved quality standards, and shrinking costs. We are definitely on the right track but of course, we still have a way to go before we reach our ambitious goals.

Schedule compression: A case history*

PETER N. WOODWARD
Consultant, Albuquerque, New Mexico, USA

1 CASE HISTORY: SCHEDULE IMPROVEMENT

I'm here to talk about a case history. This project is still in progress, so it's a little soon to celebrate victory, but the elements are all in place and we are certainly going to achieve the bulk of our goals.

I was sceptical of the statement of Howell and Ballard that most project plans contain room to improve the schedule. In my experience, construction projects always assume a momentum after the contract is entered, and they seldom ever improve. However, I had to admit that this one had an opportunity for a significant improvement in the schedule, an opportunity that could only be exploited through a contract adjustment and careful development of the project plan.

This brief talk is about how it was – and is – being done. And in the process, we'll illustrate some lean construction techniques.

2 GOAL ADJUSTMENT

The schedule improvement was in plant commissioning. In a word, what we did was advance the start of commissioning, overlapping it significantly with construction completion.

This strategy effectively involves the contractor in plant start-up, not necessarily a new idea, at least in industrial class construction. But the job we are talking about was framed contractually as a standard general contract, in which a scope of work is completed and then something else can go on.

In this case the 'something else' is starting up an industrial process. Once we realize that our immediate goal is to start an industrial process, we can view our whole plan differently. The question becomes, 'What do we have to do to start up our industrial process?' The answer to that question might be different that the answer to, 'What do we have to do to finish this contract?'

In fact, starting the process is a whole other project with a logic of its own, a logic we may be relatively unaware of. So, the task for the planner is to determine what the buyer's start-up schedule looks like.

*Presented on the 3rd workshop on lean construction, Albuquerque, 1995

In this case I went forward with my own assumptions – to see if they corresponded with the needs of the plant user. They did, with minor exceptions. The more interesting thing is that the user realized immediately what we were up to – and that there were advantages for him in the approach we were proposing – like two to three months earlier start on readying his complex equipment for production.

I argued that we did not have to achieve contract completion in order to proceed with the most essential element of this goal. In plain terms, we could start up the systems the user wanted first – before the contract was completed.

We were given the go-ahead to proceed on that basis and at the moment the user is looking forward to the prospect that some elements of the process will be available for him to start what he sees as commissioning – getting the process going – some months before they were expected. This is an almost unheard of improvement in a schedule that started out struggling against a delay and significant drawing revisions.

This project demonstrates how you have to focus on the project goals and not get diverted from those goal – even by such significant 'sub-goal pursuits' as completing the contract(s).

3 STABILIZING WORK FLOW

I was not involved with this project from the beginning but came on board around the time the building was coming up out of the ground. My task was to determine if the project could be built within the original contract schedule.

I ran some man loaded CPM simulations that showed the schedule completion date to be only marginally responsive to large man-hour additions (through change orders). However, they showed the 'permanent power' milestone – a much nearer date – to be very sensitive to changes in manning. On that basis we accelerated the electrical and mechanical subcontractors.

System start-up and commissioning is analogous to permanent power in that the same basic rough-ins are required for both. Also, the calendar dates were about the same.

You can't just advance commissioning into construction; you have to prepare the ground, in this case establish an 'inventory' of 'startable' equipment and systems. By accelerating mechanical and electrical rough-ins we established that inventory of startable systems.

Occasionally the logic of getting a particular area ready made elements of the architectural work critical. Usually we were able to deal with it simply by rearranging priorities around the systems we wanted. It was a matter of making sure that everyone was clear on those priorities. Occasionally we used spot overtime, but otherwise architectural work continued on a forty hour week.

What we needed was all that process piping, ductwork, and electrical power to start the equipment. To get it, we accelerated the mechanical and electrical subcontractors through the bulk of their rough-ins. This included probably two thirds of their total man-hour budgets. We used a 'medium level' acceleration – a fifty hour week of five tens – to reduce the 'burn out' effect on the men and excessive loss of productivity, which would be the owner's problem in a directed acceleration.

We took the mechanical and electrical contractors off overtime just as burnout was becoming noticeable. At that point we were building up our controls wiring crew, which we put on a 60 hour week. It had become apparent that we had to continue accelerating the critical work elements to maintain our inventory of startable systems.

4 FIRST RUNS OF REPETITIVE PROCESSES

Start-up and commissioning is a complex process. However, like most other construction processes, it breaks down into somewhat more simple, repetitive tasks. So a great deal can be learned from first runs of commissioning sequences.

We were fortunate that a small part of our project was a remodel that was complete long before the main part of the project. Commissioning that remodel was a perfect miniature of the full plant and it was educational, to say the least. Our first attempts to schedule commissioning of the remodel were awkward, the logic was off the mark, and it would be hard to distinguish between planning and execution failures, since there were so many of both.

But the exercise showed once again that you learn more from failure than success. We learned how a commissioning sequence was structured and more importantly, who was responsible for what, and relationships between the parties. Bear in mind that when you get into commissioning, you are dealing with a larger group of parties, some of whom are not subs to the general contractor, and some of whom have some well developed ideas about how they like to run 'their' jobs.

One of the things we learned is that you start up everything you can as soon as you can. There will always be something to modify in your equipment. And in commissioning, you have to ensure that mechanical systems are complete and operating to provide a consistent flow of work to the controls people. It is tempting to put off start-ups that seem non-critical, but there is too often mechanical modifications or missing parts when you finally start up the equipment. If you wait, your float time to order or change out missing parts, or make small modifications in the plans, is gone.

Start-up and commissioning had barely been touched on in the original schedule. We took an unplanned, 'muddle through' process, much of which was destined to take place after contract completion, planned it in fine detail, and advanced it as far into construction as possible. Because we had prepared the ground by accelerating the necessary construction processes, we were able to achieve a result far beyond the expectations of any of the parties.

5 CONTINUOUS PLANNING

It was my late entry into the project that forced me to do the detail planning of the next phase so far into construction. At the time we were looking over a three month horizon. Planning through that period made good sense to everybody. All parties were cognizant of the logic and problems governing the next three months.

I call this horizon the 'medium term' to distinguish it from 'long term' which is the original scheduling done before the start of work and 'short term' which is the weekly planning done by field forces. Long term was already complete when I got on board and I never got involved in short term until we were deep in commissioning.

However, the field forces who did the short term planning were among the primary information sources for medium term planning, so the flow of information from medium to short term plans was good, and in fact we've had good coordination and productivity throughout.

Medium term is a level you can detail with hard data. Your sources can 'see' that far ahead quite clearly. I did most of my planning on a one-to-one basis, then routed the result back to the source individual for review after it had been fitted into the overall schedule. That way everybody had the certain knowledge that they had planned their own work. They all understood what they had to do, and they delivered. You can't get that kind of response on the original planning schedule; it isn't accurate or detailed enough due to the high level of uncertainty that exists at the time it is made.

We did our medium term scheduling right in the original planning schedule. We modified it continuously, adding new logic, or replacing general logic with more detailed logic. So remember that the project schedule is evolving as the job progresses. It is important that the owner understands that the schedule will have that kind of flexibility. It isn't really compatible with a billing format and in this case the two went their separate ways.

6 INDEPENDENT SCHEDULER

When I started work the general contract had already been written. The contractual problems this created were avoided by making me a consultant to the general contractor – paid for by a change order from the owner. My independent role was written into my agreement.

If it appears that some of the work I did, such as the focus on project goals, the alignment of the schedule with the goals, and a stress on resolving conflict, was more in the spirit of partnering, that is no coincidence. I was recommended to the owner by the partnering consultant, and as a long-time disputes resolution consultant, I have a built-in predilection for partnering.

I believe that if an owner wants continuous planning – and in any complex project where the schedule is a prime concern he should want continuous planning – an independent planner scheduler is almost a necessity. Once a contract is entered, fulfilling that contract always takes over as the project goal. Lip service may be paid to improving schedule or value engineering, but in reality, the elements that need to be flexible to achieve optimum performance have been cast in concrete. Change is difficult and the parties to the contract will balk at it.

An independent scheduler is under no such constraints. His scope of work is to plan – continuously, always progressing the schedule from the general to the specific, always searching for the opportunities that may only become apparent in that medium term envelope – the visible future.

The idea of involving the partnering specialist in the actual work is new – just like the independent scheduler. But it is an idea whose time has come.

Ultra fast-track project delivery: 21st century partnering and the role of ADR*

ROBERT S. MILES
Industrial Design Corporation, Portland, Oregon, USA

ABSTRACT: Ultra fast-track design/construct methods are being developed and implemented by the leading edge of the construction industry. Facility owners, design firms, construction companies, and equipment suppliers are redefining the relationships needed to deliver built environments (some of which are among the most technically complex) in previously undreamed of short duration.

These partnering style relationships are designed to leverage the experience of past projects and to forge long term business and personal relationships. Lean construction principles are being implemented by experimentation. Project teams with members from each of the partner organizations move from project to project.

Authority levels are built into the 'systems' of the project. Since the project monetary and psychological incentives are built upon common goals, cooperation and creative problem solving attitudes are institutionalized.

Issues are resolved at the lowest possible levels. Escalation of issues is possible, but looked upon by cultural definition as failure. ADR methods 'grow' from the project team structure, personalities, and management styles. Any formal form of dispute resolution (including presently considered progressive stepped ADR) is the last resort. Each of the partners has too much at stake in the relationship to consider litigation.

However, driving decision making and issue resolution down to the individual members of the self-managing teams has created a critical need: training of parties below the project management level in issue resolution skills.

1 INTRODUCTION: 'ONLY THE PARANOID SURVIVE'

1.1 *Who will run the ASYLUM? ...and who truly are the 'sane'?*

Andy Grove, CEO of Intel Corporation, has been quoted as saying that 'Only the paranoid survive'. As major providers of design and design/construct services for the high technology industry, Industrial Design Corporation (IDC) and Technology De-

---

* Presented on the 3rd workshop on lean construction, Albuquerque, 1995

sign and Construction Company (TDC)[1] are faced with the challenge of providing facilities for our clients that enable the manufacturing of products which evolve faster than the plant can be designed and constructed. To perform such services we must be as 'paranoid' as our clients. What is sufficient today is totally inadequate (virtually) tomorrow. We cannot afford relaxing or assuming that the 'solutions' have been found, if we are to maintain a leadership position.

Manufacturing innovations in the last decade have been based upon the concepts of total quality management and now more recently on lean production[2]. Just-in-time delivery has become the standard, no longer the innovation. It has generally been the leading edge high technology manufacturing sector that has pioneered 'faster, better, less expensive' processes. No longer are these three goals seen to be mutually exclusive goals. High technology industries are now the driving force behind remolding the method of built-environment delivery. At the heart of these methods are building of multi-functional and multi-tasking working groups.

This chapter will contrast traditional, contemporary, and leading edge design/ construct processes. The discussion will lead to a requirement for a step beyond the limiting concepts of traditional alternative dispute resolution. Even current partnering concepts need to evolve from formal relationships into being parts of a living organism. We have found it to be true that 'crazy times call for crazy organizations' (Peters 1994)[3].

1.2 *So what then? ...What once was, now is not, will be again*

There was a time when the architect and the artisan worked side by side through the life of the project. Both knew each other's trade and respected each other's unique skills and talents. They complemented each other and were dependent upon each other in order to achieve shared goals for the project.

The level of quality of the design and construction was a point of pride, and all involved in the project took ownership of the finished result. While a form of division of labor existed, it was not a hindrance to the goal.

The design and construction took place simultaneously, with the architect literally living with the project from start to end. Design details were developed as a team effort, with the design evolving as the project was built. In ancient Greek construction projects, 'The contractors were responsible for cutting and shaping the blocks at the quarry and for transporting them safely to the site. There they would be trimmed

---

[1] IDC provides full services architectural, engineering, and construction management. TDC is the sister company that provides design/construct services.

[2] The concepts of Lean Production in the manufacturing sector include the following: Keep systems simple and avoid waste. Encourage the use of self-managing teams that are empowered to improve processes. Create multi-tasking and multi-lateral (cross discipline) working groups in order to expedite and simplify communications. For the fundamentals of application to the construction industry see: (Melles, undated).

[3] While the requirement for strict client confidentiality will preclude reference to specific projects, the general experiences of IDC and TDC that are driving to new methods and processes will be used to present new forms of relationships on construction projects.

down to their final surface for proper fitting. *The finishing and assembly ...were the most exacting responsibilities of the architect'.* (Kostof 1985)[4] .

The end results of these projects continue to this day to inspire awe and admiration. We wonder at the accomplishment, and marvel at the result. We cannot but question whether our major improvements in tools of the trade and technology are not wasted in the effort to find (or rediscover) an appropriate method of delivery for projects.

2 TRADITIONAL PROJECT DELIVERY 'WHEN DINOSAURS ROAMED'[5]

2.1 *Design – then – build... and watch the client's competition pass them by*

Defining the beast
Traditional delivery of the built-environment in the US has evolved into a rigid 'phased'[6] process. These are clearly defined by the American Institute of Architects (AIA). The phases are the backbone of the standard AIA and EJCDC[7] construction contract language and the CSI[8] format of the standard contract documents used in the USA. The phases are as follows:
- Programming; *then*
- Schematic design; *then*
- Design development; *then*
- Construction documents; *then*
- Bidding; *then*
- Bid analysis and award; *then*
- Construction; *then*
- Startup and commissioning; *then*
- Project turn-over (to the owner).

Perceived advantages
The traditional delivery method is often credited with providing for the greatest amount of competition, with the resulting lowest 'cost.' The production of well-defined design construction documents and a sufficiently long bidding phase purportedly allows for the 'best price' to be developed. Roles and responsibilities are arguably considered clearly defined. The design and construction schedule can be easily prepared in a linear fashion.

Each party to the project can resort to the definition of their 'responsibilities' found in the construction documents. On the surface, this seems simple and safe.

---

[4] Emphasis is mine.
[5] The reader will perhaps need to excuse the literary license of the author in utilizing some American movie titles and themes to emphasize the titles and headings.
[6] Phased delivery in the context of the Traditional Delivery process should not be confused with delivery of Phased, or Packaged, design/build. The latter form will be discussed later in this paper.
[7] Engineers Joint Construction Documents Committee.
[8] Construction Specification Institute.

2.2 *Reality always settles in... The meteorite lands – Rest in peace Tyrannosaurus-Rex*

Deficiencies come to the fore
The construction industry is well aware of the vacuous reality of the claimed advantages of the traditional delivery process. In fact, many are the seeds of the litigation explosion in the US construction industry planted in the soil of the traditional delivery process.

The seeds of discord are planted
The traditional delivery method, with its rigid definition of roles and responsibilities, provides no basis for a shared vision or shared goals. Each party has their own agenda based only on their singular interests. The legal contracts that bind the parties together become the basis for finger pointing, claims, disputes, and broken working relationships. Even at its best, this method fosters poor cooperation and resultant low efficiency.

The traditional role of the architect as the first resolver of disputes is a contractual fantasy. This *impartial* interpreter of the 'intent' of the construction documents just happened also to have prepared them.

Constructors, living in the highly competitive marketplace, discovered attorneys (or perhaps the reverse). They found that they could bid low, then recover the margin in costly change orders based upon errors and omissions in the design documents. The market competition no longer allowed constructors to bid with contingencies to cover an understandable number of missing or unworkable design details. The courts agreed that they could collect – not only for the direct cost of the added work, but also for 'damages for delays' stemming from the design deficiencies.

The owner is then in the unenviable position, contractually, of being in the middle. They hold separate contracts with the design professional and the constructor. Three-way litigation (four-way if the bonding agent became involved) often results.

Time truly is money
The Traditional process is painfully slow. As financing cost pressures increased in the US, first due to sky-rocketing interest rates in the 70's then to tight lending policies of the late 80s and 90s, any delay in the already slow process resulted is flurries of law suits due to the financing cost increase and the delay of a positive cash-flow. All these costs can and often do absorb part or all of the cost advantages of competitive.

High technology industry in particular was quick to learn that time-to-market was everything, and the sequential aspect of the traditional process could not be tolerated. The potential failure to capture the early, and most profitable, segment of sales is not acceptable. Appendix A shows a hypothetical simplified Gantt schedule of a large high technology facility project. The total duration of 47 months, even with some overlap of startup and tool install, is totally unacceptable in this time-to-market driven industry.

3 DESIGN/BUILD: 'JURASSIC PARK' – THEY LIVE AGAIN, AND THEY'RE OUT OF CONTROL

3.1 *Sole source responsibility ...the worst of both worlds brought together*

Owners, having been financially torn asunder in disputes over *ultimate* responsibility for project problems, looked for a way to have a single guilty party. They were also frustrated with poor preliminary estimates of construction costs prepared by design professionals. Design firms were not in a position to properly predict the fast changing construction market. They wanted to change the nature of construction from a process to a goal, to purchase of a product. They wanted a hard cost, known up front and fixed.

3.2 *Design/build is born ...a gene splicing experiment gone wrong*

The rise of design/build in the US was a reaction to the marketplace's dissatisfaction with the traditional process. Owners no longer accepted the concept of each project being a unique creation, something never done before. Instead they viewed the built environment as a product. However, unlike a manufactured product that could be prototyped, tested, and redesigned before it was delivered to the consumer, a building cannot. For a building the process *is* the product. So, by attempting to marry the design and construction into an unnatural union the industry bred instead an animal out of its natural environment and out of control.

Design/build removed the previous check-and-balance between the design professional and the constructor. The projects are competitively bid to design/builders, most frequently with only the sketchiest of performance specifications. Generally the only existing companies in the US able to perform this new form of work are the construction contractors. It was no surprise that the first place to cut cost to get the job and profitably produce the 'product' was in design. Next was a reduction in quality of materials that are not readily apparent to the owner. Without the traditional delivery method's *independent* design consultant to protect the owner's interests, quality suffered.

The resultant problems in the delivered project became solely the owner's problem to resolve with the design/build 'manufacturer'. They had bought a product. The sole source responsible party delivered the product: a once and forever, first time prototype, delivered *with no prior testing*.

3.3 *Fast-track ...don't get in the way of a charging dinosaur*

The concept of compressing the project schedule was the logical next response to the need to better deliver the project. There is a limit on how much a project can be accelerated, unless tasks overlap. Appendix B contains a new version of the Appendix A project schedule. In this schedule tasks are overlapped to reduce the overall duration.

The dangers of mis-executing fast-track are built into the system. The methods are much more complex. There is far greater opportunity for errors and omissions. The speed and the overlapping of design and construction make communications more difficult, but at the same time more important. Errors will require substantial re-design, re-work, and schedule delays. Unless these challenges are addressed, the re-sult is likely to be cost and schedule overruns.

The problem remains that with traditional fast-track we have only made more complex the archaic systems, relationships, and structures. Fast-track alone is not the answer. It must be coupled with entirely new philosophies on the working relationships and systems. This is the only way that disputes can be avoided, and project goals successfully achieved for all the parties to the project.

4 TWENTY-FIRST CENTURY PARTNERING: A RETURN TO RELATIONSHIPS, OR 'BACK TO THE FUTURE'

'The significant problems we face cannot be solved at the same level of thinking we were at when we created them...' (Albert Einstein).

The high technology manufacturing sector is paving the way for the demise of the traditional delivery process. The advantages of the new methods vastly outweigh the risks The perceived and real risks of progressive delivery methods will diminish with use. Costs will further diminish as smarter ways of working and long term relationships lead to new innovations in products, processes, and technology.

The remainder of this paper will present the operating philosophy and fundamental principles that IDC and TDC have implemented. These have resulted in the reduction of design and construction duration by almost one half while maintaining the highest of quality and on budget performance. This has been accomplished on projects that push the envelope of modern high technology – submicron, cleanroom, semi-conductor fabrication facilities.

4.1 *Partnering ...a modern definition*

Partnering in the context of a 21st century framework carries with it more of the soft side of business than the hard side. Systems, procedures, and methods flow from the relationships needed to perform the work – the reverse of the traditional business mind-set.

Partnering at its best is '... a long term commitment between two or more organizations for the purpose of achieving specific business objectives by maximizing the effectiveness of each participant's resources. The relationship is based upon trust, dedication to common goals and understanding each other's individual expectations and values'.[9] This is well said, but unfortunately not always well implemented. The United States Army Corps of Engineers, a pioneer in partnering, has had mixed results for this reason.[10]

---

9 The Construction Industry Institute's definition of Partnering from the Partnering Task Force Report, published in August, 1991.
10 In an interview with Keith Erickson and David Ohsiek, Kansas City District of the Corps of Engineers, March 31, 1992: The chief reason for at least one less than successful Partnering was due to poor initial alignment of the project team and the lack of true buy-in by at least one key partner.

4.2 *Lean construction ...lean production comes to construction*

Partnering cannot be effectively implemented without implementation of some key elements of lean production principles. Lean construction theory supports the practices that have been empirically discovered by IDC and TDC to correct the deficiencies in prior implementation of construction partnering. IDC and TDC have employed these for some time, without being aware until very recently of the theory that supports them.

Multi-tasking. multi-discipline, multi-functional, self-managing working groups are key lean construction concepts that are critical to the success of partnering on construction projects. In practice, these groups will evolve, and 'morph' many times between the beginning of design and the turn-over of the finished 'product' to the owner. As the needs of the project change, the same players will move seamlessly from group to group, fulfilling different roles each time. They are at one time the working group champion, the next time a support member. Working groups will overlap then disappear as their function is no longer needed. However, the key to the success is the continuity of the team members from group to group. *The process is the product.*

4.3 *Total quality management ...application to construction*

21st century partnering is Total Quality Management (TQM) and lean production for the construction industry. Prior attempts to implement TQM in the construction industry has been fragmented due to implementation only within the existing corporate and project structures. The implementation of TQM individually within each entity involved in the construction process can have only nominal success. After all, the 'product' is the result of the effort of all the parties. This is similar to the realization by manufacturing that TQM required the partnering of the manufacturer with their suppliers and distribution networkers. This included the realization that traditional competitive bidding of these important parts of the delivery system was counterproductive.

Similar to TQM efforts in manufacturing, the 'expected benefits include improved efficiencies (time to market) and cost effectiveness, increased opportunity for innovation (ncw projcct delivery methods), and continuous improvement of quality products and services (lessons learned)'.[11]

The shortcoming of TQM in the construction industry has been the attempt to implement practices designed to improve the *final product*. The production of a built environment is a unique creation, never exactly repeated again. The process is the project. Therefore the concentration must be on real-time feedback and improvement of the processes during the project. Variation can only be controlled at the micro level. Prototyping must therefore be performed on 'slices' of the project process and of the project facility itself.

---

11 Op cit. Construction Industry Institute.

4.4 *Effective implementation in practice... 'to boldly go where no man has gone before...'*

Effective implementation requires an entirely new way of thinking about projects. Notions that seem 'state-of-the-art' such as CPM scheduling quickly are proven to be inadequate to address the speed at which the modern high technology facility must be produced. The CPM chart is obsolete before it exits the laser printer. In addition the CPM tells us nothing as to *why* dates are missed and how to improve the systems and performance in real-time fashion. This requires organization and systems that respond in real-time, work at the micro level, yet communicate seamlessly.

Basic principles

Principle 1: Accountability is tied to 'buy-in' and mutual respect. Errors are treated as 'lessons learned,' with remediation *shared* by all parties to the team. The goal is for each task (and each subsequent project) to benefit from the learning experiences of the present one. Those with the most battle scares are the most honoured and most valued members of the team.

Principle 2: *Incentives* are built-in and based upon the *team* winning. All parties are members of various multi-tasking, multi-functional 'working groups,' but are also cherished members of the total project team for the full duration of the project. The working groups always include all parts of the team – design, procurement, construction, suppliers, owner, and end user.

Principle 3: Risks are fairly allocated to those best able to control them. Reasoned risks are recognized and rewarded. Innovation that benefits the *end product* is sought, not short sighted 'fixes that fail' (Senge, 1990)[12] . Efforts that do not produce lessons learned are not pushing the envelope sufficiently.

Basis of the partnership

Commitment of all parties to the shared vision and goals of the project beginning with top management, then built into all levels of the project team.

Equity: All stakeholders interests are considered in the development of the goals of the project.

Communications: Open and honest at all levels. Discussions that are 'Hard on the problem, Soft on the people' (Fisher & Ury 1983) and based upon mutual respect.

Trust: Sharing of information without fear or hidden agenda.

Continuous – real-time – measurements and evaluation of the project milestones and goals.

Timely responsiveness to issues: Working groups with the responsibility and authority to make decisions.

Issue resolution system: Processes for resolving issues quickly, at the *lowest* levels, and fairly without faultfinding or exploitation. Built-in escalation to the next higher level if not resolved in a minimum of time. A culture that encourages quick and effective resolution based upon long term personal relationships.

---

12 Definition of this Systems Thinking Archetype per Senge: 'A fix, effective in the short term, has unforeseen long-term consequences which may require even more use of the same fix'.

The team

The partnering team includes active participation of all organizations involved in the project (refer to the chart in Appendix C). This includes the following:

The client is included from the inception of the project. 'Bringing clients into the process early makes them coconspirators in the creation adventure, which often edges them toward embracing exciting (and risky) ventures that promise a wow-scale payoff.' (Peters 1994).

This must include more than just senior management. To be successful the team must include key representatives from the following client groups. These parties must be involved at the concept phase and there forward through the start-up and turn over of the facility:

– Facility construction team: The owner's construction management representatives;

– Process owners: The parties who are responsible for the design of the manufacturing processes for which the facility is created;

– Facility operations team: The owner's operating staff for the various facility systems supporting the manufacturing process.

Authorities having jurisdiction and public utilities providers are critical to the success of the project. They must be included from the beginning of the project. Experience has shown that when they are involved from the beginning, informed openly, and included on the team they become a beneficial partner in place of the hindrance and adversary that many envision them.

The design/construction team including scheduling, procurement, expediting, design services during construction (SDC), key equipment providers, as well as the standard design and construction staff. Breakdown in the project delivery systems will result if all parties are not included. In addition, the traditional adversarial relationships between these parties *must* be resolved early, monitored, and problems remediated immediately. Communications and cooperation breakdowns will otherwise cripple the project efforts.

Each team segment and working group is 'chartered' to be self-managing. This is the key and the power to the partnership. Without this there is only a team in theory, not in fact. In addition, the speed increase in the process is dependent upon timely, informed, knowledgeable, and 'bought-in' decisions at the lowest levels appropriate.

The team chartering process

Beginning with the management team, down through each working group, charter statements are written. These begin with a mission statement for the particular team segment which outlines in one or two sentences the vision they see for the best possible results of their tasks. It is followed by the specific goals of the segment.

Creation of the charters is performed in meetings attended by all parties of the team segment. Input is sought from all, and the final charter is signed by all.

While the process of creating the charters is slow, the end results of aligning the mission and goals are preeminently important to the success of the project. This is the process that creates buy-in by all parties. It is also the benchmark (self imposed) upon which performance is both self evaluated, and measured by the management team.

From time-to-time re-alignment meetings are conducted. These occur when significant redirection on the project effects the original goals. It may also occur to resolve any creeping degeneration of the team segment within, or with respect to the project team as a whole. Monitoring and immediate realignment as needed is critical to the success of the project.

Team organization
'To the outside observer, it will appear almost edgeless, with permeable and continuously changing interfaces among company, supplier, and customers. From inside the firm the view will be no less amorphous, with traditional offices, departments, and operating divisions constantly reforming according to need.' (Dividow & Malone 1992).

The organization is shaped, and re-shaped, to meet the needs of the client and the project objectives. The high technology manufacturing sector is changing at lightning speed. Manufacturing process innovation requires the capability to make changes to the project design not only from project to project, but also numerous times *during* not only design but also construction of a particular facility. The team organization must also be able to smoothly reshape when client manufacturing processes, budgets, or schedules change in the midst of design or construction.

For the partnering team, properly established and maintained, this can be accomplished quickly and with the least impact. This is because the 'organization' is already a non-hierarchical, self-managing group of team segments and working groups. Organizational charts, in the traditional sense, are no longer of value – they would be re-drawn *after* the actual re-organization, which often happens literally in hours. An attempt to draw an organizational chart at all is problematic. This is due to the many lines of both official and (perhaps more instrumental) unofficial authority and communications. For an example of a less than successful attempt, refer to Appendix D.

Management systems and project processes
Management systems and project processes must be structured to make the ultra fast-track delivery process possible. Traditional office and project methods and management styles are totally inappropriate to twenty-first century partnering. The following sub-sections will describe elements of the organization that IDC and our design/construct sister firm TDC have implemented and continue to refine.

4.4.1 *The design office*

Management structure and culture
The design office is structured around responsiveness to the client. The organization is very flat, with the minimum number of management levels. The project staff is arranged in a matrix management structure. Each technical staff member is responsible both to their technical discipline, and to the project management team of their assigned project. The management style is 'strong project manager,' consistent with putting the needs of the client first. The technical disciplines *sell* their expertise to the project.

The culture of the firm is totally client centred. The client project needs are the focus of the design of all supporting systems and infrastructure. The culture encourages a pro-active, land-on-your-feet, take-on-all-that-you-want attitude. Recognition is awarded to those at all areas of responsibility who take action. A good natured 'no whining' is a frequent response from staff who see anything less in a co-worker's approach.

Disciplines are organized into working groups and teams to address both short term and long term standards and practices development. These self-managed teams contain all categories of staff that are needed to both identify and create needed tools. These working groups and teams are the front guard of quality improvement. The formal teams are structured around technology based systems. The technology teams forge long term relationships between the staff members, as well as with respective client technologists and equipment manufacturers.

Staffing

The office is staffed with highly skilled technical and support staff. The staff is highly self-motivated and is expected to pro-actively seek solutions to problems. They are empowered to make decisions, and encouraged to innovate and take reasoned risks. The client is placed first in priority for all project and joint technology-development work groups.

The new-hire screening process is unique. Each candidate is screened first by human resources. Human resources is responsible for verifying the credentials of the candidate. Once accepted for interview, the candidate is interviewed by usually no less that five to seven members of the staff. These staff members represent a cross-section of all categories of company employees with whom the candidate will relate on a daily basis if hired. The candidate must be deemed both technically competent, and also able to work within the self-managing team environment as a pro-active problem solver and team player.

Empowerment

A first class technical library is available in-house. This includes both hard copy and on-line capability via CD ROM and links to university libraries.

The latest technology in high speed 3D CAD software is employed. In some cases, co-development of these tools is undertaken to better meet client project needs. Engineering calculation software is available in any office via the computer network. Master drawing details, and project drawing files (with appropriate security access) are accessible from any office from the network.

In-house and off-site training is tailored to the current needs of staff and projects. Every staff member is encouraged to take an active role in identifying and coordinating personal training needs for their continued growth.

Project teams

Client project work is staffed by project teams. Each of the needed technical and management discipline departments contributes dedicated staff for the duration of the project. In most cases project staff are assigned to project after project for the same client.[13] This leverages the learning process related to the needs, likes, and dislikes of the clients. More importantly, the staff at all levels of the organization develops long term working relationships with their client counterparts.

The project design team is located in a dedicated team space either in the design office or at the project site during the life of the project. Space is provided in the team area for conferences and working sessions with the client, authorities having jurisdiction, equipment suppliers, and other key parties. Once substantial construction begins, and as design efforts ramp down, key design staff are shifted to the project site for Services During Construction (SDC).

Close to the customer
Staffing, management, and infrastructure are designed to assure a close-to-the-customer environment.

– Quick and convenient communication is essential to the project processes. All offices and major project sites are fully computer networked. Every employee has a computer. In many cases eMail communications are integrated with that of the client. Lead technical and project management staff travel with notebook computers that can be connected to the network from the project sites.

– Video and audio teleconferencing are used extensively for daily meetings. However, these methods are not substituted for regular face-to-face contact with other partner parties.

– On site offices are established for all substantial duration projects. The goal is to fully integrate with the project and the client.

– Joint alignment sessions and skills training are conducted with representatives of all the partners including, and most importantly, the client.

4.4.2 *The project*

Formal dispute resolution
Since client relationships are long term, resort to formal Alternative Dispute Resolution (ADR) rarely occurs. The relationships are too valuable to all groups to not resolve differences quickly and by mutual agreement. The project relationships and the structuring of long-term working groups will almost always set the framework for resolution of issues at the lowest level of the organizations. Failure to be able to work through differences at the individual and working group level is considered a failure on all parties' parts. The issues are therefore generally resolved without escalation to formal resolution methods. While seldom resorted to, it is important that formal *last resort* ADR methods are clearly defined in advance.

Most of IDC's and TDC's clients are under long term services agreements. During the original contract negotiations dispute resolution processes are discussed and defined. While they very somewhat from client to client, they can generally be described as follows.

---

13 The majority of IDC's projects are for long term repeat clients. In some cases IDC and TDC are providers-of-choice for all design and design/build services for our clients. The highly confidential nature of our client's processes are best protected by reserving lead technical and management staff to the particular client to the highest extent possible.

The contract will stipulate traditional forms of ADR such as mediation and arbitration as the 'last resort' prior to litigation. However, direct, facilitated, negotiation is pursued as the first formal step. Each partner organization will designate a senior (decision maker) party to represent them in negotiations. A pre-negotiation conference is convened to:

1. Identify any issues that can be resolved prior to a formal negotiation;
2. Clearly define the remaining unresolved issues;
3. Agree on a facilitator;
4. Set ground rules for the negotiation sessions;
5. Agree who will attend – ensure 'comparable representation'.

It is not unusual for the issue to be resolved at step one of the pre-negotiation conference. Formal negotiation is rarely required. Arbitration and litigation are 'unthinkable.'

Project alignment

After project staffing has been assigned by all partner firms working groups are formed based upon the technological and management units. Overall facilitated alignment meetings are conducted to ensure that all parties have a part in the goal setting processes and that they understand the requirements and challenges of the project.

Joint skills training sessions are often conducted early in the project. This would include subjects such as effective communications, meetings skills, and technical sessions on any special project processes or software.

Incentive program

An incentive program for achievement of project goals is implemented. Typically the design/construct team puts a portion of their fees 'at risk' and the client will match the amount in the 'pot.' A scoring system is negotiated to measure performance that underachieves, meets, or exceeds the goals. The design/construction team can break-even, forfeit all or part of their at-risk fee, or earn all or part of the clients matching fund based upon the score achieved.

The scoring will generally include the following categories:
– Jobsite safety;
– Project schedule, measured by achievement of each major milestone;
– Budget control;
– Quality of the delivery process and the delivered built-environment.

Project scope and budget

A programming report is generated which defines the objectives of the project in their most general sense and the design criteria. This document will include detailed lists of major equipment, fundamental process flow diagrams, building floor plans with equipment layouts, and fundamental building sections. This is the result of intensive work between the design team leads and the key client parties.

Once the programming is completed, work immediately begins in producing a scope of work written document The scope of work is reviewed with the client in working sessions and amended as needed. From this, the design and construction team members work together to develop the detailed *line-item* design and construction budget. If the cost developed is in excess of that permitted by the project limits, de-scoping sessions are conducted with all partners to produce a revised scope and budget. This becomes the plan of record for the project. All future changes in scope will be baselined against this document.

All drawings produced in programming are carried forward into and through all future design deliverable drawings. Nothing is discarded; all documents are updated, enhanced and revised into the subsequent usage. Ultra fast-track delivery will not allow for waste and rework of any type. In addition, these become the living project memory.

Continual quality achievement

Quality of the delivered facility is ensured by setting a goal of zero punch-list items at the time of client beneficial use of respective portions of the project and its associated systems. There is *no* scheduled punch-list correction time built into the end of project schedule segments. This requires that the design, construction, equipment provider, and client facility owners work daily to review the in-progress work. The result is less re-work and replacement, the chief enemies of productivity.

4.4.3 Design processes

Working groups

The cornerstone of the design process is the joint working group. The project working groups are composed of representatives of design, construction, and the client's facility construction and facility operations staff. Equipment manufacturers who are part of the project partnering attend working group meetings when needed. Generally the design working groups are identified by respective facility technologies (e.g. architecture, mechanical, etc.).

Cross-coordination sessions are conducted on a regular schedule between the various technology-based working groups. This often takes the form of wall pin-up sessions. Current multi-colored CAD drawing plots are maintained on the walls by the lead designers in each technology group. These drawings serve as continual resource, a forum for informal discussion, and the subject of formal pin-up sessions. Many potential mis-coordinations are identified over coffee at the wall drawings.

The working groups make joint decisions on the design based upon constructability, achievement of the design criteria, budget, schedule, and quality. The effort is to strike a balance between these sometimes conflicting parameters. The working groups are empowered to make decisions within the plan of record scope of work and budget limits. Since design decisions are jointly arrived upon, they are not generally disputable later.

Schedule and content of work packages

The project design deliverables take the form of work packages. The work packages will include various documents issued to the construction procurement group for pricing, acquisition of the related materials and equipment, and for issue to the construction forces. These may be pre-purchase packages for early acquisition of materials and equipment that have long delivery lead times. Other packages will include the documents for installation and construction.

The schedule for delivery of work packages and the detailed limits of each package is established by the overall project team early in the project. Issues such as lead time for materials often requires that the respective providers be included in the process. As time progresses and unforeseen changes occur, the project team and the respective working groups are authorized to make necessary adjustments to work package content and schedule.

The goal is to issue documents such that procurement can obtain needed materials to the project site in a just-in-time fashion. As a result, all large equipment items, large piping, and other materials such as roofing can be off-loaded directly to the proper site location. Materials and equipment are handled only once, if at all possible.

Intelligent drawings
Complex process and mechanical piping system CAD drawings are produced in 3D utilizing intelligent software. The 3D CAD drawings are prepared by design, interference checked with other building elements, and issued to construction in both hard copy plans, and in software files. Construction electronically converts the design drawings to isometrics and fabrication spool drawings. Material lists are automatically generated by the software based upon the integrated database inputted by design. The redundant, time consuming, and costly steps of re-drawing the design documents for fabrication and of manual material take-off are eliminated.

Mock-up construction
Full size mock-ups of all repetitive, high intensity construction areas are prepared on site. Interferences, trade coordination issues, owner maintenance issues, and unforeseen design issues are resolved by perfecting the mock-up *before* full build out is started. All the client's facility owners have one last chance to see the configuration, full size, and agree to proceed. Rework, mis-perceptions, and mis-coordinations are eliminated. Since the entire facility cannot be prototyped, selected areas that are the most complex (and therefore most error prone) and most repetitive are mocked-up. This is the construction analogy to the manufacturing industry prototype.

Construction protocols
Cleanroom space construction protocols, equipment start-up procedures, and space turnover procedures are all developed jointly by working groups representing all effected disciplines. These are fully documented and signed-off *before* the respective areas and systems are ready for implementation. Strict schedules are maintained. All required efforts and staffing is brought to bear to ensure their timely completion. This is critical to avoid the typical construction symptom of requiring thirty percent of the time and budget to complete the last five percent of the work.

Design reviews
Review of the design is done continuously in real-time within the working groups. Final formal design-team and client quality-control review of the finished Issued For Construction (IFC) documents are therefore only a final confirmation. There should be no surprises in the final IFC documents or in the constructed systems.

Construction clarifications
Design clarification requests from construction are given priority over all other design efforts. No request is permitted to take longer than three days to handle from issuance to reply-received. The is essential to keeping the construction efforts on schedule in the compressed ultra fast-track process.

4.4.4 *Construction processes*

Management and supervision
It is essential that the construction site be fully staffed early with key staff. Construction engineers are needed to interpret the design documents and coordinate activities. Construction CAD detailers are required to convert 3D CAD drawings to full spool drawings. Schedulers, purchasing agents, and expediters will be fully occupied ensuring that the needed materials, and tools are available when needed. Field supervision must be in place to direct activities and coordinate with the construction engineers to ensure that skilled labor is available and equipped on time. In accordance with lean construction concepts, construction tasks are not started until labor, materials, and tools are available, and the design is sufficiently complete and verified.

Design team members now begin to transition to their new role of Services During Construction (SDC). In addition to answering construction generated design questions, they also play on-going roles in cost control, start-up and protocol planning, review of submittals, and quality control. Also, without a substantial on site SDC staff, immediate responds to during-construction design changes would not be possible.

Instead of leading the efforts, they now assume a support role to the construction forces. The design organization needs to be prepared to assign additional staff during construction, if design changes or unforeseen situations arise.

New working groups are formed that include the key construction supervision and design team members. These new working groups are established to serve various functions, but are composed of a matrix of participants from other working groups. This is the power for continuity and communications. The working groups will include: Systems start-up teams, space turn-over teams, cost control teams, and others as needed. As the working groups complete their chartered purpose, the members will transition to other tasks and rolls on the project, or ultimately move on to start the process over on the next project.

Work authorization
Clear and expeditious systems must be in place to authorize work on the construction site. The exceptional speed of construction of the ultra fast-track process will otherwise drive unauthorized work that will either be started to soon, unfunded, incorrect, or any combination thereof.

Jobsite safety
The accelerated speed of the ultra fast-track process has a greater potential for creating conditions that can result in jobsite accidents. Extreme diligence must be exercised to stress that safety always comes first. Since actions speak louder than words,

it is important that positive examples be set early and often. Safety must be stressed to take priority over schedule and budget. Every team member is responsible for safety on the job site.

Cost control

Cost control engineers are key to the feedback needed to ensure on-budget performance. The cost control systems must be operational early, fully staffed, and sophisticated enough to cope with a huge amount of input every day. The client never likes cost overrun surprises. Information that is after-the-fact is only of value to the attorneys during discovery. The timely information from the cost control working groups allows for quick changes of direction.

Cost control working groups are established when the expenditures on the site begin. Ultra fast-track schedules will not permit delayed start of cost control input and output. The working groups are broken down by technology system within the project.

Bi-weekly cost control working group meetings composed of design, construction, client construction management, and cost control are critical. The working groups review the bi-weekly reports, identify and revise any errors, and most importantly *forecast* expenditures yet to be made for the duration of the project. This forecasting, along with the actuals spent, is the basis of real-time understanding of budget status. The results of the working groups are accumulated and reported to design, construction, and the client's project management bi-weekly at a joint management meeting where any corrective measures are determined.

Independent inspection

In addition to the legally mandated code inspections, the project must implement independent quality assurance inspection. It is best that the agency for these inspections be contracted directly to the owner in order to insure true autonomy.

The design team must develop quality assurance criteria for the facility systems and materials in the design construction documents. They should work with the independent agency to establish procedures, and they will be the reviewer of the assurance reports.

Lessons learned

One of the most important functions of the design and construction team is to capture lessons learned from the project. These lessons will include both procedural and technical lessons.

Lessons are captured and used as feedback for improvement or revision of the practices and technical skills of the design/construction team. The repetition of failures or less than best successes in the next phase or project is not acceptable. In addition, lessons related to new tries that produced improvements also must be captured and instituted in future work.

This process is continuous. What worked well today may be obsolete tomorrow.

4.5 *Goals ACHIEVED ...results are the measure*

The processes described have been implemented on some of the largest and most technologically challenging project. Although there is continued room for further refinement, the results have been very encouraging.

On a typical large ultra fast-track project of the type that IDC and TDC now routinely delivers, it is not unusual to produce three or more major design work packages for construction *each week for one year*. On a recent project this amounted to a total of over 4600 drawings, plus written specifications, issued in over 150 work packages. The first construction work package was delivered three months *before* the completion of the full project design development report. During the design (and simultaneous construction), the client changed their manufacturing processes as many as three times, resulting in substantial re-design. *No schedule increase from the original plan was requested, and the project was ready for client use on schedule to the day.* We were asked to reduce the duration of the project by thirty percent, the design cost by thirty percent, and hold the construction cost equal on a subsequent project of the same scope.

The projects are being performed on budget. Scheduled critical target dates, including those established prior to substantial design changes both during design and construction, have been achieved. Some of these key dates have never before been achieved on a particular client's projects, even prior to ultra fast-track schedules.

Change orders to construction due to design omissions or errors have been less than one-third of one percent of total construction cost. The construction costs of these changes have been recovered due to the partnering agreement provisions and the exceptional overall project performance.

Minimal start-up problems occurred and all were resolved without impact on critical milestone dates.

No project issues have been escalated above the project management level. All issues have been resolved on the project site, without recourse to any ADR processes. All but a handful of issues were never taken above the field level.

Additional projects of major scope have been awarded both by present and new clients as a result of the impressive results of the ultra fast-track partnering process used on recent projects.

5 THE FUTURE – NOW: GETTING THERE IS ALL THE FUN

'The way customers judge a service may depend as much on or even more on the service process than on the service outcome'[14].

While the above statement may exaggerate the case for high technology facility design/construction (where the on-time 'service outcome' is essential), it certainly is true that new delivery processes that speed delivery of the facility (while holding the line on cost and quality) are getting the attention of owners.

The reality is that 'Change and constant improvement (kaizen, per the Japanese), the watchwords of the 80s are no longer enough. Not even close. Only revolution,

---

[14] Len Berry, Professor, Texas A&M University as quoted in Peters, 1994.

and perpetual revolution at that, will do' (Peters 1994). As Dr Deming put it: 'You don't have to do it... Survival is not compulsory.'

5.1 *Dispute resolution*

The processes required to meet the continuing challenges for better facility design/construction processes will find the need for alternative dispute resolution a last resort only. The interlocked, long term relationships required to meet the need to deliver ultra fast-track projects will create healthy inter-dependence. These relationships will be mutually beneficial and too valuable to all parties to endanger. 'Close to the customer' must become 'part of the customer.' The culture of the joint, self-managing teams required to deliver projects efficiently will create a culture of issue resolution at the fastest and lowest level.

The challenge to the ADR community is in developing flexible ADR *systems* that can be taught to those at all levels in the project team. 'People Skills' are essential in the new project delivery processes. These new processes strain the relationships skills of project staff at every level. Decisions once made at project management levels, are now being made at the senior designer or field superintendent level. Technical coordination in design is being driven down from lead engineers to the designer level. While these are the parties best able to make timely, informed decisions, they are generally not as skilled in interpersonal relationships, negotiation, and dispute resolution.

While formal ADR will continue to be an important aspect to the successful resolution of construction issues, it is this author's experience that formal ADR will soon become what litigation is today – the last resort. Litigation and arbitration will become unthinkable, as they spell death to long term relationships.

REFERENCES

Dividow, B. & Malone, M. 1992. *The Virtual Corporation. New York: Harper Collins Publishers*. pp. 5-65.

Fisher, R. & Uiy, W. 1983. *Getting to Yes, New York: Harper Collins Publishers*. New York: Penguin Books.

Kostof, S. 1985. A History of Architecture, Settings and Rituals. New York: Oxford University Press.

Melles, I.B., 1994. What Do We Mean by Lean Production in Construction? 2nd International Workshop on Lean Construction, Santiago, Chile.

Peters, T. 1994. The Tom Peters Seminar – Crazy Times Call for Crazy Organizations. New York: Vintage Books, Random House. p. 253.

Senge, P. 1990. The Fifth Discipline . New York: Currency Doubleday. p. 388.

APPENDIX A: TYPICAL TRADITIONAL DELIVERY SCHEDULE

	Jan-95	Feb-95	Mar-95	Apr-95	May-95	Jun-95	Jul-95	Aug-95	Sep-95	Oct-95	Nov-95	Dec-95	Jan-96	Feb-96	Mar-96	Apr-96	May-96	Jun-96	Jul-96	Aug-96	Sep-96	Oct-96	Nov-96	Dec-96
Programming	2 mo																							
Design Development			4 mo																					
Estimating/Budgeting					3 mo																			
Construction Documents						10 mo																		
Bidding															3 mo									
Construction																		17 mo						
Early Start-up																								
Blow Down																								
Tool Install																								
Start Production																								

Total duration: 47 months

	Jan-97	Feb-97	Mar-97	Apr-97	May-97	Jun-97	Jul-97	Aug-97	Sep-97	Oct-97	Nov-97	Dec-97	Jan-98	Feb-98	Mar-98	Apr-98	May-98	Jun-98	Jul-98	Aug-98	Sep-98	Oct-98	Nov-98
Programming																							
Design Development																							
Estimating/Budgeting																							
Construction Documents																							
Bidding																							
Construction																							
Early Start-up								4 mo															
Blow Down										4 mo													
Tool Install													4 mo										
Start Production																	8 mo					+$	

APPENDIX B: ULTRA FAST-TRACK DELIVERY PROCESS

	Jan-95	Feb-95	Mar-95	Apr-95	May-95	Jun-95	Jul-95	Aug-95	Sep-95	Oct-95	Nov-95	Dec-95	Jan-96	Feb-96	Mar-96	Apr-96	May-96	Jun-96	Jul-96	Aug-96	Sep-96	Oct-96	Nov-96	Dec-96
Programming	2 mo																							
Design Development			7 mo																					
Estimating/Budgeting				6 mo																				
Issue Design Work Pkgs									15 mo															
Construction													16 mo											
Early Start-up																4 mo								
Blow Down																			4 mo					
Tool Install																				8 mo				
Start Production																								

Total duration: 29 months

	Jan-97	Feb-97	Mar-97	Apr-97	May-97
Programming					
Design Development					
Estimating/Budgeting			→		
Issue Design Work Pkgs					
Construction					
Early Start-up					
Blow Down					
Tool Install					
Start Production				→$	

APPENDIX C: THE PARTNERING TEAM CHART

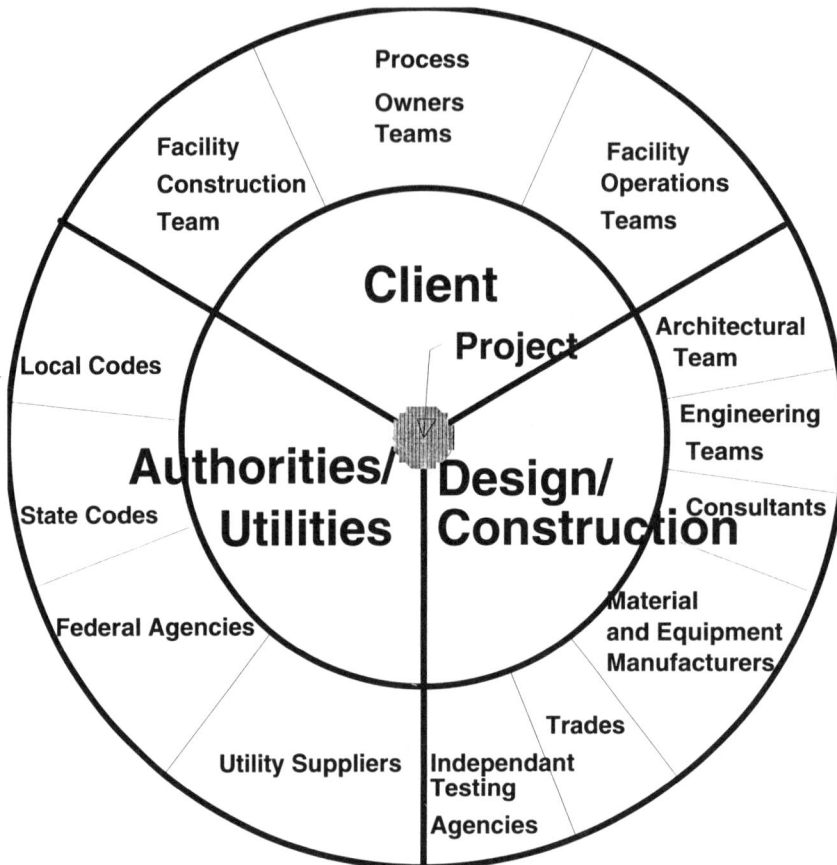

APPENDIX D: PROJECT TEAM ORGANIZATION

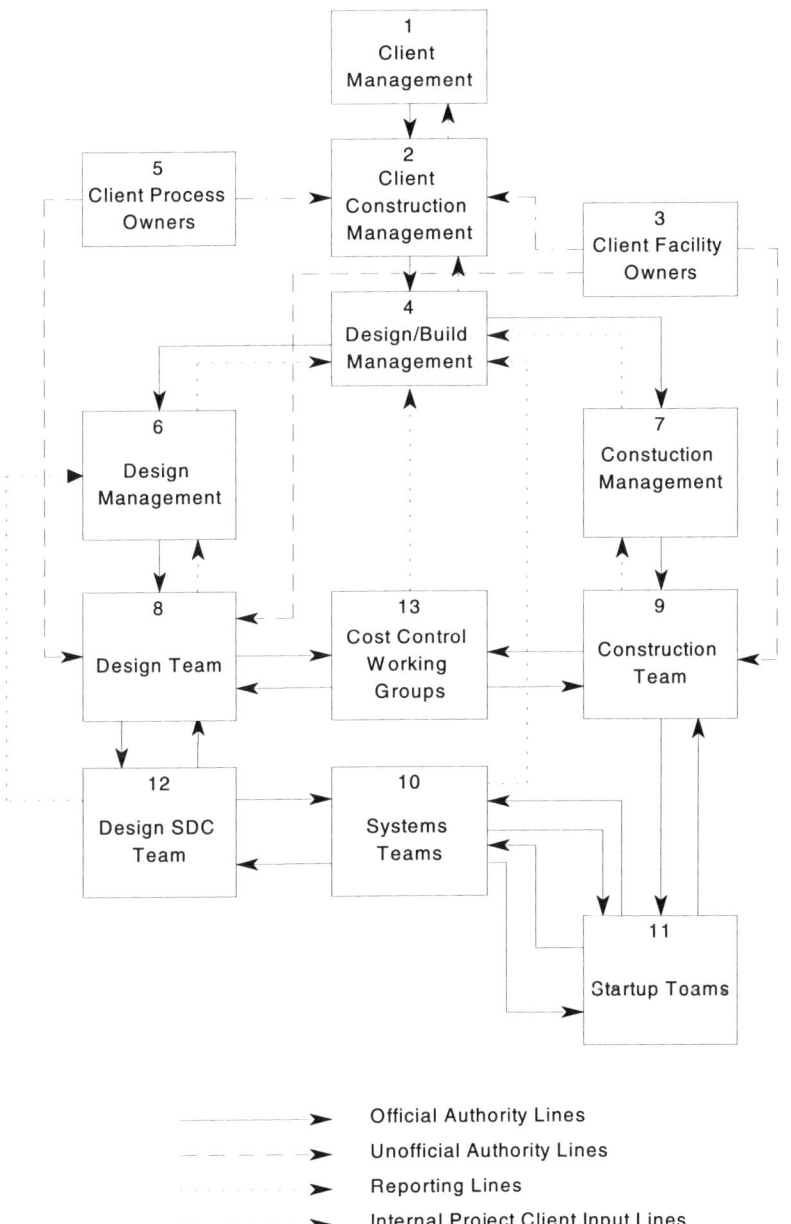

```
                                    ┌──────────────┐
                                    │      1       │
                                    │   Client     │
                                    │ Management   │
                                    └──────────────┘
```

Official Authority Lines

Unofficial Authority Lines

Reporting Lines

Internal Project Client Input Lines

New tools for lean construction*

KARI TANSKANEN, TUTU WEGELIUS & HANNU NYMAN
Helsinki University of Technology, Espoo, Finland

ABSTRACT: In this chapter we first review some of the major principles of lean manufacturing. We also discuss the situation of construction business from the lean manufacturing viewpoint. Then we define the most important requirements for tools that support lean manufacturing principles. By tools we mean both manual and computerized methods for reengineering, planning and controlling business processes. Finally, we describe two tools that are designed to support lean construction. The first one is a methodology for measuring the effectiveness of information and material flows in designer-vendor-construction site chains. The second one is a computerized tool for planning and controlling material deliveries of lean construction site.

1 LEAN CONSTRUCTION PRINCIPLES AFFECTING DESIGN OF TOOLS

When developing tools for lean construction we must consider following principles and goals that are fundamental for lean manufacturing:

1. Focus on material and information flows. The effectiveness of the whole business process that start at design and ends when the final product is handed over to the customer is more important than the efficiency of the separate units in the process;

2. Eliminate waste. Those operations in the process that does not add any value to the end product must be identified. After that the processes must be redesigned to eliminate the waste;

3. Minimize variances. Standard and simple procedures should be used as much as possible in order to make forecasteability of performance as good as possible. This way we also increase the repetitiveness of processes so that effective development can take place;

4. Consider time a key element of all business processes;

5. Focus on continuous development of the processes instead of sudden and revolutionary changes.

These principles are often argued not being applicable for construction. The main argument has been that lean manufacturing is a system for repetitive manufacturing. In construction business, products are unique and also site organization is unique for each project. Therefore lean manufacturing principles and methods cannot be applied. However, when we take process view for construction, we find a lot of repeti-

---

*Presented on the 1st workshop on lean construction, Espoo, 1993

tiveness. For example, the process chains of material deliveries can be grouped into two categories: 1) Process chain for standard materials (made to order); and 2) Process chain for customized materials (designed to order). Both of these process chains follow the same steps in all construction projects.

Another argument against applying lean manufacturing principles in construction is the 'culture' of construction industry that is against standard procedures (individual professionality is emphasized) and accepts long throughput times and poor accuracy. However, if we look the evolution of manufacturing systems of the automotive industry in Japan, we can see that the cultural aspects do not prevent developing manufacturing systems. In the 1950's, Japanese automotive industry suffered the same kind of problems than the construction companies in many countries today: poor quality, long throughput times, bad accuracy and so on. The lean automotive factories today in Japan are a result of long lasting and hard work that has gradually changed the 'culture' to support lean manufacturing. In the 1980's and 1990's the same kind of development has taken place in many industries all around the world. The leading edge companies can be found from automotive and electronic industries, but there are successful examples in almost all industries. There are no evidence that similar kind of development cannot take place in construction industry as well.

2 REQUIREMENTS FOR TOOLS THAT SUPPORT LEAN CONSTRUCTION

We categorize the tools that support lean construction into two groups:
    1. Tools that support reengineering business processes;
    2. Tools that support planning and controlling business processes.
Reengineering is clearly a management task. However, it does not mean that managers and consultants should do it and then tell workers how they should operate. Although the support of managers and consultants is valuable, it is most important that the workers who are in charge of day to day operations are involved in the development. According to our experience, the following are the key issues in the development process:
    1. The business processes to be developed are identified and the current performance level is measured;
    2. Current performance is benchmarked with 'best practices' in order to identify improvement potential;
    3. Ideal models are provided to guide the reengineering process. Although the 'ideal model' might be not applicable in all conditions, it shows the right direction.

Surveys among Finnish construction industry indicate that this is still far from the current practice: business processes are not known, current performance level is not measured, development potentials are unclear, and 'ideal models' are not known. Instead, planners and foremen either use the 'ordinary model' when designing the processes or they do not design them at all, things just happen. In the next chapter we present methods for analyzing business processes that aim at changing the current practice.

Planning and controlling business processes of construction projects is a task that can be effectively supported by modern information technology. However, there are

so many 'fuzzy' things to take into consideration that it is not reasonable to try to computerize decision making. The tools of lean construction must combine the strengths of human and computer: computers are strong in storing, sorting, calculating and transmitting big amounts of data, humans are strong in combining information, reasoning and making decisions. For human reasoning it is extremely important that information is presented and processed in such way that planner is familiar with. In practice this means that the user interface of the system must be graphical, and the system must work interactively. Also the process how we design the computer system is important. Before we specify the tool we must thoroughly understand the goals of the planning task. In lean construction this means that we don't just automate routines; the tool must also support continuous improvement of performance, and provide feedback on development of performance. The system must also be flexible enough to fit different kinds of environments. Therefore prototyping approach is the best for designing tools for lean construction. We summarize the requirements for lean construction planning and controlling tools as following:
- Graphical presentation of information;
- Interactive way to process information;
- Understand and specify the goals of planning and controlling;
- Support continuous improvement of performance;
- Provide feedback on the actual trend of the performance of planned business process.

The computerized tool for planning material deliveries of construction site, TOIMI, that is described later in this paper, has been designed according to these principles.

3 METHODOLOGY FOR REENGINEERING BUSINESS PROCESSES

In this section we describe two methods that can be used for analyzing non-value-adding activities of business process. The first method is activity and cost analyses, and the other is accuracy and delivery time analyses.

3.1 *Activity and cost analyses*

The activity and cost analyses are based on the theory of activity-based costing. The principle of activity-based costing is that the operations of the company are divided into activities, which use different resources. The same method is used here for analyzing activities in the business processes and the costs of these activities. The objective of the activity and cost analyses is to find out costs of unnecessary work and to help to remove the problems that cause these costs. The activity and cost analyses help people to focus on the most expensive activities. The analyses also show concretely potential for cost savings.

Construction projects typically involve a lot of companies: contractor, subcontractors, design offices and material suppliers. To find out waste in business processes of construction projects, all companies involved in the process must co-operate. The first step in co-operation is to improve business processes by using activity and cost analyses to find out what are the processes today and how to improve them. To start with the material group and the sites to be analysed are chosen.

The first step of the analyses is to identify all activities in the business processes. For standard materials, the business process starts with placing an order and ends when materials are assembled. For customized materials, the business process starts when architectural design starts. The focus is on non-value-adding activities (moving, storing, inspecting, sorting etc.), which are modelled in details. The value-adding activities (production, design etc.) are described on a rough level. This stage in the analyzing process is called activity analysis.

After identifying the activities, the second step is to measure the costs of non value-adding activities. For this purpose we use precalculated standard costs, for example, dollars per square meter of in-house warehouse per day. By using standard costs we better can find out the effectiveness of the total process instead of efficiency of separate activities. Finally, the costs of capital tied-up to the process is added. After summing up the costs of non-value-adding activities of the selected business process, we calculate the non value adding costs/total costs -ratio. Different kind of graphical presentations are also used for illustrating the costs of non value-adding activities.

The third step is to benchmark current performance of different construction sites and to try to find out 'best practices'. An ideal model for the whole business process is defined by combining the best practices from analyzed cases. Improvement potential can be identified by comparing the current practice to the ideal model. The redesign process aims at improving current performance towards identified ideal models by continuous development.

3.2 *Accuracy and delivery time analyses*

Accuracy and delivery time analyses have been developed to find out time lags in material and information flows. The accuracy of design offices, contractors and material suppliers can also be found out by these analyses. The objective of the accuracy and delivery time analyses is to clarify the structure of the delivery time and find opportunities to shorten it. One possibility to eliminate waste time is to study waiting times between the activities in the logistics chains.

Usually the activity and cost analyses and the accuracy and delivery time analyses are implemented at the same time, because the results of the analyses support each other. Therefore the material group and the sites that will be analyzed are generally the same in the both analyses.

Implementing accuracy and delivery time analyses is very simple. First the important time points in both the material and the information flows are defined and included to the analyses. For non-standard materials it is necessary to analyze also accuracy and lead time of the design process. It is essential to analyze both the planned and the actual points of time. This way accuracy of the material delivery process can be clarified. The ordering day, the planned and the actual day of manufacturing and the planned and the actual delivery day are some examples of these time points.

The second step is to collect the analyses data from different sources. It is necessary to use documented data from the planned and the actual time points. If documented plans are not available, the processes are probably not planned and controlled well enough. Documented data can be found for example in delivery orders, production plans, construction site diaries and installation plans.

The most informative way to present the results of accuracy and delivery time analyses is to use graphs. This way time lags and delays in the delivery processes can be easily clarified. For example the analyses show if a material delivery has arrived to the site many weeks before installation. Benchmarking is useful method also in analyzing delivery time and accuracy. 'Best practices' show improvement potential and developed ideal models guide the redesign process to the right direction.

3.3 *Experiences from practical cases*

Several material deliveries have been studied by these analyses methods. So far plasterboard, concrete element, door, window and timber deliveries have been analyzed. At the same time the problems of these delivery chains were surveyed, and the means to improve the delivery chains were presented.

Based on practical experience, the importance of the activity and cost analyses is emphasized in standard material deliveries, whereas the accuracy and delivery time analyses are most useful in deliveries with a specific plan.

4 TOIMI: A TOOL FOR PLANNING AND CONTROLLING MATERIAL SUPPLIES AT CONSTRUCTION SITES

4.1 *Objectives and basic principles*

TOIMI is a PC-based software tool for managing material deliveries of construction sites. The objective of the tool is to ease the division of material purchases into smaller deliveries based on the actual need, to support the management of incoming material flow and to assist in specifying the contents, time points and unloading places of the deliveries.

Usually more than 50% of the costs in a construction project consist of different materials. Purchasing prices are therefore subject to meticulous observation in construction companies. What is easily overlooked, is the cost of activities that occur after the purchasing agreement is signed. Several operations need to be completed before the material is ready and installed. Several studies have indicated that by systematically planning and controlling the incoming material flow, companies can achieve major cost reductions.

However, it is very wearisome to divide the material purchases into smaller deliveries when the time points and contents of the deliveries are planned accurately. As the number deliveries increase, the difficulties involved in managing the purchases increase as well. In addition, as the inventories get smaller, the risks concerning the punctuality and correctness of the deliveries become greater. At its worst, the mistakes in delivery planning may delay the whole project whereupon the total effects of the planning process quickly turn unprofitable.

TOIMI has been developed to ease the division of the purchases into appropriate deliveries according to the need of materials and to support the management of incoming deliveries. The primary user of TOIMI is the person who is responsible for the deliveries and scheduling on the construction site. The software is based on simple colour codes shown on calendar views, concepts familiar to the user, and visual scheduling which altogether make TOIMI easy and illustrative to use. The idea of the program is to provide the information essential to the timely planning and controlling of purchases to the user easily and clearly.

4.2 *Functions of TOIMI*

Following events are important when managing a single delivery:
– Rough planning of the purchase, i.e. preliminary division of the purchase into deliveries and defining the time points (delivery weeks) and contents of the deliveries;
– Scheduling of the delivery as the agreed/fixed time point draws closer, i.e., 1) Determining the date of the delivery, the unloading places on the site, and the final content of the delivery, and 2) Sending the supplier the delivery order, including the mentioned details;
– Confirmation of the ordered delivery by the supplier;
– Acceptance inspection on the site;
– Possible reclamation.
TOIMI has been designed to support these functions. Illustrative colour codes are used in managing the deliveries so that the user can directly see the state of the deliveries arriving to the construction site. The deliveries can be examined in a calendar view either relating to a particular purchase (all the deliveries of the purchase and their state in a weekly calendar) or within a chosen time period (all the deliveries coming to the site within a month and the state of these deliveries). TOIMI is a Microsoft Windows-based application. To run, it requires at least 4 MB of memory and a 386-processor.

A delivery is presented on the computer display as a colour spot at the agreed delivery date. The colours are:
– Red (the delivery has not been planned);
– Yellow (the delivery has been planned and the delivery order has been sent, but the supplier has not yet confirmed the delivery;
– Green (the supplier has confirmed the delivery).
The system changes automatically the state and the colour code of the delivery when the user sends the delivery order or when the confirmation of the supplier arrives. The colour codes are analogous to traffic lights: red gives the alarm, yellow warns, and green signals that everything is fine. When choosing e.g. a summary of all the deliveries coming to the site in present month the user can immediately see the possible problems and next tasks to perform by checking the colours on his display.

TOIMI also makes complete delivery orders and delivery reports according to the instructions of the user. The reports can be used e.g. in weekly meetings of the construction site. Furthermore, a layout of the site with unloading places marked can be enclosed with the delivery order.

The information filled in the delivery order by the inspection can be entered to TOIMI. When using the system regularly on the sites of the company, the accumulated information can be used e.g. to evaluate and to follow the performance of the suppliers.

4.3 *Further development of TOIMI*

The development of TOIMI began in the summer 1992, and the pilot version of the software was tested on a construction site at the end of the same year. The software has been further developed according to the feedback given by the users. Currently TOIMI is in pilot operation on six different construction sites. The experience gained at the pilot sites indicate that the tool is useful and deserves further development. Further development will be focused on following areas:

– Building connections between TOIMI and project management software;
– Building connections between TOIMI and cost accounting software;
– Modifications of TOIMI for small contractor companies;
– Fastening TOIMI by rewriting major parts of the code.

REFERENCES

Brimson, J.A. 1991. *Activity Accounting: An Activity-Based Costing Approach*. John Wiley & Sons, New York.
Johnson, T.H. & Kaplan, R.S. 1987. *Relevance Lost – The Rise and Fall of Management Accounting*. Harvard Business School Press, Boston.
Stalk, G. Jr. & Hout, T.M. 1990. *Competing Against Time*. The Free Press.
Womack, J., Jones, D. & Roos, D. 1990. *The Machine that Changed the World*. Macmillan Publishing Company.

Lean production as a purpose for computer integrated construction*

MARTIN BETTS
Department of Surveying, University of Salford, Salford, UK

1 INTRODUCTION

1.1 *The nature of innovation*

Technological innovation can occur through either a technology-push or strategy-pull process. These are quite distinct innovation models that have important influences on how innovation is managed. Few have addressed what these different models mean to construction innovation and with regard to strategy-pull forces, we have little understanding of what the different strategic forces are. The issue of strategy-pull is becoming increasingly complex in process and manufacturing industries generally, and in construction particularly. We are clearly moving away from the concepts of using technology solely for automation. The dynamic nature of construction, the volatile economic circumstances we face and the inflow of new ideas of production management that are increasingly divergent from mainstream construction management philosophies, give rise to a climate ripe for innovation and a paradigm shift. This paper will explore a specific example of this general phenomenon by examining the consequences of a new production philosophy to CIC. It will place this new philosophy in the context of other strategy-pull forces and developments and discuss their combined effect on the way that innovation in CIC is and should be taking place.

In doing so, the paper raises the issue of a definition of CIC and the extent to which integration is an intrinsic goal of benefit to construction organisations. The case is made for integration to be motivated by the specific improvement needs of the construction process as part of the business strategies of the participating firms.

1.2 *A systems approach to problem solving*

This paper argues for a new, alternative view of CIC based on several recent approaches stressing process improvement and business strategy. After discussing these new views generally, the implications of them for CIC are analysed. The way that this is done is by posing the question why should we pursue CIC?

In many ways this question arises from taking a systems approach to problem solving in construction as advocated by Armstrong (1985) who suggests that we

*Presented on the 1st workshop on lean construction, Espoo, 1993

343

should only move to the fourth stage of a systems approach, to operational programmes, after having: 1) Stated our ultimate objectives; 2) Identified our indicators
of success; and 3) Considered the alternative strategies for realising them. The argument here is that much of the CIC research is preoccupied with the implementation
of operational programmes. There is a need to shed light on the other three stages of
a systems approach and this paper will begin to do this by considering the first two
stages.

1.3 *Previous CIC efforts*

There is an application and vision problem with technology research generally and
construction IT research in particular. It could be said that researchers have rarely
seen their projects in the light of the way they can be applied in practice and that
most integration research provides a technology push by developing a new technology and trying to push it onto an application. The purpose of integration research in
an applications sense is often ill-defined and how well a proposed solution helps to
overcome a practical problem, or how easily the solution can be organisationally
exploited in the evolving construction context, is seldom considered or tested. Thus,
integration research has produced many solutions looking for real world problems.
An alternative approach to technology-push as a model for innovation that has
gained much acceptance in management disciplines is to follow research and development in support of a strategy pull. That is for seeing what the evolving business
needs are, and then developing a solution to the business problem.

   A number of writers have commented on how IT can be exploited by companies
to exploit strategic purposes including (Porter & Millar 1985; Earl 1989; Daniels
1991). Earl, in particular, stresses the need for the 'technology strategy connection'
to be made and advocates the use of planning frameworks to assist in this. A combination of technology-push and strategy-pull is likely to offer the best promise for a
successful implementation and application of CIC.

2 WHY SHOULD WE PURSUE CIC?

This question may equate with the systems approach activity of identifying our ultimate objectives. The answer may appear to be obvious but a justification is required
in the light of developments in management thinking. This is particularly so given
that CIC as a concept originates as a parallel of CIM which is applied extensively in
manufacturing but usually in the context of different production management and
corporate planning frameworks. A justification of why we seek CIC will have profound implications on how we attempt it and the extent to which we aim to realise it.

   For a justification for CIC then, some discussion of the recent developments in
new production philosophies, competitive strategy and strategic IT planning are appropriate. These reflect part of the body of theory of production management that we
have seldom applied to our CIC research. They are increasingly being used as a justification for technological innovation in other sectors. This may give a fresh stimulus to our consideration of the importance of CIC.

2.1 *The new production philosophy*

Many new approaches to production management have emerged within the manufacturing sectors. These include Just in Time (JIT), Total Quality Management (TQM), time based competition, value based management, process redesign, lean production, world class manufacturing and concurrent engineering. Their significance for construction has been analysed by Koskela (1992) who found that many of the approaches share a common core viewed from different perspectives. This common core is based on a conceptualisation of production activities with the perspective determined by the design and control principles emphasised. For instance JIT aims to eliminate wait times in the management of deliveries, production layouts and storage, whereas TQM aims to eliminate errors and related rework. Both apply these perspectives to flows of work, material or information.

The sum of these different management advances is a new production philosophy. Regardless of what term is used to name it, it is the emerging mainstream approach practised, at least partially, but increasingly, by major manufacturing companies in America, Europe and East Asia (Womack et al. 1990).

At the root of this new production philosophy is the concept that there are two types of activities in all production systems: conversions and flows. While all activities incur cost and time, only conversion activities add value to the material or piece of information being transformed into a product or service. Thus, the improvement of non value adding flow activities (inspection, waiting, moving), should aim to reduce or eliminate them, whereas conversion, or value-adding activities should be made more efficient. In design, control and improvement of production systems, both aspects have to be considered. Traditional managerial principles have considered only conversions, or all activities have been treated as though they were value-adding conversions. This distinction may seem slight but within manufacturing has been found to be of great consequence in effective production management in general and the successful implementation of CIM in particular. In some cases it has led to changes in business performance and strategy that have been of major significance to nations.

We can illustrate the basis of the new production philosophy and our current preoccupation with conversion activities in CIC research with a simple example drawn from Koskela & Betts (1993). This assumes a sequence of activities at the production stage of construction design to consist of those shown in Figure 1. Firstly it has to be noted, that the conventional description methods, such as generic flow-charting (used in Fig. 1), do not show attributes of flow processes important to the new production philosophy such as waste (for example, amount of rework due to errors or omissions). Taking the preparation of the brief (activity 1) as an example, making the brief is a value adding activity. The needs of the owner do not exist explicitly, but have to be established and stated. On the other hand, any requirement, which is missed in the beginning, but occurs later in the process, is costly, and produces waste. Thus, a systematised analysis of needs is extremely important; however, our conventional description methods do not suggest this in any way. The significance of avoiding having to restate needs is missing. The implications of this to CIC are that we must find ways of incorporating explicit and implicit user needs into our computer models and find ways of continuously monitoring whether they are being met as our CIC models evolve.

Secondly, value adding and non value adding activities have not usually been distinguished, as they are in Figure 1. The exchange of files (activities 5 and 6) provide an example. These are non value adding activities, which should be reduced or eliminated. If we are aiming only to increase the efficiency of these then our research may be misguided. However, this is the goal which has been primarily adopted in much CIC research. Much of the work in data exchange standards appears to be oriented towards improving the efficiency of flows. Product modelling and data integration research and innovation needs to be particularly aware of this issue.

This is obviously a very simplified example at a general level; as an example, it serves to illustrate the concepts of the new production philosophy. For a practical process improvement study involving CIC, the analysis should be taken to much

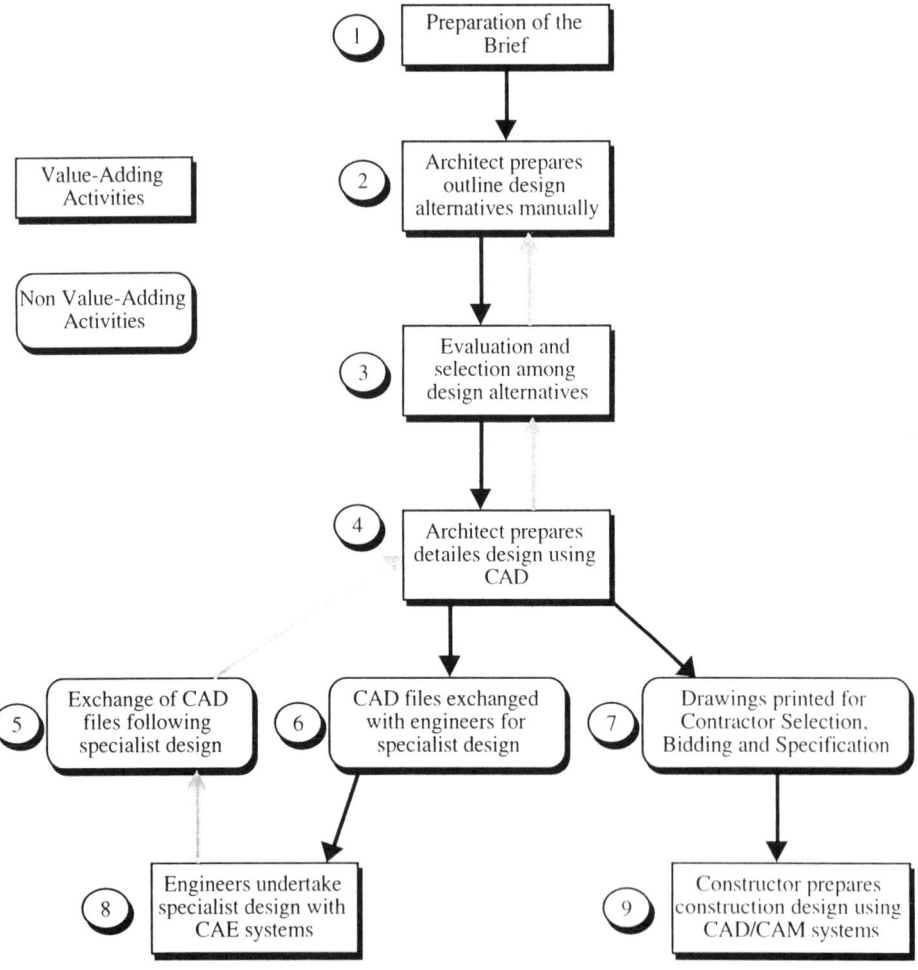

Figure 1. Construction design: An example of conventional analysis.

more detail than it is here but this illustrates the scope that lean production philosophies have as a driving purpose for CIC.

In essence, what we can conclude from applying lean construction principles to CIC is the need for integration to ensure that added value from conversion activities supported by CIC are continuously built upon and that flow activities are replaced by the automated functions of CIC. At present, automation appears to be aimed for without regard to whether activities are conversions or flows.

2.2 *Competitive strategy and strategic IT planning*

In parallel with the emergence of the new production philosophy, a second major area of development in management thinking has been the concern with competitive business strategy. Dominant in developments in competitive strategy and strategic planning in the 1980's was the work of Porter whose advancement of a five forces model for analysing industry profitability potential, value chain for exploring competitive advantage, and advocation of three alternative competitive strategies have all been popularly adopted. More recently, Porter's work in the competitive advantage of national industry segments has resulted in a diamond of four inter-related components of a competitive environment which has been applied to the construction professions by Betts & Ofori (1993).

Other strategy writers have more recently advocated other business planning techniques to be followed such as bench marking (The Economist 1991) and the core competencies approach (Prahalad & Hamel 1990). The former is also seen as one of the principles of flow process design of the new production philosophy illustrating the link between these two emerging management approaches. Core competencies address the production and process issues more directly by looking to identify a basic production skill which can be applied broadly to out-compete rivals for whom that basic skill is not available. Their significance to our discussions here may lie in whether CIC could result in a core competence that a construction organisation would use as a strategic weapon. Whether that would be CIC applied to a conversion or flow activity may arise from lean construction analysis. These strategic planning techniques and their relevance to construction enterprises have been discussed in detail by Betts & Ofori (1992).

In addition to the techniques of strategic planning, there have been other advances in techniques for identifying strategic information systems (Earl 1989). Many of these techniques pose such questions as who, why, when and for what information systems are applied. These are very similar to the fundamental questions we posed in the introduction of this chapter regarding a strategy-pull force to technological innovation. But what relevance if any do these competitive strategy and strategic IT planning techniques have to the new production philosophy and to an evaluation of the purpose of our research in CIC?

2.3 *A production and strategy synthesis*

The motivation for many process improvement activities and technological innovations is generally derived from strategic considerations, although there is some evidence for exploitation of technological opportunity. Most current strategic models do not explicitly acknowledge both the conversion and flow aspects of the new production philosophy separately. Many of Porter's early models most directly relate to conversions rather than flows and would have a different point of emphasis to the new production philosophy described. Porter argues that in particular, the three generic competitive strategies of product differentiation, low cost leadership and focus are mutually exclusive alternatives and points out the dangers of falling between two of them from a business strategy point of view. The new production philosophy in contrast advocates that both lower costs and higher quality are attainable and indeed should both be aimed for in process improvement from a production management point of view.

That is not to say that the strategic planning techniques and the new production philosophy are mutually exclusive alternatives as a model for strategy-pull forces to CIC but only that each in isolation does not show the complete picture. Applying CIC strategically using some of the above models for conversion activity improvements and flow activity automation would be beneficial. The strategic models used in isolation may not bring the order of magnitude improvements that the new production philosophy has shown to be possible in other industries which presumably is at the heart of our goals for CIC. In many ways the new production philosophy and strategic planning models can be seen as complementary and each of greater scope when applied together.

The core competence approach is also restricted from a lean construction viewpoint by virtue of its concern with only conversion activities. It is proposed by Stalk et al. (1992) that the approaches of core competence (referring primarily to conversion aspects) and process capability (referring primarily to flow aspects) could be used in parallel to chart the strategic improvement potential. Thus, in the case of building design, core competence primarily refers to the skills of designers and their design tools, such as calculating programmes, whereas capabilities refer to the properties of the actual information flows in design. Which of these do we primarily consider in CIC? It seems clear that in considering CIC there is scope for both to be embraced as a strategy.

What we are left with from this review of developments in strategic planning is an understanding that strategy and production are interdependent. To implement process improvement without a mind to the strategic objectives is unlikely to result in innovations to suit the way the construction industry will be organised in the future or that could be competitively applied to greatest advantage in the way the industry is presently organised. But by the same token, to develop competitive strategies without consideration of the way that production is carried out, and the scope for process improvement that exists, is unlikely to result in a strategy that is practically realisable. This applies most particularly to production dominated activities such as construction. Whether it also applies to other service oriented activities that form part of the total construction process remains to be established.

Our review of these two areas of management developments therefore can be combined to give a synthesised justification for technology applications. Any new developments in technology management must address the process improvement approach required by the new production philosophy and the requirements of the new thinking in business strategy. Taken together, we might describe this need for technology to contribute to the synthesis of these two approaches as 'evolving customer needs'. That is to say that the construction sector as it increasingly adopts these models of management thinking, will evolve needs to be met by technology of the type we have described. Other needs will arise from the demands made upon the sector by clients of our products and services and quality and environmental sensitivity are key in this regard.

This again brings us back to the idea of a technology strategy connection which we can illustrate here as an opportunities matrix. Organisations within construction may view a combination of technology and strategy issues as shown in Figure 2. All in construction should identify the key demand-side issues, or 'evolving customer needs', that face them in the future and combine these with a consideration of the emerging technologies of a constructional, financial, organisational and behavioural type. These may be those technologies they can acquire or gain access to rather than develop internally. They should then see where the technology-push/strategy-pull forces act together.

In terms of our desire for integration through CIC, we can use some examples from our matrix as illustrations. We may assume that one of the key emerging technologies is object-oriented systems and that an evolving customer need of an integrated AEC constructor organisation increasingly involved in Build Operate and Transfer projects would be for early buildability and maintainability advice. This may give rise to a technology strategy connection opportunity that gives a specific purpose to one particular type of CIC research. In addressing this purpose, the conversions and flows of design processes that aid buildability and maintainability must be separately considered as must the competitive consequences of developing such a technology.

A second example may see the emerging technology of broad band width data communication links and of multi-media technology with the evolving customer need for value-added early visualisation of internal office space alternatives for an owner-occupier of an office development. In addition, the traditional flow activities of the types shown in Figure 1 may need to be completely removed by automatic file transfer. Again, a specific technology strategy connection opportunity for CIC research emerges but of a quite different nature. Both examples draw from a vision of a

			Demand Side / Strategy Pull			
			Evolving Customer Needs			
			Buildability	Maintainability	Early Visualisation	Automatic File Transfer
Supply Side/ Technology Push	Emerging Technologies	Object Oriented Systems	*Example 1*			
		Virtual Reality				
		Multi-media			*Example 2*	
		Broad-band comms				

Figure 2. Opportunities matrix for an AEC organisation.

dynamic industrial viewpoint and research in response to change. Not all cells in the matrix will give rise to CIC research scope and the detailed analysis with the matrix that would be required in an individual, practical context therefore gives rise to identification of where to research and develop or generally apply CIC. All examples would further benefit from a consideration of the lean production and competitive strategy forces that affect them.

From this initial consideration of the current state of development in these dynamic fields it appears that the new production philosophy and the strategic concepts of core competence and capabilities offer scope for a combined strategy and process improvement approach which should be combined with our current technological preoccupation. It is to these ultimate objectives that our efforts in CIC may most purposefully be directed. Up till now most CIC research appears not to have been guided by any production philosophy or strategic model other than automating and integrating current practice based on an implicit assumption that automation is in itself desirable and that current industry practice is likely to remain.

The answer to our question of why we are aiming for CIC is thus not a generally applicable one but that there will be different reasons to follow different types of CIC research for different beneficiaries at different points in time and in different construction contexts. This is the range of CIC research objectives that we must consider.

3 A GENERIC DEFINITION OF INTEGRATION

Having identified the principles that explain why we are aiming for CIC in the context of emerging management philosophies we can now return to the second stage of a systems approach whereby we specify our indicators of success. This issue can be addressed by having a clearer and operationalised definition of what we mean by integration in the different contexts outlined above. Who integrates what; how and when one should integrate; and why one would choose to integrate. This leads to the development of the framework presented in Table 1. This framework was first proposed as an initial basis for discussion to define dimensions and levels of integration towards CIC (Fischer et al. 1993).

With regard to who, we can imagine integration among individuals and departments leading to the integration of entire firms and projects and ultimately to the integration of the entire AEC industry. In regard to the question of what to integrate, as an initial step we might choose to focus on sharing data. This could then be expanded to include models, such as product and process models, knowledge about decisions, and project goals. Ultimately, data, models, knowledge, and goals would all be shared. With regard to when we integrate, just a few applications within one phase and discipline might be a starting point which then could be expanded to include all applications from all disciplines and phases. Reasons why anyone might integrate or increase the level of integration are to stay in business, increase profit, market share, market size, or to enter or even create new markets. That CIC offers these opportunities has already been demonstrated by such companies as OTIS (Cash & McFarlan 1990) along with many others from outside of construction. Bradshaw (1990) also demonstrates a building materials supply company exploiting similar opportunities.

Table 1. Dimensions and levels of integration.

	Low integration $\rightarrow$		$\rightarrow$	$\rightarrow$	High integration
Who?	Individuals	Departments	Entire organisa-tions	Whole project life cycle	Entire industry
What?	Data	Models	Knowledge	Goals	All project infor-mation
When?	Islands of auto-mation	Multiple appli-cations in one discipline and phase	Multiple appli-cations for mul-tiple disciplines in one phase	Multiple appli-cations for mul-tiple discipline and phases	All applications in project deliv-ery process
Why?	Survival, stay in business	Increase profit	Increase market share	Enter new mar-ket	Create new mar-ket

Fischer et al. (1993) attempted to list the various values of these four dimensions of integration to indicate increasing levels of integration. However, they found that it is not always necessary to reach the previous step to go to the next level of integration. For example, a firm could well opt to first integrate goals and then tackle knowledge or model integration. Yet a framework such as this is important to distinguish different forms and stages of CIC and for all in practice and research to be able to cross-refer to different initiatives using a common understanding. Earl (1989) also argues that such a temporal framework is important in identifying sequential stages of progression in research and development.

This framework allows individuals, departments, companies, projects, and industries to plot their current state of integration and to indicate efforts to increase the level of integration. Thus, the framework becomes a vehicle for comparison and for focusing development and implementation efforts. For example, two departments might differ in their capabilities of sharing project information, or a company might be interested in pushing its integration capabilities from level 3 to level 4 for the how and when dimension.

This framework also provides a generic and focused definition of integration. Generically integration can be defined as the sharing of something by somebody using some approach for some purpose. Obviously this is not a very useful definition. However if one substitutes the vague expressions with values from the framework, one can create a definition that suits a particular purpose. For example, a firm might define integration as the sharing of data and models by departments using several applications pertaining to a number of disciplines and project phases for the purpose of increasing profit and market share. A government body may see integration as industry-wide data exchange between applications in multiple disciplines for industry survival. The advantage of the framework is that different definitions can be related to each other easily. The framework has five dimensions that are each presented in Table 1 independently. In reality, each dimension can be combined in a multi-dimensional way which we can illustrate by combining the dimensions why and who as in Table 2.

Through this framework and the generic integration definition that it allows, we have not specified a single indicator of success for our systems approach. We have

Table 2. An example of the interplay of two dimensions within the framework.

Who and why?				
Individuals	Departments	Enterprises	Projects	Industry
Stay in a job	Department survival	Stay in business	Complete project	Industry survival
Increase earnings	Increase profit contribution	Increase profit	Increase project duccess	Increase industry profitability
Extend job authority	Increase department political role	Increase market share	Increase project dcope	Increase industry share of economy
Change jobs	Assume additional roles	Enter new market	Make project more widely useful	Extend into other sectors
Create a new job	Create new roles	Create new market	Create project extension	Create new sectors and economic services

given the range of different indicators of success for the range of CIC research objectives that we foresaw at the conclusion of the previous section of this chapter.

4 CONCLUSIONS

This chapter began by making a case for more thought to be given to the question of why we seek CIC and what constitutes integration. By following a systems approach we first identified objectives and then specified indicators of success. Much current CIC research seeks to implement a particular operational programme without putting the work into a context of broad objectives and indicators of success. This chapter makes no such contribution and describes or advocates no particular programme.

What this chapter has shown is that the prerequisites to a programme are deciding why we are doing it and how far we want to go with it. These questions are addressed in the paper by considering our current quest for CIC in terms of recent managerial philosophies and by considering the extent in terms of a multi-dimensional generic integration definition. The contribution of the chapter therefore lies in the foundations it may give to the operational programmes research of others.

With regard to the larger picture of lean construction, the contribution of this chapter is in showing how a new production philosophy causes us to fundamentally re-evaluate our thinking in regard to improving the construction process. This is done in particular here for our consideration of how computers can be used to manage construction. Lean construction gives a focus to why we are aiming for CIC in a way that illustrates that we are looking to do more than automate. Whether we are automating flow or conversion activities will have important implications for how it is done.

There is a need for all engaged upon developments in computers for the construction sector to be aware of this if their technological improvement efforts are to fit into a broader picture of what emerging business and production developments require to happen.

ACKNOWLEDGEMENTS

This paper brings together recent collaborative work that has had a significant input from Lauri Koskela and Matti Hannus of VTT and Jarmo Laitinen of HAKA in Finland, Martin Fischer of CIFE at Stanford, George Ofori at NUS in Singapore and of Yusuke Yamazaki of Shimizu Corporation. I acknowledge and thank their contribution to many of the ideas that have been assembled here.

REFERENCES

Armstrong, J.S. 1985. *Long Range Forecasting: From Crystal Ball to Computer.* Second Edition, John Wiley and Sons, New York.

Betts, M. & Ofori, G. 1992. Strategic Planning for Competitive Advantage in Construction. *Construction Management and Economics* 10(6): 511-532.

Betts, M. & Ofori, G. 1993. Strategic Planning for Competitive Advantage: The Institutions. *Construction Management and Economics*, (in press).

Bradshaw, D. 1990. Building Blocks of Efficiency. *Financial Times*, August 30, p. 18.

Cash, J.I. & McFarlan, F.W. 1990. Competing Through Information Technology. *Harvard Business School Video Series*, Harvard, Mass.

Daniels, C. 1991. *The Management Challenge of Information Technology.* The Economist Intelligence Unit Management Guides, London.

Earl, M.J. 1989. *Management Strategies for Information Technology.* Prentice Hall, London.

Fischer, M., Betts, M., Hannus, M., Yamazaki, Y. & Laitinen, J. 1993. Computer Integrated Construction Framework. *First International Conference on the Management of Information Technology for Construction*, Singapore, August.

Liker, J.K., Fleischer, M. & Arnsdorf, D. 1992. Fulfilling the Promises of CAD, *Sloan Management Review.* Spring, pp. 74-86.

Koskela, L. 1992. Application of the New Production Philosophy to Construction. Technical Report No. 72. Centre for Integrated Facility Engineering. Department of Civil Engineering. Stanford University. 75 p.

Koskela, L. & Betts, M. 1993. Computer Integrated Construction in the Context of the New Production Philosophy. *First International Conference on the Management of Information Technology for Construction*, Singapore, August.

Porter, M.E. & Millar, V.E. 1985. How Information Gives You Competitive Advantage. *Harvard Business Review*, July-August, 149-160.

Prahalad, C.K. & Hamel, G. 1990. The Core Competence of the Corporation. *Harvard Business Review*, May-June, pp. 79-91

Stalk, G., Evans, P. & Shulman, L.E. 1992. Competing on Capabilities: The New Rules of Corporate Strategy. *Harvard Business Review*, March-April, pp. 57-69.

The Economist 1991. Competing with Tomorrow, 12 May, pp. 67-68.

Womack, J.P., Jones, D.T. & Roos, D. 1990. *The Machine That Changed The World.* Rawson Associates, New York.

Wright, R.N. 1991. Competing for Construction the World Arena. *Construction Business Review,* May/June, pp. 36-39.

Application of Quality Function Deployment (QFD) to the determination of the design characteristics of building apartments*

ALFREDO SERPELL & RODOLFO WAGNER
Department of Construction Engineering and Management, Catholic University of Chile, Santiago, Chile

ABSTRACT: A practical application of Quality Function Deployment (QFD) to the determination of the design characteristics of the internal layout of building apartments in Santiago (Chile) is described. The operational aspects of the analysis and the identification of client requirements are emphasized. The principal difficulties of the application of this method are also discussed. Finally, some results are presented to demonstrate the great potential of this tool for the solution of problems that combine demand and supply factors.

1 INTRODUCTION

When a person faces the decision of buying a home, he must include a set of variables related to the characteristics of the product in his decision. One of the most important of these variables corresponds to the internal layout of the house which is defined by the relative position of the different spaces, the form in which they are connected, and the relative size among them, given a restricted total plant area of the housing.

As a rule, experts in housing design are able of specifying – with quite precision – the characteristics that a good internal layout of a home must fulfil, but they normally don't have a detailed knowledge about the relative importance that customers assign to each one of these characteristics. This knowledge is however of great importance to offer a home that will satisfy the needs of potential customers and have thus a high selling potential, a fundamental objective of any successful real-estate project.

On the other hand, customers have limitations to express their wants in a technical and specific way, which hinders their understanding by the specialists. This situation can be explained by the fact that customers present three categories of requirements or expectations (King 1987):

1. Requirements that customers express openly which are those that a real-estate company normally satisfies. Generally this sort of requirement takes the form of general specifications that must be interpreted in a process of information exchange among customers and suppliers;

2. Requirements that customers do not mention openly but which they assume

*Presented on the 2nd workshop on lean construction, Santiago, 1994

they will obtain. These requirements fall into a category that has been nominated as 'expected quality'. These requirements are not specific satisfiers but if they are not present in the product, customers don't feel particularly pleased;

3. Requirements that customers don't expect, but people feel pleased about them. This sort of requirements has been incorporated in what is known as 'exciting quality'. These characteristics are generated by the supplier upon incorporating new ideas or innovations to his real-estate product.

The Quality Function Deployment is a method which allows to adequately incorporate customers requirements during the conceptual and design stages of a product (Bossert 1991). This method was developed in Japan around the end of the decade of 60's and it has continued its development up until now. Its application in the western world has widely expanded in the last years. This method presents a great potential of application to the design and conceptualization of houses, being one of the first applications accomplished in the construction sector (Shiino & Nishihara 1990).

This work presents an study which consisted in the application of QFD to the design of the internal layout of residential buildings apartments, emphasizing its operational and methodological aspects. Though the study also encompassed the application of this method to the design of other building zones, the presentation has been limited to the layout problem for simplicity purposes. In the next sections, a synthesis of the process carried out to achieve this application is described with particular focus on the principal difficulties found during the study.

2 OBJECTIVE AND DESCRIPTION OF THE STUDY

This study is part of a broader research work oriented to characterize the real-estate preferences of building apartment owners of the east area of Santiago. The study began with a bibliographical review that provided an overview of the potential of QFD as a method for identifying customers requirements. A Japanese application to prefabricated houses was used as a basic reference (Shiino & Nishihara 1990). Once the central objective of the study was established, 14 interviews to outstanding real estate professional and entrepreneurs were carried out with the purpose of obtaining their feedback to better focus the study and define the most appropriate methodology.

Acquisition of customers perceptions (or demand voice) was performed through an after-sale survey to apartments inhabitants. In the survey, the apartment owners were the customers and the main objective of the survey was to find out what aspects of the apartments layout were actually the most important for them. The questionnaire was designed to obtain all the information required to use QFD. A group of four real estate experts was formed to provide the study with the necessary experience and knowledge to assure success of this survey, and to perform the role of the supplier in the QFD application (supply voice).

Figure 1 shows all the stages of the QFD application as it was carried out in this study. Many different problems to obtain dependable and useful information from these sources occurred during the study, which are discussed in the following sections.

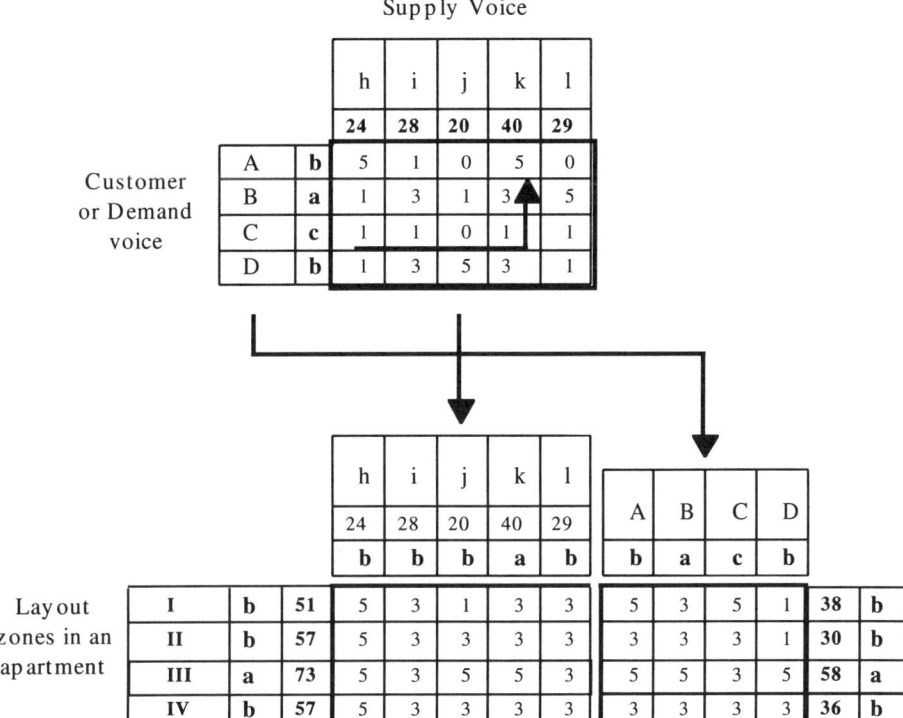

Demand voice (in the case of Middle Segment):
    A = Separation between rooms, internal privacy;
    B = Size of rooms;
    C = Functionality (relationship between the rooms and their equipment);
    D = Quality of life (acoustic and thermal isolation from outside, all rooms with natural light).
Supply voice:
    h = Separation, independence of social, service, and private zones;
    i = Design efficiency;
    j = Relative position to the sun, natural light, view;
    k = Internal privacy; l. Relative sizes of the rooms.
Layout zones in an apartment:
    I = Service zone (kitchen, loggia);
    II = Social zone (Entrance hall, living and dinning room, terrace);
    III = Main suite (bedroom and bathroom);
    IV = Other bedrooms and bathrooms.

Figure 1. Summary of QFD application (adapted from Shiino & Nishihara 1990).

2.1 *Voice of the customer or demand voice*

The voice of the customer or demand voice includes the characteristics of the layout of apartments that were considered as required/specified by their inhabitants (A, B, C, and D in Fig. 1). These opinions were obtained through two questions of the survey that was sent to 500 apartments of the east area of Santiago.

These questions were of the open type in a way that respondents could give their opinions freely. Any forced question format, for example to choose from among a set of alternatives, would introduce a bias in the customers answers. On the other hand, the use of open questions could result in a set of answers that can be too heterogeneous without indicating any trend. Because of these restrictions, a careful redaction of the text of the questions was performed. In each question, a definition of the general concept of layout of an apartment was included, to assure answers on a common basis and not in function of the definition that each one of them previously had of those terms. However, this did not suffice to obtain adequate information since it is clear that preferences change from one social segment to another. It was decided since, to stratify the demand.

It was not considered as convenient to use an economic index as a segmentation criterion, since it could produce susceptibilities upon consulting it, reducing the possibilities of success of the questionnaire. Thus, the sample was segmented according to the total area of the housings surveyed, without causing any mistrust upon asking it and being effective for stratifying purposes.

Once the surveys were back, the information was processed. It was necessary to categorize the answers using as criterion, the grouping of them according to their concepts affinity. After all the answers were grouped, the concept that was identified as the most relevant one in all the answers that constituted each group was assigned as the name of the category (group). Of course, there existed answers that did not keep similarity with any other and were let apart, without forcing their entry into a category.

The categories so obtained were also characterized by the number of answers or frequency contained by each of them. Based on this frequency, the categories received a weighting factor. Those with greater frequency received the weight 'a', the intermediate 'b' and the smaller 'c'. If the dispersion of the frequencies of the categories was small, it was decided to assign only two weights: a and b.

It is important to point out that different categorization criteria can deliver different results. In fact, an exercise was carried out with the same information which, though not useful for the QFD, delivered interesting information about demand. Instead of categorizing by concepts affinity (for example: amplitude), the information was categorized according to the type of room mentioned in the answers. As the questions of the survey did not consult about any type of room in particular, this form of categorization informed about the rooms that are of greater concern for the surveyed persons. Different, but not contradictory results were obtained from both forms of categorization, which allowed researchers to obtain a quite complete perception of demand preferences.

2.2 *Supply voice*

The supply voice corresponds to those design characteristics or parameters that define the layout of an apartment in a general form, without making specific reference to any kind of room (h-k and I in Fig. 1). Considering that this is an stage where multiple opinions can be obtained from different experts, it was apparent that the contribution of only one expert would offer a limited vision. Thus, it was determined that this topic would be covered by a multi-disciplinary team, formed by four real-estate experts, three of them architects and one civil engineer.

It is important to point out that this stage was carried out at the same time of the application of the survey. In this way, both the demand and the supply voices were obtained concurrently, without any influence between them.

This stage consulted the employment of the Delphi technique which permitted the access to an excellent group of professionals, which was not feasible to gather physically. This technique presents several other advantages, since avoids uncompromising positions and personal dominations, in addition to assuring the anonymity of the members of the team (Porter et al. 1991). The Delphi technique was not only used in this stage but also in the development of the quality matrix and in the detail phase were the same group of experts participated. In all the stages, questionnaires were used to interview team members and the group response was fed back to the specialists, exposing the arguments for and against each position.

Two rounds of meetings with team members were necessary to obtain the supply voice. This stage did not present greater problems, since the opinions of the different experts were quite coincident and the answers, more than contradictory, were complementary. A second round was used to obtain convergence. Thus, a set of design aspects represented by h, i, j, k, and I in Figure 1, were obtained. Similarly, in this same stage, the classification of the major zones of the layout of an apartment were determined (I-IV in Fig. 1). These zones are the different main areas that experts visualize within an apartment (i.e. social area, service area, etc.).

2.3 *Quality matrix*

The next step involved the construction of the quality matrix, which is used to confront the previous determined demand and supply voices (top matrix in Fig. 1). It was the labor of the team of experts to complete this matrix, defining the degree of relationship among each aspect of the demand voice and each of the supply voice. Four different levels of relationship were defined (Shiino & Nishihara 1990): 5 = strong relationship, 3 = average relationship, 1 = weak relationship, and 0 = no relationship. These levels permitted the transference of the weights of the demand voice factors to the supply voice factors. The design aspects that obtained the highest weights correspond to those that were considered critical by the customers.

This stage required three meetings with the team of experts. In this case, responses from experts were quantitative (a number, that represents the intensity of the relationship that existed among a concept of the demand voice and each one of the concepts included in the supply voice). The processing of the information was simple, using the median of the answers as the representative group answer. The first meeting with the experts brought a very high dispersion of the answers. The analysis of this situation indicated that the principal reason of this variability rested in the different meaning that each specialist attributed to each concept of the demand and of the supply voices. Though the same expert had defined the supply voice factors, after some time they forgot the meaning of the terms obtained through the Delphi technique, coming back to the daily, personal meaning of them. Something similar occurred with the names of the categories of the demand voice, where definitions as 'functionality' or 'quality of life' caused problems. In the two following meetings, the researcher had the caution of clarifying the meaning of the terms of the demand and of the supply voices to each specialist, previous to the construction of the quality matrix. Even so, the experts showed greater heterogeneity in this case than before.

This motivated the interviewer to allow the experts in some cases to qualify with 5 a relationship that other expert had qualified with 1. In any event, such situations were the exception and an adequate convergence of the answers was finally achieved. Upon ending this stage, the design phase of an application of QFD is considered complete, having constructed the quality matrix.

2.4 *Detail phase*

The lower matrix of Figure 1 shows the last stage of this application of QFD. In this matrix, the demand and the supply voices were separately confronted with the previously defined apartments layout zones. This matrix was built by the experts, something which permitted to obtain two sets of weights for the zones, one originated from the scores of the supply voice and other originated from the scores of the demand voice. Once the weights for the zones were obtained, an indication of those that required greater attention from design was achieved, in particular to those aspects that were highlighted by the supply voice according to the weights they obtained.

This grupal work in this stage did not show any serious problems, since from the beginning of the stage the research coordinator had the caution of reading to each experts, the definitions of the concepts determined in previous meetings, prior to having each expert answering the questionnaire. Taking care of this aspect yielded immediate fruit, inasmuch the opinions of the first meeting showed very small dispersion, and the convergence was easily reached in the second round of meetings.

3   EXPERIENCES OBTAINED

In the first place, it seems necessary to comment the excellent response level obtained in the survey, receiving 139 correctly answered surveys from 509 questionnaires (27.3% of the sample) initially delivered to 10 buildings. A careful preparation of the survey and of the presentation letter, in addition to the use of incentives, would explain such a high response rate.

With respect to the problems that occurred during this application of the QFD method, a comparison among the five principal problems evidenced in a study performed in 1986 by Yoyi Akao (King 1987) and the problems found in the present application is presented below:

1. Matrices too big, and as such, very high handling complexity: In this particular case, the biggest matrix had 4 rows and 5 columns, and did not present managing problems. In any event, the size of the matrix is a variable that cannot be handled without influencing the results since in this case, the number of rows and columns results from the number of concepts or categories that are determined by the demand and the supply voices, and the layout zones;

2. Difficulty in obtaining the requirements of the demand: this aspect did not present large difficulties, due apparently to the careful design of the survey questions, which had the valuable contribution of 14 real-estate professional and entrepreneurs interviewed previous to the application, with the purpose of perfecting the form and the fundaments of the survey;

3. Difficulty in the categorization of answers of customers: This was a complex process, that began by the definition of the categorization criterion. It was already indicated that two different categorization criteria can deliver different results based on the same group of answers. The criterion used here consisted in grouping the requirements with some similarity or affinity. To apply this criterion, the method of the affinity diagram was used. This technique allows a team of experts to organize a set of not-structured ideas into a sound group of categories (Bossert 1991);

4. Difficulty to determine if the answers of demand were appropriated for the study: this aspect did not present difficulties in this application due apparently to the careful design of the questions, which permitted to direct the answers toward the sought objectives;

5. Difficulty in determining the degree of relationship among each one of the terms of the supply and of the demand voices: It was one of the greatest complexities of this application. Some times there was considerable divergence from one expert to other. It was appreciated that in this case, the supply was quite less homogeneous than the demand in their preferences.

Any opinion survey must take into account that people, when consulted about requirements of quality, name specifications and don't make references to characteristics of quality of other type, like those mentioned before in the introduction of this paper. For example, in an apartment most of the customers expect that some sort of floor cover will be delivered. It is not thinkable that someone would deliver a naked concrete slab. Now, if this expected quality is not provided, probably the product will not be sold, but if it is provided, people will not indicate any particular satisfaction. They are simply obtaining what they expected to receive. In summary, the characteristics of the expected quality do not call the attention if they exist, but most of the customers would resent the lack of any of them.

There are also new characteristics that designers confer to their products. Generally, clients do not expect to get them but will eventually please them if these characteristics constitute improvements (i.e. they have value for the customers). That sort of quality characteristic has the purpose of motivating people to acquire the product.

It is important to take into account these concepts when analyzing the results of a market survey. Upon analyzing the answers obtained in this study, it can be seen that when the persons were consulted about important aspects of the apartments layout they answered, as a rule, in function of what they had noted as not present or that was lacking in their particular case. It was observed clearly that at a greater economic level, a greater number of aspects become 'expected' and, as such, they disappear from among the explicit quality requirements of these persons. This was seen in several different results of the study, like the one that is presented in Table 1.

In Table 1, the aspects of the demand voice that were named more frequently by people from the economic segments middle-high and high are shown. It can be observed that in the middle-high segment it exists 5 categories and in the high segment exists only three which are exactly the same three first categories of the segment middle-high and with the same weights. The fact that the high segment does not have the others two categories could probably obey to the fact that the design of the high level apartments, as a rule, include an adequate functionality and the existence of a good entrance hall. The people of this stratum do not request these characteristics,

Table 1. Comparison of the demand voice of economic segments middle-high and high.

Segment middle high	Score	Segment high	Score
Separation among rooms, interior privacy	a	Separation among rooms, interior privacy	a
Amplitude	a	Amplitude	a
Direction, natural lighting, view	c	Direction, natural lighting, view	c
Existence of an entrance hall, adequate distribution spaces	c		
Functionality	c		

since in their particular case these design problems are generally well solved and therefore, do not call their attention.

Figure 1, though is only presented in order to show the stages that were followed in this methodology, contains real results from apartments of the middle level. It is important to look at the case of zone III in the lowest matrix. The obtained weights from both the supply and the demand voices are the highest for this zone. This situation confirmed this zone as a critical one that requires the greatest design attention, in particular, with regards to that aspect that was emphasized by the supply voice with the highest weight (in this case, the aspect k). In this study, always existed coincidence between the supply and the demand voices with respect to critical layout zones.

4 CONCLUSIONS

The application of QFD described in this chapter has allowed researchers to obtain experiences that can be useful in future applications. Some of them have been discussed already. However, one of the main problems to deal with when using QFD, is the retrieval of the information required by this method. This information should be reliable and complete to assure good results.

The research work demonstrated that the technique is fully applicable to the sort of problems described here, and concluded that the QFD is a technique that uses the demand opinions to weight the design aspects that the supply itself has defined. This affirmation is extremely relevant since a common survey of opinion can obtain the requirements from the clients but expressed in their own terms. This might not be very useful for the supply due to the fact that he probably uses special variables in the design of its products, that can be difficult to relate to the quality requirements stated by customers. In QFD, on the other hand, it is the supply the one which determines the variables and the design aspects as it likes it, and customers are entrusted with saying what of those aspects are the most relevant for them.

ACKNOWLEDGEMENTS

The authors wish to thank the collaboration of all the experts and professional that provided their knowledge and experience in this study and the permanent support of the Chilean Construction Research Corporation.

REFERENCES

Bossert, J.L. 1991. Quality Function Deployment, to Practitioner's Approach. Marcel Dekker, USA.

Porter, A., Roper, T., Mason, T., Rossini, F., Banks, J. & Wiederholt, B. 1991. *Forecasting and Management of Technology*. John Wiley & Sons, Inc., USA.

King, R. 1987. Listening to the Voice of the Customer: Using the Quality Function Deployment System. *National Productivity Review*, Summer, USA.

Shiino, J. & Nishihara, R. 1990. Quality Deployment in the Construction Industry, In: Yoji Akao (ed.), *Quality Function Deployment: Integrating Customer Requirements I intuit Product Design*, cap. 10, Productivity Press, USA.

Tools for the identification and reduction of waste in construction projects*

LUIS F. ALARCÓN
Department of Construction Engineering and Management, Catholic University of Chile, Santiago, Chile

ABSTRACT: A discussion of the importance and impact of introducing the new production philosophy in the field of construction is carried out, based on experiences in other industries and recent applications in construction companies. The importance of the heuristic principle of reducing non value-adding activities is emphasized, as the fundamental focus for achieving improvement. Different definitions and classifications of waste are compared which extend this concept beyond its traditional meaning. Within this context, examples are presented of tools which can be used to identify and reduce waste, such as: Work sampling, resource balance charts, and a waste diagnostic survey.

1 INTRODUCTION

Unlike manufacturing activity, where the rhythm of production is fundamentally governed by the machines used in the manufacturing processes, construction depends on the management of information and resource flows. This is due to its large field work component, the provisional nature of some of its organizations, and its intensive use of labor and non-stationary equipment. The organization, planning, allocation and control of these resources is what finally determines the productivity that can be achieved. In spite of this reality, until now the conceptual model used either implicitly or explicitly to analyze construction is that of conversion of inputs and outputs, which ignores important aspects of the flow of information and resources. For many years, the utilization of these models has helped to emphasize the differences between construction and stationary manufacturing activity, and has limited the spread of the new production technologies and philosophies which have been generated in the latter field. However, the recent advances developed by Japanese industry and currently being disseminated in Europe and the United States, are based on production philosophies which explicitly consider information and resource flows. Additionally, they are focused on productive processes and are perfectly adequate for application in construction, in spite of their singularities.

Construction is a key sector of the national economy and faces contingent problems which must be resolved. Increasing foreign competition, the scarcity of skilled

*Presented on the 2nd workshop on lean construction, Santiago, 1994

labor, and the need to improve construction quality are the challenges it faces today. Responding to these challenges imposes an urgent demand to raise productivity, quality, and to incorporate new technologies. A lack of responsiveness can hold back growth and the development of the needed infrastructure for the construction industry and other key activities in the country.

In an attempt to respond to these challenges, several parallel research efforts are currently being developed in the Department of Engineering and Construction Management of the Catholic University of Chile, in direct collaboration with companies within the sector. The general aim of this research is the development, adaptation and application of methodologies for the analysis and improvement of productivity and quality in construction projects. A few examples of tools which have been adapted or developed are shown in this article to illustrate one of the lines of work presently being carried.

2 THE IMPACT OF THE NEW PRODUCTION PHILOSOPHY

The introduction of a new production philosophy, which has been applied with so much success in Japanese industry, has been principally promoted by administrators and engineers in the daily practice of their professions, with almost total independence from the academic world. The revolutionary ideas that serve as the foundation of this new philosophy evolved from different starting-off points. Terms such as: 'Total quality control', elimination of waste through the 'just in time' philosophy, 'simultaneous engineering', and 'continuous improvement', describe different forms of implementation of these ideas. Only recently the academic world has begun to study the concepts underlying this philosophy. Particularly, some researchers have recognized the applicability of these concepts to different activities of stationary manufacturing. The new philosophy promotes a continuous improvement in the productive processes through a reduction of 'waste' (time, unnecessary or excessive processing, resources, etc.) and an increase in 'value' (quality, improvements, finished products, etc.).

Construction processes are characterized by high contents of non value-adding activities leading to low productivity. Thus, the development of methods of analysis and process improvement, as well as the introduction of new production philosophies which contemplate continuous improvement, can have an important impact on management, productivity, quality, and current construction technologies. Likewise, it is possible to visualize a direct transfer of the benefits to other activities with similar characteristics, such as industrial maintenance and agricultural exploitation. The potential for improvement is tremendous, and proof of this are the results obtained in the automotive industry, where impressive reductions have been achieved in the use of each of the components of the productive processes: 50% in human labor, 50% in assembly space, 50% in investment in tools, 50% in engineering hours, plus the development of new products in half the time that was necessary before. Similar orders of magnitude of benefits have been documented in other industries. If just a minimal fraction of these benefits were achievable in construction, the impact would be tremendous.

3 CONCEPTS OF LEAN PRODUCTION

The conceptual foundations and principles of the new production philosophy are substantially simple. The flows of information and materials are the bases for analysis, different combinations of which lead to three different points of view on production:

– Production is a conversion of inputs into products (traditional approach);
– Production is a logistical flow (just-in-time approach);
– Production is a generation of value through the satisfaction of the client's requirements (quality approach).

The new philosophy considers production as being essentially a flow of materials and/or information, from raw material to final product (Fig. 1). In this flow, the material is processed (converted), inspected, experiences wait periods, or is transported. All these activities are inherently different among themselves. The various processes represent the different aspects of conversion encompassed within production, while inspection, transport and waiting time represent its flow aspects.

The flow processes can be characterized according to their duration, cost and value. Value refers to the satisfaction of the client's requirements. Only the processing activities add value to the product. For example, transporting materials to a site where by they can be converted into forms that clients can assign value to, in a certain sense adds value. However, this occurs because the materials are found in an appropriate place and at the precise moment, which theoretically could occur without need of transport. In fact, the client will not perceive any difference if the component materials are transported a few meters or several kilometres. Similarly, inspection is necessary at certain stages of the production, but can be eliminated with further improvements, such as precision tools or error-proof procedures.

For the flow of materials, processing activities produce variations of form or substance, assembly or disassembly. Something similar occurs with the flows of information. However, how could the processes of process flows be designed, controlled and improved in practice? In several sub-fields of the new production philosophies, a significant number of heuristic principles have evolved (Table 1).

Within the above, principle no. 1, 'reduce the share of non value-adding activities' is among those which exert the greatest impact.

Value-adding and non value-adding activities can be defined as follows:

– Value-adding activities: Those which convert the materials and/or information in the search to meet the client's requirements;
– Non value-adding activities (waste): Those which are time, resource or space consuming, but which do not add value to the product.

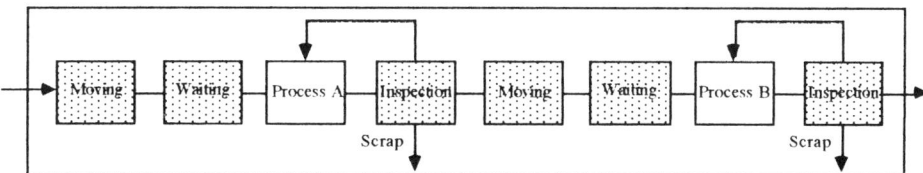

Figure 1. Production as a flow of processes: Simple illustration. The shaded boxes represent non value-adding activities, in contrast with processing activities, which do add value (Koskela 1992).

Table 1. Principles of improvement of production processes (Koskela 1992).

0. Increase the efficiency of the value-adding activities.
1. Reduce the share of non value-adding activities (also called 'waste').
2. Increase the product's value through the systematic consideration of the client's requirements.
3. Reduce variability.
4. Reduce cycle time.
5. Simplify through the minimization of steps, parts and the need to reconcile information or unions.
6. Raise the flexibility of the outputs.
7. Increase process transparency.
8. Focus process controls on the totality of the process.
9. Introduce continuous process improvement.
10. Establish permanent references in the processes (benchmarking).

Experience shows that non value-adding activities predominate in the majority of processes: Usually just 3% to 20% of the steps add value (Ciampa 1991), and their share in total cycle time is negligible, from 0.5% to 5% (Stalk & Hout 1990). In construction, it is possible to immediately identify innumerable non value-adding activities: inspections, moving, waiting, deficient or incomplete information, etc. Besides these, there are a large number of activities which apparently add value but which, on closer inspection, can be completely eliminated. For example, if reasonable finishings and plumbs could be obtained in the building process, it would eliminate the need to chip and stucco the walls. In this light, reducing non value-adding activities offers great potential for development in most construction processes.

4 FORMALIZATION OF THE CONCEPT OF WASTE

4.1 *A broad definition of waste*

The focus of the new production philosophy presupposes a broader concept of 'waste' with respect to the usual meaning given to the term. For example, Toyota defines waste as: 'Anything that is different from the minimum quantity of equipment, material, parts and labor time that is absolutely essential for production'. A Western definition using similar terms would be the following: 'Anything that is different from the absolute minimum of materials, machines and labor necessary to add value to the product'.

In general, waste is defined as the loss of those activities which, while they produce a cost, whether direct or indirect, do not add value or help in advancing a project. These waste categories are measured as a function of their costs, including opportunity costs.

Another type of waste is related to the efficiency of processes, equipment, and personnel. These are harder to pinpoint and measure, since knowledge of the optimum attainable efficiency is required, and this is not always possible.

4.2 *Waste classification*

Shigeo Shingo, in his book *Study of Toyota Manufacturing System* (1981), proposes the following waste classifications:
 1. Waste due to overproduction;
 2. Waste due to wait periods;
 3. Waste due to transport;
 4. Waste due to the system itself;
 5. Waste due to stock;
 6. Waste due to operations;
 7. Waste due to defects.

George W. Plossl, in his book *Managing in the New World of Manufacturing* (1991) adds three more categories:
 8. Waste due to time;
 9. Waste due to people;
 10. Waste due to bureaucracy.

By examining the focus of construction-productivity improvement, one may conclude that the concept of *Lean production* is not completely new in construction. This approach, from the 1970s, added new elements of performance for the study of productivity, upon recognizing that the traditional control systems in construction were exclusively aimed at information on cost and schedule, and did not allow for identifying the performance elements that reduce productivity.

In 1986, Borcherding proposed an interesting qualitative model to identify the causes of reduced productivity in construction. He postulated that loss of productivity in large and complex constructions is explained by the use of five categories of non-productive time:
 1. Waste due to waiting or idle;
 2. Waste due to travelling;
 3. Waste due to slow work;
 4. Waste due to ineffective work;
 5. Waste due to rework.

These categories correspond to the definition of waste within the terminology of the new production philosophy, and it is interesting to note the correlation between these definitions of waste and the others enunciated above.

5 DIAGNOSTIC AND IMPROVEMENT TOOLS

Almost all the 'waste categories' are invisible within traditional control systems. Nevertheless, the productivity approach proposed new tools for diagnosis, measurement and improvement for this purpose. Delay surveys for foremen, work sampling methods, records of materials and other tools have been adapted or developed to allow decision making for the improvement of productivity in construction (Oglesby et al. 1987). The main objective of these tools is to reduce delays, interruptions and to improve storage of resources, coordination, and planning on the construction site.

In general, actions based on information provided by these tools have been directed at eliminating 'the organization's restrictions'. For example, reducing transport time for materials supply, or storing tools close to the construction site; modifying the distribution of installations; providing cranes or elements of materials transport to eliminate transportation and transfer time. All these partial measures seek to reduce or eliminate non value-adding activities. Additionally, keeping records of the evolution of the different performance elements is directed toward measuring 'variability', which has been suggested as a necessary measure for improvement in construction (Koskela 1992). A description of some of these tools and examples of their application follows.

5.1 *Diagnostic and improvement survey*

To try and identify waste categories in construction works, a survey was created that was very simple to use and which has to be very useful as a communication tool among site supervisors, to enable them to generate their own mechanisms of waste identification and reduction. The survey, whose format appears in Appendix A, makes it possible to identify the most frequent waste categories and their most frequent sources. It also provides a mechanism for investigating the causes of waste.

The survey can be created by the members of the project team themselves, based on some examples provided in a basic format. The information required for preparing the survey consists of two main parts: a list of waste categories, which introduces a common language among the members of the project team to enable their identification; and a classified list of the potential sources of waste, according to some kind of criterion that would allow the identification of the areas with the greatest deficits. Tables 2 and 3 give examples of the information used in some recent cases.

Each of the preceding waste categories generates multiple impacts whether in the project's cost, schedule or quality, and has its origin in some source, as in those listed in Table 3.

The survey showed in Appendix A, once processed, makes it possible to pinpoint the most frequent on-site waste categories, according to the project team members' per

Table 2. Examples of waste in construction projects.

1. Work not done
2. Re-work
3. Unnecessary work
4. Errors
5. Stoppages
6. Waste of materials
7. Deterioration of materials
8. Loss of labor
9. Unnecessary movement of people
10. Unnecessary movement of materials
11. Excessive vigilance
12. Extra supervision
13. Additional space
14. Delays in activities
15. Extra processing
16. Clarifications
17. Abnormal wear and tear of equipment

Table 3. Examples of waste sources according to area.

A. Administration
    1. Unnecessary requirements
    2. Excessive control
    3. Lack of control
    4. Poor planning
    5. Excessive red tape

R. Use of resources
    1. Excessive quantity
    2. Insufficient quantity
    3. Inadequate use
    4. Poor distribution
    5. Poor quality
    6. Availability

I. Information systems
    1. Not necessary
    2. Defective
    3. Slow
    4. Unclear

ceptions, and to simultaneously identify the most frequent sources of waste. Appendix A shows the results in an industrial assembly project where 18 supervisors took part in a survey. The necessary first step for the improvement of a construction project is to remove the causes of waste; a cause-effect matrix is used to capture the perceptions of the project team of the present causes and the existing waste potential. In this way it is possible to adopt immediate improvement measures. For example, the results show that materials waste is the most frequent on that site. If one wishes to attack the causes of this waste, one can examine the matrix indicating that the principal causes of this problem are 'lack of control' (mentioned 6 times), 'excessive amount' (mentioned 3 times), and, later, several others of similar importance. The project team can now focus its attention on determining actions to improve control and reduce the amount of materials, if this is possible. Other actions over this waste, as well as over others, could be aimed at finding out more about the problem, using more specific measurement tools that would allow measuring future performance for improvement purposes.

The same matrix can be used to measure the potential for waste in each source, adding up the columns in order to achieve a better control over the sources showing the greatest potential for waste. The potential impact of each source can also be analyzed by examining each column individually. The usefulness of this type of analysis is that it makes it possible to discover, in a systematic way, the general perception of the project team members. This would allow to focus the supervisors' attention on the most pertinent processes and to create – while all this is done – a mentality of improvement, through the use of a tool that promotes participation and commitment on the part of the participants.

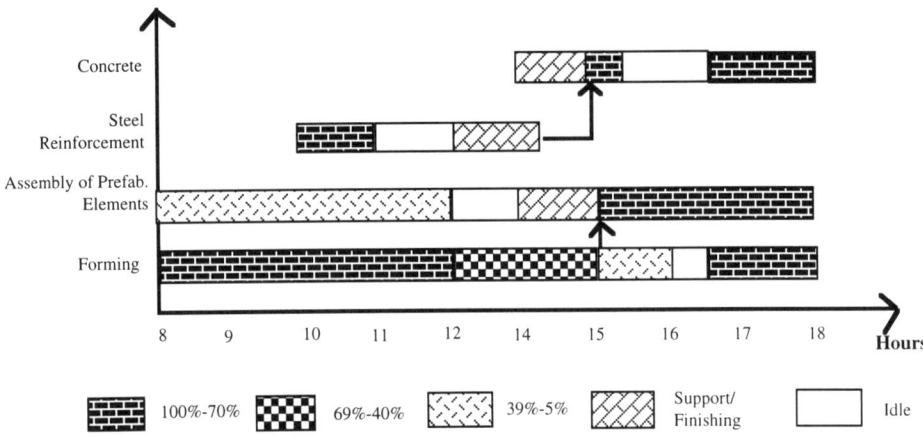

Figure 2. Balance chart of an integrated process.

5.2 *Balance chart of a multi-operational process*

The balance chart is a classic method of operational analysis which makes it possible to do a detailed follow-up on the use of resources within an operation in order to identify opportunities for improvement. Traditionally, this tool has been applied to the analysis of a specific operation, to improve the efficiency of the operation. Nevertheless, an alternative application was recently developed to carry out a less detailed but more complete follow-up on the use of resources on a construction site. Figure 2 shows this application, featuring an interesting use of the balance chart of an integrated process as a tool for the identification of waste categories, especially those derived from the interrelationship between operations. Figure 2 shows the activity developed by each of the operations observed, indicating an estimate of the degree of effort carried out in the operation. The notation makes it possible to identify the existing interrelationships between the activities, pinpoint bottlenecks, and achieve a global vision of the integrated process of the operations.

Focusing attention on the complete process makes it possible to analyze global improvements, which is not always possible to do with merely an individual vision of each operation. This analysis, for example, suggests the need to balance the production rhythms through multi-functional crews. The application of solutions of this type allows the reduction of waiting times and to better balance the use of resources, significantly increasing the project's productivity.

Obtaining a balance chart of this type requires constant observation of the process, which implies an important expenditure of resources for gathering information. Nevertheless, given its minimal requirements for detail, this can be done by video filming at certain time intervals. This would reduce the time for gathering information and would provide an additional record of tasks for their subsequent review.

5.3 *Work sampling*

Traditional work sampling is a very attractive option as a tool for waste detection.

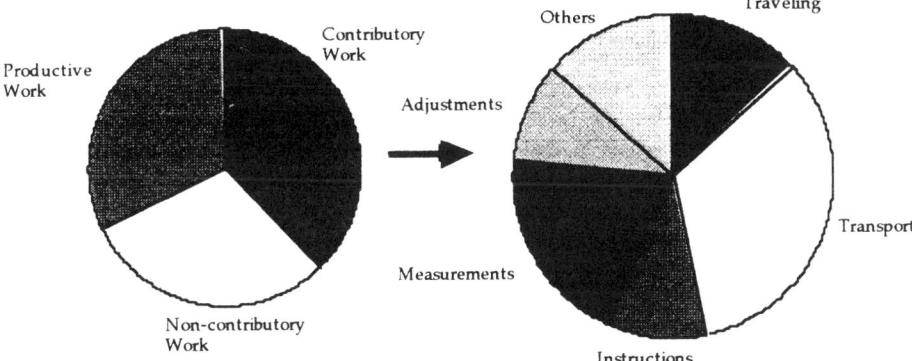

Figure 3. Use of work sampling to identify waste categories.

Through random observations, it is possible to estimate, with statistical validation, how labor time is being used in a construction site. Figure 3 shows a classic example of how sampling can be used to identify productive, contributive and non-contributive time in a construction project.

The information from the work sampling points out to unusual situations or abrupt variations in the use of time. This information, supported by the comparison with standards for the same or similar projects, focuses attention on identifying the causes of the detected waste. Intelligent use of this tool can be of enormous help in the works' management.

6 CONCLUSIONS

Initially, *Lean production* has been approached through the application of several methodologies in vogue, such as total quality management, just-in-time production, concurrent engineering, process re-engineering, benchmarking, and others. These methodologies are inherently partial approaches, for they originate around one or more central principles. For example, quality approach is based on variability reduction. Nevertheless, instead of implementing just one methodology, it seems more effective to adopt base principles and methodologies which provide the greatest potential for results in each particular case; this is the direction of the *Lean construction* approach.

The evolution from the traditional control of construction schedules and costs toward a more comprehensive measurement of performance, implies a change similar to the move from uni-dimensional to bi-dimensional vision. Adopting the paradigm of *Lean construction* constitutes an even more radical change involving the adoption of a multi-dimensional vision: A wide scope for the effort of improvement that is not just focused on construction productivity, costs or schedules, but on the reduction of a wide spectrum of waste. A broad definition of waste in the construction processes adds a global dimension to the improvement effort, and an integrated vision of all phases of the construction project is achieved, including design, materials supply, construction and subsequent operation. The tools shown here can help in the task of creating a mentality of improvement, which will make it possible to gradually create the conditions for adopting the principles of Lean Construction.

ACKNOWLEDGEMENTS

Several of the examples presented in this work arose from a joint project with other colleagues and students. In particular, the diagnostic survey was prepared in a project with students Francisco Lowener, Francisco Lira and Marcelo Beratto, which was supervised jointly with Prof. Carlos Videla. The extended concept of the use of the balance chart was developed in a joint work with Prof. Hernán de Solminihac.

   The author very sincerely thanks the Chilean Construction Training Corporation for its collaboration and the sustained support for this line of research that it has extended to the Executive Research Board of the Catholic University of Chile (DIUC).

APPENDIX A: WASTE IDENTIFICATION SURVEY

The following survey is designed to identify the most frequent waste and its sources according to your perception of your work environment. Please, answer carefully the questions below.

---

1. Name:

---

2. Position:
– Resident engineer
– General foreman
– Foreman
– Procurement
– Administration
– Other (specify):

---

3. Company:

---

4. Type of project:
– High rise building
– Building
– Highways and roads
– Mining
– Industrial
– Civil
– Others (specify):

5. Identify with a tick mark the waste sources present in your work environment:

Administration
1. Unnecessary requirements
2. Excessive control
3. Lack of control
4. Poor planning
5. Bureaucracy

Use of resources
1. Surplus
2. Shortage
3. Misuse
4. Poor distribution
5. Poor quality
6. Availability

Information systems
1. Unnecessary
2. Defective
3. Late
4. Unclear
Others (specify):

6. Select the five most frequent waste types that you perceive in your work environment:

1. Work not done
2. Rework
3. Unnecessary work
4. Errors

5. Stoppages
6. Loss of materials
7. Deterioration of materials
8. Unnecessary movement of people
9. Unnecessary handling of materials
10. Excessive vigilance
11. Extra supervision
12. Additional space
13. Activity delays
14. Extra processing
15. Clarifications
16. Abnormal wear of equipment
Others (specify):

Rank order according to their impact on overall performance:
1.
2.
3.
4.
5.

Sources of Waste

Waste	Administration	Unnecessary Requirement	Excessive Control	Lack of Control	Poor Planning	Bureaucracy	Use of Resources	Surplus	Shortage	Misuse	Poor Distribution	Poor Quality	Availability	Information Systems	Unnecesary information	Defective information	Late information	Unclear information	Others (specify)
1.- Work not done		1	2	5				1	1	1	2				1	0	1	2	
2.- Rework			4	4						2	2					1	2	2	
3.-Unnecessary work			1	4						1	2					1	2		
4.- Errors			3	2	1					2		2					1	4	
5.- Delays	1		1	2	1					1			1		2	3	1		
6.- Loss of materials	1		6			1	3			2	2						2	1	
7.- Deterioration of materials	1		3							2	2	4	0		0	1	0	1	
8.- Unnecesarry movement of people			1	5						2	3					1	1	3	
9.- Unnecesary handling of materials			2	2					1	1	4		1		2	1	2		
10.- Excessive vigilance		1		2						1					1	1			
11.- Extra supervision	4					1	1	0	1	2	1				3	1		2	
12.- Additional space		1		2					2	1		1			2				
13.- Delays in operations	1		3	2	1					3	1	2			1	1	4	1	
14.- Extra processing	1				1					1	1				1			2	
15.- Clarifications	1			2			1			1					1	2	0	2	
16.- Abnormal wear and tear of equipment			4						1	4						1		2	
Others (specify):																			

Figure A1. Waste according to source of waste.

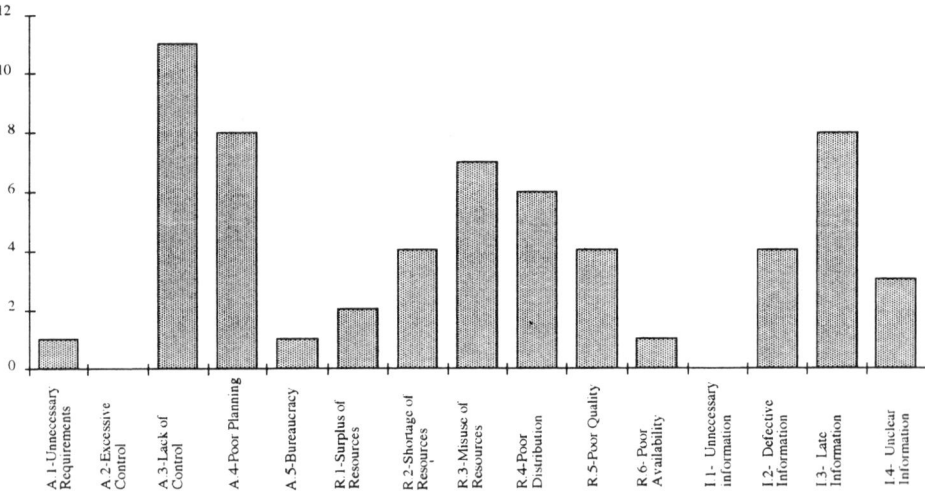

Figure A2. Main sources of waste according to survey results.

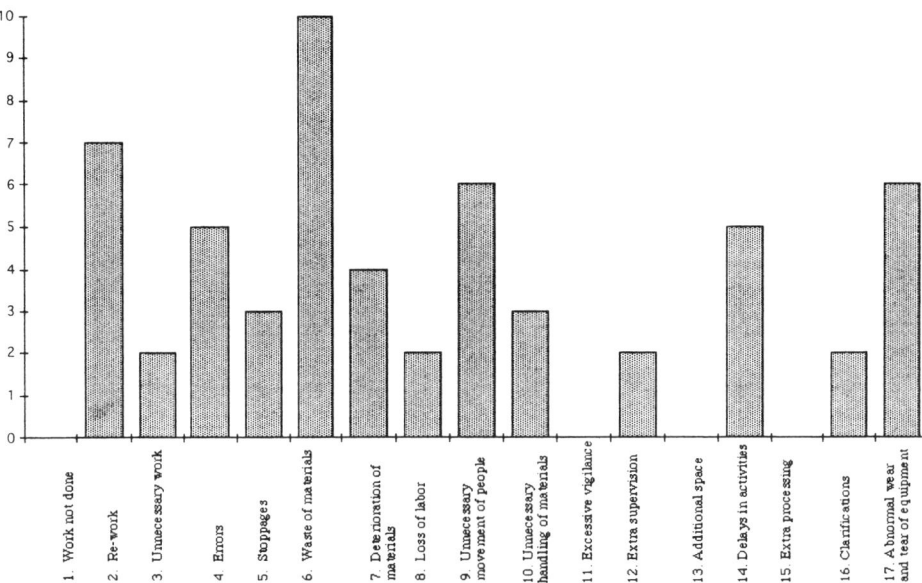

Figure A3. Waste frequency according to survey results.

REFERENCES

Alarcón, L.F. 1993. *Modelling Waste and Performance in Construction*. Presented in the Work-shop 'Lean construction', August 11-13, 1993, Technical Research Centre of Finland, Espoo. 23 p.

Borcherding, J.D., Alarcón, L.F. 1991. *Quantitative Effects on Construction Productivity*, The Construction Lawyer, Vol. 11, No. 1, January.

Koskela, L. 1992. Application of the New Production Philosophy to Construction. Technical Report 1972. Centre for Integrated Facility Engineering. Department of Civil Engineering. Stanford University. 75 p.

Lowener, L.F. 1994. Aplicación de Benchmarking a Procesos Constructivos, Informe de Investigación. Depto. Ingeniería y Gestión de la Construcción, UC. 40 p.

Plossl, G.W. 1991. *Managing in the New World of Manufacturing*, Prentice-Hall, Englewood Cliffs. 189 p.

Shingo, S. 1988. *Non-Stock Production. Productivity Press*, Cambridge, Ma. 454 p.

Construction models: A new integrated approach*

SAIED KARTAM, GLENN BALLARD & C. WILLIAM IBBS
Department of Civil Engineering, University of California, Berkeley, USA

ABSTRACT: Modelling any system is a critical step for understanding it and improving its performance. The thrust of this research is the belief in the development of valid credible models as a logical precursor to automation. This chapter reviews the key models used to represent construction work processes. Through an examination of these models, this chapter makes an important distinction between process and system modelling concepts. This distinction is the basis for the conclusion that no single tool, by itself, is accurately capable of fully modelling the construction system. This is why this research integrates a set of descriptive tools to allow the development of construction system models. This chapter presents one of these tools as a new system modelling concept, called 'workmapping', that overcomes the deficiencies in the current modelling approaches. The power of this new model is illustrated in a detailed comparison among the key system modelling concepts.

1 INTRODUCTION

Paradigms are the fundamental assumptions, principles, and models that serve as the basis against which choices and decisions are made. Due to the lack of a proper model, people in construction plan using individual paradigms derived only from personal experience rather than some set of first principles. There are problems associated with such a heuristic approach. It is difficult to test, verify, and teach. Moreover, the individual paradigms governing this approach are often situation dependent and inappropriate to apply in a new situation (Koskela 1992).

Modelling any system is a critical step for understanding it and improving its performance. Also, the development of valid credible models should be a logical precursor to automation. Those principles illustrate the value of developing valid models. This chapter identifies a major problem in the current modelling concepts. This problem is based on the distinction, made in this chapter, between the objectives of process and system modelling concepts.

This chapter first reviews the key models that have been used to represent construction system. Through this review, this chapter demonstrates the advantages and disadvantages of each. Then, it presents a new system modelling concept, called 'workmapping', that overcomes the deficiencies in the current modelling approaches.

*Presented on the 2nd workshop on lean construction, Santiago, 1994

The power of this new model is illustrated in a detailed comparison among the key system modelling concepts. Henceforth, the article makes an important distinction between process and system modelling concepts. This distinction is the basis for the conclusion that the existing modelling concepts, in isolation, are insufficient to re-engineer the construction system. In response, this research utilizes a new integrated set of descriptive tools that can be used to re-engineer the construction system. The authors have validated the appropriateness of these tools through applying them in various case studies and demonstrating significant benefits in each one. These tools and case studies are published in a subsequent paper (Kartam et al. 1995).

2 A PROCESS VERSUS A SYSTEM DEFINITION

A process can be defined as a set of consecutive steps or activities with an end product or service being delivered. These activities can be identified as value adding or non-value adding activities. Non-value adding activities are those activities that add no value to the process, but add cost. Thus, waiting for materials, waiting for instructions, rework, and inspection are considered non-value adding activities in the construction process. On the other hand, a system can be defined as a process with its surrounding environment, i.e. inputs, outputs, directives, feedback loops, and interactions with other processes. Thus, a process is one component of a system.

3 SYSTEM MODELS

3.1 Existing key models

Many system models exist in the general management literature. The next few pages summarize key models which relate most closely and recently to the construction industry.

The conversion model and Walker's adaptation
The conventional model dominating the manufacturing process view is the conversion model. It is the bedrock for subsequent construction models. According to this model, production is understood as conversions of materials and labor inputs into product output. Furthermore, production processes can be hierarchically divided into sub-processes which, in turn, are also conversion processes. Walker applied this conversion model to the construction process as shown in Figure 1 (Walker 1985). He developed an input-process-output model for the process of providing a project.
To an extent this model and its corollary principles are acceptable. For instance, it is generally consistent with estimating practice whereby total project cost is computed by adding the estimated cost of individual components. However, applying this conversion model to the construction process to analyze and manage productive operations, is misleading and often false (Shingo 1988). One of the key problems in the conversion model is that it does not differentiate between processing activities (value adding activities) and flow activities (non-value adding activities). Indeed, it considers all activities as value adding.

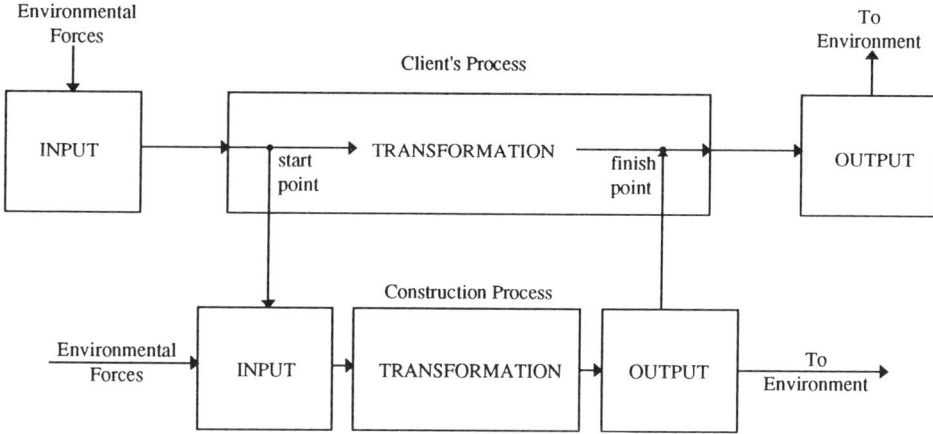

Figure 1. An input-process-output model of the process of providing a project.

Another problem in the conversion model is that it neglects important aspects of resource flows. For instance, one of its key premises is that the total cost of a process can be minimized by minimizing the cost of each sub-process independently. This ignores synergistic and interdependence effects. In addition, it neglects output variation and rework. It assumes that work passes linearly and sequentially through a system. Moreover, the conversion model neglects the impact of poor quality inputs as well as the impact of variability and uncertainty.

The above mentioned deficiencies in the conversion model, which has been the dominating model in manufacturing, explains the industry's misleading focus. Its focus has been on making sub-processes (islands) more efficient, rather than more effective, through changing technology (automation).

The dynamic model and Sanvido's adaptation
Alexander proposed a model to represent a dynamic system that static models have failed to represent (Alexander 1974). An example of such static models are PERT diagrams, bar charts, organization charts, and decision trees. Since construction is a dynamic system, Alexander's model is more accurate to view the process than those static models. The dynamic characteristic is captured through using output feedback for inputs control. This characteristic is an important advantage of this model over the conversion model.

Sanvido realized the adequacy of Alexander's dynamic model to construction and thus developed his overview model of construction as shown in Figure 2. This model identifies eight activities that constitute an on-site construction process. Sanvido developed this further until he reached his consolidated control model that follows the common hierarchical organizational structure (Sanvido 1984).

Sanvido's model still encounters all the deficiencies that Alexander's model has. One deficiency is the model's inability to differentiate between value adding and non-value adding activities. It also doesn't differentiate between inputs that are resources and those that are constraints. Moreover, it doesn't incorporate the future learning capability into the model.

382 *S. Kartam et al.*

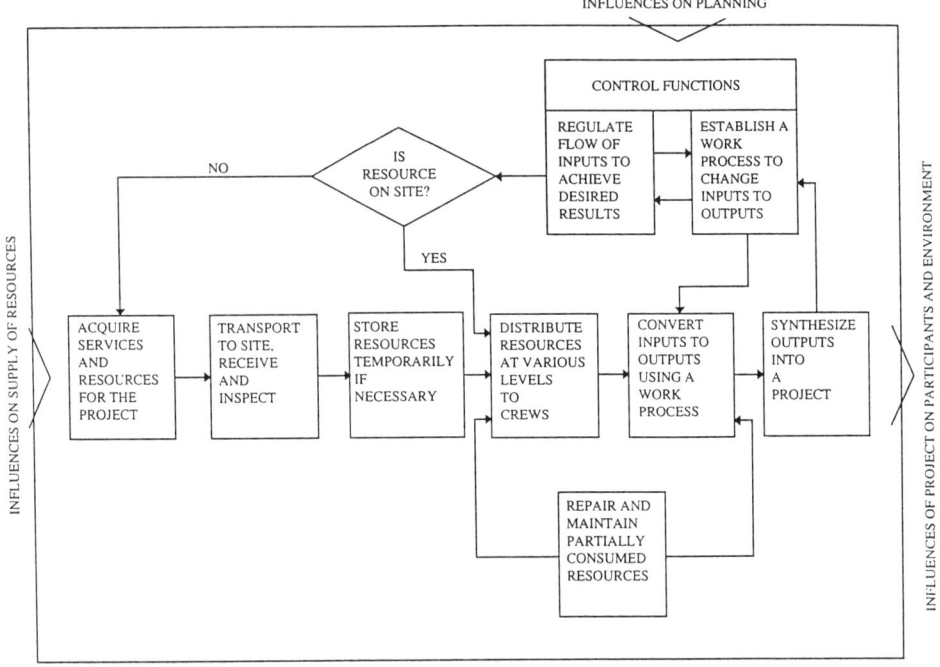

Figure 2. Sanvido's overview model of the construction process (Sanvido 1984).

The structure analysis and design technique (SADT) and Chung's adaptation

The structure analysis and design technique (SADT) is a graphics language that can be used to model a system and its environment. This technique, developed by SofTech Inc., is one of several modelling tools developed in the software engineering discipline (SofTech Inc. 1979). The SADT consists of a hierarchical series of diagrams. Each one consists of a set of boxes, each of which represents a transformation function. The functions have four types of data: input, output, control and mechanism.

Chung developed an integrated building process model (IBPM) from the perspective of the owner of the facility. His model is considered an extension of Sanvido's model. In doing so, he used the structure analysis and design technique (SADT) as his basic framework.

Chung's IBPM is a generic hierarchical representation of the construction process from conception to demolition. The model begins with an overview diagram of the primary process, which is 'provide facility'. Then it breaks this process into five major processes as shown in Figure 3. Those processes are plan for a facility, design a facility, construct a facility, operate a facility, and manage a facility. Each process is further broken down into several sub-processes. Each of these sub-processes is then decomposed to show greater detail. This results in a hierarchical breakdown of the process until the basic process elements are defined (Chung 1989).

This modelling technique has two major advantages over the previous models. One advantage is this model's ability to differentiate between inputs that are resources and those that are constraints. The second advantage is the model's ability to

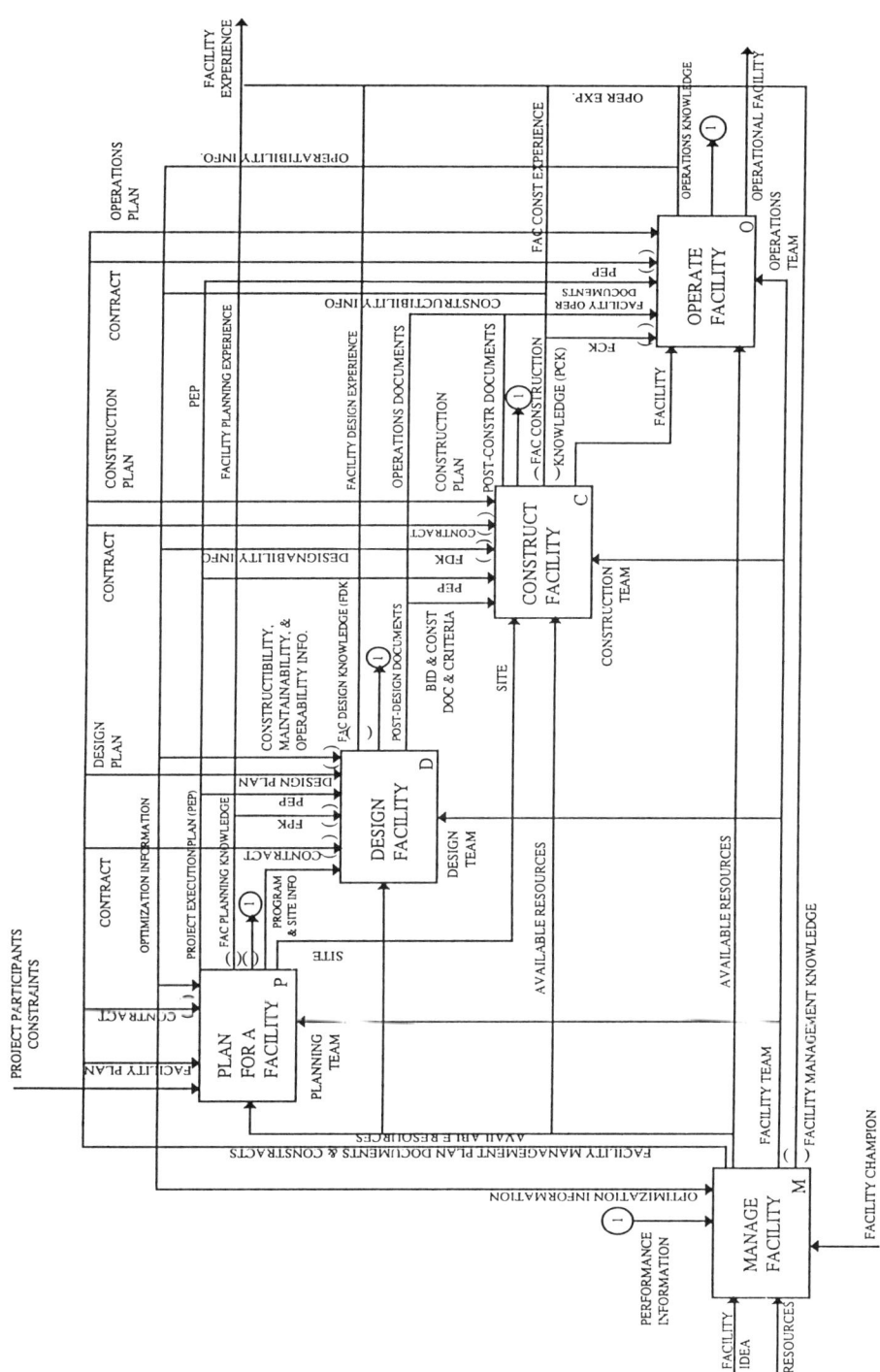

Figure 3. Detailed diagram of the IBPM (Chung 1989).

incorporate the future learning function. On the other hand, evolving from the conversion model, it still encounters its deficiency in concealing the value adding concept. Moreover, Chung's model is implicit in its structure making it difficult to interpret. That's because the model uses lines to represent inputs, outputs, constraints and feedback; i.e. everything, but processes, are represented by lines.

3.2 *The new system modelling concept, 'workmapping', and its application to construction planning*

The new model is rooted into the conventional conversion model, in which outputs are provided by processing inputs. Also, it is rooted into the structure analysis and design technique, in which data is divided not only into inputs and outputs but also into mechanisms and control.

This concept represents the process in a rectangular box, the input, output, and directives in a circle, and the feedback for control and breakthrough purposes and future learning in a diamond shape. Figure 4 illustrates the general format of the workmapping model. It describes the processes with their outputs, the directives that drive them, and the resources they use. Moreover, it portrays the relationship between different types of systems, including production, resource management, planning, contracting, and control or breakthrough.

The best strategy for introducing and using the workmapping concept is to develop it more methodically. That means to first apply the concept to model the production process, then extend its application to management processes (i.e. administratively-driven production processes). Thus, when applying the workmapping concept to the construction process, inputs are resources, outputs are completed work or deliverables, and directives are the plans produced by the planner. Then, applying this workmapping concept to the planning process, inputs tell the planner what work can be done while directives tell him what work or results are supposed to be accomplished. Those directives that the planner receives may range over a spectrum from simply specifying goals to specifying the actions to be executed in achieving those goals. These directives are the output of multiple processes, such as estimating,

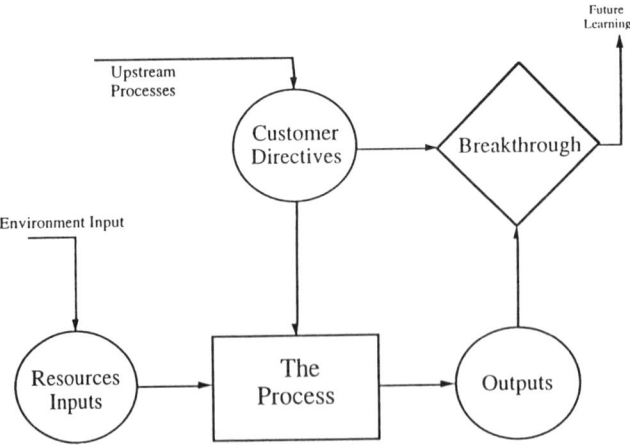

Figure 4. The general format of the workmapping model.

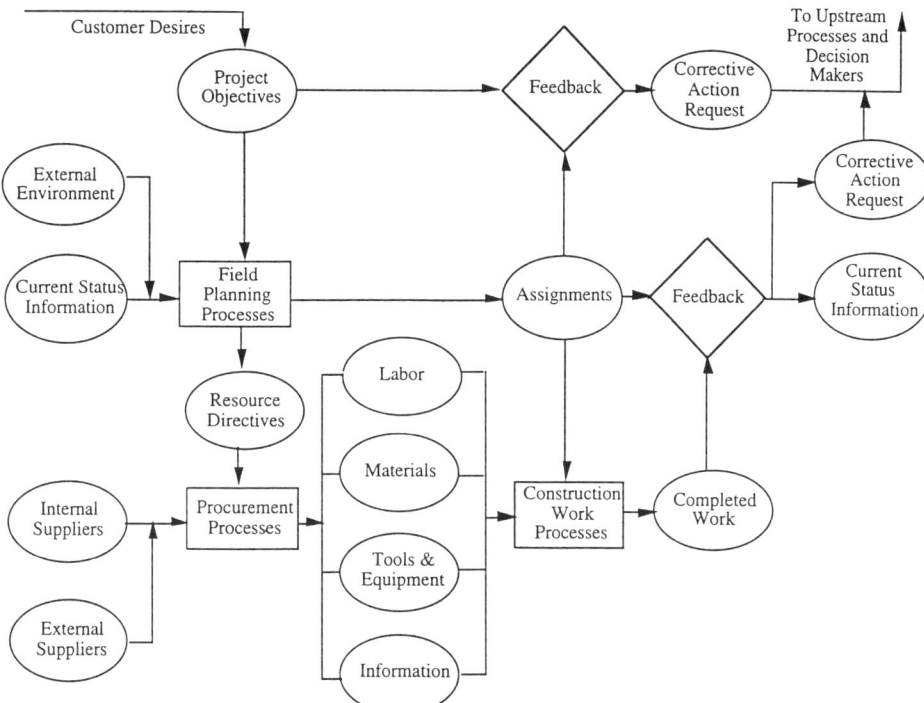

Figure 5. The generic field planning workmap.

from simply specifying goals to specifying the actions to be executed in achieving those goals. These directives are the output of multiple processes, such as estimating, bidding, scheduling, budgeting, quality assurance planning, policy making, etc. The outputs of the planning process are the directives to the construction process.

Construction resources are the inputs into the planning process. Those resources primarily consist of the planners themselves, their tools, materials and information. These resources are the output of other processes, such as hiring, training, construction equipment leasing, materials fabrication, purchasing, and transportation, information collection, report preparation, etc.

In the workmap's control process, data is collected on quality, schedule, cost, safety, and other criteria and then they are compared with the directives. The situation to be analyzed, at this control process, is whether the system work processes and outputs confirm to the system directives. It then, generates both feedback into the planning process regarding current status of production and corrective action requests to the appropriate upstream decision makers. This may result in changes to the directives governing the field planning process as shown in Figure 5. Thus, the control process is better named the control/breakthrough process since it will be used not only to control the process, but also to allow breakthrough in the system performance. This function represents the future learning capability inherited in the system.

3.3 *Comparison among the key system modelling concepts*

Since the conversion model portrays a process in terms of transforming inputs to outputs only, it completely neglects the processes' interactions in terms of their inter-dependencies and feedback information, i.e. the big perspective. The dynamic model starts to resolve this deficiency by portraying the process's feedback. The workmap model and the structure analysis and design technique (SADT) contribute more to portraying the interrelationships and interdependencies among processes through differentiating between inputs which are directives that drive the production process, and inputs which are resources that are converted in whole or part. Moreover, these two models highlight the use of feedback loops not only for control purposes but also for breakthrough and future learning.

Thus, the workmap model and the SADT both posses important modelling characteristics that make them better than the rest. The major advantage of the workmap model over the SADT is the use of explicit graphic symbols for easier understanding and interpretation. In the SADT, resource inputs, directives, outputs, feedback, control, breakthrough and future learning are all represented by lines. On the other hand, the workmap model portrays inputs, outputs and directives in circles, control and breakthrough in a diamond shape, future learning in an arrow and different processes with different four-sided shapes. This characteristic makes the workmap an explicit graphical model that is easy to understand, interpret, update and verify. Moreover, the detailed graphical vocabulary of the workmap gives it the ability to be used as an evaluation tool. In this research, for example, the workmap was applied to the field planning system and served to indicate appropriate system evaluative criteria (Ballard et al. 1994). Thus, it is clear that the workmap is graphically superior to the SADT, and thus, easier to understand and use.

Table 1 summarises the comparison among the various system modelling concepts presented earlier with respect to important modelling graphical characteristics. Two major conclusions can be drawn from this comparison:

1. The workmapping modelling concept overcomes the deficiencies in the current system modelling concepts;

2. All of these models lack the value adding concept. Thus, to fully model any system, both a system model and a process flow model need to be utilized.

4 A PROCESS MODEL

Modelling the flow of resources is essential to indicate the non-value adding activities and thus eliminate waste. Although, this type of waste is widely recognized, very little has been done to eliminate it. It is clear that speeding the flow of resources saves time, reduces cost, and thus satisfies customers. The traditional production philosophy is based on the conversion model. It conceives production activities as sets of operations. These operations are controlled, operation by operation, for least costs and improved periodically in regard of productivity by implementing new technology.

Koskela has realized this lack of unified conceptual framework for construction. He proposed a new conceptual model for the direct production process as a generalization of different models suggested in different fields, such as the just in time (JIT) movement and the quality movement. The new production philosophy conceives production activities as materials and information flow processes. These processes

Table 1. Comparison among system modelling concepts.

Model/ characteristic	The conversion model (Walker's basic model)	The dynamic model (Sanvido's basic model)	The structure analysis and design technique (Chungs' basic model)	The workmap model (the authors' basic model)
Portraying the interrelationship among various processes (the big perspective)	None*	Limited**	Exists***	Exists***
Input differentiation	None*	None*	Broken into resource inputs and constraining inputs***	Broken into resource inputs and directives (constraints)***
Graphic symbols for easier understanding	Lines and squares**	Lines and squares**	Lines and squares**	Lines, squares, circles and diamonds***
Feedback loops	None*	Only for control**	For control and future learning***	For control, future learning and breakthroughs***
Processes differentiation	None*	None*	None*	Uses different 4-sided graphical shapes for different processes***
The ability to be used as an evaluative tool	Doesn't exist*	Doesn't exist*	Doesn't exist*	Serves to derive appropriate system performance criteria***

*Disadvantage; **Okay; ***Advantage.

are tightly controlled for minimal variability and cycle time and improved continuously in regard of waste and value and periodically in regard of efficiency by implementing new technology.

The new production theory seeks cycle time reduction, total waste elimination, zero defects, and flexible output. According to Koskela, the model adequate to the new theory is a flow process model, in which production is conceived as a flow of materials and information through four types of stages: transport (moving), waiting (delay), processing (conversion), and inspection as shown in Figure 6. This model has an advantage over all of the previous models. It differentiates between value adding activities and non-value adding activities. It also concentrates on the process flow rather than the exchange among the processes. As a rule in this model, only processing activities are value adding activities. Reducing the share of the non-value adding activities is the target for continuous improvement (Koskela 1992).

Koskela believes that the new production theory applies to construction, and further proposes that the dominance of the old paradigm (which in his opinion is the

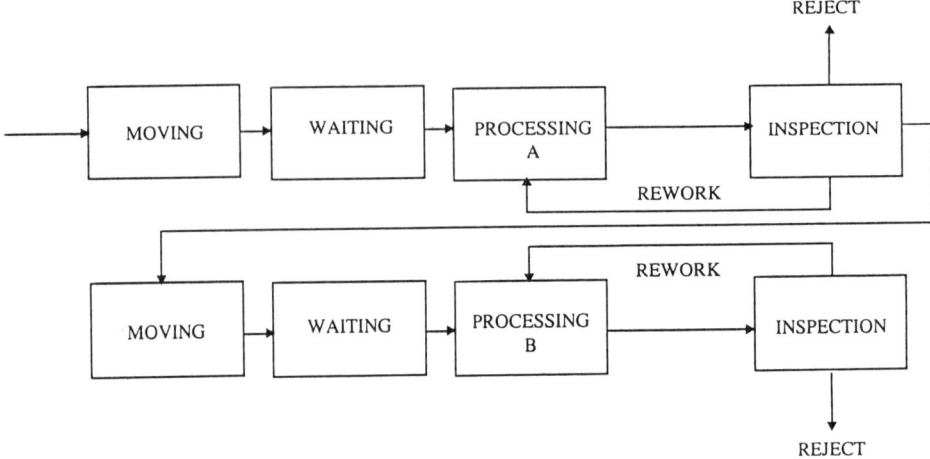

Figure 6. Koskela's production model (Koskela 1992).

conversion model) leads to poor management practices. This creates many fires to fight and accordingly becomes another obstacle to embracing the new philosophy.

Koskela flow process model is a new contribution to the construction industry. It is the only model that realizes the value adding concept. On the other hand, this model is still lacking the full picture. It doesn't realize the system concept. It ignores the processes interactions and interdependencies. It models the production process but doesn't model the management processes that affect the production process. Moreover, it is essential to differentiate between production processes that are administratively-driven and those which are machine-driven. Contrary to manufacturing, construction has more administratively-driven production processes than machine-driven ones. This is due to the fact that construction is dynamic in its work place and accordingly depends extensively on manpower and non-stationary equipment. The construction work processes are dominated by the mobilization of resources to the site. The planning and control of resources are what ultimately determine construction production. In addition to these differences, construction differs from manufacturing in the degree of impact of uncertainty.

5 THE NEED FOR A NEW INTEGRATED MODELLING APPROACH

It can be seen from this article that the major problem in the current modelling concepts is based on the distinction between the objectives of process versus system modelling concepts. Distinguishing between the value adding and the non-value adding activities is the major objective of modelling a process, while highlighting the various processes' interactions with their inputs, outputs, directives and control is the major objective of modelling a system.

Given this distinction between process and system modelling concepts, no single tool, by itself, is fully capable of accurately modelling a system. Therefore, there is an extreme need to develop an integrated set of descriptive tools that can be used to model not only the construction processes, but also the construction system. These models will serve as theoretical foundations that can help understand the construction system and contribute to the paradigm shift in the theory of construction.

One of these tools is the new system modelling concept mentioned above, i.e. workmapping. It is considered not only a modelling tool, but also a theoretical foundation by what it presents of values and concepts. Other tools utilized in this research are used to portray the processes' perspective rather than the system perspective (Kartam et al. 1994). This new integrated modelling approach overcomes the deficiencies in the current modelling concepts

6 CONCLUSION

While Walker, Sanvido, and Chung highlight the need for more accurate graphical system models, and Koskela highlights the need for a process flow model, there should be a bridge between these two to integrate them in a fundamental foundation of construction. This bridge is necessary given the important distinction made in this chapter between process and system modelling concepts. This is why the authors felt an extreme need to develop an integrated suite of descriptive tools that can be used to model not only the construction processes, but also the construction system. This new integrated modelling approach overcomes the deficiencies in the current modelling concepts.

REFERENCES

Alexander, M.J. 1974. *Information Systems Analysis*. Science Research Associates Inc., Kingsport, Tennessee.
Ballard, G., Howell, G. & Kartam, S. 1994. Redesigning Job Site Planning Systems. *Proceedings of the ASCE First Congress on Computing in Civil Engineering, Washington, D.C.*
Chung, E.K. 1989. A Survey of Process Modelling Tools. Technical report No. 7, Computer Integrated Construction Research Program, Department of Architectural Engineering, Pennsylvania State University.
Kartam, S. 1995. Reengineering Construction Planning Systems. Ph.D. Dissertation, Department of Civil Engineering, University of California at Berkeley, Berkeley, CA.
Kartam, S., Ibbs, W. & Ballard, G. 1995. Re-engineering Engineering and Construction Planning Work Processes. *Project Management Journal*, Project Management Institute.
Koskela, L. 1992. Application of the New Production Philosophy to Construction. Technical report No. 72, Centre for Integrated Facility Engineering, Stanford University.
Sanvido, V.E. 1984. A Framework for Designing and Analyzing Management and Control Systems to Improve the Productivity of Construction Projects. Technical report No. 282, Department of Civil Engineering, Stanford University, Stanford, CA.
Shingo, S. 1988. *Non-Stock Production: The Shingo System for Continuous Improvement*. Productivity Press.
SofTech Inc. 1979. An Introduction to SADT: Structure Analysis and Design Technique.
Walker, A. 1985. *Project Management in Construction*. Collins P. and T. Books.

Training field personnel to identify waste and improvement opportunities in construction*

LUIS F. ALARCÓN
Department of Construction Engineering and Management, Catholic University of Chile, Santiago, Chile

ABSTRACT: Generally, construction field personnel tend to overlook many opportunities of improvement in their everyday work. One of the reasons is that they don't realize that many activities they carry out do not add value to the work. Even managers tend to think that if they keep everybody busy they are being 'productive'. This chapter proposes a shortcut to introduce concepts of lean construction in the field using a simple approach which consists of short training and motivation sessions for the field supervisors and foremen, the application of a survey to identify waste and its causes and team work sessions to analyze the survey results and propose improvements. The results of the application of this approach to the field personnel of four industrial projects and the home office of a construction company are presented and compared with a diagnosis of the firm carried out simultaneously. One application to a housing project is used to illustrate some of the actions taken to reduce waste and implement improvements.

1 INTRODUCTION

Performance improvement opportunities can be addressed by adopting waste identification/reduction strategies in parallel to value adding strategies. However, concepts such as waste and value are not well understood by construction personnel and they need to be educated to be able to adopt such strategies. In particular, waste is generally associated with waste of materials in the construction processes while non value-adding activities such as inspection, delays, transportation of materials and others are not recognized as waste.

In lean production waste is defined in broad terms. Toyota defines waste as 'anything different from the minimum amount of equipment, materials, parts, or time absolutely essential for production'. An equivalent definition says 'anything different from the absolute minimum amount of resources of materials, equipment and manpower necessary to add value to the product'.

In general, all those activities that produce cost, direct or indirect, but do not add value or progress to the product can be called waste. Waste is measured in terms of costs, including opportunity costs. Other types of waste are related to the efficiency

---

*Presented on the 3rd workshop on lean construction, Albuquerque, 1995

of the processes, equipment or personnel and are more difficult to measure because the optimal efficiency is not always known.

This chapter summarizes a recent experiment carried out by the author visiting several construction sites, in different locations, carrying out training sessions to teach construction field personnel to identify waste and value in their own processes. These sessions were developed using a very simple methodology to structure the discussion and the identification of performance improvement opportunities.

2 METHODOLOGY DESCRIPTION

The methodology shown in Figure 1 includes a brief training in concepts of lean construction, with emphasis on waste and value concepts, followed by brainstorming sessions to define waste categories and causes of waste. After these preliminary activities, the participants have gained sensitivity toward all types of waste, in broad terms, which exist in their own and surrounding processes and they are ready to answer a survey prepared with the information collected in the previous brainstorming sessions. The purpose of this survey, which format is shown in Appendix A, is to identify the most frequent waste categories, according to the participants' perceptions, and to identify the most critical causes of waste.

The use of this methodology is illustrated later using two cases, one case, involving several projects and the home office of an industrial construction contractor, where the methodology was applied as a diagnosis tool, and another case of a housing contractor where the methodology has been fully applied.

3 INTRODUCING WASTE AND VALUE CONCEPTS

Educating or sensitizing construction personnel to identify waste is an interesting experience. 'waste' is an ugly name, especially if you suggest to somebody that most of their everyday activities are 'waste'. It is usual to find negative reactions when you suggest that activities such as inspection or transportation of materials do not add value to the product and should be minimized or eliminated. These two activities are considered inherent to construction work and therefore people think they cannot be eliminated. It takes some examples of how they can be reduced, or in some cases

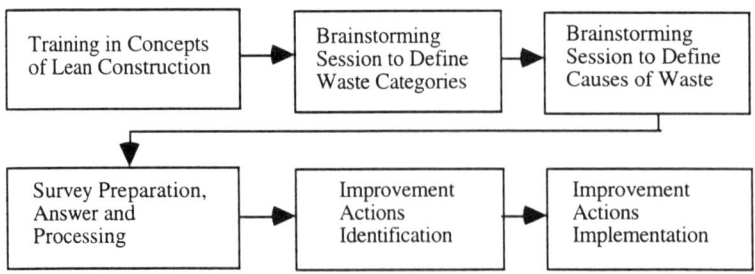

Figure 1. Proposed methodology.

eliminated, before people start to believe this can be done and reluctantly accept these waste categories. However, carrying out this exercise makes the people aware of a broad range of improvements which can be carried out in their work by reducing waste and increasing value.

Value concepts at the field level are useful to satisfy internal customers, however, the concept of value for the entire product, requires attention beyond the construction site. Sales and design personnel, who interact directly with customers, have opportunity of incorporating value in the product from its conception. Tools like Quality Function Deployment have proved to be very promising in this task. Nevertheless, a new categorization of work, which incorporates value concepts, is discussed below as it could be useful for discussion by field personnel.

Work study or labor productivity studies in construction usually break down work in three main categories, for work sampling purposes:

1. *Effective work*: Activities directly adding progress to a unit being constructed such as placing bricks, painting walls, placing concrete, etc.;

2. *Contributory work*: Activities not directly adding but required to finish the unit such as reading plans, receiving instructions, moving outside the work position, etc.;

3. *Non contributory or idle*: All other activities.

For construction people, familiar with these concepts, contributory work and idle categories can be immediately categorized as waste according to lean production concepts, and there is a natural tendency to associate the effective work category with value-adding activities. However, this seems to be incorrect because the traditional effective work category does not incorporate the concept of value and there are many examples in construction where activities considered as effective work do not add value to the product. In order to introduce a more comprehensive evaluation of waste and value present in the production process, Orbeta (1995) has suggested to further breakdown work into five work categories:

1. *Obviously valued effective work*: Activities directly adding a characteristic valued and generally expected by the customer. For example, painting walls or placing bricks;

2. *Competitively valued effective work*: Activities adding a characteristic valued by the customer, which differentiate the product from others adding competitiveness. For instance, a special floor or wall finishing;

3. *Non value-adding effective work*: Activities producing a transformation in the product which is not valued by the customer. For instance, extra processing or activities which are not rework but are required because of poor quality internal processes;

4. *Contributory work*: Activities not directly adding but required to finishing the unit such as reading plans, receiving instructions, moving outside the work position, etc.;

5. *Non contributory or idle*: All other activities.

These categories are not appropriate for work sampling, but they can be used to motivate people to think harder to identify waste and value. This breakdown of work can force people to question whether many activities which are usually accepted as effective work really add value to the product and to find new opportunities for improvement.

4 CASE 1: A COMPANY DIAGNOSIS

This experience was carried out in four construction projects, located in copper and nitrate mines, and in the home office of a construction company which works mainly in industrial construction. The participants in the construction sites were almost all the field supervisors, from engineers to foremen, and some administrative personnel. In the home office the participants were construction professionals and administrative personnel in charge of key support functions for the construction work such as procurement, cost estimating, and safety management. All these people participated in a twenty hour training program in productivity and quality improvement which was carried out in five sessions of four hours each. However, the methodology was applied only in one of the sessions with an estimated duration of two hours, plus another period of thirty minutes to answer the survey.

The answers of the survey were processed for each location with the purpose of using this information for diagnosing waste and its causes and to select improvement projects to be carried out by designated teams. The results of the five locations were also grouped to help identify company wide opportunity improvements and are shown in Appendix B. Figure B1 shows the most frequent waste categories perceived by the participants; the ordinates show the percentage of respondents which mentioned that waste category and the number above the bar indicates the importance assigned by the respondents to each type of waste. The four most frequent waste categories were *Excessive space requirements*, *Excessive surveillance, Damaged material,* and *Excessive supervision.* Among these four, *Damaged material* was ranked higher in importance. It is important to note that the following two waste categories *Uncompleted work* and *Extra processing* were perceived less frequent but were ranked as the most important.

Figure B2 shows the most critical causes of waste, including a judgement of the frequency of its occurrence. The first cause is *Late information*, the second is *Poor planning* and the following three are all related to resources: *Shortage of resources, Resources unavailable,* and *Poor distribution.*

The matrix in Figure B3 shows the participants' perceptions about the causes associated with each waste category, where the numbers indicate the number of respondents who selected the association between the corresponding waste category and its cause. For instance, the matrix suggest that 'Excessive space requirements' is caused mainly by 'Poor planning' (16), 'Poor resource distribution'(12), and 'Excessive amount of resources'(11). On the other hand, the matrix suggest that 'Uncompleted work' is caused mainly by 'Poor planning' (27), 'Late information' (20), and 'Shortage of resources' (16). Managers or project team members can use this information to take immediate actions to remove the most critical causes of waste.

The surveyed contractor had requested a more exhaustive study from the consulting team of the Catholic University of Chile to obtain a diagnosis of the company before starting a company wide improvement program. This study consisted of an extensive number of interviews with key company personnel and a comprehensive survey to almost all the company permanent staff. Part of this study was an assessment of the efficiency of key company functions, where each function was rated according to the following scale: very inefficient (1), inefficient (3), efficient (5), and very efficient (7). Ten out of twenty one functions were evaluated with grades below (4).

Figure B4 compares the grades of the company functions with the results obtained from the Waste Identification Survey. These results were obtained by adding the columns of the cause-effect matrix shown in Figure B3 for the main waste categories, defined as those that were mentioned by at least 20% of the respondents. The results represent most frequent causes perceived for the main waste categories. Figure B4 suggest that the results of both studies are consistent and that most of the causes are related to the poorly graded functions of the company.

5 CASE 2: A HOUSING PROJECT

A recent experience in a housing project in Santiago has shown the effectiveness of the methodology for diagnosing and reducing waste. The team consisted of all the field supervisors for the project, including the field engineer and the foremen. The methodology was applied in two team work sessions. The first session lasted about one hour and ended with the survey response. The second session lasted another hour and was a review of the survey results and a brainstorming of ideas for removing the sources of waste identified in the survey and during the team work session.

Figure C1 shows the results of the Waste Identification Survey (WIS) applied to this project. The results highlighted rework, that was attributed to deficiencies in the quality assurance system, especially in the interaction with subcontractors. Activity delays and interruptions were also important and they were attributed mainly to deficiencies in the material distribution process. The material process was also responsible for waste of materials. The analysis of these results during a problem solving session lead to suggestions to remove the causes of waste: 1) The general foreman was asked to develop a planning system for the distribution of materials with several suggestions to improve the current system; 2) A quality assurance system for individual contractors was proposed and a more strict quality control of materials was also implemented.

A Foremen Delay Survey (FDS) was carried out after the preliminary work session. The results of this survey during the first week are shown in Figure C2. The analysis of these results was carried out one week after the WIS had been discussed and they confirmed the analysis of the first survey. The results were completely consistent and no new actions were necessary from the results of the FDS. This experience showed the effectiveness of the methodology validating the results and showing similar or superior waste identification capabilities than a FDS, with a reduced effort and a more timely response.

Figure C3 shows the evolution of man-hours lost after the first team work session. Even though no specific actions were in progress during this period, some of the reduction could be attributed to the motivational impact of the training session. In fact, the application of this methodology has shown an educational effect in the teams where it has been applied and it seems to be an effective communication tool for identifying improvement opportunities.

6 SUMMARY AND CONCLUSIONS

This chapter presents a methodology which provides a shortcut to introduce concepts of lean construction in the field using a training session, a waste identification survey and a problem solving session. This methodology can be applied periodically to institutionalize a team work effort in the project team and to keep the team focused on detecting improvement opportunities.

The results of the application of this approach to the field personnel of four industrial projects and the home office of a construction company were compared with a diagnosis of the firm carried out simultaneously using a more extensive methodology. The comparison showed that this methodology can be also effective in diagnosing company processes or functions with a limited effort. The application of the methodology to a housing project illustrated the ability of the methodology to identify waste and its causes in a timely fashion, with additional gains due to the educational and motivational effect of the team work sessions. Until now, the use of this methodology has been in experimental stage. However, after evaluating the results of recent applications, this methodology will be adopted as part of the standard tools applied by the Construction Consulting Group of the Catholic University of Chile.

REFERENCES

Alarcón, L.F. 1994. *Herramientas para identificar y Reducir Pérdidas*. 2nd International Workshop on Lean Construction, Santiago, Chile.
Orbeta, F. 1995. Aplicación de herramientas de lean production en la manufactura. Progress Report, Catholic University of Chile, Department of Construction Engineering and Management.
Plossl, G.W. 1991. *Managing in the new world of manufacturing*. Prentice Hall, Englewood Cliffs.
Shingo, S. 1988. *Non-Stock Production*. Productivity Press, Cambridge, Massachusetts.

APPENDIX A

1.- Please, make an assessment of the frequency of the following causes of waste :

	Frequent	Sometimes	Seldom	Never
Management				
1- Unnecessary Requirements	☐	☐	☐	☐
2.- Excessive Mgmt. Control	☐	☐	☐	☐
3.- Poor Mgmt. Control	☐	☐	☐	☐
4.- Poor Planning	☐	☐	☐	☐
5.- Bureaucracy, paperwork	☐	☐	☐	☐
Resources				
1.- Excessive amount	☐	☐	☐	☐
2.- Shortages	☐	☐	☐	☐
3.- Misuse	☐	☐	☐	☐
4.- Poor Distribution	☐	☐	☐	☐
5.- Poor Quality	☐	☐	☐	☐
6.- Availability	☐	☐	☐	☐
Information				
1.- Unnecessary	☐	☐	☐	☐
2.- Defective	☐	☐	☐	☐
3.- Late	☐	☐	☐	☐
4.- Unclear	☐	☐	☐	☐

Other (specify):

2.-Identify the five most frequent waste categories in your work environment:

1.- Uncompleted work ☐
2.- Rework ☐
3.- Ineffective work ☐
4.- Defects ☐
5.- Interruptions ☐
6.- Material wasted ☐
7.- Damaged Material ☐
8.- Unnecessary labor movement ☐
9.- Unnecessary material handling ☐
10.- Excessive surveillance ☐
11.- Excessive supervision ☐
12.- Excessive space ☐
13.- Delays ☐
14.- Extra processing ☐
15.- Clarification needs ☐
16.- Abnormal equipment wearing ☐

-Rank in order the waste categories, according to their
importance: 1)
                    2)
                    3)
                    4)
                    5)

Others (specify):

Figure A1. Waste identification survey.

Causes of Waste →	Management	Unnecessary Requirements	Excessive Control	Poor Control	Poor Planning	Bureaucracy	Resources	Resorce Surplus	Resource Shortages	Misuse of Resources	Poor Distribution	Poor Quality	Availability	Information	Unnecessary	Defective	Late	Unclear		
Waste																				
Excessive space																				
Excessive surveillance																				
Damaged Material																				
Excessive supervision																				
Uncompleted work																				
Extra processing																				
Clarification needs																				
Defects																				
Rework																				
Unnecessary material handling																				
Material wasted																				
Ineffective work																				
Interruptions																				
Delays																				
Unnecessary labor movement																				
Abnormal equipment wearing																				

For each waste category, identify the causes according to your perception

Figure A2. Cause-effect matrix.

APPENDIX B

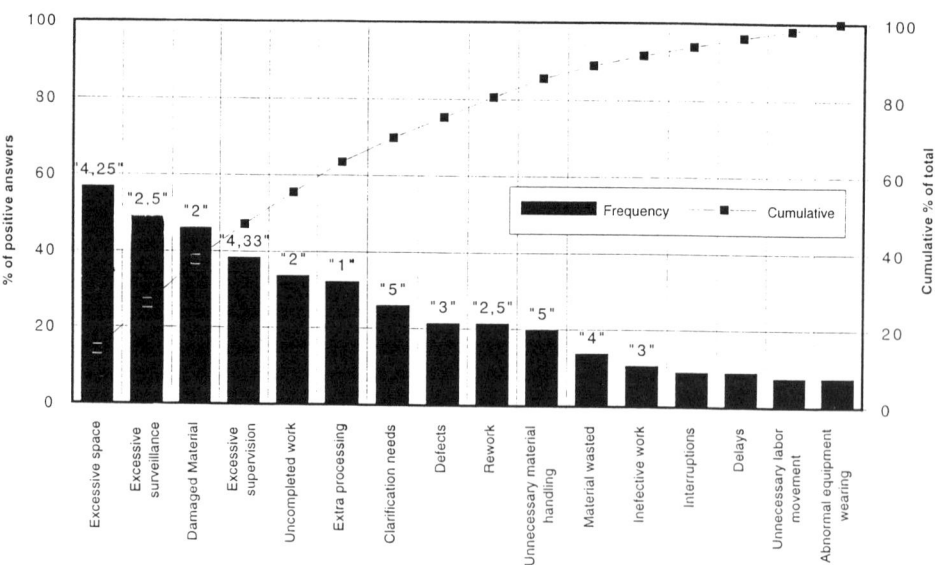

Importance: "1" (most important) - "5" (least important).

Figure B1. Waste frequency.

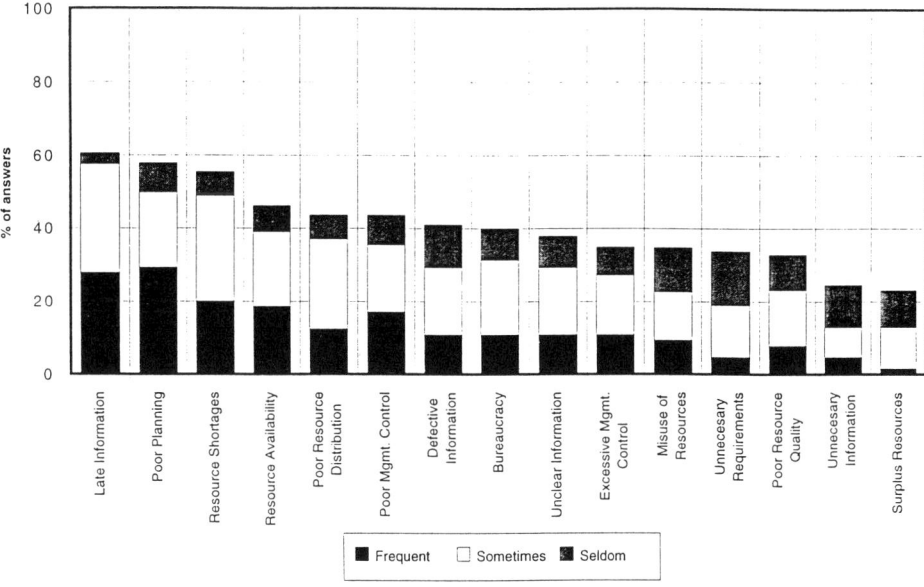

Figure B2. Causes of waste.

Causes of Waste → Waste ↓	Management					Resources						Information			
	Unnecessary Requirements	Excessive Control	Poor Control	Poor Planning	Bureaucracy	Resorce Surplus	Resource Shortages	Misuse of Resources	Poor Distribution	Poor Quality	Availability	Unnecessary	Defective	Late	Unclear
Excessive space	7	1	2	16	2	10	3	4	11		2	9	1		3
Excessive surveillance	8	13	4	5	3	8	1	7	5	1	1	4	1	2	3
Damaged Material		1	22	5		5	1	22	4	5	2	4	4	3	7
Excessive supervision	7	13	2	9	2	8	1	3	4	5		6	4	3	2
Uncompleted work	4	2	15	27	2	1	16	5	10	1	14	4	6	20	13
Extra processing	5	5	8	8	1	3	1	8	5	6	1	5	11	6	9
Clarification needs	1		3	4	2	1		1	2	3	1	4	9	6	7
Defects	3	2	24	22	4	1	5	10	5	10	2	3	24	8	20
Rework	3		16	22			3	12	1	24	2	2	21	18	20
Unnecessary material handling	10	1	5	22		4	3	5	21	3	1	7	6	11	2
Material wasted	7	2	23	12		17	3	22	7	12	3	3	11	4	8
Ineffective work	11		14	23	2	5	1	10	9	2	2	6	8	8	10
Interruptions	2	5	8	17	9		13	8	14	6	8	2	9	17	8
Delays	3	3	6	24	10		15	7	14	7	8	3	12	20	14
Unnecessary labor movement	9		6	26	2	3	9	6	18	4	4	10	9	5	6
Abnormal equipment wearing	3	1	18	7	3		1	16	2	9		2	14	6	11

Figure B3. Cause-effect matrix.

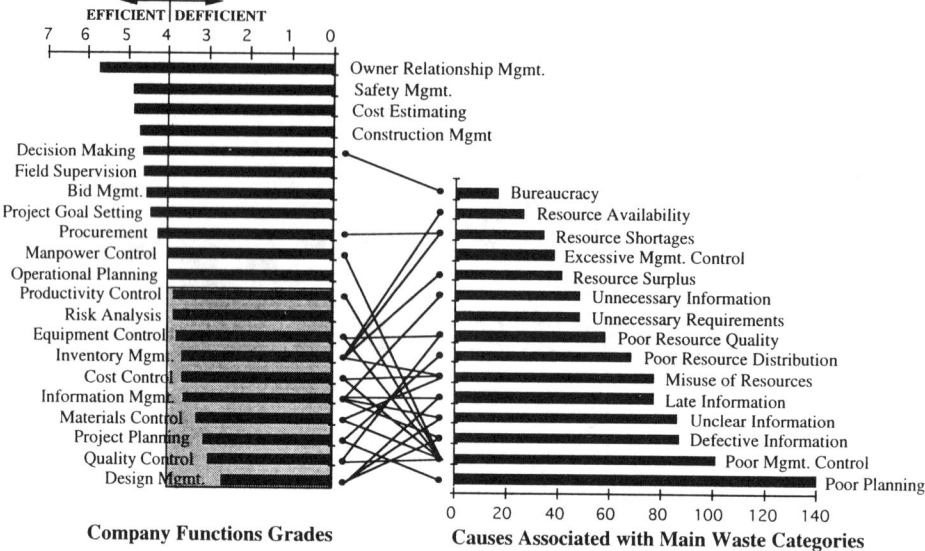

Figure B4. Relationship between organizational study and waste identification survey.

APPENDIX C

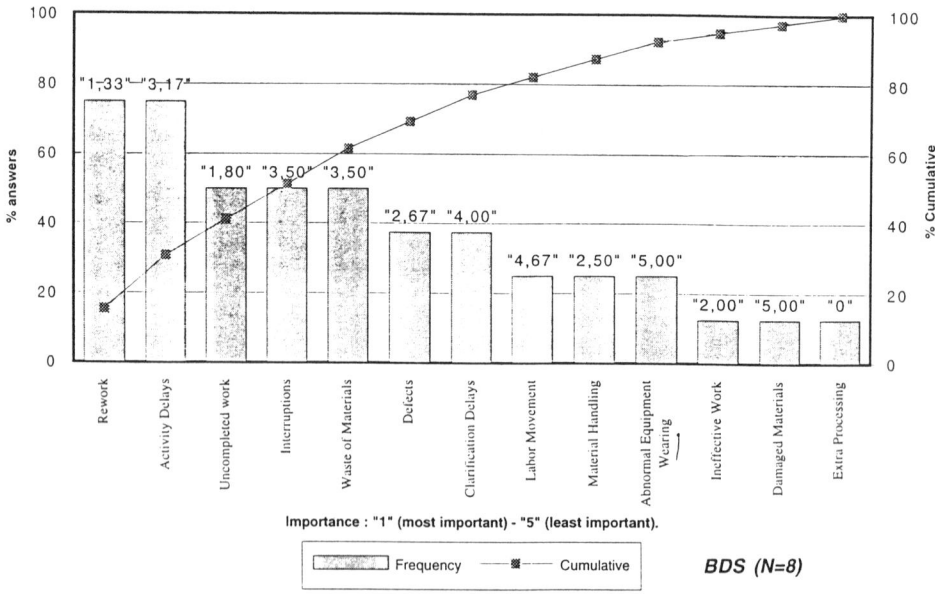

Figure C1. Waste frequency in housing project.

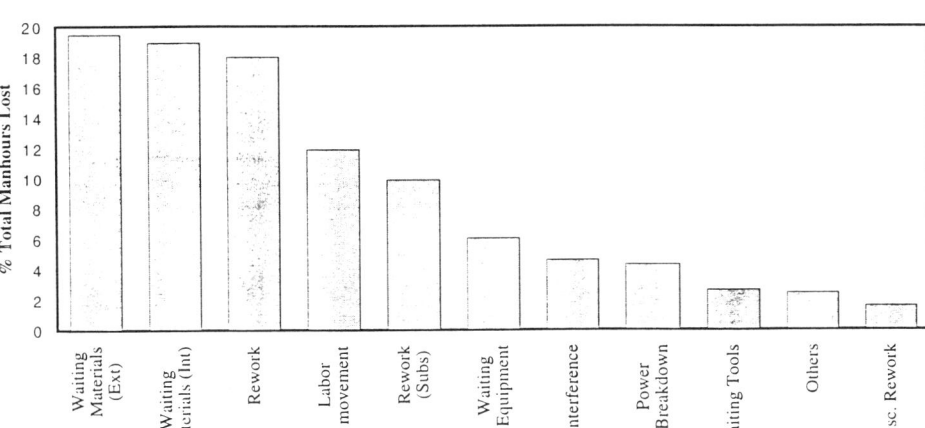

Figure C2. Foreman delay survey results housing project.

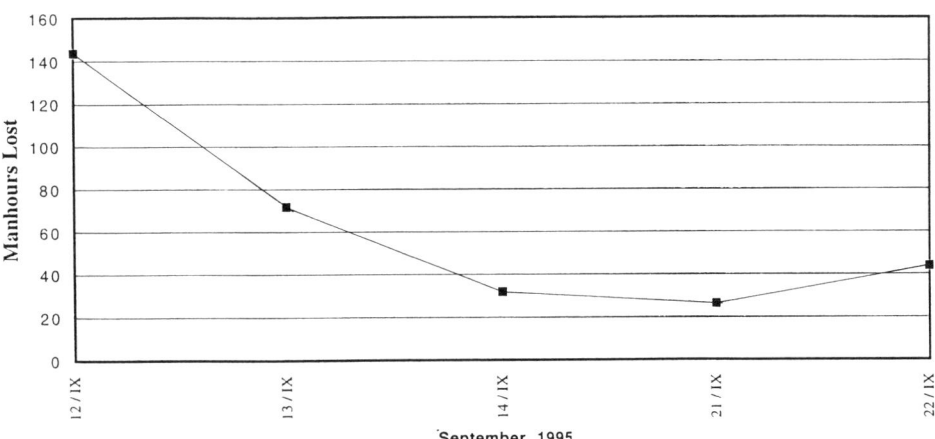

Figure C3. Evolution of man-hours lost.

Involvement of customer requirements in building design*

PEKKA HUOVILA, ANTTI LAKKA, PETRI LAURIKKA & MIKKO VAINIO
VTT Building Technology, Espoo, Finland

ABSTRACT: Customer orientation is becoming an important competitive factor also in construction industry. Satisfying of varying needs of customers (clients, end-users or internal customers) is setting new challenges also for building design. Success in making designs for complicated 'high tech buildings' by temporary project organizations may require systematic working procedures and appropriate tools. Applicability of Quality Function Deployment (QFD) method, that has been successfully practiced in industrial product development projects, has been tested in a technology transfer project for building design.

This chapter presents essential findings of that project (Lakka et al. 1985, 1995). QFD was applied in three construction projects as a team decision making tool to listen to the voice of customer to achieve common understanding, consensus and commitment in design objectives and design solutions. The results were encouraging: QFD, although requiring some 'extra work' compared to the tradition of having little customer involvement, provided a systematic method for the analysis of the customer demands. It also resulted in some design changes that were appreciated. Finally, some recent examples of further QFD development and QFD integration are, as found from literature, briefly raised for discussion: job sharing between the project team and functional departments (QFD & QFD), functional decomposition and planning of design (QFD & DSM), and strategic justification of computer-integration technologies (QFD & IDEF0).

1 INTRODUCTION

Historically, designs have been transmitted between the craftsmen in the form of the product itself. Basic need for separating design arose first in the architecture and civil works, which involved teams of people and took decades to complete. The earliest known design document is a plan of a building on a stone tablet in 2900 BC (Dixon 1988). The development of *design methods* has been quite recent in that time perspective if we think that the subject received substantial academic recognition only in the 1960's (Dasu & Eastman 1994). Still it is widely argued that engineering design lacks sufficient scientific foundation, and that without an adequate base of scientific

*Presented on the 3rd workshop on lean construction, Albuquerque, 1995

principles, engineering design practice is too much guided by specialized empiricism, intuition, and experience (Cross 1993).

Today, buildings are often complicated structures consisting of tens of thousands of parts and sophisticated technology. They are described in one-of-a-kind designs, plans and specifications that are produced by design teams which may be formed for each project, and they are constructed rapidly in a changing environment that doesn't offer factory-like working or management conditions (Huovila & Serén 1995). Design is also a compromise between what is desirable and what is possible. Design process is said to be exploring of what is possible under a specific design context and adjusting performance criteria accordingly, since what is desirable is not always possible (Papamichael & Selkowitch 1991). Important decisions are often made already at early stages of the design process – not always fully realizing their influence in final performance of that building.

It is then maybe no wonder, that buildings don't always meet the needs of their users. Some of the problems identified in the design process are:
 – Finding out what clients really need (*even if it is not expressed, but only expected*);
 – Documenting of requirements (*not only solutions*);
 – Selecting of essential requirements (*trade off for what can be afforded*);
 – Planning of the design process (*reduce late iterations = waste*);
 – Managing conformity of chosen solutions (*tracing to requirements*);
 – Realization of designs (*constructability*).

This chapter is discussing the first half of these problems: how to find out what is needed, how to document it, and how to select what is important and can be afforded. Some tools that can be applied in finding out what clients need are check lists, such as ISO6241 performance standards (ISO 1984) for user requirements or CIB master list (CIB 1993) for performance (behaviour of product or service in use), and design work (technical and economic suitability). Such tools are not widely used in Finland. They are found too heavy to use with no computer integration and either to consist of too general (self-evident) data, or too much data making it hard to find what is sought. One documented design method is problem seeking (Peña 1987), that is aimed at architectural programming. The method does not solve the problem of selecting the essential requirements, and may have some disadvantages in the documentation phase being a manual method.

In industrial engineering, manufacturing companies have successfully applied concurrent engineering tools (Huovila et al. 1994), e.g. Quality Function Deployment (Akao 1990) to determine customers' needs for the features of the product into design at its early stages of development, to integrate concurrent design of products and their related processes, and to consider all elements of the product life cycle. Customer-oriented 'champion products' may also be priced higher than their competitors, and still become as market leaders. In spite of its 'success stories' in other industries during the past decade, QFD has been little applied in construction. Examples from Japan (Oswald & Burati 1993), United States (Andersson & Jacobsson 1993), Sweden (Serpell 1994) and Chile (Tsuda 1995) show, however, its potential also in building design.

QFD provides an empty matrix (House of Quality) to be filled (Fig. 1): customer requirements and their importance in the rows along the left hand side, properties of

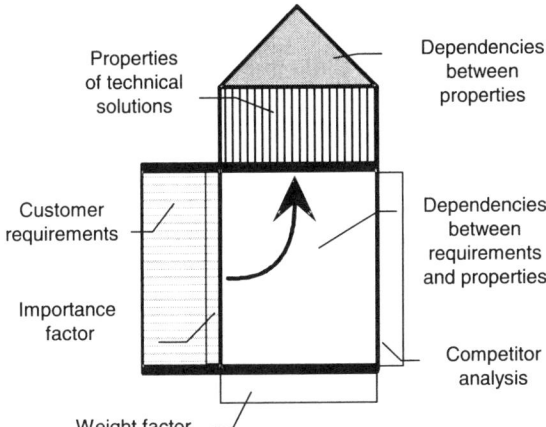

Figure 1. House of quality.

the solutions in the columns along the top portion with their dependencies in the roof of the house. The centre describes the matrix-relationship of requirements and corresponding solutions. The importance measures (weight factors) are at the bottom, and the right hand side of the box shows the evaluation of competing alternatives.

2 APPLICATION OF QFD TO CONSTRUCTION

In our research project, QFD was applied to construction (Lakka et al. 1995) together with a structural design firm and two contractors in three construction projects. Our objective was to test its applicability to construction. The following purposes were identified most potential for its implementation:

1. *Project programming.* Setting objectives and quality requirements for a project;

2. *Innovative design.* Meeting with an unusual demand by customer, where routine procedures cannot be applied;

3. *Strategic planning.* Planning of company's products to bring a strategic competitive advantage;

4. *Development of company's operations.* Analysis of requirements for company's processes and development of these processes.

The following phases of a construction process were identified potential for QFD implementation in construction (Fig. 2):

1. *Programming.* Customers' requirements for the building and design objectives;

2. *Design.* Design objectives and construction drawings;

3. *Production.* Construction drawings and production plans;

4. *Construction.* Production plans and construction phases.

Different kinds of ongoing construction projects were finally selected by participating companies for different purposes of QFD implementation as follows:

1. Programming of an apartment block by the main contractor;

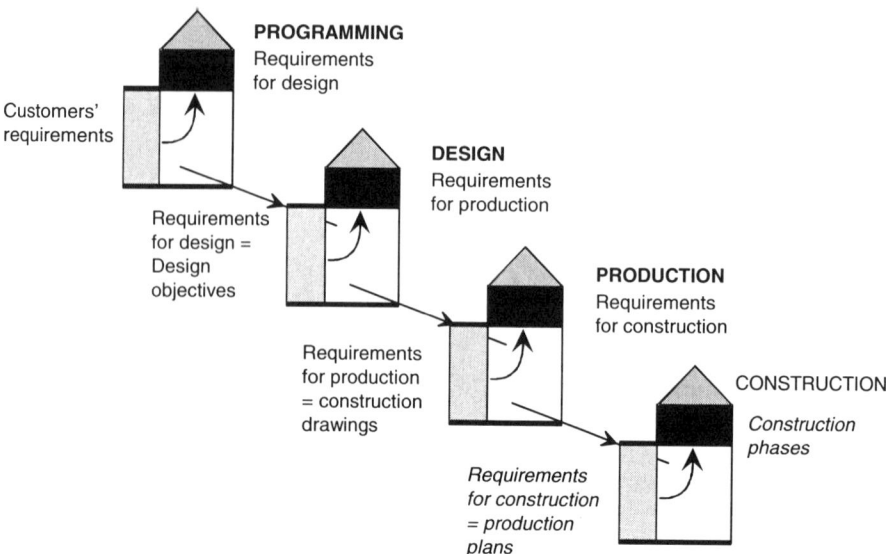

PROGRAMMING
Requirements
for design

Customers'
requirements

Requirements
for design =
Design
objectives

DESIGN
Requirements
for production

Requirements
for production
= construction
drawings

PRODUCTION
Requirements
for construction

CONSTRUCTION

*Construction
phases*

*Requirements
for construction
= production
plans*

Figure 2. Applicability of Quality Function Deployment in construction process.

2. Lay out design of a restaurant in a turnkey project of an office building by the main contractor;

3. Structural design of an industrial building by the structural designer.

2.1 *QFD software*

To facilitate the House of Quality matrix operations, a simple spreadsheet application was developed using the macro language of Microsoft Excel 4.0. The application can be used to process the QFD matrix and to compute the weight factors of various properties. In addition to applying it to the basic matrix, it can be used to set a new target goal, to compare one property against another and to prepare a new QFD matrix for the next phase. The application built on top of the spreadsheet program proved to be an excellent tool for controlling the QFD matrix.

3 QFD FOR PROGRAMMING OF AN APARTMENT BLOCK (SKANSKA OY)

The first case was to test suitability of QFD for programming of an apartment block (Fig. 3) by the main contractor. The emphasis of the work was on an inquiry which attempted to identify the key properties of the building from the point of view of customers and their demands or expectations for the building to be constructed on a new residential area at seaside, south of Helsinki. In fact, the building was already designed and its marketing was under way when QFD came into the picture. Advance marketing, based on preliminary designs, had raised interest of 160 persons,

Figure 3. QFD case 1: Programming of an apartment block.

who were selected as a target group for this inquiry. 53 of them responded to this in-quiry and their answers were analyzed using QFD against existing design solutions.

In the inquiry, requirements of customers were analyzed both for the whole building, and for the apartment they were interested in. The survey targeted at gen-eral features, such as architecture of facades, courtyard, safety, surfacing materials, natural lighting etc. Moreover, the demands on various individual spaces (living room, kitchen, bedroom, bathroom, sauna and balcony), common spaces (staircase and storage facilities), and the courtyard, related primarily to technical solutions, were charted in more detail. Since the plans were ready, it was possible to compare the needs and expectations expressed by customers in the inquiry to implemented design solutions and used sales arguments.

3.1 *Essential results*

A QFD matrix was prepared on the basis of the inquiry. Use of a QFD in that case was found to contribute little added value to the inquiry since the basic idea of QFD on the separation of demands and their technical solutions had already been incorpo-rated into the structure of the questionnaire. Thus, the summary of the inquiry results itself contained all essential information. The inquiry (Table 1) showed that most of the customer demands had been taken into account in the building design. Some de-sign solutions did not, however, conform to the needs and demands expressed by customers.

The results of the study offered to the company, however, a good starting point for assessing its marketing strategy vis-à-vis the project. The inquiry revealed the prop-

Table 1. Examples of similarities and differences between inquiry results and implemented design solutions.

	Indicated customer demand	Design solution
Average apartment size	3,2 rooms + kitchen	2,9 rooms + kitchen
Security	General security quite important: weight 3,6-3,8 (max. 5)	Fire alarms, door intercom, no burglar alarm system
Parking space	Heated parking facility or garage: 52%	Some kind of parking space: 63 %
Individual room planning	Important	Possibility of connecting rooms horizontally and vertically
Fireplace in the balcony	No, or maybe not: 53 %	Yes
Sound proofing	Very important: weight 4,4 (max. 5)	A floating floor (mineral wool insulation), and other solutions
Windows	Larger than normal	French balconies, glazed wall between living room and balcony
Storage space	Walk-in closets: 59 %	Separate closets

erties and accessories that dwelling buyers do not want if they have to pay something extra for it.

3.2 *Experiences from QFD*

+ Systematic approach to mapping and understanding of customers' needs.
+ Inquiry provided direct feedback from customers to company.
+ Questionnaire was found efficient for prioritizing customers' needs.
+ Spreadsheet application facilitated the processing of QFD matrix.
– Processing of QFD matrix was not found very useful in analysis of inquiry results.
– Processing of matrix was laborious.
– Difficult to differentiate between dwelling buyers' needs/demands and building's properties (technical solutions) in a residential building.

4 QFD IN LAY OUT DESIGN OF A RESTAURANT IN AN OFFICE BUILDING (PUOLIMATKA-YHTYMÄ OY)

The second case was to test applicability of QFD in lay out design of a restaurant of an office building (Fig. 4) in Espoo by the main contractor. The area located near the Helsinki University of Technology and the Technical Research Centre, consist of several office buildings aiming at gathering of international, growing, high tech companies together. The objective was to discover key properties of the space in question from customer's point of view to be used as design guidelines.

QFD was applied first to the kitchen and then to the dining room of the restaurant. QFD working groups included members of the contractor with the restaurateur's staff, users, external clients, the developer as well as the researcher as an observer. The groups started off by defining their customers. They were found to be diners, cook, cleaners, waiters, dishwashers, hygienic responsible, suppliers, maintenance

Figure 4. QFD case 2: Lay out design of a restaurant (in Building 3).

personnel, work safety delegate, restaurant manager and central administration staff for the kitchen. Customers of the dining room were lunchers, visitors of companies, cafeteria customers, conference attendees, hosts, cleaners, waiters, the restaurant manager, the cafeteria keeper and the cashier.

The customer demands discussed within the groups were categorized and entered in the QFD matrix (Fig. 5). The weights were computed using the developed spreadsheet application. Most important properties were selected for further processing. Selected properties were divided further into more detailed parts. It was kept in mind that the group was not intended to devise ready-made solutions but to provide the designers with important properties for basic data.

4.1 *Essential results*

Design brief for the kitchen and dining room was complemented through QFD meetings. Some design solutions of preliminary designs proved to deviate from needs and demands expressed by customers. The designs were then changed and complemented after QFD analysis (Fig. 6). The following list sums up some changes and additions resulting from QFD analysis:

– Change in kitchen layout: visibility and traffic routes were improved;
– Soundproofing of dishwashing station was improved: the cost was small, but important to users;
– Kitchen was provided with a space for receiving goods;
– Cold storage facilities were rearranged;
– Storage spaces were redesigned;
– More reservations for equipment were made; at this stage it was easy to reserve space for equipment that might be procured later;

Requirements \ Properties	modifiability of lighting	colours, textiles, plants	acoustics; no echo	different access for personnel	sound reproduction possibility	no fixed installations	collection of dishes and garbage	no tresholds	extra utility points	dirt-resistant joints of tiles	sufficient number of trash bins	stackability of chairs	clearance in front of windows	quidance "start here"	peaceful spots	continuous counter line	buffet service option	interior decoration possibilities	electric points in floor	tables around columns	Importance/Weight factor (P1)	Comparison: Current situation (Innopoli)	Comparison: Competitor (Spektri Pilotti)	Comparison: Competitor (PM Head quartersi)	Comparison: Desired situation (P2)
COMFORTABLE	9	9	9	1	3		3			1	1		1	1	9					3	4	3		3	4
MODIFIABLE	9	3			9	9		1	9			3					9	9	9		5	1	2	4	3
FUNCTIONAL	1			9	1	9	9	3	9		3	3	1	9		9	3	1	3	3	4	2	2	4	5
FLEXIBLE				3		3	9		9		1	1		3		3	9	3	9	1	3	2		3	5
ATTRACTIVE	3	9	3		3		3			1				3	3					1	4	3	4	2	5
AMPLE CAPASITY				1			9		9					3		3				3	5	4	2	4	3
CLEAN, CLEANABLE	3						9	9		9	9	9	9								4	4	2	4	4
UNCOMPLICATED		1					9							9		9					5	4	2	4	4
INDIVIDUALISTIC	3	9			3										3		3	1		3	2	4	1	2	3
ADAPTABLE	9	3			3	9		3	9		1	3			3		9	9	9	3	3	4	3	3	3
Weight factor (P1)	142	119	48	54	88	117	213	62	180	44	58	75	44	121	63	105	117	87	111	61	1909				
Weight factor %	7 %	6 %	3 %	3 %	5 %	6 %	11 %	3 %	9 %	2 %	3 %	4 %	2 %	6 %	3 %	6 %	6 %	5 %	6 %	3 %	100 %				
Weight factor (P2)	131	130	51	67	77	114	216	63	171	45	63	74	45	124	69	105	123	77	114	64	1923				
Weight factor %	7 %	7 %	3 %	3 %	4 %	6 %	11 %	3 %	9 %	2 %	3 %	4 %	2 %	6 %	4 %	5 %	6 %	4 %	6 %	3 %	100 %				
Selected	X						X		X					X			X		X						

Figure 5. QFD matrix of the dining room.

– The service counter line was altered and separated by a back wall from the kitchen;

– The coffee dispensing point was downsized and oriented differently;

– Preparations were made for possible future items of internal decoration in the suspended ceiling and walls of the dining room (signs, decorations, etc.);

– More electrical outlets were installed in the dining room floor.

The contractor measured customer satisfaction during the design phase by an inquiry directed at the tenants of that building. According to customer feedback, the restaurateur was clearly more satisfied than customers on average.

4.2 *Experiences from QFD*

+ Customer is happier since his voice is systematically listened to with QFD;

+ QFD helps customers to focus on their needs;

+ QFD helps clients (no construction professionals) to develop their own future space needs;

+ QFD meetings deepen developer's understanding of clients' demands;

+ QFD is a good means of management by which the person in charge of planning determines customer needs and guides the design process;

+ QFD produces a document on which the decisions are based;

– Processing of a QFD matrix is laborious;

– When properties are translated into demands and divided into more detailed

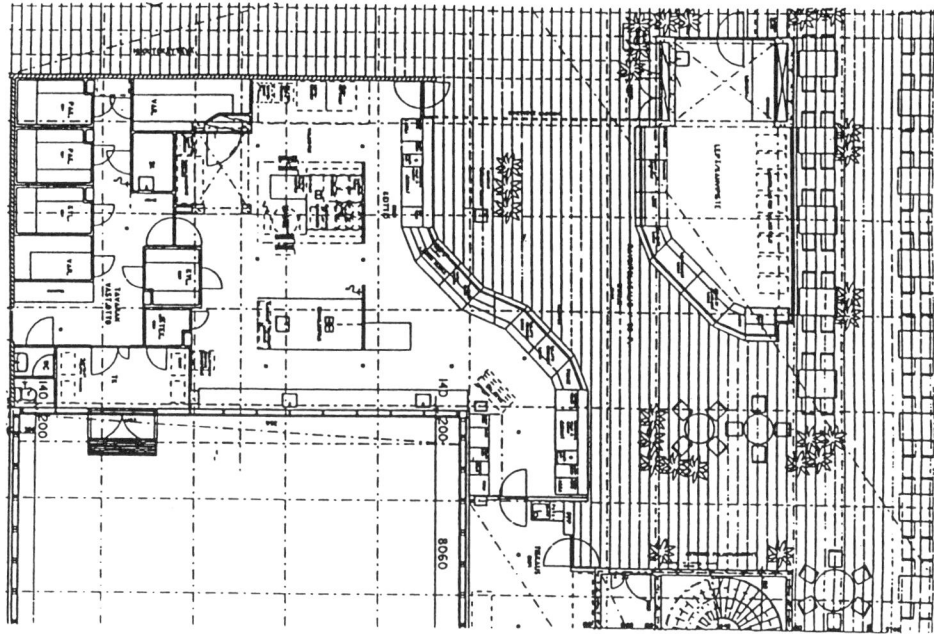

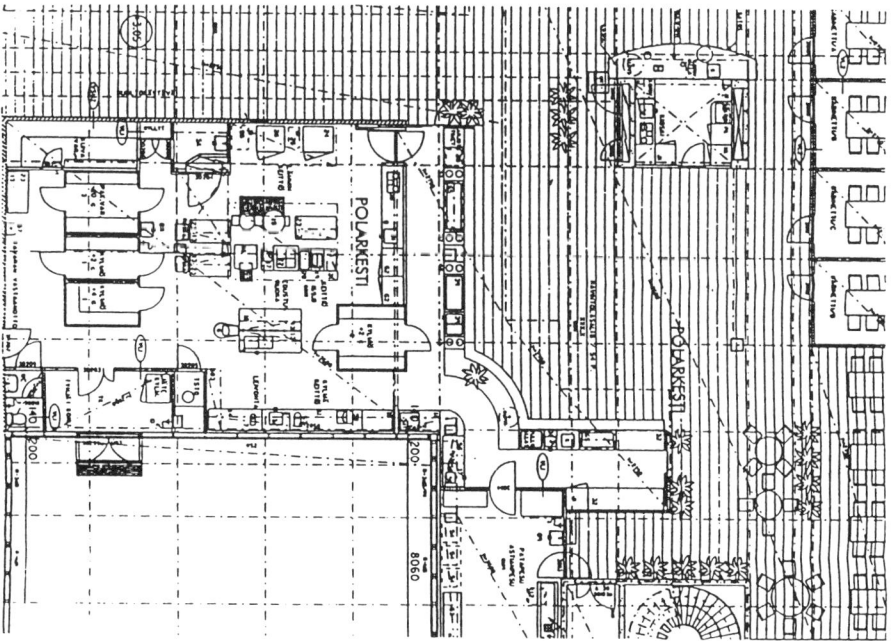

Figure 6. Part of the restaurant lay out before (top) and after (bottom) the use of QFD.

Requirements	foundations	ground floor	vertical frame	intermediate floor	external walls	windows	top floor	ceiling	surface of floors	internal walls	doors	Importance/Weight factor (P1)
LOADBEARING CAPASITY OF FLOOR	3	9	9	9								5
DURABILITY OF FLOOR		9		3						9		5
MINIMUM NUMBER OF COLUMNS	9		9	9	1		9			1		4
POSSIBILITY OF HAVING HOLES IN BEAMS				3			3					3
NATURAL LIGHTING						9	3	9				2
SMOKE ABATMENT				1	1	9	3	9		9		3
ATTACHMENT POSSIBILITIES			3	3	3		3	9		3		3
ELEGANCE			1	3	3	3	3	9	9	3	3	3
NOISE ABATMENT				1	1		1	9		3	3	2
VIBRATION RESISTANCE			3	9								3
FIRE SAFETY			9	9	3		9	9	1	9	3	3
GOOD ELEMENT DESIGN	3		3	3	9	1	3				1	4
RESIDUAL VALUE	3	9	3	9	9	3	9	1	3	3	3	2
LOW ENERGY CONSUMPTION		1			9	9	9				3	3
CLEANABILITY		1	1	1	1	3	1	3	9	1	3	3
OFFICIAL REGULATIONS			9	9	9	9	9	3	3	3	9	3
Weight factor (P1)	69	114	177	242	147	127	194	164	117	100	79	1530
Weight factor %	5 %	7 %	12 %	16 %	10 %	8 %	13 %	11 %	8 %	7 %	5 %	100 %
Selection			x	x	x		x					

Figure 7. House of quality: Customer requirements and structural system.

parts, the expertise of the group may become exhausted (e.g. with respect to practical work);

– It is difficult to differentiate between customers' demands and properties (solutions);

– The sensitivity of the method and minor variations in weight percentages cause uncertainty;

– Selling of a new method to customers requires special abilities from one's own organization.

5 QFD IN STRUCTURAL DESIGN OF AN INDUSTRIAL BUILDING (A-INSINÖÖRIT OY)

The third case intended to test QFD in structural design of expansion of an industrial building in Tampere by the structural designer. The aim was to identify important properties of structures from customers' point of view focusing especially on structural design. Customer needs and product properties were analyzed using QFD matrix and results were compared to existing plans.

Customer demands and their importance were determined by interviewing the technical director of the manufacturing company using the industrial building, who was responsible for the building project. At the first phase of QFD, subsystems of the structural system were selected as properties, and four most important of them were chosen for further processing on the basis of the results. At the second phase, the properties of the subsystems to be measured were defined; demands were same as in phase 1. Correlation between demands and properties was determined by two designers working together, and who combined their views in the final matrix. QFD processing yielded the most important factors to be taken into account in structural design.

5.1 *Essential results*

Most important factors to be considered in structural design found out through QFD processing were somewhat surprising to designers and deviated in some solutions from the emphasis of design. Structural designers decided to pay special attention to flawless element and to detailed design. It was neither possible, nor was there any desire, to affect to selection of the roof slab beam-type after QFD. The chosen roof slab meets fire safety rules that require sprinkler systems for all spaces as well as compliance with fire endurance class. Structural designers decided to pay attention to the details of smoke abatement in further design.

Special attention had been paid to the intermediate floor solution during design. For instance, high loads from the machinery, their positions and penetrations as well as requirements concerning the floor's levelness, fracture-proofness and mechanical and chemical endurance were considered. The structural designers decided to pay special attention in their later work to the structural solutions vis-à-vis the intermediate floor.

5.2 *Experiences from QFD*

+ Systematic way of compiling the customer's needs and wishes into basic design data and, especially, for prioritizing them;

+ A separate customer inquiry form is needed for each building type, but once it has been made, it helps determine the needs of many customers and their order of importance;

– Definition of the properties of structures is laborious;

– Filling in of the QFD matrix is laborious;

– Personal differences in filling the matrix lead to different emphases.

6 CONCLUSIONS

QFD is a tool of, especially, a project manager when applied to construction projects. Its greatest advantage is that it provides the project manager with a systematic method of compiling and analyzing the customers needs. The problem is, generally, that customer himself does not know his needs. Application of QFD forces the customer to focus on his future situation and the related demands.

The most critical phase of application is definition of customer demands. Various methods are available according to the needs. They include interviews, teamwork, inquiries and various combinations of them. Filling of the matrix is just a mechanical procedure required to organize the data.

QFD is not a substitute to expertise. It provides added value to an expert to distinguish most critical customer demands and to discover the corresponding properties. On the other hand, the matrix is a document that enables tracing of information behind decisions afterwards.

This study did not aim at developing a complete system to cover the entire building. The experience from these three application projects supports understanding that it is reasonable limiting matrix size to less than 30-40 rows or columns when applying the method. A modern advanced spreadsheet application is suitable for controlling of the QFD matrix as it is: user-friendly and flexible.

6.1 *Further development of Quality Function Deployment*

QFD, as such, seems to be applicable for different tasks in building design. In industrial engineering, where it has already been practiced for years, some topics have been recently raised for further development and integration. Three recent examples are shortly presented as picked from literature. In its home ground, in Japan, a two-storied version of QFD (Fig. 8) has been tested for job sharing between the project team and functional departments (Tsuda 1995).

Another potential technology transfer tool for design, more precisely for planning and management of design process, is Design Structure Matrix (DSM) (Eppinger 1991). Creating of a Concurrent Engineering toolbox for Customer-Oriented Design (QFD, DSM, etc.) is underway in our future research. Use of QFD and DSM (Fig. 9) for functional decomposition and planning of design (Christiansen 1994) is presented for load distribution in virtual teams.

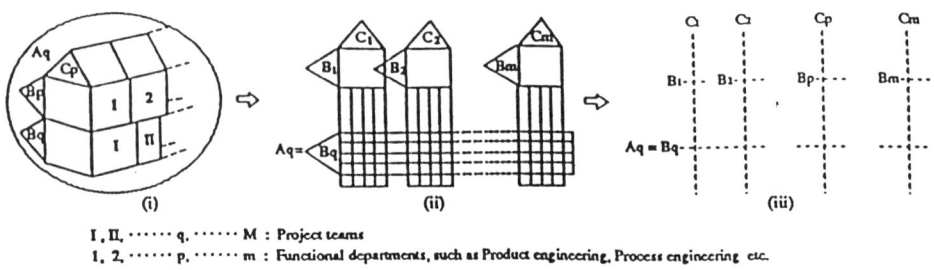

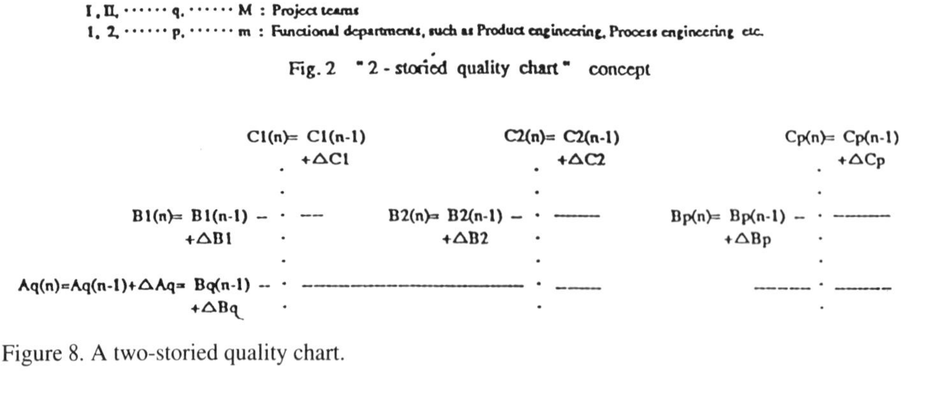

I, II, ······ q, ······ M : Project teams
1, 2, ······ p, ······ m : Functional departments, such as Product engineering, Process engineering etc.

Fig. 2 "2 - storied quality chart" concept

$C1(n)= C1(n-1) +\triangle C1$

$C2(n)= C2(n-1) +\triangle C2$

$Cp(n)= Cp(n-1) +\triangle Cp$

$B1(n)= B1(n-1) -- \cdot --- +\triangle B1$

$B2(n)= B2(n-1) -- \cdot ------ +\triangle B2$

$Bp(n)= Bp(n-1) -- \cdot -------- +\triangle Bp$

$Aq(n)=Aq(n-1)+\triangle Aq= Bq(n-1) -- \cdot -------------- +\triangle Bq$

Figure 8. A two-storied quality chart.

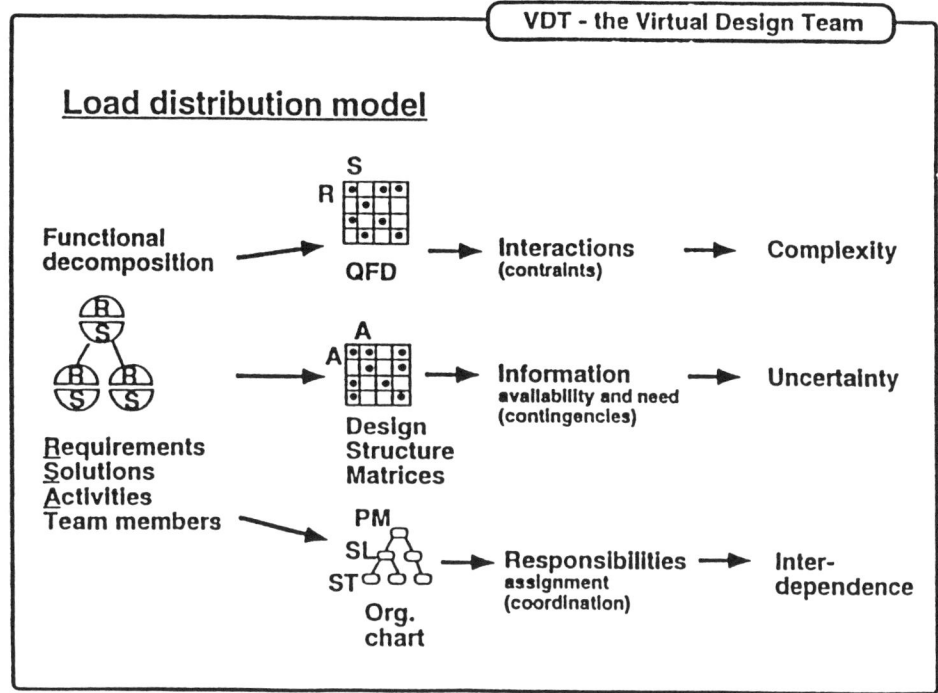

Figure 9. The coordination load model.

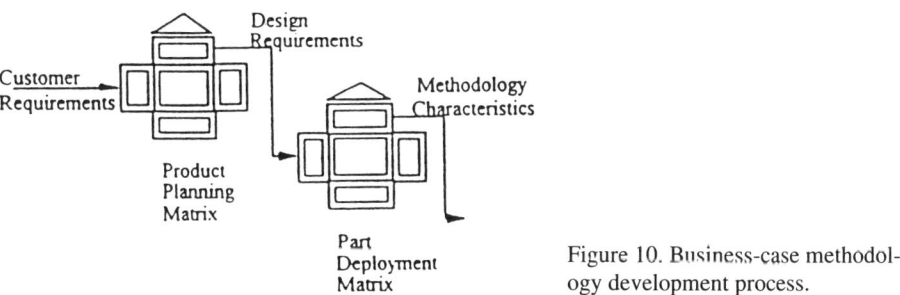

Figure 10. Business-case methodology development process.

Integration of QFD and IDEF0 (NIST 1993) has been developed (Fig. 10) as organizational decision support methodology. The aim in that case is strategic justification of computer-integrated technologies (Sarkis & Liles 1995).

REFERENCES

Akao 1990. Quality Function Deployment (QFD), Integrating Customer Requirements into Product Design. Productivity Press, Cambridge, Massachusetts, USA.
Andersson & Jacobsson 1993. QFD för byggbranchen, anvisningar och tillämplingserfarenheter SBUF (in Swedish). FoU-VÄST projekt rapport 9306, Göteborg.

Christiansen 1994. Describing Coordination Load Distribution in the VDT – a Practical Guide to Modelling for Organizational Engineering. DNV Research Report No. 94-2000, Det Norske Veritas.

CIB 1993. CIB Master List, Headings for the Arrangement and Presentation of Information in Technical Documents for Design and Construction CIB Publication 18.

Cross 1993. *Science and Design Methodology: A Review. Research in Engineering Design* 5. Springer-Verlag London Limited.

Dasu & Eastman 1994. *Management of Design, Engineering and Management Perspectives*. Kluwer Academic Publishers, Boston/Dordrecht/London.

Dixon 1988. On Research Methodology Towards A Scientific Theory of Engineering Design. *Artificial Intelligence for Engineering design, Analysis, And Manufacturing* 1(3).

Eppinger 1991. *Model-Based Approaches to Managing Concurrent Engineering*. International Conference on Engineering Design, Zürich, August 27-29, 1991.

Huovila & Serén 1995. *Customer-Oriented Design Methods For Construction Projects*. International Conference on Engineering Design, Praha, August 22-24, 1995.

Huovila, Koskela & Lautanala 1994. *Fast or Concurrent – the Art of Getting Construction Improved*. International Workshop on Lean Construction, Santiago, 28-30.9.1994.

ISO 1984. Performance Standards in Building – Principles for their Preparation and Factors to be considered ISO6241, 1984.

Lakka, Laurikka & Vainio 1995. *QFD:n soveltaminen rakentamiseen* (in Finnish). VTT Research Notes 1685, 1995.

NIST 1993. Integration Definition for Function Modelling (IDEF0). FIPS Publication 183, Computer Systems Laboratory, December 21, 1993.

Oswald & Burati 1993. Adaptation of Quality Function Deployment to Engineering and Construction Project Development. CII Source Document 97, Bureau of Engineering Research, University of Texas at Austin.

Papamichael & Selkowitch 1991. *A Computer-Based Building Design Support Environment*. First International Symposium on Building Systems Automation-Integration, June 2-9th 1991, Madison, Wisconsin, USA.

Peña 1987. Problem Seeking, An Architectural Programming Primer. AIA Press, Third Edition, Washington.

Sarkis & Liles 1995. Using IDEF and QFD to Develop an Organizational Decision Support Methodology for the Strategic Justification of Computer-integrated Technologies. *International Journal of Projects Management* 13(3).

Serpell 1994. Despliegue de la Functión Calidad para la Determinación de Características de Diseño de Viviendas (in Spanish). International Workshop on Lean Construction, Santiago, 28-30.9.1994.

Tsuda 1995. A Consideration of Applying QFD to Concurrent Engineering Processes of Automobile Development. *Proceedings of the 1995 Design Engineering Technical Conferences, ASME United Engineering Centre, New York.*

Use of the design structure matrix in construction*

P. HUOVILA, L. KOSKELA, M. LAUTANALA,
K. PIETILÄINEN & V.P. TANHUANPÄÄ
VTT Building Technology, Espoo, Finland

ABSTRACT: The Design Structure Matrix (DSM) is a novel, powerful method for analyzing and improving design processes, used successfully in product development projects. For evaluating its usability in construction, the 'as-is' design process of a fast track office building project was modelled using the DSM representation. The method provided a new, more efficient 'should-be' sequencing of design tasks. When analyzing the problems having occurred in design (monitored independently), it turned out that the majority of problems are located in process parts less effectively sequenced, as pinpointed by the method. Thus, the 'should-be' solution would probably have prevented a large share of problems. This chapter ends with a discussion of potential uses of DSM in construction.

1 INTRODUCTION

1.1 *Problems of the design process in construction*

Building designs are produced by temporary project organisations. These design teams consist of different skills, firms and individuals, that may have never worked together beforehand. Their design procedures can be company or individual related. In practice, planning of the design process is, in many cases, managed without systematic working procedures, and with few (if any) appropriate tools. Building designs consist of specific information at a stage, when it is not always known who will be applying that information: which are their products and processes.

At early stages of the design process, clients often have difficulties in expressing formally their real needs or requirements. Neither do they get input of the priorities of information that should be available to the other designers and to production. Designers have constant problems in getting reliable information in time. Clients, or architects, may change their minds during the design process, other designers may start their work too late, contractors may have certain preferences in construction methods to be applied, estimated investment costs may rise too high, the business environment may rapidly change, etc. All that mess affects in numerous changes to designs that may be already completed – or even constructed.

---

*Presented on the 3rd workshop on lean construction, Albuquerque, 1995

1.2 *Use of DSM in other industries*

In manufacturing industries, as reducing of lead time in product development projects has been emphasized, design procedures have become more complex. Concurrent engineering requires a framework to determine which tasks should begin early in the development process, which tasks should be left for later, and which tasks have strong interdependencies needing so a special attention to be paid in order to ensure that the coming tasks will have reliable input data in time. There has been initiatives to employ the Design Structure Matrix method (DSM) to manage the product development processes.

DSM was developed by Steward (Steward 1981) and recently it has been applied by several authors (Eppinger 1991; Eppinger et al. 1994; Kusiak et al. 1994; Austin et al. 1994; Austin et al. 1995). DSM is aimed to study the design processes which include iteration (or circuits). Circuits can not be handled with the critical path methods, and DSM serves as a useful tool to analyze, re-engineer and schedule the design processes.

2 BASICS OF DSM

2.1 *Theory*

Systems can be analyzed from two points of view: structure and semantics. For example, a graph or a matrix can be used to describe the structure of a system, by indicating which parts of the system affect which others. The structure does not tell how or why the part affects the others. The semantics of the systems answer to questions of 'why and how the parts affect' (Steward 1981).

Directed graph (Fig. 1) is used to represent the structure of a simple system. Vertices stand for the parts of the system, for example, tasks of a design process. An arc, between the two vertices, shows direct effect between respective parts of the system. *A*

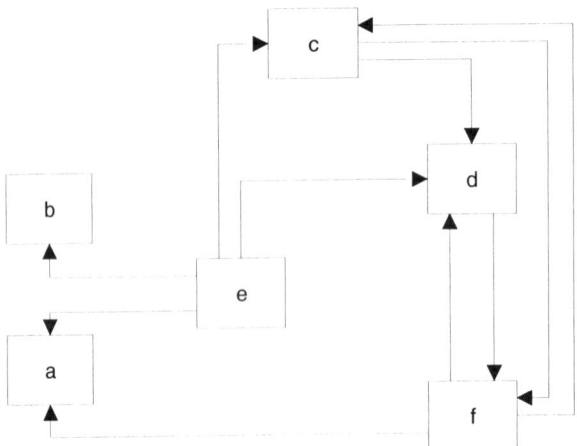

Figure 1. Structure of a simple system, described with graph.

simple path is a list of vertices in which successive vertices are connected by edges in the graph. A simple path of two or more arcs, e.g. (A, B, C), shows indirect effects.

A *cycle* is a path that is simple, except that the first and last vertex are the same. Cycles are the central concern of DSM. In our example systems, there is one cycle, (c, d, f, c). To solve a cycle in design process guesses are made for preliminary design, and then iterated to obtain improved designs.

Design structure matrix is a tool to find the cycles (or strongly connected components to use the language of graph theory) and to order the subprocesses so, that every vertex not belonging to any cycle, is listed before all the vertices it points to (Sedgewick 1988). *Partitioning* is a simple algorithm used to assign the subprocesses to blocks and to assign distinct integers to the subprocesses such that following conditions hold:

1. Subprocesses within a block are assigned contiguous integers;
2. No arc goes from a high-numbered subprocess in one block to a low-numbered subprocess in another block;
3. Given any two subprocesses x and y in the same block, there is a path from x to y and a path from y to x. (Steward 1981).

In Figure 2 is the structure of our simple system described with precedence matrix. Subprocesses are assigned to a row and corresponding column. Direct effects, or relations, are marked with 'x' in the matrix and the diagonal is marked with 'o'. A 'x' in column i row j represents an arc from vertex i to vertex j. Reading rows indicates on which other subprocesses the respective row depends on, and reading columns indicates other subprocesses which require input for the respective column.

After the execution of the partitioning algorithm, the marks appear either below the diagonal or within square blocks on the diagonal. In partitioning whenever reordering of the elements is done, the same reordering is made for the rows and columns. The block is the smallest possible such that all the variables that occur in a cycle will be found in the same block. When the system contains no circuits it is possible to reorganize the matrix into lower triangular form, i.e., no marks appear above the diagonal. Design processes usually contain iteration. This means that during partitioning blocks will be found and marks, illustrating feedback, will exist above the diagonal.

	a	b	c	d	e	f
a	o				x	x
b		o			x	
c			o		x	x
d		x		o	x	x
e					o	
f			x	x		o

	e	f	d	c	b	a
e	o					
f		o	x	x		
d	x	x	o	x		
c	x	x		o		
b	x				o	
a	x	x				o

Figure 2. Precedence matrix representations for the structure of the simple system. Matrix on the right is reordered with the partitioning algorithm and the block found is shaded.

2.2 *Practical point of view to DSM*

To decompose the design process into (sub-)tasks, and to determine the relations between the tasks usually enhances considerably the understanding about the whole process. DSM offers a simple and powerful method to model the structure of a system and to study it. Understanding the structure of the design process offers several practical opportunities to improve the process.

Blocks found with the DSM illustrate the tasks, that are coupled, and have to be done jointly. Concurrent engineering techniques can be used within the blocks to achieve needed co-operation between the tasks, persons and organizational units.

The right order is easily found for the sequential and parallel tasks. This information can be used, when the design processes are planned and scheduled.

The design structure matrix illustrates the information flow between different tasks and persons. A model of information flows can be used to identify which other persons have to be informed when design changes are made.

Tasks producing information for several other tasks can be identified. These tasks have significant role in the whole process and special attention has to paid to their performance (correctness of information, timing).

The design structure matrix models the connections between tasks and helps to manage the whole process. On the other hand, DSM gives a clear indication how the work of an individual designer affect on the whole process. This helps the designers to understand the requirements of their work and to communicate with the right people, when problems occur or changes have been made.

3 FRAMEWORK OF EXPERIMENTATION

3.1 *Summary of the construction project*

Design Structure Matrix method was tested ex post in a fast track construction project, where briefing phase was started already in 1991, and re-started in December 1994. Structural design of that 7.000 m^2 and 25.000 m^3 office building was started in the beginning of January 1995 and construction at the end of the same month. The five-floor building is planned to be finished by the end of November, and to be completed monthly, floor by floor, starting from the third floor at the end of July. The construction time of the building will be thus approximately 11 months. The previous phase of the same complex, slightly smaller than this one, was completed 1991. The construction time was 16 months, which is a standard construction time for a building of the size.

3.2 *Method of experimentation*

The construction project was followed and evaluated by a researcher who first collected all interesting documents of the project: drawings, minutes, time-tables, etc. Secondly, he observed and documented the ongoing design tasks: information flows between different project actors, task dependencies, their duration, and occurred problems. Thirdly, gathered information was completed and checked by interviewing the designers.

Once the design task information was filed and analyzed, it was put in a matrix form, where inputs and outputs between design tasks were identified and marked. After that, the matrix was reorganized using the DSM method. Main idea of the experimentation was to see afterwards, if the needs for changes, found through DSM method in the design process (changed order of design tasks, special attention to be paid, or specific methods to be applied in critical coupled tasks, etc.) would correspond to the problems that were actually identified when following the design process.

4 RESULTS

4.1 *The Design Structure Matrix*

This exercise was performed to test the feasibility of the DSM method in construction. The actual process was simplified to focus our efforts on testing the method. There are some tasks missing and the division into tasks could be more fine-grained. The tasks of the actual design schedule (Fig. 3) were selected as the tasks in the matrix. Still, we believe that we have enough data to test the feasibility of the method.

Collected data was input in the form of matrix presented in Figure 4 describing the as-is situation. Matrix was re-arranged using the partitioning algorithm and the result is presented in Figure 5. Majority of the tasks were either sequential or parallel. Only one major loop was found, forming a block of 15 tasks.

A large block of coupled tasks in the beginning of the design process indicates the difficulty in freezing the information in the early phase of design. This phenomenon should be familiar for people involved in large design processes. Design Structure Matrix does not offer any solutions for the problem, it only identifies it. However, solutions can be proposed only when problems are identified and their structure is known.

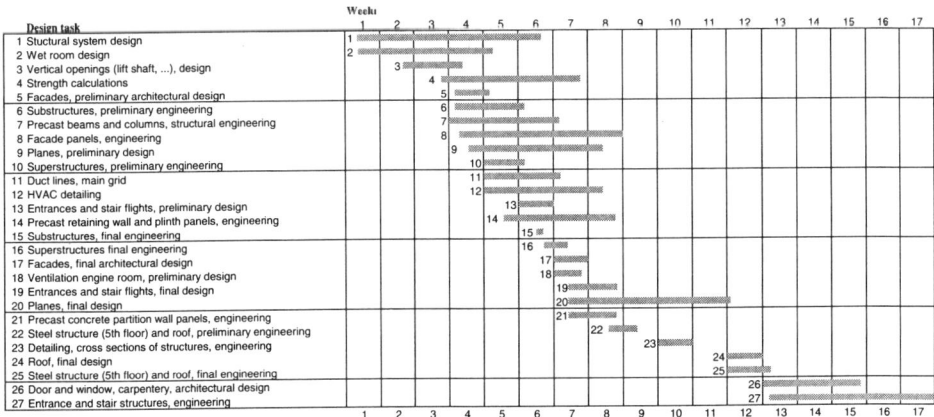

Figure 3. The actual design schedule.

		1	2	3	4	5	6	7	8	9	10	11	12	13	14	15	16	17	18	19	20	21	22	23	24	25	26	27	Inputs
Stuctural system design	1	1			x	x				⊗			x																4
Wet room design	2	x	2	x																									2
Vertical openings (lift shaft, ...), design	3		x	3																									1
Strength calculations	4	x			4	x																							2
Facades, preliminary architectural design	5	x		x		5																							2
Substructures, preliminary engineering	6	x			x	x	6																						4
Precast beams and columns, structural engineering	7	⊗		x	x			7					x	⊗															5
Facade panels, engineering	8	x			x	x			8				x	x			⊗												6
Planes, preliminary design	9	x	x		x					9			x																4
Superstructures, preliminary engineering	10	x	x		x	x					10		x	x															6
Duct lines, main grid	11	x			x				x		x	11											x						6
HVAC detailing	12							x	x	x	x	x	12																5
Entrances and stair flights, preliminary design	13													13															1
Precast retaining wall and plinth panels, engineering	14	x			x	x							x	⊗	14														5
Substructures, final engineering	15						x	x							x	15													3
Superstructures final engineering	16							x	x		x		⊗				16												4
Facades, final architectural design	17			x	x								x					17											3
Ventilation engine room, preliminary design	18				x				x										18										2
Entrances and stair flights, final design	19												x					x		19									2
Planes, final design	20								x				x					x			20								3
Precast concrete partition wall panels, engineering	21	x			x								x								x	21							4
Steel structure (5th floor) and roof, preliminary engineering	22				x				x														22						2
Detailing, cross sections of structures, engineering	23									x											x			23		x			3
Roof, final design	24	⊗										x						x	x						24				4
Steel structure (5th floor) and roof, final engineering	25											x									x		x		x	25			4
Door and window, carpentry, architectural design	26																	x	⊗	x							26	x	4
Entrance and stair structures, engineering	27													x							x					x	x	27	4
Outputs		12	3	5	8	10	1	4	2	6	4	3	9	8	1	0	0	4	2	2	5	0	1	0	2	0	2	1	

Figure 4. As-is matrix. Design tasks in chronological order. Identified major problems are marked with a bold X.

		1	2	3	4	5	7	8	9	10	11	12	13	17	18	20	6	14	15	16	19	21	22	24	25	27	26	23	Inputs
Stuctural system design	1	1			x	x			⊗			x																	4
Wet room design	2	x	2	x																									2
Vertical openings (lift shaft, ...), design	3		x	3																									1
Strength calculations	4	x			4	x																							2
Facades, preliminary architectural design	5	x		x		5																							2
Precast beams and columns, structural engineering	7	⊗		x	x		7					x	⊗																5
Facade panels, engineering	8	x			x	x		8				x	x	⊗															6
Planes, preliminary design	9	x	x		x				9			x																	4
Superstructures, preliminary engineering	10	x	x		x	x				10		x	x																6
Duct lines, main grid	11	x			x			x		x	11												x						6
HVAC detailing	12						x	x	x	x	x	12																	5
Entrances and stair flights, preliminary design	13		x										13																1
Facades, final architectural design	17			x	x							x		17															3
Ventilation engine room, preliminary design	18				x			x							18														2
Planes, final design	20							x				x		x		20													3
Substructures, preliminary engineering	6	x			x	x						x					6												4
Precast retaining wall and plinth panels, engineering	14	x			x	x						x	⊗					14											5
Substructures, final engineering	15						x										x	x	15										3
Superstructures final engineering	16						x	x		x		⊗								16									4
Entrances and stair flights, final design	19											x		x							19								2
Precast concrete partition wall panels, engineering	21	x			x							x				x						21							4
Steel structure (5th floor) and roof, preliminary engineering	22				x			x															22						2
Roof, final design	24	⊗									x			x	x									24					4
Steel structure (5th floor) and roof, final engineering	25										x					x							x	x	25				4
Entrance and stair structures, engineering	27												x			x									x	27	x		4
Door and window, carpentry, architectural design	26													x	⊗						x					x	26		4
Detailing, cross sections of structures, engineering	23								x							x									x			23	3
Outputs		12	3	5	8	10	4	2	6	4	3	9	8	4	2	5	1	1	0	0	2	0	1	2	0	1	2	0	

Figure 5. Should-be matrix. Design structure matrix arranged using partitioning algorithm. Blocks are marked with border. Identified major problems are marked with a bold X and a circle around them.

4.2 *Comparison between the actual structure and suggested structure: how could the design have been improved?*

The first finding to point out is that many down stream tasks (17, 18, 20) belong to the main block. The DSM analysis suggests to move these tasks earlier in the design process. We did not, although, consider the importance of each input and output. In case these tasks will provide output which is not imperative to the other tasks within the block, significant problems might not emerge.

The main block consists of tasks from all major design disciplines: architectural design, structural engineering and HVAC engineering. This emphasizes the need of team work within the design team.

The tasks within the main block have notably more outputs that the tasks which are not in the main block. Thus, successful execution of the block tasks is essential to the rest of the design process. Management of the design process should focus on the main block.

There are few tasks which have notably many outputs (task numbers 1, 4, 5, 12, 13). Delays or changes in these tasks have obviously high potential to generate more problems later in the design process. Special attention should be paid on management of these tasks. Especially tasks 12 (HVAC detailing) and 13 (preliminary design of entrances and stair flights) are problematic, because they are late in the process and they have many outputs to the prior tasks. Still, it is obvious that the outputs of task 12 are mainly needed to finalize the other tasks and it is not of crucial importance. So, the main focus should be on task 13, and any action to advance the execution of this task should be taken.

It is obvious that problems could be encountered in the cases when output is created to the tasks before and especially when the delay is long. This criterion points out the tasks 20, 17, 13, 12 and 9. Team work in the respective tasks is suggested.

4.3 *Identified problems in the actual design process*

Comparison of the re-arranged matrix with the encountered problems revealed that the majority of the problems were somehow connected with the tasks within the block. They are either within the block or they get input from the block. Having input from the tasks in the block is problematic, because the input has a tendency to be continuously changing. All the tasks in the block depend on the other tasks and a minor change in one task can affect all the other tasks in the block. There is only one major problem (marked with bold X in the table) which did not meet this condition (output from 19 to 26).

There were two problems that seemed to be above others. First the architect could not prepare drawings for entrances and stairs. This was also crucial input for many other tasks including engineering of precast beams and columns, which, furthermore, provided HVAC engineering with necessary input. The problem was recognized during the design process, but the architect and the client could not agree on a satisfying concept, and the start of the design task was delayed. The DSM analysis pointed this potential problem clearly: task 13 is in the block, it has many outputs also to preceding tasks. With the information provided by the DSM the task could have been started earlier and with more determined attention.

The other major problem was related with the structural system design (task 1). The client could not define the user requirements and the architect could not prepare preliminary plans. Furthermore he/she was unable to define the structural system (input from task 9 to task 1) and its requirements (fire classes, etc.).The DSM analysis identified also this problem clearly. Task 1 has many outputs and there is input from following task to previous task with significant delay between the tasks (from 9 to 1, and from 12 to 1). Still, the seed of the problem was the client's incapability to define the user requirements.

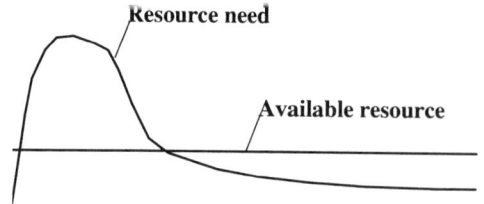

Figure 6. Needed capacity and allocated resources in architectural design.

The block is composed of 15 tasks, eight of which are tasks of architectural design. Additionally, in the beginning of the design process there were also severe problems in the definition of user requirements, which required plenty of the architect's time. Thus the architectural design has a load peak in the beginning of the design process. However, the architect did not have the capacity available (Fig. 6). The DSM analysis clearly pointed this potential problem and it might have been possible to assign more resources to the architectural design.

The comparison of the identified problems and the DSM analysis revealed that it is possible to find the critical elements of a design process. The DSM can as well be used effectively during the project to study the effects of possible changes to the design process.

5 CONCLUSIONS

The case study described proves that the design structure matrix can effectively be used in construction for finding better sequences of design tasks, as in other industries. We envisage that this method will be used in construction for following purposes:

– Planning and management of design. The matrix in itself (without resequencing) is a powerful tool for planning and communication, and can be used in the overall management of design;

– Fast tracking analysis. An improved and speeded design sequence may be searched by means of the DSM method. Through careful analysis, it is possible to keep the inevitable added costs of fast tracking to a minimum;

– Effect of changes. The schedule effects of changes, initiated, say, by the client, can be visualized through the matrix.

The DSM method can be used manually, even if it is laborious. We anticipate a more widespread use of DSM after related design management software has been commercialized. Now such software exists only as research prototypes.

All in all, the DSM method seems to provide a rigorous and effective conceptual basis for construction design management.

VTT Building Technology will continue this work. We have just started a project where the DSM method will be used in design planning to analyze and avoid potential problems. Some other DSM implementation and development projects are currently under planning.

REFERENCES

Austin, S., Baldwin, A. & Newton, A. 1994. Manipulating the flow of design information to improve the programming of building design. *Construction Management and Economics* 12: 445-455.

Austin, S., Baldwin, A. & Newton, A. 1995. A data flow model to plan and manage the building design process. *Journal of Engineering Design* 6(3).

Eppinger, S. 1991. *Model-Based Approaches to Managing Concurrent Engineering.* International Conference on Engineering Design, Zürich, August 27-29.

Eppinger, S.D., Whitney, D.E., Smith, R.P. & Gebala D.A. 1994. A Model-Based Method for Organizing Tasks in Product Development. *Research in Engineering Design* 6: 1-13.

Kusiak, A., Larson, N.T. & Juite, W. 1994. Reengineering of Design and Manufacturing Processes. *Computers & Industrial Engineering* 26(3): 521-536.

Sedgewick, R. 1988. *Algorithms,* 2nd ed., Addison-Wesley Publishing Company, 657 p.

Steward, D.V. 1981. *Systems Analysis and Management: Structure, Strategy and Design*, Princeton, NJ: Petrocelli Books, 287 p.

Benchmarking, best practice – and all that*

SHERIF MOHAMED
CSIRO Division of Building, Construction and Engineering, Highett, Victoria, Australia

ABSTRACT: Benchmarking has been in use in various industries for many years. While there have been many successes, its use has been rather limited in the construction industry. This chapter addresses the benchmarking concepts and its application to construction and presents a three-level – internal, project and external – framework for benchmarking current practice. Each benchmarking level is examined in detail with an illustration of the need to adapt to improve the construction output. A generic definition of benchmarking is used throughout this chapter to ensure applicability to the different and many aspects of the construction process.

1 INTRODUCTION

Over the years, the Australian construction industry has made a notable contribution to the national Gross Domestic Product (GDP). The size of this contribution was mainly determined by two factors – demographic development and national economic growth. A third factor is now emerging in the form of the fast-growing construction activity in the region of East Asia, as leading Australian construction firms become more active in targeting large projects in this region. In the process of competing for such projects, these forms are finding themselves in fierce competition with their international counterparts. As a result, some firms are starting to critically assess (benchmark) their output against what is known as best-practice performance.

The successful implementation of benchmarking in the manufacturing industry is reflected in the large number of publications which address the concept, application and limitations of benchmarking. Benefits of benchmarking are still largely unrecognised in the construction industry despite the fact that the best-practice concept has been indirectly investigated by both practitioners and researchers. Practitioners demonstrated how best can be achieved through the presentation of case studies where construction time and/or cost had been substantially reduced. Researchers, on the other hand, continued to propose theoretical measures as a remedy for improving the construction output.

*Presented on the 3rd workshop on lean construction, Albuquerque, 1995

2 BENCHMARKING CONSTRUCTION

To date, researchers are arguing about differences and similarities between construction and manufacturing. There are those who claim that both industries are quite different (Groak 1994) mainly due to the lack of 'one roof' organisations which characterise production processes in manufacturing. On the other hand, there are others (Ayres 1989; Koskela 1992) who are looking for the 'applicability' of manufacturing production tools in construction. This chapter views construction as a unique industry with a significant potential to develop its own tools based on imported concepts which have been successfully used elsewhere. This chapter demonstrates a typical example to support this viewpoint – the application of benchmarking, which is a manufacturing managerial concept, to construction.

Benchmarking is essentially a goal-setting procedure to compare one's current performance against the best practice in any particular area. Typically, it is applied to value-adding activities, i.e. activities which affect timeliness, cost or quality. Once a firm's performance is assessed against the best, those selected activities are redesigned (re-engineered) according to the strategic goals of the firm. In most cases, re-engineering an activity may extend well beyond its working procedure to cover associated managerial and non-technical issues.

The focus of benchmarking in manufacturing is the ability to meet customer requirements and to adopt innovative practices regardless of their source. Data availability has a major contribution to the success of benchmarking in manufacturing. Over the years, manufacturing firms have developed measures to assess their performance. Typical measures were components assembly or unit production. These measures usually involved time as the measuring basis. In so doing, firms collected data regarding the different procedures performed to deliver the output to be measured. In some cases, and to account for any data unavailability, data was obtained from other forms with common process steps, e.g. assembly, packaging, storage (Macneil et al. 1994).

In construction, however, benchmarking is not a straightforward task due to both the very nature of the industry which lacks solid data gathering and the remarkable fluctuation in productivity. Collected construction project data usually lacks consistency in structure and compilation (Choi & Ibbs 1994). Those who attempt benchmarking in construction are certain to be faced with difficulties such as incomplete or non-existent data. Even if data is well recorded and retrievable, it would be highly dependent on the special characteristics of the project, e.g. size, type and budget. Therefore, it is difficult to effectively use it as a basis for comparison. The structure of the industry with its temporary nature in organising the construction process, where a number of firms get involved in designing and constructing a single project, adds to the complexity of the benchmarking task.

Benchmarking only works in an enlightened environment where consistent methods of measuring the performance of operations can be developed and introduced. Currently, such methods do not exist. The lack of methods, the comparative information, and the systems to measure it, prevent the national industry from assessing its position relative to international best practice.

In the face of the above, i.e. the absence of data and measuring methods, it becomes clear that another dimension has to be given to the benchmarking form, as currently applied in manufacturing, before it can be applicable to construction. This dimension is presented and discussed in this chapter, in the form of the following three different levels of benchmarking:

– *Internal benchmarking*. Where a construction firm aims towards identifying improvement areas within its organisation through comparing its business operations with those of others who do it better, thus setting new targets to meet;

– *Project benchmarking*. Where a construction firm assesses the performance of projects in which it is involved with an aim to meet customer requirements, measure productivity rates, and validate and maintain its estimating databases;

– *External benchmarking*. Where the industry attempts to increase its productivity through making tools and techniques, developed and successfully used by other industries, applicable to construction.

For a proper benchmarking application in construction, all proposed benchmarking levels should be considered as discussed below.

3 INTERNAL BENCHMARKING

Internal benchmarking is the examination of an individual firm's current processes and practices. In other words, it is the measurement of the firm's performance. Naturally, performance indicators depend upon the firm's line of business, but useful indicators can include: Customer perspective (service, cost, quality); business evaluation (market share, successful/failed tenders, conflicts); and financial stability (turnover, backlog). These measurements can be used in identifying the firm's strengths and weaknesses before proceeding with the comparison of its performance to that of other direct competitors.

A firm launches its benchmarking campaign to improve its performance and business competitiveness. As in any business process, there would be a number of non-value-adding activities embedded within the process, i.e. activities that consume time and/or cost, without adding value to the process. Through benchmarking, these activities can be identified and, at a later stage, eliminated, leading the way to a more efficient business practice. Processes targeted by the internal benchmarking exercise must be in line with the firm's overall strategic business objectives to avoid wasting benchmarking efforts.

A critical prerequisite to successful benchmarking is the commitment to the exercise from all levels of the firm, especially the top management level who have the authority to implement and integrate the benchmarking results into the firm. Other levels have to be flexible in the preparation for changes in job functions and resources allocation. Those most affected by the benchmarking results must be empowered to use these results as a positive change agent (Almdal 1994). This is a necessary step to achieve an essential change to processes and optimum performance improvement.

This level of benchmarking represents a major step towards establishing well-defined measures and measuring systems within the firm to better understanding its own processes, resources (human and technical) and functions (operational). Once these measures are established, the firm embarks upon the second phase of benchmarking – the comparison phase.

After determining where the areas for greatest improvement are, the issue of selecting a 'best-practice' firm becomes the target; who is best and what do they do.
The selection process is outside the scope of this chapter. However, and as a word of
caution, ultimate care should be taken to account for the differences in operating environments, levels and quality of service between firms. This is a fundamental point
to avoid being overwhelmed, if the firm taking up benchmarking does not enjoy a
high calibre of performance on the national level.

For those firms who are considered to be the national industry leaders, they have
to measure their performance against world-class competitive standards. Otherwise,
any chance of improving national industry standards will diminish. These firms
usually have a reasonably effective system that can guide and support the benchmarking exercise. Nevertheless, the system must be maintained in a way that allows
continuous benchmarking, ensuring the firm's competitive advantage.

The road to best practice is not a short one. It starts with the firm exploring its
system. Then it proceeds to benchmark its performance against that of another firm
who does it best, in order to generate improvement opportunities to increase the
firm's productivity and competitiveness. A more advanced stage along this road is
identifying innovative practices which can secure a firm's place amongst the best-
practice forum.

4 PROJECT BENCHMARKING

The second level of benchmarking is the measurement of performance of projects in
which the firm is involved. Traditionally, measurements have been in terms of com-
plying with completion time and allocated budget. This benchmarking exercise is
desirable, but is, however, highly dependent on the project characteristics, e.g. degree
of complexity, performance of other key participants, and most importantly, the
trends of the industry, i.e. the performance of the industry as a whole.

Current Australian attempts at benchmarking construction have focused on measuring the time performance of projects. In a particular comprehensive investigation
by Walker (1994), useful indicators for benchmark measures were identified and four
major issues found to affect construction time performance: 1) Construction management effectiveness; 2) The sophistication of the client in creating and maintaining
positive relationships with the construction management and design teams; 3) The
effectiveness of the design team in communicating with client and construction management teams; and 4) A small number of factors describing project scope and complexity.

The need for taking construction benchmarking a step further became clear by the
early 1990s, when it was noted that they key objective was to develop practical
means by which firms can compare and assess their own performance and the overall
project performance. A positive step in this direction was taken by the Construction
Industry Development Agency (CIDA) which collected data on the performance of
more than 50 national projects completed between 1988 and 1993 (CIDA 1993). Despite the reported significant improvement in project cost and time performance, this
survey fell short of meeting the requirements of benchmarking due to the following
reasons:

– The survey reported the current performance of projects in Australia, but it did not search for reasons underlying this performance – this opens the door for construction firms to speculate reasons for their project success or failure as the survey outcome did not help identify improvement opportunities.

– Performance in the survey was measured in terms of meeting planned project costs and schedules as defined by contractual details; other critical factors such as quality and resource availability were not accounted for.

– Performance was measured internally against the expectations specified in the contract – once again, construction firms found themselves questioning the credibility and reality of these expectations.

– The survey was carried out through quantity surveying firms only, thus presenting a partial view on current performance and violating one of the major requirements of benchmarking, i.e. the surveyed sample must be representative of the process. In this survey, the process lacked the input of other key players in the industry.

To lay out the foundations for benchmarking construction, the Construction Industry Institute Australia (CIIA) focused its research efforts towards the preconstruction phase in an attempt to isolate the most important aspects that determine performance. These were, according to CIIA, project management and project communication. A survey carried out by the CIIA Benchmarking Task Force (1995) examined a number of broad issues relating to project management teams, team performance and communication. The principal finding of this survey was that an excellent project outcome can be expected from a management team if the team is cohesive (operates as a team), customer-focused, and goal-oriented. However, this finding raises some important questions, such as the following:

– When should the team building process begin? (before or after defining the project scope);

– Who is directly responsible for creating this quality project management team? Is it the client, architect, project manager or someone else?

– Is team building a progressive (dynamic) process in which the members joint in the team only when their skills/expertise are needed?

– How would the quality project management team be successful in overcoming the rules of thumb that overwhelm current practice?

– What are the measures/actions that could be taken to avoid the existing sources of adversarial relationships among project participants, thus allowing team cohesiveness to develop and sustain throughout the project?

For an efficient team-building process leading to a much improved performance, as that suggested by the survey, these questions remain to be addressed in a clear and systematic manner to create the environment in which an unambiguous team-building framework can be produced independently of the project size, budget, type, location, etc.

The process of project benchmarking is essential for supporting the firm's internal benchmarking exercise; the performance measurements(s) carried out throughout the lifetime of various projects would serve as an effective gauge for the success/failure of the actions taken in the light of the internal benchmarking recommendations.

Irrespective of its line of business, what should a construction firm consider when taking up project benchmarking? To answer this question, let us assume that the firm under investigation has been awarded a contract for a project that aligns with its overall strategy. Throughout the different stages of the project, the firm attempts to perform a twofold task: first, to carry out the assigned 'to be done' work to meet customer expectations as well as contract specifications (being on schedule, on budget, etc.); and second, to deliver the 'already done' work whilst meeting its own profit goals and without the need for pursuing legal claims after the completion of the project.

For a proper benchmarking of the firm's performance during the execution of the first part of the task, each value-adding activity within the process has to have a well-defined customer, i.e. the next user, who sets the measures for the process performance. For each selected activity, relevant customer requirements must be clearly identified at the earliest possible stage so that sub-activities can be designed, managed and performed with the aim of satisfying these requirements. Once the customer is satisfied with the outcome, work then proceeds to the next activity.

Adopting such a sequence will not only ensure systematic customer involvement, but will substantially simplify the task of project benchmarking, as it will give rise to simple and clear measures for the firm to meet. Whenever practical, these measures should cover both quantitative and qualitative issues. For example, and in addition to the well-known time, cost and quality measures, there can also be additional measures such as the number of queries raised by the customer, technical changes after final submissions, design variations to suit site conditions or disputes at the hand-over stage.

Efficiency of site operations is a good example for project benchmarking from the design team's perspective. The more construction knowledge and contractor experience that is implemented into the design stage, the higher the designer's performance rate is. Therefore, performing a constructability analysis can be used in benchmarking the designer's performance against the measurements/criteria specified by the next user, i.e. the contractor.

To benchmark the second part of the task, the firm should establish a set of project benchmarking measures to trace the performance against its own expectations and initial estimations. Provided the firm carried out its precontract estimations correctly, extra care should be taken to detect any 'holes' that may arise during execution resulting in the reduction of the assumed profit margin. One of the most common holes that exists in construction practice is reworking incomplete tasks – redoing activities. This undesirable feature puts considerable pressure on the firm in the form of direct additional costs and lost time, which could have been avoided if things were done correctly in the first place.

Any designed set of measures, as required and needed by the firm's discipline, should be simple enough to be 'built-in' within the process allowing accurate and representative measurements to be taken. These measures would initially help detect the errors which lead to rework and ultimately lead to their prevention in similar future projects. It is worth noting that a proper quality assurance (QA) application in construction does help reduce design and construction errors. However, it does not allow the performance measurements to be taken. To illustrate this type of benchmarking, the size of the rework cycle experienced in current practice, as well as the role of design and construction teams in its detection, measurement and correction, are discussed below.

As far as it is known, there has never been any systematic attempt to observe rework cycles in construction. Sources of quality deviations have been investigated by Burati et al. (1992) who indicated that rework cost is a significant portion of the total project cost. Cooper (1993) reported that the design of large construction projects may require up to two and a half cycles of rework to 'get it right'. These findings clearly give an indication of the large amount of time and cost associated with re-work in construction. Furthermore, it is widely believed that current quality measures are not adequately implemented within construction firms. Consultants and contractors who use inspection to achieve desirable project outcomes, aim at detecting and correcting deficient results. Inspection not only consumes resources and time but also does not improve working procedures. Therefore, inspection can be seen as a non-value-adding activity which ideally should be eliminated from the construction flow process (Koskela 1992).

Design and construction teams have a thorough understanding of the technical, managerial and operational aspects related to their project areas. The basis of their judgement, during the project lifetime, is highly dependent upon their personal knowledge and past experience. This experience is not being captured 'as it happens' or documented into the firm records for future use. That is where project benchmarking should come in recording the performance in terms of productivity rates, allocated resources and cost analysis. A feedback link should be established by the firm undertaking benchmarking to examine how the actual performance-cost rates compare to what was actually assumed at the estimating/planning phase. This is an essential link for assessing current performance during construction, thus enabling the firm to decide on effective actions to be performed to maintain the planned schedule and cash flow. This proposed link would also provide invaluable information to the estimators/planners.

The above discussion highlights the deficiency in current understanding of project benchmarking. It demonstrates that project benchmarking should not be merely focused on investigating the project progress and then comparing it to the planned schedule and allocated budget, as widely believed and practiced by many researchers in this area. however, it calls for construction firms to take the initiative in monitoring their own performance during the time of their involvement in the projects, in both a quantitative and qualitative sense, so that a true performance picture can be obtained well beyond that of team building and project management team performance.

5 EXTERNAL BENCHMARKING

This level of benchmarking is mainly concerned with the selection and implementation of managerial and technological breakthroughs, developed by other industries, to generate significant improvement in the construction productivity. Although external benchmarking may not provide the construction firm with immediate benefits, it is still considered as an important tool for identifying improvement areas in performance standards that constitute the firm's business framework. In external benchmarking, there is an obvious opportunity to introduce innovative processes into traditionally performed operations.

Productivity is the most important issue in any industry. In the last decade, industries like manufacturing and electronics have achieved remarkable high productivity through integrating design and production processes. To a large extent, design and construction processes are still carried out in what can be described as 'the way they have always been' with noticeable disunity between them. In the majority of construction projects, engineers are consulted once the architectural plans are completed, while contractors become involved after most of the drawing details and specifications are set. The temporary relationship between the client and architect extends to a large degree covering other project key players. This type of short-lived relationship does not help create a stimulating environment for increasing productivity and enhancing value (Kubal 1994). Furthermore, existing bidding systems, post-tendering negotiations, industrial relations and cultural attitudes towards change, limit the chances of attaining substantial improvement in construction productivity (Ireland 1995).

By comparison, other industries have succeeded in establishing a much healthier and more competitive environment in which companies endeavour to produce what seems to exceed customer requirements. These companies also strive to optimise their production activities to gain a larger share in the market. Thus it is clear that there is a need to benchmark other industries in an attempt to incorporate, subject to practicality, the best in the construction industry. Many attempts have been made to acquire and adapt the management tools and process techniques used successfully in manufacturing, to construction, e.g. prefabrication, standardisation, automation, computer integration, quality control and many more.

While some of the above-mentioned tools and techniques have enjoyed a reasonable degree of success in construction, many others have not. This can be mainly attributed to the high uncertainty factor usually associated with construction operations. Uncertainty is regarded as a fact of life in most construction projects rather than an isolated problem of limited importance (Laufer & Howell 1993). Sources of uncertainty in construction are too many, and are outside the scope of this chapter. Simply by ignoring this factor, tools imported from manufacturing or any other industry will have difficulty in delivering what has been promised. Practitioners and researchers have to acknowledge that this uncertainty factor is, and will always be, a feature of construction operations. Therefore, external benchmarking should target tools and techniques that can either cope with such a high degree of uncertainty or help reduce it.

Upon the selection of a tool or technique, its implementation should be investigated under the current features of construction practice to determine its impact upon productivity. Impact assessment in an assumed 'perfect' environment will almost certainly receive a negative reaction from the industry, especially when the implemented tool or technique requires major changes in the associated features of the process. Practitioners tend to be more receptive to new ideas and their adoption when these ideas do not originate in their own industry (Gable et al. 1993).

Benchmarking is a typical example of managerial tools imported from manufacturing. In the first instance, it is quite an appealing tool for those who are interested in improving the construction output. Yet, construction circles are not susceptible to the idea due to the following:

– Misunderstanding of the benchmarking concept; for many it simply means measuring everything;

– Confusion surrounding what is required to take up a benchmarking exercise: what to measure? how is it measured? against what should these measurements be compared?

– Unavailability of data mainly because the structure of the process does not allow data associated with field-based operations to be collected readily.

– Application of benchmarking seems to require changes in the way information is handled and documented.

– Lack of relevant conceptual models to support and guide data collection.

The above demonstrates that considerable work is required to fully utilise an imported tool in construction. Nevertheless, the well-demonstrated effectiveness of such tools in other industries necessitates external benchmarking. An appropriate level of external benchmarking is bound to give rise to effective tools that my solve, or at least ease, some of the inherent problems the construction industry is facing.

6 CONCLUDING REMARKS

Construction output has been characterised by fluctuation in productivity. This represents the major drive behind the need for benchmarking construction as other industries have proved to be well ahead of construction in measuring their performance. In order to improve the construction output, it is essential to have accurate and representative measurements reflecting current practice, trends and productivity. This chapter has examined the application of the benchmarking concept to construction through three levels – internal, project and external.

On the road to imitate best practice, construction firms should be effectively taking up internal benchmarking to gain a thorough understanding of how they do business and how their customers evaluate their services. For efficient internal benchmarking, allowances has to be made for the differences in operating environments, levels and quality of service, to avoid wasting effort. Firms are urged to be actively involved in project benchmarking to assess their performance, measure their productivity rates and validate their cost-estimation databases. Also, firms have to be more open to benchmark what has been successful in other industries and assess if it can be adaptable to construction.

Finally, benchmarking should be seen as an integrated part of an ongoing process aiming at improving construction productivity. A fully implemented benchmarking environment within the industry, as suggested herein, will significantly improve the quality of decision making related to design.

REFERENCES

Almdal, W. 1994. Continuous improvement with the use of benchmarking. *CIM Bulletin*, 87(983): 21-26.

Ayres, R.U. 1989. Construction as manufacturing. *Proc. CIB Workshop on the Future of Construction, Haifa, Israel*, January 1989, pp. 1-16.

Burati, J.L., Farrington, J.J. & Ledbetter, W.B. 1992. Causes of quality deviations in design and construction. *ASCE Journal of Construction Engineering and Management* 118(1): 34-49.

Choi, K.C. & Ibbs, C.W. 1994. Functional specification for a new historical cost information system in the concurrent engineering environment. *The International Journal of Construction Information Technology* 2(2): 15-35.

CIDA 1993. *Project Performance Update – A report on the Time and Cost Performance of Australian Building Projects Completed 1988-1993.* Construction Industry Development Agency, Sydney, Australia.

CIIA 1995. *Benchmarking Engineering and Construction – Winning Teams.* Construction Industry Institute of Australia.

Cooper, K.G. 1993. The rework cycle: Benchmarking for the project manager. *Project Management Journal* 24(1): 17-22.

Gable, M., Fairhurst, A. & Dickinson, R. 1993. The use of benchmarking to enhance marketing decision making. *Journal of Consumer Marketing* 10(1): 52-60.

Groak, S. 1994. Is construction an industry *Construction Management and Economics* 12(4): 287-293.

Ireland, V. 1995. The T40 Project: Process re-engineering in construction. *The Australian Project Manager* 15(1): 31-37.

Koskela, L. 1992. *Application of the New Production Philosophy to Construction.* Technical Report No. 72, Centre for Integrated Facility Engineering, Department of Civil Engineering, Stanford University, California, September.

Kubal, M.T. 1994. *Engineered Quality in Construction: Partnering and TQM.* McGraw Hill Inc., New York.

Laufer, A. & Howell, G.A. 1993. Construction planning: Revising the paradigm. *Project Management Journal* 24(3): 23-33.

Macneil, J., Testi, J., Cupples, J. & Rimmer, M. 1994. *Benchmarking Australia: Linking Enterprises to World Best Practice.* Longman Business & Professional, Australia.

Walker, D.H.T. 1994. An investigation into factors that determine building construction time performance. Ph.D. thesis, Department of Building and Construction Economics, Royal Melbourne Institute of Technology, Melbourne, Australia.

Assessing quality control systems: Some methodological considerations*

DAVID SEYMOUR
School of Civil Engineering, University of Birmingham, Edgbaston, UK

ABSTRACT: The chapter reports a study that was designed to establish the circumstances in which different levels of quality with regard to a feature of steel reinforced concrete structures were achieved. A procedure in the research process is discussed in order to consider ways in which the research paradigm used influences findings and the practical inferences that may be drawn from them. Four research paradigms are considered and contrasted, two of them are referred to as 'applied' and two as 'critical'. It is argued that if developments like that of lean construction are to realize their potential benefits, 'critical' research has an important contribution to make.

1 INTRODUCTION

My contribution to this workshop may seem rather esoteric in view of the workshop's essentially practice orientation. However, I hope to show that as researchers who aim to make a contribution to the improvement of practice, we ignore at our peril the question of how come we know what we think we know.

One reason that I, personally, feel forced to ask this question is the relative failure of QA under the aegis of BS 5750/ISO 9000 in the British Construction Industry. A superficial explanation of this failure is that, by and large, the standard is fine in principle but that it has been received within a set of attitudes or culture which remains unchanged; that it has therefore been distorted, subverted and discredited. Though it contained some good sense, so the explanation goes, many of it proponents have either been naive or opportunistic, ignoring the reasons for opposition which are deeply rooted in the attitudes, culture and institutions of the industry and has, therefore, largely failed as an instrument of change. There is something in this explanation and it needs to be looked at more closely because even now, the more sophisticated of the quality praxes, like TQM, benchmarking and even lean production, are receiving the same comments and stand to suffer the same fate.

An easy response to this kind of analysis is that if you sufficiently fine tune your instrument of change, you can change attitudes and culture, so, what I propose to do is consider what researchers and would-be contributors to the industry's efficiency might mean by this. In particular, I wish to consider *our* attitudes and culture as researchers and the role they play in mediating our perception of what should be

*Presented on the 2nd workshop on lean construction, Santiago, 1994

437

changed; whether, perhaps, they need to be changed quite as much as the culture and attitudes of 'those-out-there' who are frequently blamed for the lack of improvement, as they certainly are, in the UK, at least. For what I conclude from the study reported here is that any of the quality initiatives must be much more carefully studied and reported as to why they work or why they do not work.

In this chapter I will draw attention to some research problems, initially identified as methodological in nature, that were encountered in an empirical research project concerned with the achievement of quality in concrete structures. I will then suggest that these problems, in fact, raise some much more fundamental questions about standard assumptions that are made about the aims of and the relationship between researchers and practitioners.

Firstly, I will present some of the findings of the study which was designed to establish the extent to which the quality achieved, with respect to a specific structural feature in a sample of twenty-four projects, conformed to the specifications provided in the design and what organizational features, work practices, and the like, correlated with the quality achieved. The purpose, in short, was to describe the contexts in which given levels of quality were achieved. The research was carried out by a team of engineers and a social scientist and was designed to satisfy the expectations of the sponsors who appointed an experienced civil engineering research project manager and closely monitored progress. The research was therefore carried out within what may be called the conventional construction management research paradigm, which, as I will argue, comes firmly within the rationalist tradition. The manner in which I present these findings also conform to what I take to be the standard conventions for reporting research within this tradition.

Secondly, I will conduct what might be called an 'auto-critique' of a feature of our research practice, suggesting that it carries certain rationalist assumptions which influence the analysis, the inferences that may be drawn and the practical consequences that are implied. I will, therefore, thirdly, contrast the rationalist approach with three others and consider in turn the inferences that may be drawn from them. Finally, I will suggest that despite appearances, many rationalist assumptions or ways of thinking are deeply embedded in the culture of the industry and are reflected in both the practice of and the commentaries on quality management and that this compromises much of the promise of the quality management initiative and developments like lean construction.

2 THE STUDY

2.1 *Data collection*

The purpose of the study (as noted above), which we have been carrying out over the last two years, was to record the quality achieved in a sample of structures and to relate these findings to the circumstances in which the work was carried out. To date, twenty four sites have been investigated, varying in type of structure, size, contract type and located throughout the UK. A sample of structural elements on each site was selected and the construction process monitored through its various stages. The findings on twelve of the sites have been analysed, to date, and are discussed in this chapter. Ten of the twelve contractors who participated in the study were quality assured with BS5750. The completed structural elements were subjected to a series of tests and measurements. Photographs were taken of work in progress and informal interviews conducted with operatives, foremen, engineers and managers during the course of the work, about what they were doing, why they were doing it and so on. The period of investigation for each site was, on average, one month.

It was established that, consistent with normal practice, quality control was carried out by both the contractor's management staff and the client's representative who undertook procedures of inspection and approval.

Contractors' staff monitored the work at various stages in its progress against standard checklists. Where defects were observed, rework was ordered and when deemed satisfactory, a request sheet was submitted to the client's representative for final inspection and approval before concreting. When the client's representative was satisfied, permission to proceed was given. Though there were variations in the checking procedures followed by contractors and client representatives on each site, they generally conformed to a common pattern.

Defects, that is, features of a structural element that have a direct negative effect on the achieved quality of a particular feature of construction were identified by the researchers by monitoring the complete construction process of the structural elements. The monitoring process took three to four days for each site.

2.2 *Analysis*

Following discussion with an expert panel of structural engineers, defects were distinguished into two classes on the basis of the origins of the defects identified. Within the first class, referred to, for the purposes of this chapter, as *management-controllable* defects were placed those defects which were judged to have originated in:

1. Impractical *design* such as a clash in the position of components;
2. Components having been despatched to site by *suppliers* wrongly dimensioned or out of tolerance;
3. Inappropriate methods of construction adopted by *contractors* or the supply of inappropriate materials.

These defects are referred to as management-controllable defects since they could have been *prevented* from occurring in the built structure, by prior management action.

A second class of defects, called *operative-controllable defects* because they are judged to result directly from operative action, include such features as misplacement or omission of components.

In summary; defects were classified as those which appear during construction but originate before it (management-controllable) and those which occur during it (operative-controllable).

The quality management systems that were observed in the research were only capable of detecting the defects that occur *during* construction and are directed at the operative-controllable defects. Since the research does not address whatever quality control systems that may have been in place at the design stage or within the suppliers' organizations, I cannot comment on them. However, I note that management activity (time and effort) may be directed at influencing others' input (over which it has no direct responsibility) to those processes for which the contractor does have direct responsibility. At this stage of the research I can only comment on; a) efforts made by contractors in the sample to effect this 'external' influence and report that it appeared minimal, and b) the *relative* emphasis contractors placed on the management controllable defects which directly came within their control (i.e. the adoption of certain construction methods) and, what we have called, the operative-controllable defects. The comment is that by far the greater emphasis was placed on the operative-controllable defects which, the findings showed, constituted only about half of the defects.

How effective were the procedures in controlling the 'operative controllable defects' i.e. those to which the controls were primarily directed? The procedures were investigated as follows. On each site, the checklists used by site personnel were examined and the site engineers were asked about the items they check, the purposes and activities of the on-site researcher having been explained to them. The checklists, completed by site staff, and claims about what had been checked were then compared with the actual condition of what had been checked or allegedly checked.

Items appearing on the checklists related, for example, to the use of correct bar size, correct number, properly cleaned formwork. Six items, common to all the checklists, were selected on the basis of an engineering judgement, provided by a panel of experienced engineers, that they were of special significance to the achievement of the required quality in the particular feature of construction under consideration. The six items were also claimed to have been checked by those interviewed. On each site, a sample of structural elements (minimum of five) were monitored throughout the construction process (as described earlier) and the degree of conformance of each item was established by comparing them with the design drawings, standard recommendations and specifications. Each site was given 'quality performance' scores of 1, 2, 3 or 4, for each of the six items as evidenced in the sample of structural elements that was monitored on that site. Scores were given on the following basis:

1. The item is generally properly carried out in all the structural elements in the sample, monitored by the researcher, according to specifications and method statements;

2. The item is generally properly carried out on most of the elements in the sample but not properly on one or two of them;

3. The item is generally not properly carried out in most or all of the structural elements in the sample;

4. The item is not properly carried out and likely to impair to a serious extent all the structural elements in the sample.

2.3 *Results*

The 'quality performance' score for each item, as it appeared in the sample monitored for each site is shown in Table 1. Entries in the Table thus provide a basis for comparing the relative effectiveness of the procedures intended to control operative-controllable defects.

As shown in Table 1, it was found that 57% of the items that were claimed to have

Table 1. 'Quality performance' scores for six items.

Site No.	Item					
	1	2	3	4	5	6
1	3	4	3	1	1	1
2	2	4	1	1	2	1
3	2	2	2	1	1	1
4	2	2	2	2	1	2
5	1	2	1	1	1	1
6	2	2	3	1	1	1
7	1	4	4	3	2	1
8	1	2	1	2	1	1
9	1	1	1	1	1	1
10	3	4	4	4	2	1
11	1	4	1	1	1	1
12	1	4	1	1	1	1

been checked scored 1, that is, were properly constructed according to the specification. The incidence of different kinds of defects varied from site to site though there were some common patterns. Except for Site 9, Item 2 showed a high incidence of defects. In contrast, Item 6 was found to be correct on all the sites except Site 4. For Item 1 and 3, the score of 1 was achieved on 60% of the sites, For Item 4, 66%, and for Item 4, 75%.

Of the defective items on all sites, 55% of them scored 2, that is, the items were generally correct on most of the structural elements monitored on a given site but defective on one or two of them; 16% were of class 3, that is, most of the items were not carried out properly and 29% were of class 4, that is, the items were not carried out properly on all structural elements monitored.

2.4 *Discussion*

Remembering that the items monitored were considered, from an engineering point of view, to be of high significance in the achievement of the required quality, these findings indicate that the checking systems falls some way short of achieving its intended purpose. Many defects, it would seem, remain undetected by site engineers even if the checking is carried out as thoroughly as they deem reasonable or are capable of effecting. It also appears that engineers make arbitrary choices, from a strictly engineering point of view, about what to check and what not to. In sum, about a third of the defective items fell seriously short of the requirements which is, we suggest, a significant percentage.

Why did it happen? From the interviews, it would appear that 10% to 30% of site engineers' time is allocated for checking the work on site. From the findings this would seem to be insufficient. The findings also indicate that on some sites, there are different emphases regarding how limited time should be spent, some concentrating on this item and some on that. This was evident on sites 11 and 12, where Item 2 does not seem to have been checked compared with other items which were more effectively checked. In contrast, on sites 1, 7 and 10, 50% of the items that were supposed to have been checked were either not checked or the checking procedure was not properly carried out. A possible explanation, inferred from what was said during the interviews, is that the site engineers may settle for less than the required standards in favour of other priorities such as time or cost. Furthermore, the site engineers interviewed on seven of the sites did not consider the achievement of the particular feature of construction under investigation to be a problem that requires special attention despite the fact, as noted at the outset, that there is considerable engineering research evidence to suggest that it does.

Of the operatives and working foremen interviewed, many showed great awareness of the structural significance of the feature in question. When asked why it had not been achieved on particular occasions, typical answers were: 'They (contractors' personnel) don't trust us; they treat us as pig-ignorant. Why should we bother?', 'They're paid to check us', 'We're screwed down to a price (inference: what we are paid is unfair. Of course, we cut corners!)', 'Even when we point things out they don't listen'.

2.5 *Conclusions*

If the quality control adopted on site allows 43% of the items to fall short of the required quality standard, it would seem that something is severely amiss. Certainly, a quality problem is recognised, hence all the attention it is attracting. However, findings suggest that control of quality lies exclusively with site engineers and management and the client's representative and not with the operatives who do the work. This practice accords exactly with the principles of scientific management enunciated by F.W. Taylor one hundred years ago, where a distinct separation between supervisors (controlling) and workers (doing) was recommended.

The questions are asked; given the mass of evidence accumulated in this period and the modest addition that this study makes to the evidence, that it does not work:

1. Why does this approach to the control of quality persist?;

2. Why isn't a prevention rather than detection mode adopted consistent with the QA principle which, at least, the ten QA'ed firms in the sample are supposed to be following?

Before proceeding with an attempt to answer these questions, I wish to draw attention to a feature of the way the research just reported was conducted, and, indeed, reported, that is, within the rationalist paradigm. I will then develop a fuller critique of the rationalist paradigm contrasting it with three others and will suggest, in each case, how the above questions concerning resistance to change would be answered.

3 AUTO-CRITIQUE

Categorization of defects into the four types of cause, as described above, was carried out by an expert panel comprising four structural or civil engineers. It is to be noted that while the panel members were able to arrive at a consensus, though not without lengthy discussion in some instances, the procedure was suspect in a number of crucial respects.

Firstly, there is a vast difference between a group of people making technical decisions in the abstract with no real involvement in the situation, where no real consequences, especially contractual and financial ones, attend them and, in contrast, people making such decisions whilst being very much aware of possible consequences.

Secondly, even to the extent that they were able to empathise, there is the problem of indexicality where the meaning of something is necessarily provided in the context in which it occurs. To 'remedy indexicality' (Garfinkle & Sachs 1970) they had imaginatively to reconstruct the situations by drawing on their own experience.

Thirdly, accepting that they were able, on matters of pure engineering judgement, to accomplish this without too much distortion, i.e. were able to identify defects, categorize them, allot scores and identify the immediate antecedent 'technical' causes of the defects, however, to establish the sequence of decisions and events which had led to the technical defects and causes presented a different order of problem. Reconstructing what had happened took the panel far beyond purely engineering considerations into the realms of organizational practice, contractual liability, human motive and so on.

The panel's task, then, was to reconstruct what had happened on the basis of the field researcher's measurements, observations and reported interviews/discussions with participants. For example, a typical event needing to be evaluated was the construction of a reinforced concrete lift shaft on a building project. The reinforcement box was constructed with the appropriate placement of spacers. The form fixers (who provided the plywood for the forms) had difficulty fixing the form because the box was slightly over size (designers' fault? supplier's fault? Contractor's fault for not checking?). In the process they knocked off most of the spacers in order to fit it in place. The result did not meet the specifications. Poor workmanship, ignorance or carelessness on the part of the operatives? There was no apparent consultation between or attempt to co-ordinate the steel fixers and form fixers. It was also observed that although the site agent saw the event which led directly to the defect no corrective action was taken.

It was found later through interviews with site staff that achieving the specifications with regard to the particular feature under investigation was not seen as a priority. Greater priority was given to finishing the job in as short a time as possible. This was attributed to the fact that site staff were under pressure from head office and that, anyway, they didn't have enough engineers. Further, it was found that steel fixers were employed on an individual basis as labour-only and that far from there being consultation between the two trades, there was no possibility of it. By the time the form fixers started work the steel fixers had left for another site. Finally, it was alleged that even if it had been the right size it was a 'damned stupid' design which would have been difficult/impossible to comply with.

This is a familiar enough story and equally familiar, no doubt, is that depending on whom one talks to, different accounts, explanations and rationalizations are offered. People blame each other; accusations are made about incompetence, laziness, dishonesty and so on.

While it was possible at times to verify the factual accuracy of what a respondent claimed had occurred, why it had occurred and whose responsibility it was supposed to be, for much else of the time, there was no way in which the accuracy or truth of a given account could reliably be established. Besides, the whole texture of informal arrangements and tacit agreements that characterize a workplace and particularly a construction project were largely unknown to the field researcher and entirely un-known to the panel.

The troublesome questions raised, therefore, are: how far could viable assess-ments of these matters be made without the in-depth, situated knowledge of those actively engaged and what, in fact, was the status of the panel's judgement? What kind of fact was established? Did consensus guarantee truth? Did the panel's un-doubted competence and experience as engineers increase their credibility as moral adjudicators?

I conclude that this was not just a research problem but, firstly, one intrinsic to the situation where it is inevitable that there will be a range of explanations as to why something had happened and where there is no single, objective or neutral account to be had.

Secondly, accounts and explanations of events are provided for specific purposes within specific conventions and that they are only judged accurate or true 'for all practical purposes' (Schultz 1967) within the situation in which they are called for.

Thirdly, the situation in which they are called for in this case is one where the re-search sponsors want unambiguous findings expressed in the rationalist terms that they have learned to expect, from which specific remedies to the problems identified can be formulated.

Fourthly, there is what Becker (1970) has called a 'hierarchy of credibility' in which certain kinds of information and certain kinds of informant are more likely to be believed than others. I take it that the judgement of the engineering panel is likely to be believed because they are engineers, members of an occupation who, through their skills in the application of instrumental rationality, have played a central role in creating the technology and institutions of the construction industry.

Fifthly, engineers' roles and their self-conception in UK construction has been so shaped that they either distance themselves from the practicalities of construction or align their interests with the contractors who employ them. Either way, to expect them to provide an 'objective' account of a process in which members of their pro-fession are variously involved as active participants seems to be, at least, highly questionable.

In short, the engineering panel, in all conscience, tried to provide an objective ac-count of what happened and why, so as, thereby, to be in a position to propose remedies. However, the procedure was inevitably selective and adopted an engineer-ing perspective. Answers to the questions regarding diagnosis and remedies therefore are biased by this perspective and effectively foreclose possible alternatives.

4 PRACTICAL IMPLICATIONS OF THE RESEARCH DILEMMA

At this point during the course of our research, although I knew of the work of many of the quality gurus like Deming and Juran and was aware of their arguments con-cerning 'self-control' (Juran & Gryna 1993), providing the workforce with the means to effect their own control (Deming 1986) and so on, I was introduced to Womack et al.'s *The Machine that Changed the World*. In particular, I was struck by the de-scription of Ohno's insight regarding the consequences of engineering-dominated, mass-production thinking (Womack et al. 1990). This thinking results in a production system where no reference is made to those who operate it apart from how they have to be controlled in order to fit it. Only the supervisor was authorized to stop the production line, on the assumptions that the system as a whole could only be over-seen from a synoptic point of view and that if the workers were allowed to stop it they would do so for various personal and nefarious reasons. Recognising how disas-trously inefficient this was, with errors piling upon errors till the system boundary was reached, Ohno introduced a facility for stopping the line at each work station so that problems could be remedied as they occurred. He went further, instituting the Five Why's so that workers could trace back each error to its ultimate cause.

It seemed to me that there was a clear parallel here to the research problem de-scribed above and the way we tried to deal with it. No engineer or panel of engineers were in a position to adjudicate on why defects occurred. It required the situated knowledge of those actually on the spot. Was our research procedure, therefore, simply an endorsement of the engineering mode of thinking whose negative conse-quences Ohno had so astutely noticed?

I provisionally conclude, firstly, that indeed it was. The research procedure was perfectly respectable within the rationalist paradigm. So long as due care is taken ac-cording to its conventions, the researcher is expected to establish the facts of what happened. That is what research is supposed to be for. However, it was critically flawed since it could not, in fact, address the question of what was going on and, consequently, directed attention away from a conclusion like Ohno's, which, signifi-cantly, involved a great reduction in the number of engineers required and who now worked alongside operatives as part of a collaborative team.

Secondly, I conclude that though Womack's report of the Ohno example is highly suggestive, we need to know much more about the circumstances in which it oc-curred before making inferences about practice. For example, was the pursuit of causes so firmly directed at the technical, as is implied, or were there instances of buck-passing, covering of backs, denials of alleged causes and so on, such as were present in the cases our research considered? If so, how were they adjudicated? Given the probability that many answers are provided in the mass of research carried out by the Womack team and point, say, to a highly cohesive corporate culture (Dore 1973) where there is great incentive to look only for technical causes or where fin-ger-pointing is untainted by the suspicion of personal or political motive, there re-mains the problem of explaining the existence and evolution of such a culture. How far can it be manufactured or grafted on to cultures with different traditions, expecta-tions about work, democracy, individual freedom? (White & Trevor 1983). One un-derstands, for example, that there is a less acceptable face to Japanese industry where dissent is met with intimidation, and harmony masks resentment and jealousy of fel-low workers (Junkerman 1982). Thus, thirdly, I conclude that the style of research used to produce answers to such questions directly influences conclusions and per-ceived practical options.

5 CHOICE OF RESEARCH PARADIGM

What, then, are the alternatives? What styles of research are there that might reveal a different, perhaps fuller, more accurate picture and therefore choice of alternatives? I will describe four in broad outline, at the risk of caricaturing them, suggesting how, in each case, they might be expected to answer the questions posed above concerning reasons and remedies for the situation as described in the study. The first is the rationalist, whose methods and assumptions, as I have said, have been taken for our research. The second is what I call the culturalist, third, the institutionalist, and fourth, the interpretive.

5.1 *Rationalist*

The rationalist approach derives from the assumption made in the natural sciences that objective reality can be described from the neutral observation of it. Using this assumption, the intention is to describe social systems, as, for example, a construction firm or a project, as though they were natural systems operating according to some inherent logic that research attempts to reveal. In this attempt, a key question is: how are various parts or functions of the system inter-related? Then, to the extent that this is accurately answered, it becomes possible to ask the practical question how can it be better co-ordinated? This is readily translated into: how does the firm or project work and how can it be better managed or controlled? Given that the approach postulates an objective reality 'out there', distinct from the observer as neutral scientist, it further becomes possible to distinguish (sub)systems within the system and a conventional distinction has become that between the technical and social or human.

In construction, two so-called systems that have received a great deal of attention are the technical and the contractual. On the technical front, the problem is defined as the need to articulate in exact and comprehensive detail the intrinsic technical logic of the construction process. The attempt is to construct a blueprint of how the process can be made to work most efficiently. The blueprint is intended to provide management with a control tool in two senses. It provides detailed description of exactly what goes where, when and how, against which implementation can be checked and the necessary remedial action taken. In the more ambitious versions of this approach, the intention is so perfectly to hone the tool that nothing does go wrong, very much as one might engineer the manufacture of a motor car. Secondly, it acts as a device to legitimate management control. Offered as the pure expression of instrumental rationality it brooks no objection. Since the approach makes an essential separation between technical system and human system, separates, as Scarbrough and Corbett (1992) put it 'the dancers from the dance' (that is, despite choreographers dances do not exist without dancers) the intention is to perfect the (abstract) technical system and then fit the human system to it. Answers within this framework to the questions posed (why does checking persist and why not alternatives?) are that there is nothing wrong with trying to control quality through checking since engineering the system is an essentially esoteric activity and that there is no reason to expect those subject to it to understand it. If the system designer works hard enough he can and should aim to make it 'idiot-proof'. Thus, the search for 'efficient' checking techniques and tools is still very much part of the agenda. (see, for example, Parsons (1972), Chase & Federle, (1992)).

This brief description is offered in a fairly stark way and though proponents of it make concessions concerning the completeness of the system which they can construct and the social or human systems that the technical system must accommodate to, nonetheless, a great many commentators in construction management research seem to hold this view. In its extreme form we may call it the 'Engineering Fix'. It holds that, given the vagaries of human nature and the uncertainties of human conduct, it is possible, in principle, to understand and describe the inherent characteristics of the technical system. With more and more sophisticated tools to help, it becomes possible to get at least *that* right. This image of the fully articulated cognitive map of the technical system seems still to be a holy grail for many researchers.

If concern on the technical front is to devise a rational system describing what should happen, on the contractual front the concern is with devising ways of controlling relationships so as to increase the chances that what should happen will happen. As is frequently noted, contract in construction assumes special significance because of the size, cost, complexity, uncertainty, vulnerability, prototype nature and so on, of many projects and, therefore, the necessary involvement of many parties experiencing different levels and kinds of risk. A central issue becomes; how should rights and responsibilities be allocated or, in other words, who legitimately controls whom? The point of reference seems to be the idealized, unitary organization with a centralized source of authority and control, more or less completely responsible for designing and manufacturing an identifiable product and offering it to the market. The problem is conceived of in terms of how, through contractual obligation, the many parties to a construction project can be similarly tied to the common undertaking. Examples of participants in this perennial debate are Lombard (1975) and Breakwell (1985) who argue that it should be the contractor who is in control; Isaak (1982) who argues for the design professionals and Bubshait & Al-Musaid (1992) who say that it should be the client. The comment is well taken that any analysis of the shortcomings of the industry from Banwell (1964) to Sir Michael Latham's most recent contribution (Latham 1994), looks to new forms of procurement contract for a solution. While, clearly, contract is important, the comment here is that research on contract in construction can easily become dominated by the rationalist mode of thinking where answers to our questions about resistance to change centre on tools of control.

I suggest that the point to note about this explanatory framework, in its two variants, is that it is a *constituent* of the current situation; it represents a crucial ingredient of the ideology which underlies practice and contributes to the situation which we are trying to explain. From within this framework, answers to the questions: Why does the practice persist and what are the remedies? one would expect them to be: because of incomplete control systems, inadequate supervision or weak management (technical) and contracts which do not adequately specify requirements and obligations and are not adequately enforced. Thus, the remedies are tighter more elaborate regulatory technical systems and tighter contractual regulation.

5.2 *Culturalist approach*

The culturalist approach recognises the weaknesses of the technical system-human system split. Developing the insights of, for example, Simon (1947), March & Simon (1958) and Weick (1979), the emphasis is shifted to the subjective experience of participants.

The approach recognises and erects as its cornerstone the principle that people think and feel, need meaning in their lives and work and have far more to offer than they are conventionally allowed to. They must be empowered, given practical autonomy, but empowered to act in ways which promote the interests of the corporation. Herein lies the crucial dilemma of control.

Now while, in my view, the culturalists are right to reject the 'Engineering Fix' of the rationalists, and to look to the affective domain as the key for improvements in quality, efficiency and productivity, they risk treating this domain quite as instrumentally as do the rationalists. In other words, the general goal and the general good, as implied by such general terms as improving quality, efficiency and productivity, become particular goals and specific objectives, as defined and set by corporate management. A strong corporate culture is seen as an instrument to these ends. The intention is that employees will identify with corporate goals, gain satisfaction from trying to achieve them, give willingly of themselves, their hearts as well as their minds, in return for the sense of community that sharing of the corporate values brings. As Peters and Waterman put it, it gives them meaning as well as money (Peters & Waterman 1982).

In short, culture is undoubtedly important but how far is it possible or morally defensible to control it, indeed manufacture it, in the pursuit of specific objectives? There are two questions here regarding the dangers which, in my view, are present in the culturalist approach. Firstly, if you intend to change people's values and beliefs, is it not necessary to know what they currently are? This raises methodological issues. Secondly, is it intrinsically manipulative, and therefore wrong, to set out to change people's values and beliefs in order to secure one's own definitions of what is good or desirable? This raises moral issues. As to the first, in the concern with meaning, the culturalist approach draws heavily on the Interpretive tradition (see below) but with the notable difference that it shelves methodological and, indeed, moral concerns in just the same way as does the rationalist approach, while the interpretive approach, in contrast, continues to recognise them as problematic.

For, as noted above, the rationalist approach claims that neutrality can be assured by observing certain methodological routines based on those of the natural sciences. The culturalists adopt a similar standpoint. The interpretive approach recognises that the standpoint and values of the researcher easily become the occasion for bias and special pleading and must be kept continually in view. Subjectivity cannot entirely be remedied if for no other reason than that the very capacity to understand the meaning of others is contingent upon being a social being with values and beliefs of one's own. (Weber 1966). However, the researcher can go a long way in remedying it and thus increase the accuracy of his/her account. Because, although meaning is indexical i.e. what is said or done requires a context with which to interpret it, it is not indeterminate. Given the necessary contextual information, the correct interpretation of meaning is usually unproblematic 'for all practical purposes' for those involved. So too, given knowledge of the practical purposes and the necessary contextual information, a correct interpretation is available to the researcher.

What prevents this from happening, I suggest, is that the practical purposes of participants are simply assumed without reference to those they have in fact, and in the worst forms of culturalism, ignored or rejected. There is both a methodological and moral dimension to this. Methodologically, there is failure to achieve an accurate interpretation of what is going on which, I believe, has practical implications for realizing their own agendas. It becomes a moral and, indeed, political issue insofar as the failure to achieve this understanding is the result of seeing the purposes of others as an obstacle to one's own.

Insofar as the culturalists put themselves in the service of the corporation and take as unquestioned givens the corporation's objectives of, say, profitability and competitiveness, even though they may justify this endorsement on the grounds that successful corporations promote the general or even global good, then they obscure the substance of what they need to know from their own purely practical point of view.

It seems to me that the culturalists compromise the development of a methodology which might genuinely concern itself with meaning and which may well, therefore, lead to improvements in quality, productivity and efficiency (which are not necessarily the same as profitability), because they are not willing to face up to, for some, the unpalatable and unavoidable egalitarian implications of such a methodology. The methodology, in other words, rests on the principle that essentially every point of view is valid.

To see people's attitudes, priorities etc. as merely obstacles to the achievement of a cultural consensus around an already decided set of values is to 'ironicise' those attitudes and marginalize them, to exclude vital information about what needs to be understood. The most fully articulated versions of the interpretive perspective rely upon 'bracketing' the researcher's attitudes just as the natural sciences do. The difference is, however, that the latter do it in the pursuit of 'objectivity' and the former to achieve a fuller understanding of subjective understandings other than one's own.

Even if it is possible to offer plausible reasons for the need to change these viewpoints, you must in the first place know what they are. Vaill (1975) very persuasively argues the consequences for organizational change initiatives that ignore or misconstrue the perspectives of those involved. Any changes achieved are skin deep, short lived and invariably fail.

In summary, it seem to me that three questions must be asked of the culturalist approach:

1. *Methodological.* Does the drive to build strong corporate cultures and change people thereby lead to a failure to see what is actually going on; to see what one chooses to see, quite as much, as I have tried to show, the approach adopted in my own study initially prevented us from seeing what is going on?;

2. *Practical.* Does the strong corporate culture ideal actually work? There is already much evidence to suggest that, to the extent that it is developed, it results in ritualism, fulfilling public expectations and manipulating images but retreating into private worlds behind the masquerade. Its effects can be quite the opposite of that supposed by the likes of Peters and Waterman. Instead of individualism, dynamism and innovation, there results totalitarianism, conformism, the absence of diversity and the very fear of failure that they, along with Deming, say has to be driven out;

3. *Moral.* Beside the negative practical consequences of trying to manufacture culture, are there absolute grounds on which to reject manipulation and totalitarianism?

Applying the culturalist viewpoint to construction, the diagnosis of what is wrong and remedies for it centre on the quality of relationships and the need to change confrontational attitudes, encourage trust and teamwork which will make the work both more satisfying and financially rewarding for all involved. The emphasis is on strong positive leadership, frequently involving a plea to clients to set the example, which will demonstrate the practical benefits of more collaborative forms of working. In this way, attitudes will be changed and people coaxed into rejecting the old ways.

5.3 *Institutional approach*

This approach treats the experiential reasons given in the reported study as symptoms of much more radical causes built into the whole structure of relationships that have developed in construction and which, if the explanation is being offered by a Marxist, are particular expressions of more global contradictions and conflicts. The concern is to 'situate' the phenomenon in question in its historical context. So, for example, according to Ball (1988) the entire contracting system as it developed from the early 19th Century in Britain, was instigated by a few large contractors with the connivance of powerful clients and the emerging professions, like architecture and civil engineering. According to this analysis, in pursuit of greater profits they took control of the labour process forcing economically independent craftsmen into the role of wage labour. The professions acceded to these developments because they achieved considerable status and economic benefit as design specialists and honest brokers. A consequence of this 'carve-up' – the separation of design and implementation, was an inefficient system which stifled technological innovation and more efficient and productive working practices at the expense of labour, clients and society as a whole.

The essence of this kind of research is to explain cultural phenomena in 'relation to its conditions of possibility' (Willmott 1993). That is, in addition to trying to establish what a phenomenon consists of experientially for those involved, to ask: why here and why now? With respect to corporate culturalism, Willmott (1993) suggests: '... the move towards more flexible forms of accumulation, the Japanese challenge, a resurgence of economic neo-liberalism and the reassertion of managerial prerogative in the governance of employee values'.

The 'Quality Movement', so this kind of analysis would go, is the latest attempt designed to shore up an essentially illogical system, and that not surprisingly it is treated with suspicion by everybody, apart from those who see it as a means of retaining their control or who see a role for themselves in effecting this form of control. In light of this analysis, remedies are described in terms of restructuring the entire contracting system. The impact on practice of Marxist versions of such analysis have been minimal and not surprisingly the remedies offered have be roundly rejected.

However, recently this perspective, with a non-marxian agenda, has been brought to bear on the 'Quality Movement', which has largely remained the domain of the gurus and their committed followers. My view is that the institutionalist approach poses some serious questions that must be answered. Amongst these is a fuller understanding of the circumstances in which, for example, the lean production methods described by Womack took root. Plausible suggestions include the fact that post-war Japan was significantly feudal in many of its institutions and had not fully embraced Cartesian forms of Western rationality and, indeed, rejected much of Western individualism. Extensive concessions were made to labour in order to defeat militant trades unionism. The fact that it was able to rely on distinctive forms of reciprocity and collectivism in the development of its technologies and that attempts to transplant them have had variable success has forced attention to the fact that technology is itself a cultural phenomenon, a fact that the rationalists have difficulty in accepting.

As to the culturalists, who do indeed accept it, they are in equal danger, given their confidence in being able to manufacture corporate culture, of making equivalent mistakes to those of the rationalists. This to say that the rationalists sought to control employee behaviour treating them as one-dimensional utilitarian decision-makers. While the rationalists reacted to, and to some extent accommodated to, the fact that people are moved by needs and desires that cannot be reduced to simple instrumental rationality, the culturalist seek proactively to control the non-rational. They seek to control feelings, emotions and value. As Willmott (1993) puts it, they seek to colonize the affective domain as well as the instrumental.

The institutional approach provides a healthy counterbalance to this colonizing intent. For example, in his comparison of lean systems with the Saturn and Uddevalla projects, Rehder (1994) writes:

'While the (Womack) study presents a most convincing case for the comparative advantages of the Japanese lean system over traditional mass production and management system, it does not give equal attention to the perceived significant human costs and retaliated political and environmental problems which many union members, employees and scholars in North America, Europe and Japan attribute to it'.

There needs to be, he suggests, an on-going debate about the trade-off between a manufacturing organization's efficiency and the quality of work life. In order that this debate should not be dominated by the interests of powerful corporations the institutional and interpretive perspectives provide a research base from which evidence germane to this debate can be gathered.

5.4 *Interpretive approach*

In the fourth, the interpretive approach, the researcher's aim is to report the perspective or worldview of the participants in particular settings. The concern is with how and why *they* do things as they do. It does not presume to report any single, overarching truth but assumes that any particular report or account of how and why things happen is produced for particular purposes, audiences and circumstances and follows the conventions of those circumstances. This assumption holds as much for the accounts of a researcher as the people s/he is researching. The accounts that the researcher offers are intended to reveal how people construct their world, what meaning they attach to things and events. As Vaill (1975) puts it s/he is concerned with what people *think* they are doing, not what they may be said to be doing from any, claimed neutral standpoint. What people think they are doing, so this viewpoint holds, is the very crux of what is meant by culture, so any talk of changing or adjusting to it must recognise this fact. Culture is, first and foremost, a method used by people with reference to other people (Eglin 1980). You can treat it as some organic totality or system as the culturalists do, and try to gain control of it as they try to, but the fallacies and dangers associated with this attempt have been noted.

Thus, the interpretive approach may observe, for example, how, in construction, engineers assert or fail to assert the right to make decisions on engineering quality, the circumstances in which and the conditions on which this right is legitimated or contested, on the grounds, for example, of individual competence or experience, how that competence should be assessed and so on. They may note that engineers, through their various institutions, have tried, more or less successfully, to regulate entry into the profession, control standards of practice and the criteria concerning what an engineer is expected to know and be able to do; the fact that engineers are conscious that there is a formal aspect to their knowledge, which is set out in text-books, codes and the like, but that there are also indefinable aspects that they talk about in terms of skill, flair, habits of mind, intuition or experience. They may note that some engineers experience this as a source of frustration; on one hand they are expected to be upholders of pure instrumental rationality, dealers in pure objective fact, but on the other hand, have moral qualms about what they do, are conscious of uncertainty and experience doubt just like anybody else.

The interpretive researcher is interested to know how they resolve these dilemmas and what effects this has in their dealings with others. The researcher may also report the view that there are different kinds of engineer, that there are, for example, those who think up 'smart solutions' to practical problems that 'don't have a cat's chance in hell' of getting off the ground because they don't take into account day-to-day realities; that there are those engineers 'on the other side of the contractual fence' who provide unworkable details, or, expressing the other viewpoint, are always looking for reasons for claims and extras.

Equally, the approach may be applied to steelfixers or formwork carpenters, in which case, engineers would appear as a feature in their landscape, given authority over them which, they feel, is often undeserved, and thus, to be dealt with, be given instructions by, negotiated with and so on. The answers to the questions (diagnoses and remedies?) would take at face value the existential force of what participants say. A number of such explanation were noted in the foregoing study. Head office re-quires site engineers to carry out checking in a certain way. Head office also requires them to do other things. They make a trade off. This is a fact that it is the re-searcher's task to reveal for the very reason that this is the stuff of their attitudes and culture, which, the simple diagnosis has it, needs to be changed. As noted above, if you do not take such things into account efforts to effect change 'backfire'.

5.5 Summary

At the further risk of distortion through oversimplification, Table 2 indicates the main differences and tendencies of the approaches identified. The horizontal axis concerns the domain of interest, that is, with either the societal, system or macro level – the 'objective', or with the meanings that processes, events, etc., have for people involved in them – the 'subjective'. In the former, the aim is to provide ex-

Table 2. Differences and tendencies in research approaches.

	Applied	Critical
Objective	Rationalist	Institutionalist
Subjective	Culturalist	Interpretive

planations or accounts of the circumstances which can be shown to impact particular processes or events; it tries to describe general patterns and dynamics that underlie and may be used to explain them. The other is concerned to describe and sometimes explain the relationship between values and beliefs on one hand and patterns of conduct and human institutions on the other. On the vertical axis, the distinction is between the assumed purposes of enquiry and the ethical justification for it. The 'applied' justification is that there is practical benefit to be obtained insofar as immanent causes can be established. The 'critical' justification is that understanding and trying to report the truth is a value that needs no further justification.

These methodological alternatives affect the knowledge we accumulate about construction practice and the inferences we make about how to improve that practice. In particular, the suggestion is made that: 1) The rationalist approach is an extremely potent force in the industry; that it is rooted in a mind-set, set of attitudes, worldview, explanatory framework, ideology, call it what you will, and that to challenge it and the diagnoses and remedies that it contains, is to challenge the role of those who uphold it, and; 2) The culturalist approach is rightly concerned with the role of human meaning but that it is uncritical of its own methods and assumptions. Thus, 3) Research into the vitally important role of human meaning will benefit from the contribution that the interpretive and institutional approaches can make to our knowledge and understanding of it.

6 CONCLUSIONS

The arguments or, perhaps, the doubts expressed in this chapter have turned on a biographical experience -reading Womack's book, a key text in the development of lean construction praxis, at a juncture when I was wrestling with a methodological problem. While the book deeply impressed me it also worried me for reasons I do not fully understand. This chapter is, in a sense, an exploration of these doubts.

I have tried to explain my thinking with reference to a research problem and the Ohno example discussed above. It begins with the importance of situated knowledge for both research and practice. For practice, Ohno's conclusion was that a production method should be devised which located the rights to diagnosis and remedy at the point of production. The implications for research are that someone without that situated knowledge is in no position to offer either diagnosis or remedy.

The question then arose: How do research and practice relate? My answer is the one given by Max Weber (1919) many years ago: they are crucially different. The objectives of practitioners, for example, quality, efficiency, productivity or profits, cannot be taken to be self-evident by the researcher. An essential purpose of research is to establish what participants in the situation under study, managers, engineers or steelfixers, mean by them and what values and beliefs underlie such meanings. The researcher may well share the meanings of the participants but s/he has, so to speak, to hold them at arms length, and there are no simple recipes for achieving this. In describing the interpretive and institutional perspectives I have briefly tried to indicate how this can be done and have suggested that the key to such a method is that the researcher take seriously and at face value the worldviews of the people concerned and does not attempt to explain them away according to some presumed over-arching, 'real' version of the 'facts', Again, the principle was stated many years ago, this time by W.I Thomas: 'If men define situations as real they are real in their consequences'(Thomas 1964).

The problem that I see with both the rationalist and culturalist approaches is that in pursuit of this over-arching view they do not relativize their own position but hold that, given the use of certain methodological routines, either their own values are irrelevant or essentially benign or self-evidently in tune with what any 'right-thinking' person would hold. In my view, this lack of self-consciousness results in a failure to relativize their own knowledge, producing methodologically unsound research and therefore findings which are radically incomplete and, therefore, any action programmes based on them of questionable practical value.

Thus, while my conclusions essentially concern methodological issues as the basis for establishing practical and useable findings, at the centre of the methodological issues there seems also to be a moral one. This is that researchers must take seriously the culture and attitudes of others and not see them as simply needing to be changed in order to accommodate their own.

Finally, it seems to me that Deming's career can be seen as a struggle to come to terms with these issues and thus provides a highly significant comment on them. A lukewarm response to his ideas at home, a warm reception in Japan, then, given the reputation his success had brought, an initial clamouring for his ideas and an attempt to appropriate them, then once again, a cooling off as the radicalness of his ideas became apparent. Deming's increasing attention to the fact of organization as human process, conflict and contradiction and, in all this, the centrality of meaning and ways of thinking, were simply not possible to absorb or appeared too threatening to Western conceptions of organization and management's role in it. This has been attributed to the power of vested interest and, while I agree with this diagnosis, I would add that there is a more subtle, deep-rooted reason. This is that the very methods we use to investigate and understand organization dispose us to certain conclusions about them. I have suggested that we must ask genuinely empirical questions about what is going on in situations we wish to understand and, particularly, about those admired innovations we wish to learn from.

Thus, looked at from the perspective of the rationalists and the culturalists who both hold to the effective primacy of a single rationality, Deming's Fourteen Points are a series of contradictions. The fundamental propositions are not a set of procedures or algorithms; they do not offer a clear set of instructions. They are, rather, a set of heuristics expressed as paradoxes. Thus, for those brought up with the ideology of the control loop where objectives are clearly stated, methods monitored and shortfalls remedied, albeit the loop is never closed, constancy of purpose set against a philosophy of change are difficult to reconcile. So too are *obsession* with quality and a scientific approach. And so it continues. Is pride in workmanship consistent with the collective pressures of continually refining procedures? If we encourage education do we mean indoctrination and if not, education in what, for what? How do we drive out fear and break down barriers at the same time?

In short, Deming, it seems to me, was forced to address the multiple, perceived realities that constitute organizations and the intrinsic contradictions that this view reveals. His advice is in terms of paradoxes because that is all it sensibly can be. How far Deming was himself constitutionally capable of pursuing the research agenda to which his experience pointed it is not necessary to speculate. I conclude that his later American experience caused him, at least, to widen his perception of what was at issue; to look beyond the more operational aspects of his work which Japanese receptivity to them had not required him to do and to reconsider exactly what it was about the Japanese situation that had led to this receptivity. The simplistic answer to all this, given by many of the followers of Deming and the other gurus, is that it lies in culture which can be managed and controlled just as it was once believed that organizations could be controlled like rationally designed machines.

ACKNOWLEDGEMENTS

Parts of this chapter have been published in Shammas-Toma, M., Seymour, D.E. and Clark, L.A., 'The Effectiveness of Formal Quality Control Systems in Controlling Structural Quality', in Proceedings of ARCOM Annual Conference, Loughborough, September, 1994. I am grateful to John Rooke who provided valuable comments on an earlier draft of this chapter.

REFERENCES

Ball, M. 1988. *Rebuilding Construction*. Routledge, London.
Becker, H. 1970. *Sociological Work*. Aldine Publishing Co. Chicago.
Breakwell, N. 1985. Applying Quality Assurance to Civil Engineering Work on the Sellafield Site. *Quality Assurance* 11(3).
Bubsait, A.A. & Al-Musaid, A.A. 1992. Owner Involvement in Construction projects in Saudi Arabia. *ASCE Journal of Management in Engineering* 8(2).
Chase, G.W. & Federle, M.O. 1992. Implementation of TQM in Building Design and Construction. *ASCE Journal of Management in Engineering* 8(4).
Deming, W.E. 1986. *Out of Crisis*. MIT Centre for Advanced Engineering Study: Cambridge, Mass.
Dore, R.P. 1973. *Japanese Society*. Sussex Publications, Devizes.
Eglin, P. 1980. Culture as method: Location as an Interactional Device. *Journal of Pragmatics* No 4.
Garfinkle, H. & Sachs, H. 1970. On formal structures of Practical Actions. In: McKinney, J, & Tiryakin, E. (eds), *Theoretical Sociology*. Appleton-Century-Crofts, New York.
Isaak, M. 1982. Contractor Quality Control: An Evaluation. *Proceedings of the American Society of Civil Engineers, Journal of the Construction Division*, Vol. 108.
Junkerman, J. 1982. We are Driven. In: *Mother Jones*. August.
Juran, J.M. & Gryna, F.M. 1993. *Quality Planning and Analysis*. 3rd Edition McGraw Hill, New York.
Latham, M. 1994. *Constructing the Team*. HMSO, London.
Lombard, M.A. 1975. A Contractor's Need for Quality and Economy Assurance. *ACI Journal*, Title No. 72-22.
March, J.G. & Simon, S.G. 1958. *Organizations*. Wiley, New York.
Parsons, R.M. 1972. System of Control for Construction Quality. *Proceedings of the American Society of Civil Engineers, Journal of the Construction Division*, Vol. 98.
Peters, T. & Waterman, R.H. 1982. *In Search of Excellence*. Harper and Row, New York.
Rehder, R.R. 1994. Saturn, Uddevalla and the Japanese Lean Systems: Paradoxical Prototypes for the Twenty-first Century. *British Journal of Management*.
Scarbrough, H. & Corbett, J.M. 1992. *Technology and Organization*. Routledge, London.

456 *D. Seymour*

Schutz, A. 1967. *The Phenomenology of the Social World.* North-western University Press, Chicago.

Simon, S.G. 1947. *Administrative Behaviour.* Wiley, New York.

Thomas, W.I., quoted in Coser, L. and Rosenberg, B. 1964. *Sociological Theory.* Macmillan, New York, page 232.

Vaill, P.B. 1975. Practice Theories in Organization Development. In: Adams, J.D. (ed.), *New Technologies in Organization Development.* University Associates Inc., La Jolla, Cal.

Weber, M. 1919. 'Politik als Beruf', Duncker and Humblodt, Munich, published as 'Politics as a Profession'. In: Gerth, H.H. & Mills, C.W. (1962), *From Max Weber.* Galaxy Books, New York.

Weber, M. 1966. *The Theory of Social and Economic Organization.* The Free Press, New York.

Weick, K. 1979. *The Social Psychology of Organizing.* Addison-Wesley, Reading, MA.

White, M. & Trevor, M. 1983. *Under Japanese Management.* Heinemann, London.

Willmott, H. 1993. Strength is Ignorance: Slavery is Freedom: Managing Culture in Modern Organizations. *Journal of Management Studies* 30(4).

Womack, J.P., Jones, D.T. & Roos, D. *The Machine that Changed the World.* Rawson Associates, New York.

Quality assurance and partnering: A lean partnership*

IAN M. EILENBERG
Construction Management Unit, Department of Building & Construction Economics, RMIT University, Melbourne, Australia

ABSTRACT: Over the years there have been many systems introduced to improve various aspects of the construction industry. In most cases these have been aimed at curing a problem, rather than preventing it in the first place. Two management systems have been developed over recent years which are aimed at removing the causation of problems, especially disputes, before they arise. One development is 'Quality Assurance', which focuses on providing a series of steps to ensure that the item being assured is done correctly and at the best possible level, the first time.

The other system is that of 'Partnering' which was developed to minimise the number of disputes arising, rather than trying to resolve them once the disagreement has arisen.

A combination of these separate systems, if adopted by the construction industry can produce a better quality product as well as a better and more amicable working environment both on and off site.

1 INTRODUCTION

In the 1950's and 1960's 'systems engineering' was the preferred management tool for streamlining company operations. The method of implementation was to employ a 'consultant' to analyse the company, the way it worked and to provide a 'system' for each aspect. The results tended to be a piece of paper for everything and eventually a store room full of unused paper. The problem with the system was that the number of forms that had to be filled in by various people became so great that the person responsible for checking these documents became overwhelmed by the paperwork, and simply found it all 'too hard'.

Added to this was the cost of printing. At that time business did not have photocopy machines suitable for bulk copying and all the forms had to be printed, usually by external companies. Any changes to the form as a result of business growth, meant that any unused forms became scrap. The author worked for a medium sized construction company in the late 1960's which had over fifty such forms, yet hardly any were actually used for their original purpose. Changes to contract conditions and forms made many of the forms redundant. Change of management personnel made others useless. One set of forms had four different sheets per set, yet only the original

---

*Presented on the 3rd workshop on lean construction, Albuquerque, 1995

was still in use, the other three ending up in the rubbish bin at every use, having printed information on the back, ruling out using them for recycling as scribble pads.

Systems engineering had a serious downside. It looked at what was and decided what should be, but did not consider what might happen later. It was not dynamic and so eventually fell into disuse, at least in this arena.

2 QUALITY ASSURANCE

Whilst the formal name of the system is comparatively recent, this in itself is not a new concept, having been widely adopted by the Japanese after the second world war. Dr Deming, an American, took the concept to the Japanese who adopted it with the well known, and documented, results. The idea was also adopted by various European companies and provided the means for the great success of so many of our competitors. The picture began to change in the 1980's, very largely as a result of the huge success of the Japanese automobile companies in the lower and middle price range, in direct competition with the major American automobile manufactures, Chrysler, GM and Ford.

The origin of the formal title of 'Quality Assurance' and in particular of the ISO 9000 document dates back to the joint development between France and England of what has been described as one of the most successful products every produced, namely the Concorde aeroplane. The need for a set of control documents to be used on each side of the English (French) Channel to ensure commonality of production, led to the creation of a set of documents setting out the why, how and when, that formed the basis for ISO 9000.

In Australia (as well as many other places in the world) implementation of quality assurance documentation conforming to ISO 9000, is now generally a necessity before a company can tender for government work. This applies as much to the construction industry (if not more so) as to any sector of industry.

Not all quality assurance implementation is well done. However many lessons have been learnt since the first quality manuals were written in the late 1980's. The 'golden rule' quoted by quality assurance consultants is that a good system as documented in the quality manual, should use what the company already has, not invent new ways of doing things. Improvements will come out of this by review and by the very nature of the quality assurance system. This is a case of 'less is more'.

One outcome from of the companies that have adopted quality assurance has been an improvement in communications within the company. By necessity departments within a company now have to be able to relate to ensure that what they do relates correctly to what others will do with what you produce. In effect implementing the quality circle, often without being aware of what they are actually doing.

The down side of the 'Quality Assurance' is that whilst the system itself is excellent, the emphasis is on the system, not on the objective of the system. Each step is detailed with its objectives and reporting system clearly defined. The argument is that people implementing quality systems place too much emphasis on the 'system' and insuring that the manual is followed, and not enough on the quality of the actual product being produced - in this case the building. The 'system' and in particular the recording itself has become paramount rather than the product.

3 DISPUTATION

Whilst internal management procedures are undergoing improvement, the construction industry is still suffering from arguments between the parties which should never have occurred. the industry.

As the author writes this chapter, two unions are on strike over which of them should have the right to sign up members working in the pre-mixed concrete plants. This is brought all construction in Melbourne to a halt. Nobody will win – the unionists won't get paid, the contractors won't be able to build, the owners won't get their buildings. Loss of revenue at all levels. The matter is before the Commonwealth Arbitration Commission and will end up with an order to return to work, whilst the matter is 'negotiated'.

At any given time there will be contractors arguing with architects about the cost of variations (change orders), head contractors with their sub-trades, clients with contractors about completion dates and so forth. Many of these arguments will end up in formal hearings – arbitration if not the courts. There will be winners – the arbitrators, the lawyers (attorneys) and the various experts who will give opposite opinions fully justifying their own stance. The losses will involve the amount of time it will take to even get to the hearing of the dispute, much less the time involved in the hearing itself. Periods of weeks are minimum, periods of years not unheard of. By then who can remember what the argument was really about in the first place? Written notes are not the same as fresh memory.

In America the cost of these disputes was becoming so large that the Corp of Engineers decided to develop a better way of doing business in an attempt to minimise the number, size and length of disputation. The system developed was termed 'Partnering'. The concept is to improve communication and attempt to overcome the 'them and us' syndrome that occurs between the players in any major construction project.

The method of achieving this is to gather all the players together in a 'retreat', with the people who will actually make the decisions and to develop a series of strategies to maximise communication between these parties. A 'charter' of aims is usually developed with the overall aims expressed. These include simple statements such as: 'To complete the project ahead of time and ahead of budget', and usually contain references to occupational health and safety; ' we aim to have an accident free project' and so forth.

However research by the author has shown that whilst the 'charter' is a useful tool it is the communications network that develops during the 'retreat' that is of most value. The people who have to make decisions at the various levels, in different companies, get to meet each other as people, not just voices on the telephone line or names on the end of a letter. The whole concept of 'partnering' depends upon the active participation of top management, not just their moral support.

The dispute resolution matrix (Table 1) is decided at the 'retreat'. It establishes the type of dispute should be handled at a given level and by whom. A time frame is also attached at each step, so that a dispute cannot be effortlessly put in the 'too hard' basket. If nothing transpires, then the dispute escalates to the next level at the expiration of the stated time.

Table 1. Issue resolution matrix – typical. Note: under each heading a specific person or persons is/are nominated.

Level	Client	Contractor	Consultant	Time frame
1st	Works manager	Foreman/Leading hand; superintendents		1/2 day
2nd	Supervising engineers	Superintendent; site engineer	Project manager	1 day
	Works manager	Project engineer; quality assurance rep		
3rd	Team leader	Project engineer Project manager	Project manager	*Construction*: 1 day *Design*: 2 days
4th	Manager construction	Project manager		2 days
5th	Project manager	Project manager; construction manager		4 days
6th	Director major projects	Area manager		

Ground rules: 1. Resolve issue at lowest level; 2. No jumping levels of authority; 3. Ignoring the issue of 'No decision' is not acceptable; 4. Unresolved issues to be escalated upwards by both parties, in a timely manner prior to causing project delays or costs; 5. Decision time frame may be extended for data gathering, by mutual agreement, of both parties.

Again research has shown that this system works well at the lower end of the management team. Those disputes with readily determined outcome and/or with minimal cost implications will be resolved. Those of a more complicated nature and involving more money will not be so readily resolved at the lower management level and will tend to still proceed to the formal resolution system preferred by the antagonists.

4 THE QUALITY DISPUTE BOARD

The aim is to maximise the benefit of each system in such a way that the result is better than if either or both systems were adopted independently. In Hong Kong on the Queen Elizabeth Hospital a series of dispute resolution systems were adopted starting with partnering and proceeding through to the dispute resolution board. This board is made up of a representative from the owner/client, the head contractor and a third party skilled in dispute resolution – often a retired legal person – judge or similar.

The use of a quality manual will improve the quality of the work, but does little for the dispute resolution process for the construction. The aim then is to extend the partnering retreat to include the quality implementation processes so that the participants not only declare to build a 'quality assured' building but have a series of steps built into the systems to ensure that any problems related to questions related to quality are resolved in the same way as any other dispute that may arise.

To achieve this a 'quality reference' is required. This will include all the currently supplied samples of the various parts of the building, as well as the assurance system itself, but will go one step further. This will be the 'quality panel'. A group of personnel who will determine what is, and what is not, acceptable. Again, like the dispute resolution board it would have a representative from the owner/clients, the contractor and a suitably qualified neutral – possible an architect or engineer specialising in construction materials and systems.

Like the disputes resolution board, the members would be kept informed as to the progress of the building, the current assurance records and of any area that may be causing some concern. Should a question arise with respect to quality – is it good enough? Does it meet the specification, if not, who is responsible. This board would meet to make an early decision and ensure that the minimum of interruption occurred on the work site itself.

Either of the boards could call for members of the other (probably the third party neutral) to attend the other board's meeting, should this assist in resolving the matter under discussion.

5 CONCLUSION

The use of the principles of 'partnering' and 'quality assurance' in a combined system will have several outcomes. It will improve the final quality of the building, a better working atmosphere between each of the participants, both on and off site and should disputes arise a system of resolution engineered to resolve them with minimal expenditure of money and time.

REFERENCES

Cowan, C. 1992. *Partnering, A Strategy of Excellence*. MBAV Workshops.
Deming, W.E. 1986. *Out of the Crisis*. Cambridge University Press, UK.
Patching, A. 1994. Partnering and Personal Skills for Project Management, APA, Australia.
Reilly, N.B. 1994. *Quality, What Makes it Happen*. Van Nostrand Reinhold, N.Y.

TQM the Nordic Way: TQMNW*

AXEL GAARSLEV
The Technical University of Denmark, Department of Construction Management, Lyngby, Denmark

1 INTRODUCTION

In 1992 a Nordic group was founded. The group consists of thirteen contractors from the Nordic countries – among these the largest firms – and four universities/research institutes, one from each of the following countries: Norway, Sweden, Finland and Denmark. The goal was to cooperate on improving the standing of both the companies and the research organizations involved.

In the last three years four main activities have been under way:
– Preparing a general development strategy for the companies involved;
– Ways to measure quality standing in the companies;
– Procedure for reengineering through benchmarking;
– Ways of measuring customers' satisfaction.

At the moment the two first activities are documented and will be reported in this paper. The last two listed are still in progress and the results of the work will probably be published in the spring of 1996.

The group is at the moment discussing the areas of cooperation for the next two years and the following topics seem to be of common interest:
– Networking/partnering;
– Environmental management.

The work is performed in a number of small, mixed groups with contractors and researchers reporting to a plenum at seminars.

2 PREPARING A GENERAL DEVELOPMENT STRATEGY FOR THE COMPANIES INVOLVED

As a starting point the top management of the companies involved were very confused about developing trends for their companies. Reading modern management literature did not really clarify the ideas. They were confronted with a number of fancy terms like: Kaizen, lean production, world class manufacturing, total quality management, double-loop learning organizations, anthropocentric management, team building etc. Difficult words and difficult to see any real substance in after studying a book or two on the topic. They felt, that it was just what they always have been do-

*Presented on the 3rd workshop on lean construction, Albuquerque, 1995

ing but now just put in a fancy term. Other more concrete terms like: Time-based management, JIT, benchmarking, reengineering etc. gave meaning but did they fit construction? – and understandable procedures for implementation were needed. So the companies' motivation for cooperation was on one hand frustration and on the other hand a feeling of obligation to jump on the band wagon. For the research organizations the motivation for cooperation primary was the chance to get access to real world data.

All the companies were at that time heavily involved in quality work so this was defined as the starting point of cooperation. In the same way total quality management was associated with the work as a meaningful scenario. The Nordic Way was used to signal that the companies involved had some kind of special feature based on location, culture etc.[1]

The first job of the group was to cope with the confusion described. To show a meaningful way from the quality work the companies were doing at that time to the fancy terms – an overall plan for a development strategy for the companies involved – from confusion to TQM (Sjøholt et al. 1995).

First of all it was decided that the strategy should be based not on revolution but on evolution. A strategy slowly building up competence on the premises at the moment in the companies. Figure 1 shows the general strategy developed.

Phase 1: Quality assurance – building the basic system
Quality assurance and quality control is aimed at satisfying clients' demands to a specific quality standard. Often the ISO 9000 series is the base. The focus is product quality. The goal is zero defects at delivery, no repair in the guaranty period and costs as planned.

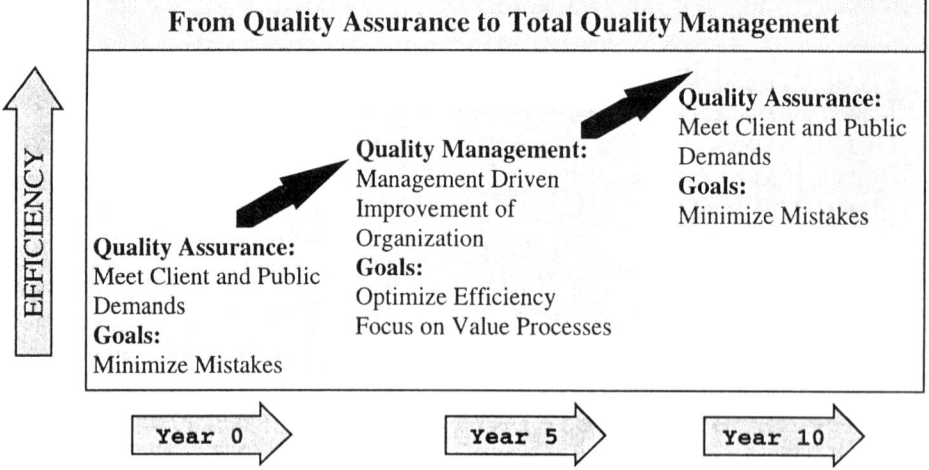

Figure 1. Three phase strategy.

---

[1] It has later been very difficult to give this idea any substance and meaning in a marked perspective.

Normally the following components are used in the basic system:
– Quality assurance ISO 9001/9002;
– Quality revision ISO 10011;
– Quality handbook ISO 10013;
– Quality plan ISO 9004-5;
– Quality certification EN 45012;
– Health and safety. EU directives. Internal control procedures;
– Environmental control ISO 14001;
– Building codes. Internal control procedures.

The feelings about certification are mixed in the group. But again it is important that the improvement work not is started just to obtain certification but really to improve quality awareness and standing in the company. Certification might be the final result but not the starting motive. Again we will focus on evolution and not on revolution.

This client motivated quality assurance system is normally the starting point for most quality work in the companies. In many ways it's a difficult start when only motivated by external demands. It's demotivating to ask people to fill in more forms to satisfy external demands only. It's in this way very important at an early stage for management to start also an internally motivated process using the resources of all employees in an ongoing improvement process – phase two.

Phase 2: Quality management – widen the scope of the basic system
The goal is efficiency through a determined management effort and with a special focus on satisfying customers' demands. The focus is not on products' quality but on the quality of the processes involved removing all no value activities.

This management driven improvement work will normally be based on the following parts:
– Quality management ISO 9004-1(-2);
– Quality improvements ISO 9004-4;
– Process cost analyses BS 6143;
– Time-focused improvement work;
– Deming's steps 1-14;
– Juran's steps 1-12;
– Measuring process efficiency.

Phase 3: Total quality management – the final stage
The strategy for improving efficiency is now based on a concept with a focus on total aspects and analyses and on new and alternative cooperation forms.

The following ingredients will normally be used:
– Measuring customers' satisfaction;
– Quality function deployment;
– Concurrent engineering;
– Time-based management;
– Logistics management;
– Partnering;
– Team building;
– Benchmarking;

– Business process reengineering;
– Measuring company total quality standing.

3 WAYS TO MEASURE QUALITY STANDING IN THE COMPANIES

The measurement of results is the driving force behind the process of improvement in a company. Planning improvement work without measuring is of no meaning. The companies in the TQMNW group found this an important issue to address (Sjøholt et al. 1994).

Planning is first of all a process. You make the best possible initial plan, you control the progress, compare the actual results with your target and produce a new plan, adjusting your old plan and producing a new one in accordance with actual progress and actual circumstances.

Secondly, the process is controlled by your objective. The 'best plan' had to be measured on a scale – a goal, a criterion. This process is illustrated in Figure 2.

In this context the interesting thing is that the goal always has been discussed on just one dimension, and furthermore on a changing dimension. This is the reason for the fragmentation we see so often in construction management.

The first topic in focus was probably budget control, then schedule, resources (men, machines etc.), etc. And the 'hot topics' today are probably product quality, risk control, technology control and coming up seems: working conditions, human resources' management, controlling use of scares natural resources and environmental impact.

In the literature the problem is always treated as an one dimension goal problem. Basically the goal changes according to society attention and fashion and the old one is forgotten in the headlines. And the management consultants can sell their old tools again just under another heading.

But this is not reflecting reality for construction planning in practice. The manager cannot forget all about schedule control because environment or something else has

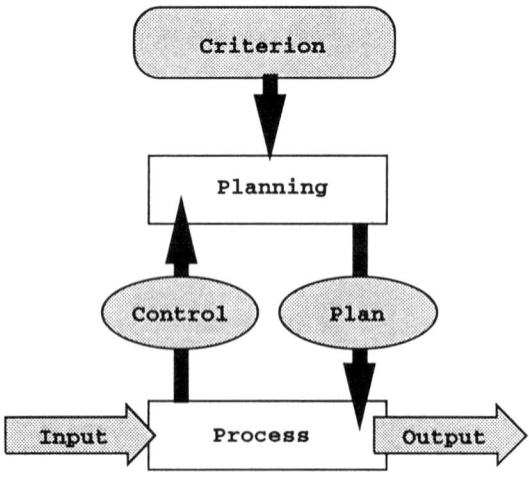

Figure 2. The planning/control process.

got all the attention. Often these goals are in conflict and he has to make the trade-off. Real life construction management do not need a specialist in schedule control or ecology, we will have a number of knowledge-based systems and other tools available for giving the right advice, but we need a very qualified person able to span the entire spectrum of problems, making the optimum trade-offs based on a solid general knowledge.

Despite this fact, the organizational structure of most construction companies and the basic managerial culture of the company does not support this point of view but show a fragmented reality without coordination. Specialization and formalization and the resulting hierarchy have created a reality far from our real needs. This is due to a number of factors as: size of organization, imitation of the stationary industry, lessons taught at traditional management schools and courses etc.

In the first and second phases of the strategy described in the previous section measuring can be meaningful on one dimension, e.g. reliability of deliveries, fulfilling of internal planning requirements, handover of dwellings free from failures, budget and schedule goals, customer satisfaction etc.

But in the third phase when the company wants to assess its own TQM level new tools are needed and Sjøholt et al. (1994) recommends the use of TQM quality award systems for this purpose. As discussed in Hyldgaard et al. (1993) the TQM philosophy was introduced to a Danish construction company (Rasmussen & Schiøtz Øst A/S, also the first ISO 9001 certified construction company in Denmark).

The goal was twofold: to improve overall control of the company and to run for the Danish quality award. For that reason the criterion from this award[2] was applied when designing the system. Figure 3 shows the criteria and the weights adopted. A system was developed on a case basis and the results and the principles of the system are shown in Figure 4.

We will recognize the nine criterions and the total score available for each item from the award system.

Two types of auditing are performed. The head office (HO) is audited and a selected number of the sites or them all. The head office score and the average site score are combined and as a result a total score of the company can be calculated and compared with the total score available equal to 1000 points[3].

As a planning and control device this concept is superior to normal procedures. All criteria for planning and control are observed in the same picture headlining priorities and trade-offs in contrast with normal procedures with separate planning and control systems for each criterion often separated in independent organizational units. Management issues are in focus. It is a valuable tool to communicate priorities and goals for the company and to measure the actual performance on site, head office and in total for the company. Doing this over a period of time furthermore gives an indication of the trend in performance. The tool can also be used for benchmarking with other companies in the trade.

---

[2] The award system used is identical to the assessment model used by the European Quality Award.

[3] It should be mentioned that the company was awarded a second price in the Danish quality award competition only surpassed by Rank Xerox.

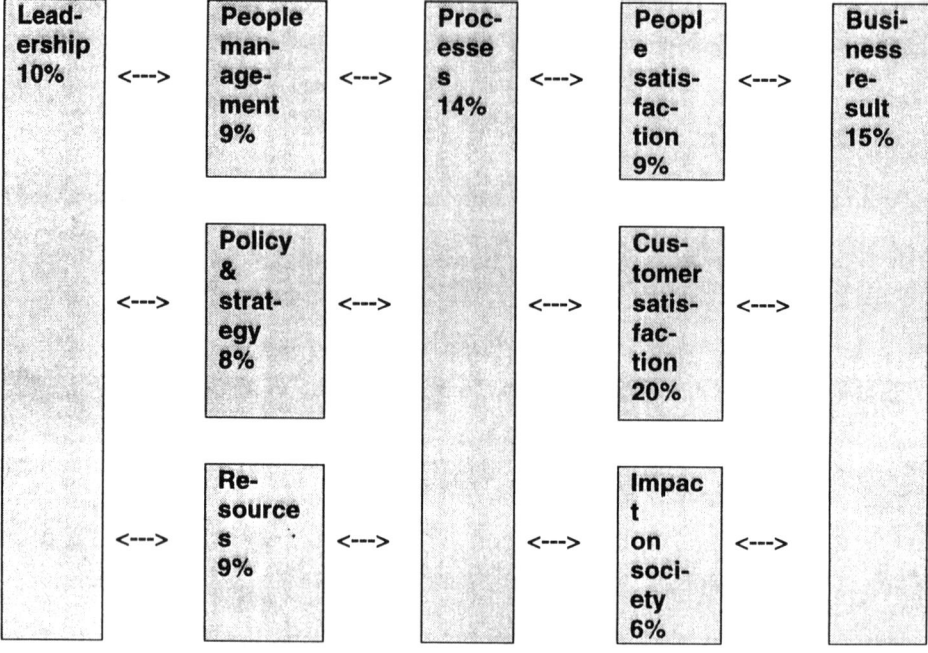

Figure 3. The European quality award assessment model.

4 CONCLUSION

There is really no conclusion to this story. The case is still in progress. So far so good. But working that close with industry has been a rewarding but often also a frustrating experience for the research institutes involved.

REFERENCES

Sjøholt, O. et al. 1995. From quality assurance to improvement management, BYGGFORSK, Oslo, Project report 06198.
Sjøholt, O. et al. 1994. Measuring the results of quality improvement work, BYGGFORSK, Oslo, Project report 0155.
Hyldgaard, U. et al. 1993. TQM in construction, Report from the Technical University of Denmark, Department of Construction Management.

I-tem	Scope	Total score avail.	Avail. score split HO/ sites	Sum HO/ sites eval.	HO score	Aver-age sites score		Sites 1...n
1	Lead-ership	100	HO1 S1	P1	% of HO1	% of S1	<-->	Item 1 s11...s1n
2	People man-age-ment	90	HO2 S2	P2	% of HO2	% of S2	<-->	Item 2 s21...s2n
3	Policy & strat-egy	80	HO3 S3	P3	% of HO3	% of S3	<-->	Item 3 s31...s3n
4	Re-sources	90	HO4 S4	P4	% of HO4	% of S4	<-->	Item 4 s41...s4n
5	Proc-esses	140	HO5 S5	P5	% of HO5	% of S5	<-->	Item 5 s51...s5n
6	People satis-fac-tion	90	HO6 S6	P6	% of HO6	% of S6	<-->	Item 6 s61...s6n
7	Cus-tomer satis-fac-tion	200	HO7 S7	P7	% of HO7	% of S7	<-->	Item 7 s71...s7n
8	Impact on soci-ety	60	HO8 S8	P8	% of HO8	% of S8	<-->	Item 8 s81...s8n
9	Busi-ness re-sults	150	HO9 S9	P9	% of HO9	% of S9	<-->	Item 9 s91...s9n
			Avail. total score 1000	Sum= act. score				

Figure 4. Example of implementing the European quality award system.

Limitations of the use of tolerances as a means of stating quality requirements in reinforced concrete*

DAVID SEYMOUR, MAZIN SHAMMAS-TOMA & LESLIE CLARK
School of Civil Engineering, University of Birmingham, Edgbaston, UK

ABSTRACT: The chapter reports an empirical study designed to establish the extent to which adequate concrete cover to reinforcement in a sample of structures was achieved. It was, found that the standards fell significantly short of those specified. Two kinds of explanation are considered to account for these findings. The first accepts as given the existing conventions for specifying quality and looks to identify the reasons for non-compliance. The second proposes that the conventions used for specifying the required cover are inappropriate to the conditions of variability and uncertainty standardly met with in construction. An alternative approach based on the concept of continuous quality improvement is described and discussed.

1 INTRODUCTION

One of the problems of achieving quality in construction is being able to specify the level of quality that meets the design requirements and in such a ways that the design is also buildable. Seymour & Low (1990) suggested that there are two schools of thought on the matter. There are those who argue that it must be made possible to state the required level of quality in exact, unequivocal and numerical terms: 'If you can't quantify it, don't specify it' (Pateman 1987). On the other hand, there are those who argue that, for many purposes, this is not possible and that required quality can only be achieved through more adaptable, collaborative and negotiative processes which allow for uncertainty and variability.

Achievement of the necessary cover to the steel reinforcement in concrete structures is vital. Inadequate cover allows the penetration of water and salts with serious negative consequences for the quality of the concrete. The assumption in design practice for specifying the required cover has been that there are consistent patterns in the distribution of cover achieved. Based on this assumption, cover surveys have been conducted and their findings used to establish values and tolerances that need to be complied with on site (Clear 1990). This chapter examines the suitability of this assumption for in situ concrete construction.

The chapter reports a study, the purpose of which was to establish how far requirements, as stated in terms of nominal cover and specified tolerances, were in fact achieved in samples of reinforced concrete structural elements on twenty-five pro-

*Presented on the 3rd workshop on lean construction, Albuquerque, 1995

jects. The results suggest that the convention of using specified tolerances is not effective. We will propose that in view of the great ambient and circumstantial variation that is met with on construction sites, alternative methods of communicating and achieving required quality levels need to be developed. The framework for such a method addresses the total design and implementation process and emphasises the importance of co-ordination in adapting the communication of design requirements to project circumstances. Ultimately, such a framework must locate construction in a much broader process governed by the client in the context of his evolving business needs and objectives. This holistic conception of the process is at odds with the current situation where a set of fixed and finite requirements is expected to be passed from client to designer, from designer to contractor, from contractor to suppliers and subcontractors. This latter conception does not match the reality of meeting client needs and focuses attention more on issues of contractual liability, thereby undermining the possibility of co-ordination and teamwork.

2 THE STUDY

The study was carried out through field investigation of twenty-four construction sites variously located throughout the UK. Projects ranged in type from buildings, water retaining structures to bridges. The principal characteristics of the projects are set out in Table 1. The sample was selected on the basis of availability since the number of projects underway at the time of the study was extremely limited. The only criteria imposed for inclusion in the sample were that a sample of about six walls or columns should be under construction; that permission to monitor construction, to measure the cover achieved and to interview those involved be obtained. Despite these limitations of the sampling procedure, we consider the projects fairly representative of current UK construction practice. The period of investigation for each site was, on average, one month.

On each of the sites the cover achieved on a sample of vertical structural elements (walls and columns) was measured to establish whether it complied with the specified values. A cover meter with an accuracy of ± 2 mm was used to measure the depth of cover to the steel reinforcement bars after concreting. The method of measurement was to scan the structural element from top to bottom and record minimum readings indicated on the cover meter as the search head was moved downwards.

On each site, at least five structural elements were measured up to a maximum of eleven. The type and number of elements available for investigation, accessibility, and time available to conduct the investigation dictated the choice and number of elements surveyed.

In addition to the survey of the physical structures, interviews were conducted with clients' representatives, site engineers, managers, foremen and operatives both during and after construction about what were doing (or did), why they were doing it, and so on. Photographs were taken of work in progress. In the initial phases of the study a camcorder was used but this practice was found to be impractical, 'off-putting' for those involved and, in the event, found to be not very useful. The practice was therefore abandoned. The purpose and reasons for the study, as outlined above, were explained to all respondents.

Table 1. Characteristics of the investigated sites.

Site no.	Type of structure	Contract type	Value £	Client	RE/CoW	Concrete work subcontracted	Location
1	Water tank	ICE	2.5 m	Private	RE	S/F & FIF	Midland
2	Building	GCIW	8 m	Public	CoW	All	Midland
3	Building	Mang. contract	1 bn	Public	RE/CoW	All*	South England
4	Building	JCT	4.5 m	Private	–	S/F & F/F	South England
5	Reservoir	D&B	1 m	Private	–	All	Midland
6	Building	D&B	17 m	Private	–	All*	Midland
7	Building	JCT	3.5 m	Private	CoW	All*	Midland
8	Water tank	ICE	0.25 m	Private	RE	All	Midland
9	Bridge	ICE	60 m	Public	REICoW	S/F &F/F	Midland
10	Building	JCT	4 m	Private	RE	All	Midland
11	Bridge	ICE	36 m	Public	RE/CoW	All*	Wales
12	Bridge	ICE	37 m	Public	RE/CoW	All*	Wales
13	Bridge	ICE	10 m	Public	RE/CoW	All*	Midland
14	Bridge	ICE	0.5 m	Public	CoW	All	Midland
15	Bridge	ICE	52 m	Public	RE/CoW	All*	North England
16	Bridge	ICE	8 m	Public	RE/CoW	All*	Midland
17	Bridge	ICE	18 m	Public	RE/CoW	S/F & F/F	Scotland
18	Bridge	ICE	23 m	Public	RE/CoW	S/F & F/F	Scotland
19	Building	JCT	8.4 m	Private	RE	All*	North England
20	Rail track works	D&B	240 m	Private	RE/CoW	S/F & F/F	North England
21	Bridge	ICE	16 m	Public	RE/CoW	SIF & F/F	Midland
22	Building	Mang. contract	46 m	Private	RE/CoW	All*	North England
23	Building	D&B	4 m	Private	RE/CoW	All*	North England
24	Sewerage treatment works	ICE	20 m	Private	RE	All	East England
25	Building	Mang. contract	22 m	Private	–	All*	West England

Key: S/F = Steel Fixer subcontractor, FIF = Form fixer subcontractor, All = S/F, FIF and concreting are subcontracted to different companies, All* = S/F, FIF and Concreting are subcontracted to the same company, JCT = Standard form of building contract, ICE = Institution of civil engineers conditions of contract, D&B = Design and build contract, GC/W = General conditions of works contract, RE = Resident engineer, CoW = Clerk of Work.

3 FINDINGS

Before presenting the findings, it is necessary to make explicit an assumption on which the use of specifications rests. The aim of designers is to convert product features into finite product characteristics. This is usually achieved by: 1) specifying a nominal value for that feature; and 2) specifying tolerances. It is expected that the tolerances will reflect the functional needs of the product and be realistic, that is, achievable in the circumstances in which the product will be produced.

A statistical analysis was carried out to establish the difference between specified nominal cover and the cover actually achieved as revealed by the cover meter readings. Specified nominal cover is defined as the design depth of concrete to the outside of all steel specified by designers.

Table 2 shows 1) the specified nominal cover for each site (column 1); 2) the number of elements measured on each site (column 2); 3) the mean and standard deviation for each of the elements on each site (column 3); 4) the mean and standard deviation for the site as a whole (column 4); 5) the minimum and maximum cover achieved on each site (columns 6 and 7). Also shown (column 5) is whether the distributions are normal (N). A few of them deviate from normal (N*) but approximate to normal sufficiently for parametric tests to be carried out.

Table 3 shows the percentage of the achieved cover deviation below and above the specified nominal cover at -5 mm, -10 mm, $+5$ mm and $+10$ mm for each site. The table shows that the specified nominal cover has not been achieved by wide margins. 69% of the achieved covers were higher than the specified cover compared with 31% lower.

It was also found that there was significant variation in the cover achieved from the nominal cover within the sample of elements on each site and significant variation between sites. Table 2 shows that the mean covers of many of the sites were much higher than the specified nominal cover and that there was also wide variations of the standard deviations.

We will now propose that there are two kinds explanation available to account for these findings which, we infer, indicate that the quality achieved falls some way short of the desirable. The first looks for causes within the existing conventions for specifying the required cover and the second queries the conventions themselves.

4 FAILURE WITHIN THE EXISTING CONVENTIONS

The standard explanation, we suggest, would distinguish design defects originating in the design office and construction defects originating in contractors' construction methods, poor workmanship, the use of incorrect materials and components and so on. Such distinctions are, for example, the basis of the NEDO (1987) study into quality in construction. The diagnostic framework it implies may be illustrated as shown in Figure 1. Thus, any given defect may be attributed to poor design: for example, the tolerances were physically impossible to achieve for some reason like a clash of bars at an intersection. The findings suggest that such design problems contributed to about 30% of the defects identified in the study (Shammas-Toma 1995). However, the buildability problem can itself be construed in two ways: as the failings of an individual designer or of the practice of the particular firm which produced the design or, on the other hand, as a failing in the conventions s/he is working within and which all design firms, by and large, use. Whilst not denying the impact of, say, inexperience, carelessness or poor checking procedures on the part of the designer and/or design firm, for the reasons set out below, we suggest that a more systemic explanation should be inspected.

Secondly, it might be that there was a failure of communication between designer and site personnel. Thus, it could have been that an individual designer's stated requirements were simply ambiguous. Alternatively, it might have been that for reasons of inexperience or carelessness, stated requirements were misinterpreted on site.

Table 2. Statistical analysis of cover readings.

Site no.	Specified nominal cover (mm)	No. of elements	Mean/SD (mm)	Site mean/SD (mm)	Min. cover (mm)	Max. cover (mm)
1	50	1	49/3.7	50/5.7	35	62
		2	47/5.8		38	70
		3	55/6.1		31	71
		4	47/6.6		36	63
		5	52/3.9		39	59
2	20	1	30/6.7	33/6.9	7	46
		2	31/6.0		15	41
		3	37/5.2		31	50
		4	31/5.9		18	47
		5	42/3.7		36	49
3	35	1	35/6.5	38/7.3	25	56
		2	38/6.1		22	51
		3	40/4.6		31	48
		4	34/7.3		15	51
		5	38/5.8		19	55
		6	38/10.9		20	59
4	20	1	22/6.1	20/6.1	4	38
		2	22/6.8		10	39
		3	21/4.5		14	32
		4	22/4.0		13	28
		5	18/3.4		14	27
		6	18/6.1		2	24
		7	16/3.9		10	24
		8	18/3.0		12	24
		9	18/4.1		10	26
		10	17/8.0		8	34
		11	20/5.0		11	36
5	40	1	48/7.0	46/6.7	35	62
		2	48/6.8		20	58
		3	44/5.3		34	55
		4	43/5.4		29	53
		5	46/7.8		21	58
6	35	1	38.11.4	41/8.2	18	62
		2	43/6.2		31	58
		3	45/8.0		28	59
		4	37/6.1		24	52
		5	41/5.3		28	57
7	30	1	35/4.1	36/14.1	10	68
		2	37/6.1		21	47
		3	35/5.9		20	46
		4	36/8.6		14	53
		5	46/9.5		36	67
		6	30/5.8		17	38
		7	35/7.1		17	46
		8	24/12.8		0	50
		9	28/6.4		18	40
		10	73/8.0		57	90
		11	42/4.0		31	49

Table 2. Continued.

Site no.	Specified nominal cover (mm)	No. of elements	Mean/SD (mm)	Site mean/SD (mm)	Min. cover (mm)	Max. cover (mm)
8	40	1	40/4.1	44/9.5	31	49
		2	37/7.1		22	52
		3	37/6.7		22	51
		4	48/7.3		36	62
		5	47/9.0		34	65
		6	46/11.0		17	80
		7	44/6.4		35	59
		8	45/11.7		27	64
		9	48/8.6		31	64
9a	40	1	44/7.2	47/6.3	21	53
		2	45/2.9		38	53
		3	46/9.1		30	59
		4	45/2.4		39	49
		5	52/4.0		42	59
9b	50	6	73/11.5	67/9.3	61	99
		7	65/3.7		59	73
		8	63/9.9		53	88
		9	67/1.7		64	70
		10	62/3.3		56	66
10	35	1	47/6.8	42/9.8	29	59
		2	45/8.8		26	62
		3	35/5.2		26	46
		4	41/8.2		20	57
		5	42/5.6		29	52
		6	34/16.7		3	61
11	50	1	50/6.1	50/9.4	44	68
		2	52/6.2		39	64
		3	45/16.8		13	74
		4	48/6.8		35	65
		5	51/6.2		36	62
		6	54/4.3		44	62
12	50	1	48/3.1	45/3.6	39	53
		2	45/4.9		35	55
		3	45/3.1		37	51
		4	45/2.6		39	52
		5	45/2.6		37	50
13	35	1	40/6.2	38/6.0	24	58
		2	41/5.2		30	55
		3	36/5.2		18	47
		4	41/5.5		32	55
		5	34/4.8		26	52
14	55	1	51/2.9	49/6.1	46	57
		2	55/3.1		49	59
		3	56/2.6		48	59
		4	43/2.8		37	47
		5	44/4.8		37	55
		6	45/3.7		40	52

Table 2. Continued.

Site no.	Specified nominal cover (mm)	No. of elements	Mean/SD (mm)	Site mean/SD (mm)	Min. cover (mm)	Max. cover (mm)
15	40	1	51/4.3	57/13.1	45	59
		2	49/3.8		42	55
		3	66/18.6		41	88
		4	63/17.4		41	94
		5	55/4.7		44	66
16	35	1	41/4.6	44/9.2	29	56
		2	47/5.8		36	58
		3	51/9.3		40	73
		4	38/8.1		13	48
		5	38/5.9		26	47
		6	56/6.2		48	67
17	35	1	50/8.8	49/8.5	35	61
		2	55/6.6		42	67
		3	47/3.3		42	54
		4	43/10.9		21	59
		5	49/5.5		41	58
18	35	1	50/9.6	59/11.7	39	83
		2	59/9.6		46	75
		3	71/9.9		51	83
		4	46/6.2		36	56
		5	57/4.6		50	66
19	30	1	31/6.4	34/12.7	18	39
		2	25/13.4		4	50
		3	42/7.0		23	53
		4	39/19.6		8	73
		5	35/5.9		25	47
21	40	1	43/5.1	52/9.6	34	53
		2	45/6.6		29	57
		3	52/3.6		46	64
		4	58/8.3		40	75
		5	61/8.6		29	72
22	40	1	45/15.1	45/16.7	10	73
		2	49/15.9		28	85
		3	37/13.4		0	61
		4	50/20.3		3	83
		5	44/15.2		14	85
23	40	1	41/8.7	43/9.3	13	58
		2	42/6.3		22	51
		3	51/9.7		34	71
		4	42/11.5		18	61
		5	43/4.5		36	57
		6	44/11.0		31	61
		7	47/7.6		35	60
24	40	1	52/7.5	51/11.9	41	70
		2	50/7.2		39	66
		3	50/12.2		33	77
		4	49/7.6		35	69
		5	70/19.6		42	103
		6	48/6.0		40	61

Table 2. Continued.

Site no.	Specified nominal cover (mm)	No. of elements	Mean/SD (mm)	Site mean/SD (mm)	Min. cover (mm)	Max. cover (mm)
25	35	1	35/6.4	37/7.6	25	56
		2	40/7.7		25	57
		3	37/7.2		24	58
		4	37/6.4		24	52
		5	37/9.3		21	69

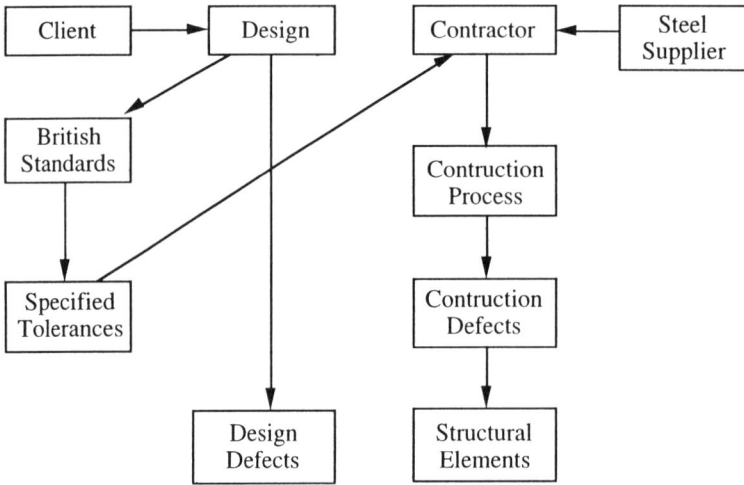

Figure 1. Relationship between defects and cover tolerances.

Table 3 shows the tolerances specified by designers compared with the tolerances that site staff said they were trying to achieve. The fact that, as shown in Table 3, all but three of the site engineers accepted and said that they worked to the tolerances given suggests that the problem is not to be located in ambiguity or communications failure. Furthermore, it might be said that, in a sense, the communication failure diagnosis is irrelevant within the present contractual arrangement since contractors are bound to work to the specifications as given.

Thirdly, and this is the most likely candidate within this kind of explanation, it might be that site personnel simply failed to achieve the tolerances because of inefficiency, poor workmanship, poor supervision, inadequate controls and so on. There is evidence to support this diagnosis. About half the defects could reasonably be ascribed to operative errors while site staff held that there was insufficient time and staff available to check and have faulty work rectified. Thus, it might be concluded that the specification convention is essentially viable, and that the failure lay with operatives guilty of poor workmanship and/or site staff who did not exercise proper supervision and checking.

Table 3. Percentage of cover achieved on site below and above the specified value.

Site no.	Cover <			Cover >		
	Spec.	Spec. −5 mm	Spec. −10 mm	Spec.	Spec. +5 mm	Spec.
1	51	20	6	49	16	3
2	5	2	0.5	95	86	70
3	36	18	5	64	22	15
4	52	20	5	48	17	9
5	22	7	3	78	57	25
6	22	9	4	78	51	30
7	36	16	10	64	48	32
8	42	15	5	58	40	29
9a	12	7	2	88	59	23
9b	0	0	0	100	98	83
10	23	13	5	77	59	42
11	57	20	13	43	30	10
12	93	54	11	7	0	0
13	38	7	2	62	40	18
14	80	59	38	20	0	0
15	0	0	0	100	61	40
16	8	6	2	92	80	50
17	10	2	1	90	82	65
18	0	0	0	100	95	92
19	38	20	17	62	50	36
21	24	16	5	76	45	39
22	40	21	17	60	45	40
23	40	19	10	60	40	20
24	12	2	0	88	79	50
25	41	18	2	59	37	18
Av.	31	15	6	69	50	33.5

Though this explanation cannot be discounted, the fact that there was a consistent pattern across all sites and contractors, most of whom were highly reputable, national firms, suggests, at the very least, that there is some systemic problem at issue. Thus, we propose to inspect the alternative explanation which is that the present convention is fundamentally flawed.

5 FAILURE OF THE SPECIFICATION CONVENTIONS

In order to consider this alternative explanation, which centres on the impact of the use of the specification conventions themselves, we need to adopt a systemic approach and reconsider the way factors identified above (design, supervision, etc.) affect the process of an element's construction and may have a negative effect on the achievement of the required cover.

Quality standards are decided at the design stage in response to the client's brief, cost limits and so on. They are passed on to the contractor who will build the structure according to the design and the quality standards set for it. The quality standards set at the design stage are intended to act as the reference point against which the construction process is controlled. The relationship may be illustrated as in Figure 1.

The factors identified in the study as having a negative impact on the cover achieved may be distinguished using the concepts developed within the analytical framework known as lean construction (Koskela 1992).These are 'resourcing' or 'flow activities' and 'conversion' or 'value-adding activities'. The latter are activities which contribute directly to the fabric of the artefact being constructed. Defects that occur here are, for example, non-alignment of rebars, inadequate ties, misplacement of spacers, poor compaction.

The former relate to the supply of all the resources necessary for the value-adding activities to be carried out and include information, materials, equipment and manpower. Thus, considered in these terms, specifications as provided in the design are a key resource. Inadequate resourcing can be traced to a number of participants in the process; the designer may produce an impractical or unbuildable reinforcement design where, for example, there is a clash of slab and column reinforcement at their intersection. Reinforcement may have been dispatched to site by suppliers wrongly bent or out of tolerance. The contractor may have adopted inappropriate methods of construction or supplied inappropriate materials, wrong-sized spacers, etc.

However, within the present conventions the only form of defect that is recognised to originate in the design process is what we have called a buildability defect where it is physically impossible to comply with the specifications. Therefore, it becomes simply a matter of contract that any other kind of defect will be the fault of the contractor.

The distinction that we are pointing out here is that between a source of error that may be located anywhere in the system as a whole (possibly with the designer) and a source of error which can only, from a contractual point of view, be located in those phases or constituents of the process for which the contractor has accepted responsibility. The effect of this is that the contractual system inhibits the necessary co-ordination within the technical system. The contractor is forced into a mode of quality control through checking, with reference to the standards provided by the designers, firstly, at the beginning of the conversion activity where the specifications are accepted and on its completion where a check is carried out with reference to the specifications. We have argued elsewhere (Shammas-Toma et al. 1994) that this is ineffective, as revealed in the resultant quality, and will argue that the necessary condition for developing an alternative strategy would be co-ordination between design and construction which would allow specifications to be set with reference to the conditions encountered on site. (see Fig. 2).

6 THE CONVENTION OF USING TOLERANCES TO SPECIFY REQUIRED COVER

On the projects investigated the standard convention of specifying nominal cover with tolerances to communicate design requirements to the site was used. By and large, the designers were happy to comply with the codes of practice except on those occasions where it was felt that there were ambiguities, in which case, an in-house convention was adopted. In all but three cases site personnel proceeded with refer-

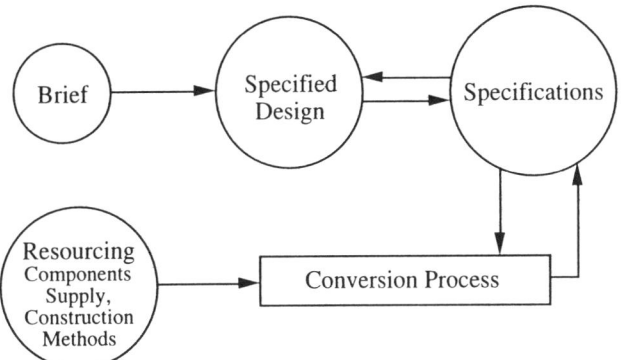

Figure 2. Information flow necessary for viability of tolerances.

Table 4. Tolerance specifications for cover.

Site no.	Designers tolerance		Site staff tolerance	
	Min. (mm)	Max. (mm)	Min. (mm)	Max. (mm)
1	–5	+10	–5	+10
2	–5	+ 5	–5	No limit
3	–5	+ 5	–5	+ 5
4	–5	0	–5	+ 5
5	–5	+ 5	–5	+ 5
6	–5	+ 5	–5	+ 5
7	–5	+ 5	–5	+ 5
8	–5	+ 5	–5	+ 5
9	–5	+20	–5	+20
10	–5	+ 5	–5	+ 5
11	–5	+20	–5	+20
12	–5	+20	–5	+20
13	–5	+20	–5	+20
14	–5	+20	–5	+20
15	–5	+20	–5	+20
16	–5	+20	–5	+20
17	–5	+20	–5	+20
18	–5	+20	–5	+20
19	–5	+ 5	–5	+ 5
21	–5	+20	–5	+20
22	–5	+ 5	–5	+ 5
23	–5	+ 5	–5	+ 5
24	–5	+10	–5	+10
25	–5	0	–5	+ 5

ence to them. As shown in Table 4 the cover tolerances specified for bridge structures (Nos. 9, 11, 12, etc.) gave a positive tolerance greater than that given for the others which were building and water retaining structures (20 mm as opposed to 5 mm). All were given the same (5 mm) minimum tolerance.

Asked about the procedure adopted for choosing a tolerance, the consultant engineers interviewed said that each contract has its own specifications. The company has a set of master specifications, based on the relevant codes of practice, which is continually revised in the light of feedback from site and the development of new techniques. This up-dating procedure is carried out by a small, standing committee in each design company. The practice implies two things. Firstly, each project is viewed as unique and the tolerances given variously differ from the master. Secondly, it implies that although based on the same codes of practice, companies differ in the master they use.

The question raised is: what is the rationale or reasoning that designers apply when they depart from the codes? Consultants interviewed confirmed that they were satisfied with the −5 mm tolerance given in BS 81 10 for durability and complied with it. There was some dissatisfaction regarding maximum cover, the suggestion offered being that the wording in the relevant clauses in BS 8110 is ambiguous. Some, therefore, felt that it is appropriate to exercise their judgement. One consultant (site 6) said, for example, that he chose a +5 mm tolerance and inserted a specific clause in consideration of the control of the lever arm to the reinforcement. The clause that he inserted stated: 'The effective depth shall not be reduced by more than 2.5% subject to reduction of 15 mm in any member'. He said that this clause had been inserted following a training day which he had attended and at which the issue had been discussed.

We do not know the substance of this discussion. However, in general terms, we suggest that the major concern of a designer, within the limits of the contractual arrangement, is to try to ensure that the specifications given will not be misinterpreted. The accepted way to do this is to specify with reference to the codes. Where specifications depart from those of the codes, the reasoning is based on any number of unverifiable assumptions that may be made about the circumstances that will arise, the capabilities of the people who will interpret the specifications and so on. This may be referred to as 'closed-system' reasoning in the sense that for any particular project only very general assumptions can be made about 'downstream' activities in the project. Effectively, they are treated as constants that do not enter the system or frame of reference within which the designer works. In such discussions as the one referred to above and in the general acquisition and sharing of experience that goes on amongst any professional group, the feedback of problems met with in the past and solutions devised for them provide a reservoir of general knowledge that may be used. It is this use which is recognised in the notion of the professional discretion that the professional is expected to exercise.

Another way of putting this is that consideration is given to the general case and not the specific because the designer is in no position to know, and, anyway, contractually that is not his/her job. Thus, general rules, distilled from particular cases encountered on any number of past projects, are sought that will be applicable to particular cases. The approach is non-holistic in the sense that the particular cases, which form the basis for formulating the general rule, are abstracted from the contexts in which they occurred. The significance of this point will be taken up below when we propose an approach to managing quality more suited to the variable conditions of construction.

7 LIMITATIONS OF THE USE OF TOLERANCES

The view will now be elaborated that the first set of diagnoses are inadequate and that the problem lies with the form of specification itself. That is, though it might be possible to give designers better general training so that they do not commit buildability errors and/or rewrite the codes with the view to eliminating ambiguity or tighten up on checking procedures on site, the problem is one of an inappropriate conception of the entire process where circumstances largely unknown to the designer are excluded from consideration. In other words, the preconditions which would justify the use of tolerances to specify requirements are absent in construction so long as it produces in situ, 'one-off products subject to the ambient variations of different sites and has to meet the unique requirements of different clients.

An attempt is made to address some features of this issue in BS 5606 (1990). In recognition of the distinctive features of the construction process, whilst preserving the logic of specification through the use of tolerances, it states that tolerances should be specified which reflect the following:

1. The dimensional tolerance needs of the design;
2. The particular requirements of the various elements and components of the construction.

According to these criteria, the tolerances for concrete cover are to be determined from the assessment of combined deviation limits. In other words, the criteria are intended to take into account the deviations that occur at the various stages in the construction of an element. Deviations may occur as follows:

1. The accuracy of setting out and positioning the kicker. This was specified on some sites as 'the position of structural elements shall be accurate to ±3 mm';
2. The accuracy of the thickness of the structural element was stated in many specifications to be within ±4 mm;
3. The plumbness of the structural elements was stated to be within ±6 mm per storey;
4. The bending of reinforcement is recommended to be ±5 mm for closed dimensions.

Many respondents expressed their concern about the incompatibility of the cover tolerances with other construction tolerances of the structural element;

'Over-tightening of bolts of the forms that is within tolerance but thinner than the specified thickness of the wall. We can over-tighten the bolts by 5 mm or something like that and this reduces the cover'.

'We have tolerances on the kicker. We set out the kicker separately from the starter bars. We sometimes end up cranking bars to get the cover on the kicker'.

BS5606 provides an equation to calculate the combined tolerance of the total deviation of a specific design or interface;

$$DL_t = \sqrt{(DL_1)^2 + (DL_2)^2 + ... + (DL_n)^2}$$

where DL_t = Total deviation limited, DL_n = Component deviation limits.

Using this equation to calculate the cover tolerance, the tolerance that needs to be applied is ±10 mm for vertical structural elements. However, this does not take into account the movement of the shutters or reinforcement during placement and compaction of the concrete. Further, the tolerance will change depending on the changes of the individual tolerances of the component features introduced in particular contract specifications. Theoretically, then, the construction process should comply with a cover tolerance of ±10 mm for vertical structural elements, assuming that it is physically possible for the design to be erected within the tolerances specified.

8 THE APPROPRIATENESS OF USING TOLERANCES

We take the above to be a special instance of a yet more fundamental problem. As stated by Juran & Gryna (1993) it is that two criteria must be met for a specified constant tolerance to be used as an effective quality standard. The two criteria are:

1. Product repetitiveness, that is, the production of identical products;
2. Tolerances must be based on the variability of the process. This requires the use of statistical analysis for the designer to quantify variability.

To the extent that construction uses in situ production methods, the first criterion is not satisfied. Structural elements differ in type, shape, size, design complexity and so on. This makes the application of a constant tolerance value as a quality standard for all these structural elements within a project impractical.

There is then the possibility of specifying a fixed tolerance for each identical group of structural elements in a project assuming that the designer would have the necessary field data to specify such a tolerance. This would require the establishment of a degree of consistency in the construction process which would be difficult to meet.

The second criterion was considered in the study in relation to cover depth as revealed in the survey of elements described above. The mean and standard deviation of the cover readings for each structural element on each site were calculated. The structural elements for each site were then grouped according to their size, reinforcement design, shape and the gang of operatives who constructed them. Parametric tests were applied to establish the homogeneity (consistency of cover readings) among identical structural elements on each site. The test used to compare the homogeneity of means is the One-Way Analysis of Variance (Paradine 1970), and the Bartlett test (Wetherill 1981) was used for the standard deviations. The assumption was made that the cover-meter readings' means and standard deviation for identical structural elements were equal for each site. The tests were intended to establish the significance level at which the assumption was validated. This would give an indication of the extent of variability in the construction process of identical structural elements on each site by keeping all other factors (shape, size, reinforcement design, operatives involved) constant.

The findings suggest that there is variation in the consistency of the process of constructing identical elements. Thus, being able to determine a constant specified value that would reflect the variations in the construction process would seem highly unlikely. It is concluded that the two criteria proposed by Juran and Gryna are not satisfied, given the present and normal circumstances of construction.

Firstly, the manufacturing analogy which is implied in the use of tolerances is intrinsically inappropriate for most of construction activity, though some aspects may better fit the analogy – off-site manufacture of components, industrialized housing constructed in tightly controlled conditions, for example. In the main, however, 'Construction is best conceived as a product development process, extending from product design through process design to facility (the manufacturing process tool) construction, the end result of which is readiness for manufacturing' (Ballard 1994).

Ballard is here describing the extreme case of developing a manufacturing facility, however, we suggest that this analogy better captures the critical, developmental and uncertain features of constructing a 'one-off facility, whether an airport terminal, bridge or water retaining structure, than the manufacturing analogy upon which the present specification convention is largely based.

Secondly, given this developmental nature, the process is heavily labour intensive. Prototypes are produced in variable locations and subject to the evolving needs and requirements of clients. These are not the conditions of manufacturing where machines perform the conversion process on a throughput to predetermined standards and where outputs can be set and regulated in a controlled manner. In these circumstances, a machine can be monitored, calibrated and adjusted to achieve the required specifications of a run of items. In construction, it is, by and large, operatives who directly perform the conversion process – fixing rebars, placing concrete and so on. Certainly, these processes can be much more effectively routinized but to do so it requires the establishment of a basis on which this can be effected. Insofar as the present conventions direct attention to remedying sporadic defects, or, in other words, 'fire fighting', little attention is directed at identifying chronic defects and therefore laying down more appropriate criteria against which a process of continuous improvements can be built into the system.

In short, in conditions of uncertainty and variability, any quality control process must look to managing uncertainty as a necessary precondition for improving the efficiency of conversion processes. It is suggested, that the current convention, featuring specified tolerances, supply unrealistic standards for the conversion process and so no benchmark against which improvement can be made.

9 QUALITY IMPROVEMENT

The alternative approach to ensuring quality is based on a rejection of the closed system logic which, it is suggested, characterizes the current approach to achieving the required quality. To repeat, the conventional approach aims to secure compliance to a stated measured quantity, specified by a designer, on the assumption that corrective action will be taken to what is characterized as a sporadic occurrence, e.g. replacing a wrong size reinforcement bar with a correct one. In contrast, the concept of quality improvement is concerned with addressing long-standing or chronic problems which are frequently accepted as inevitable or as an unavoidable feature of 'the system'.

Quality improvement thus rests on the principle of tracing the origin of these systemic or chronic problems and remedying them and/or of reducing their impact on the conversion or value-adding activities. In order to do so the theory of variation needs to be understood. The theory is described with reference to Figure 3. The figure illustrates three typical situations. In Situation A the process is in statistical control. This means that variations are the result of a multitude of unassignable causes, so tolerances are set in order to accommodate them. Such is the situation standardly to be found in a machine production line set-up. In Situation B, occasional variations

Situation A

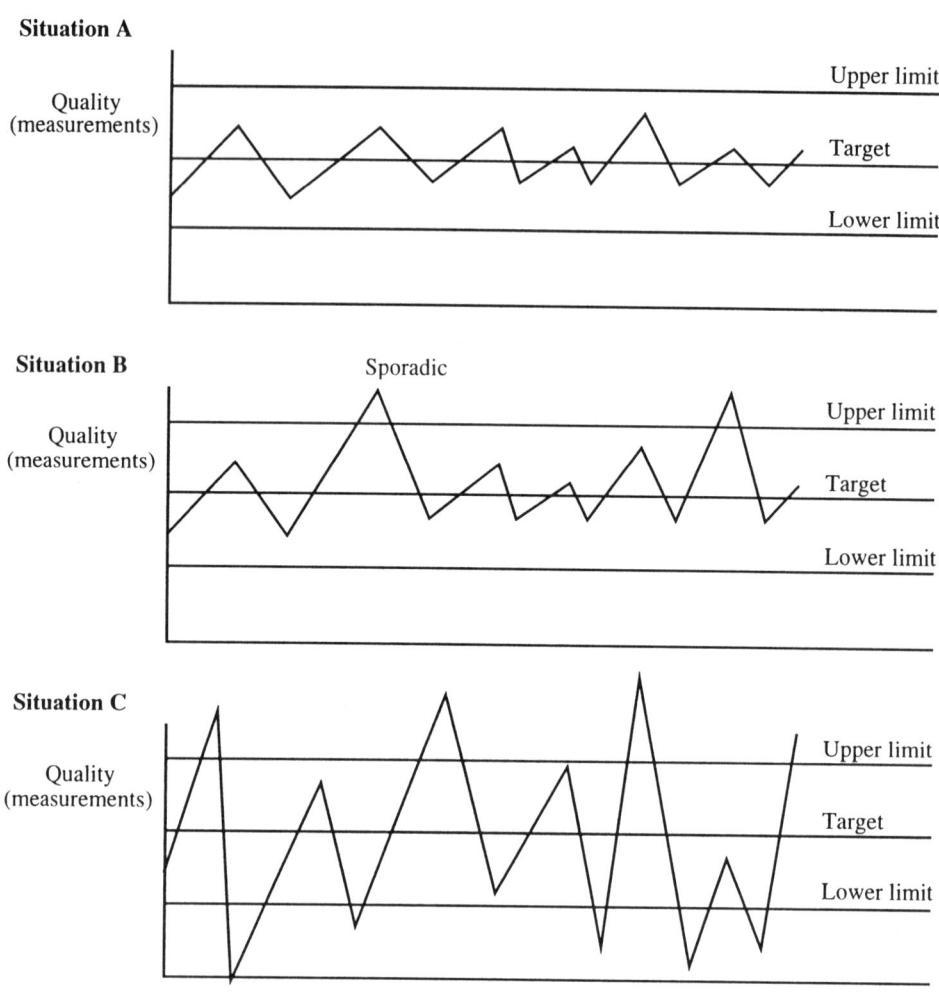

Figure 3. Sporadic and chronic defects.

may be traced to specific, sporadic or assignable causes. So for example, on a production line, an oil leak develops. The fault may then be identified and eliminated. However, in Situation C, the frequency of variation from the tolerances set is such that the tolerances are failing to act as a standard against which variations can be controlled. In these circumstances it is necessary to establish the reasons for this variation and not treat each variation as a 'one-off and deal with it as though it were.

The situation that the present research reveals is that of Situation C. Defects are being treated as sporadic when their frequency, in fact, shows that many of them are chronic. Thus, for example, a defect which occurs as the result of using an item of damaged formwork in the context of a generally acceptable standard being achieved, identifies the defect as sporadic and the faulty item can be replaced. In contrast, where the tolerances are regularly not being achieved, it points to the need to revise unrealistic specifications.

The practical implications of this are, firstly, that the aim could be to bring the process under statistical control. This was the conclusion of earlier thinking in the development of lean construction theory. However, as important as this aim may be ultimately, a prior concern is to manage the uncertainties (variable site conditions etc.) of construction. In fact, Ballard (1994) goes so far as to suggest that variation in construction has a quite different significance from what it has in manufacturing – the model against which the early lean construction thinking was developed.

'... variation has a different significance in construction than it does in manufacturing. In manufacturing, variation is to be eliminated because the standards against which variation is measured are themselves invariant. In construction conceived as a product development process, we are producing those standards. Variation cannot be eliminated, but variation can be managed' (Ballard 1994).

Ballard and Howell have developed several methods and techniques showing how this can be done. These methods and techniques have relevance far beyond the scope of the present chapter, being concerned with all aspects of resourcing value-adding or conversion processes.

One is the principle of the 'last planner' which will be briefly considered here to illustrate the deficiencies of the use of tolerances. The last planner is the person whose 'planning process is not a directive for a lower level planning process, but results in production' (Ballard & Howell 1994). The last planner makes a commitment that the planned work can actually be done rather than, as is standardly the case, accepting a work assignment in the hope that it can be done. To the extent that the work done falls short of what is planned, as usually happens, the plan is revised.

The principle is illustrated in Figure 4 (ibid.). The figure shows the complete planning process beginning with the project objectives. At each stage, initially formulated objectives (*should*) are adjusted to what is actually achievable (*can*). On the basis of this, a firm commitment (*will*) can realistically be made. In turn this firm

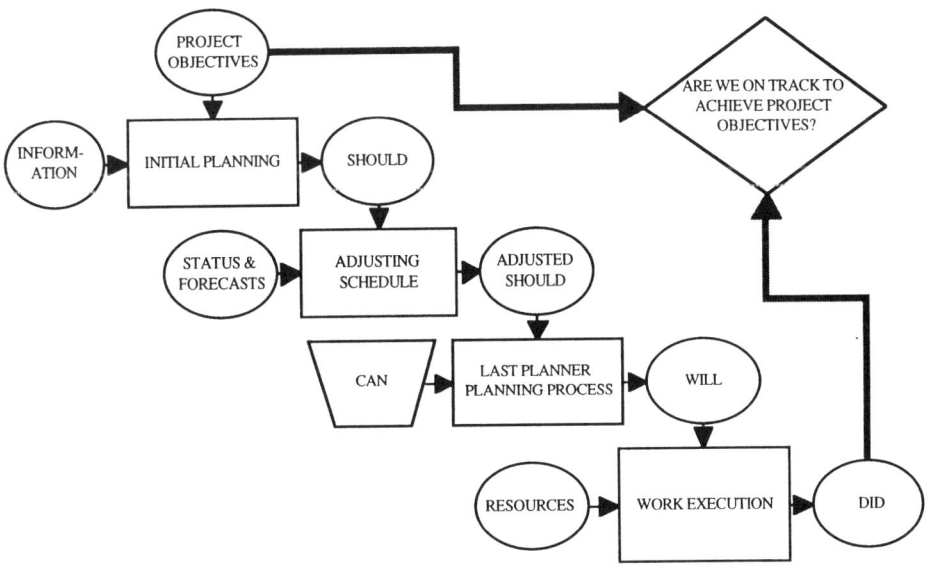

Figure 4. Measuring and improving performance (reproduced from Ballard & Howell, 1994).

commitment can provide a benchmark against which what was actually achieved (*did*) can be measured. This lays the ground for identifying the reasons for shortfalls in performance and remedying them.

With respect to the present findings, the tolerances provided are in the nature of '*shoulds*'. Without reference to the circumstances in which they are supposed to be achieved, there is no way in which to establish performance criteria on the basis of which improvements can be made.

To develop an approach in construction which facilitates continuous improvement, faces many problems. Principally, because the construction process standardly involves many separate parties, it requires a level of co-ordination and collaboration underlain by a spirit of mutual trust not commonly found. It is therefore one thing to trace the origin of chronic defects and another to get them remedied. However, what we have suggested here is that the current convention of specifying requirements is unsuited to the variable and developmental nature of construction. These conventions imply the need for a level of certainty for the most part unattainable in construction though it is the formal expectation of certainty which acts as a kind of fiction on which the traditional contractual system is based. It will be seen below that a number of practitioners do recognise the limitations and are prepared to act in the spirit of collaboration but it is also to be noted that present contractual considerations do not favour it.

It is beyond the scope of this chapter to consider all those relational and contractual issues, admirably described by Latham (1993, 1994), which centre on this issue, but, clearly, the need to provide a contractual framework and procurement procedures which encourage and reward collaboration is vital. However, we also see the danger that the emphasis on contract can easily draw attention away from the spirit of collaboration which new forms of contract may well help to foster but will certainly not guarantee so long as the intrinsic limits of exact specification are not recognised.

10 PROFESSIONAL DISCRETION

A major concern of this chapter has been to show how one vital resource to the value-adding process of constructing a concrete element -the level of cover required- is standardly assumed to be capable of being communicated in absolute and non-negotiable terms. We have argued that in the normal conditions of construction such an assumption is not warranted. We suggest, therefore, that while it is recognised that some features of relationships in the construction process are subject to ambiguity and that there is, therefore, need for greater collaboration in order to resolve them equitably, even with regard to a material like concrete, whose characteristics are subject to finite engineering criteria, there is still need for collaboration and mutual adjustment. The suggestion is therefore made for the need for professional discretion in order to address this issue.

The present conventions specifying required quality already acknowledge two principles associated with professionalism. The first is the attempt to ground practice in rational technique and to seek formulae to which precise values may be attached and which can be expressed in standardized codes. The second lies in the notion of discretion whereby the limits of precision and codification are recognised and the need for judgement to be exercised in response to particular cases is acknowledged. Over reliance on the former, it is suggested, leads to the closed system mode of specification, as described, and encourages a control mentality where it is assumed that finite measures can be provided, against which a series of checking procedures can be carried out to ensure compliance. The alternative is to give more scope to the discretionary aspect of professionalism.

The research revealed that the two models of professionalism, implied above, do co-exist. However, within the current contractual frameworks and financial climate, the scope for REs to exercise discretion in the interest of collaboration and quality is constrained. Thus, on the one hand there were those REs who tended to the former model: 'There is no compromise when it comes to achieving cover. We [REs] have specified a certain tolerance and the contractor has to comply to that. This is what he had agreed when he tendered for the job'.

And those who tended towards the latter; 'The specifications should not be read and taken as an absolute bible. We [REs] have got to apply common sense to them. The actual site conditions may preclude the achievement of the specifications.'

While the intransigent attitude, expressed in the first quote, might be justified if the specified cover were achieved, the evidence, however, suggests that it is not.

It is also to be noted that there was some evidence to suggest that where contractors' formal quality control system complied with BS 5750 (ISO 9000) it had a detrimental effect on the quality of the relationship between the REs and site management with a tendency for the RE to adopt the non-compromising view illustrated above.

Discussions with respondents suggest that two broad strategies are envisaged for the future. Increasing competition and the need to reduce overheads are interpreted to mean a reduction in management time available for quality management. A solution is looked for in tighter more elaborate models and control systems made possible by sophisticated information processing hardware. The alternative would be the recognition that the capacity to control, in this bureaucratic sense, is limited and therefore an acceptance of the need to delegate control to site, foster on-site collaboration and pursue a strategy of continuous improvement. On the evidence of this study, perceived competitive pressures make the latter seem only a distant possibility.

11 CONCLUSIONS

It is suggested that the conventions for achieving quality in the construction process, involving, as it does, many autonomous parties, have evolved on the expectation that a balance needs to be achieved between efforts to specify construction requirements exactly at the design stage and the need to apply professional discretion in the light of circumstances as they emerge. However, because of the increasing technical complexity and scale of many projects and particularly because of the commercial pressures that all participants are experiencing, these conventions are under considerable strain. We have noted a tendency for the specifications to be taken at face value with control systems directed at trying to achieve compliance through control systems based upon checking. We have also noted that this approach is failing to deliver the necessary quality.

To overcome these failings it is proposed that, consistent with the precepts of continuous quality improvement, attention be directed at ways of preventing defects, rather than, as at present, relying on detection through checking as the means of remedying them.

It is recommended that every effort be made to develop the collaborative approach. This will require acceptance, in the first place, of the fundamental principles of quality improvement where it is stressed that the reasons for a defect's occurrence must be understood. In this chapter the basic distinction between sporadic and chronic defects has been used to argue that current methods of specifying the required quality make it difficult to address chronic defects. It has been suggested that the principal reason for the inadequacies of the present systems is the lack of coordination between design and construction. The process is conceived of in terms of a series of discrete phases where the method of communicating requirements 'downstream' is unsuited to the reciprocal interdependence characteristic of the uncertain, developmental nature of construction. In this chapter we have focused on one relationship or interface in this process – that between design and construction, but note that there are similar considerations to be addressed at all the other interfaces.

The conditions which would allow these problems to be addressed are not fostered by the litigious and confrontational attitudes identified in the Latham Report 'Constructing the Team'. The authors therefore strongly endorse the Report's finding that procurement methods which foster collaboration must be developed. Such changes are a necessary precondition for implementing the recommendations which now follow.

12 RECOMMENDATIONS AIMED AT IMPROVING THE ACHIEVEMENT OF COVER

To illustrate ways of implementing quality improvement, the following recommendations are made:

– When preparing bar-bending schedules, consider the practicalities of bending the bars;

– BS4466 currently states bending tolerances for bars less than 1000 mm of +5 mm and –5 mm. However, it has been observed that closed dimension bars, such as links, tend to experience an increase in their length rather than a decrease. A recommendation is made, therefore, of a change in the tolerance to +0 mm and –10 mm for closed dimensioned bent bars;

– Wherever possible avoid bar shapes with fixed dimensions. For example, use two L-bars instead of one U-bar;

– Provide more sections in drawings, particularly sections at intersections between different structural elements;

– Avoid specifying different covers for the same structural element and state clearly to which bars the cover refers;

– Draw the reinforcement to scale at complex intersections in order to identify buildability problems at the design stage and, also, assist the steel fixer in his work;

– Contractors should have an input at the design stage to comment on buildability. For example, the contractors could provide reinforcement detailing since, it could be argued, they are better equipped than designers to understand the buildability implications of it;

– Spacers to achieve cover are part of the permanent work of the structure. However, they have not been included in the design requirements. Hence, it is suggested that they should form part of the design. For example, it could be that designers specify the type and location of spacers to suit the design loading imposed on the reinforcement, instead of the current practice of placing upon the contractor the responsibility of specifying them;

– Contractual arrangements need to be revised within the framework of a 'co-ordinative' approach and amended accordingly. The New Engineering Contract, for example, with its emphasis on clarity and simplicity and aimed at minimizing disputes and conflict may be capable of providing such a framework.

Clients' procurement policy is a key to bringing about a transition towards a co-ordinative approach. Appointment on the basis of lowest tender bid is not conducive to construction quality. Value for money is more likely to be achieved if the co-ordinative approach is adopted and if more stable relations with designers, contractors and suppliers are sought in order to achieve this end.

REFERENCES

Ballard, G. 1994. *Re: Definition of Lean Construction.* Unpublished Memo to Members of International Group on Lean Construction.

Ballard, G. & Howell, G. 1994. *Implementing Lean Construction: Stabilizing Work Flow.* Unpublished paper presented at the International Seminar on Lean Construction, Edefica, Pontificia Universidad Catolical, Santiago de Chile.

Clear, C.A. 1990. *A Review of Surveys of Cover Achieved on Site.* Building Research Establishment, Garston.

Juran, J.M. & Gryna, F.M. 1993. *Quality Planning and Analysis.* Third Edition, McGraw-Hill, New York.

Koskela, L. 1992. *Application of the New Production Philosophy to Construction.* Technical Report No. 72 Centre for Integrated Facility Engineering, Department of Civil Engineering, Stanford University.

Latham, M. 1994. *Constructing the Team.* HMSO, London.

NEDO 1987. *Achieving Quality on Building Sites.* HMSO, London.

Paradine, M.A. & Rivett, B.H.P. 1970. *Statistical Methods for Technologists.* Sixth Edition, The English Universities Press, London.

Pateman, J. 1986. There's More to Quality than Quality Assurance. *Building Technology and Management,* Aug/Sept.

Seymour, D.E. & Low, S.P. 1990. The Quality Debate. *Construction Management and Economics* Vol. 8.

Shammas-Toma, M.G. 1995. The Specification and Achievement of Cover to Reinforcement in Concrete Structures. Unpublished PhD Thesis, University of Birmingham.

Wetherill, B.G. 1981. *Intermediate Statistics.* Chapman and Hall, London.

Author index

Subject index